*High-Pressure Shock Compression of
Condensed Matter*

Springer

*New York
Berlin
Heidelberg
Barcelona
Hong Kong
London
Milan
Paris
Singapore
Tokyo*

High-Pressure Shock Compression of Condensed Matter

L.L. Altgilbers, M.D.J. Brown, I. Grishnaev, B.M. Novac, I.R. Smith, I. Tkach, and *Y. Tkach:* Magnetocumulative Generators

J. Asay and *M. Shahinpoor* (Eds.): High-Pressure Shock Compression of Solids

S.S. Batsanov: Effects of Explosion on Materials: Modification and Synthesis Under High-Pressure Shock Compression

R. Cherét: Detonation of Condensed Explosives

L. Davison, D. Grady, and *M. Shahinpoor* (Eds.): High-Pressure Shock Compression of Solids II

L. Davison, Y. Horie, and *M. Shahinpoor* (Eds.): High-Pressure Shock Compression of Solids IV

L. Davison and *M. Shahinpoor* (Eds.): High-Pressure Shock Compression of Solids III

A.N. Dremin: Toward Detonation Theory

R. Graham: Solids Under High-Pressure Shock Compression

J.N. Johnson and *R. Cherét* (Eds.): Classic Papers in Shock Compression Science

V.F. Nesterenko: Dynamics of Heterogeneous Materials

M. Sućeska: Test Methods for Explosives

J.A. Zukas and *W.P. Walters* (Eds.): Explosive Effects and Applications

Vitali F. Nesterenko

Dynamics of
Heterogeneous Materials

With 162 Illustrations

WITHDRAWN

 Springer

Vitali F. Nesterenko
Department of Mechanical and
    Aerospace Engineering
University of California at San Diego
La Jolla, CA 92093-0411
USA
vnestere@mae.ucsd.edu

Editors-in-Chief:
Lee Davison
39 Cañoncito Vista Road
Tijeras, NM 87059
USA

Yasuyuki Horie
Los Alamos National Laboratory
Los Alamos, NM 87545
USA

Library of Congress Cataloging-in-Publication Data
Nesterenko, V.F.
    Dynamics of heterogeneous materials/Vitali F. Nesterenko.
        p. cm. — (High-pressure shock compression of condensed matter)
    Includes bibliographical references and index.
    ISBN 0-387-95266-7 (alk. paper)
        1. Inhomogeneous materials.   2. Materials—Dynamic testing.   I. Title.   II. Series.
    TA418.9.I53 N47 2001
    620.1′1—dc21   2001020051

Printed on acid-free paper.

Production managed by Francine McNeill; manufacturing supervised by Erica Bresler.
Camera-ready copy provided by the author using Microsoft Word.
Printed and bound by Maple-Vail Book Manufacturing Group, York, PA.
Printed in the United States of America.

9 8 7 6 5 4 3 2 1

ISBN 0-387-95266-7   SPIN 10491732

Springer-Verlag   New York Berlin Heidelberg
A member of BertelsmannSpringer Science+Business Media GmbH

To my family

Preface

This monograph deals with the behavior of essentially nonlinear heterogeneous materials in processes occurring under intense dynamic loading, where microstructural effects play the main role. This book is not an introduction to the dynamic behavior of materials, and general information available in other books is not included. The material herein is presented in a form I hope will make it useful not only for researchers working in related areas, but also for graduate students. I used it successfully to teach a course on the dynamic behavior of materials at the University of California, San Diego. Another course well suited to the topic may be nonlinear wave dynamics in solids, especially the part on strongly nonlinear waves. About 100 problems presented in the book at the end of each chapter will help the reader to develop a deeper understanding of the subject.

I tried to follow a few rules in writing this book:

(1) To focus on strongly nonlinear phenomena where there is no small parameter with respect to the amplitude of disturbance, including solitons, shock waves, and localized shear.

(2) To take into account phenomena sensitive to materials structure, where typical space scale of material parameters (particle size, cell size) are presented in the models or are variable in experimental research. This book considers many aspects of "highly" heterogeneous materials in the sense that the typical scale of parameters variation (shock front, soliton width, shear band thickness) is comparable with particle size, and their spread from the mean value is comparable with its value itself.

(3) To concentrate attention on phenomena and models that are not widely represented in other books. In some cases, these phenomena do not have well-established mechanisms, and their analysis can be considered as a recommendation for more detailed study. This should be especially useful for young researchers.

(4) To present experimental material in a way that allows others to repeat it, or to use it for validation of analytic or numerical modeling.

(5) To introduce equations and models that allow for experimental check. I think that D. Bernoulli's remark made more than two and a half centuries ago (1738) is still valid: "To find the solution I used new principles, and therefore I wanted to

confirm the theorems with experiments, so people would have no doubt about their truth."

The phenomena considered in this book (solitons, shock waves, shear localization zones, nonequilibrium heat release) are typical during dynamic loading. At the same time, these results can be of broader interest, for example, in the area of strongly nonlinear wave dynamics of microstructured materials, nonlinear signal transformations in electronic devices, and material fabrication. The results obtained with simple models may inspire other scientists to look into the behavior of more complex media, and serve as guides for experimental research.

One of the important points for me was to pay tribute to those who in different countries contributed to this area, and later on were forgotten for various reasons. Some journals where important papers were published are no longer in existence, some results were published in technical reports that were available only inside institutes or in conference proceedings with only a small number of copies printed, and some results were not available to the public at all. I tried to follow the chronological order of publication of the most important results in all areas, no matter where the result was published.

A short description of the content of this book may help its readers to concentrate on special issues. Each chapter has a separate list of references and problems.

Chapter 1 describes the behavior of strongly nonlinear waves in discrete materials and represents the first attempt to give a comprehensive review of the area. It demonstrates the importance of new types of strongly nonlinear solitary waves under the impulse loading of ordered heterogeneous systems. Chapter 1 represents these waves using granular materials as an example, but the same type of disturbances can play a crucial role under ultrashort intense (like laser-induced) dynamic loading of monocrystals and polycrystals. It was shown that these strongly nonlinear waves (shocks and solitons) exist at any nonlinear interaction potential between elements (atoms, granules, layers) of discrete systems. Another important aspect considered in this chapter is connected with stationarity conditions of shock waves in discrete systems and comparison of different disturbances (shocks, solitons, nonstationary "shock-like" waves).

Chapter 1 also represents the natural step from the weakly nonlinear case to the strongly nonlinear case, important for intense dynamic loading of ordered systems. It places this research in the general context of wave dynamics, especially because qualitatively new strongly nonlinear waves (nonexistent under a weakly nonlinear approach) were discovered analytically, in numerical calculations and in experiments.

The main emphasis is placed on uncompressed or slightly compressed granular materials, where the linear part of the interaction law is absent or does not play a significant role. This is called a "sonic vacuum," to emphasize new media where the traditional wave equation cannot be applied. In this case, the new wave equation is based on only one small parameter—ratio of particle size to wave length. Such an equation is more general than a weakly nonlinear Korteweg–de Vries equation that is based on two small parameters (for a 1-D chain, they are the ratio of wave amplitude to the initial stress and the ratio of particle sizes to wave length). This more general wave equation has a stationary solution with unique properties. A similar situation can be expected in other strongly nonlinear cases of wave motion, for example, waves in water channels, electrical lines, and plasma.

Periodic waves, compression solitary waves, and shock waves for these materials are qualitatively different from a weakly nonlinear case, and cannot be obtained in the frame of this approach. They have unique features; that is, the spatial extent of compression solitons does not depend on amplitude, initial sound speed does not determine the soliton parameters if strain in the wave is much greater than its initial value, and the initial impulse is split into a soliton train quickly on very short distances from the entrance.

In a case opposite to the "sonic vacuum," strongly nonlinear materials with abnormal behavior, there are periodic waves, rarefaction shock waves, and solitons. They are supersonic, relative to the initial state. No stationary compression waves are allowed in such materials.

The existence of these new types of wave disturbances is confirmed through an analytic approach, through numerical calculations, and in experiments. Potential applications are not only in areas traditional for granular materials but also in other areas, like delay lines or signal processing (generating) devices that allow for a strong modification of properties by external electrical fields or in other areas. The new types of wave disturbances also represent a very compact way for transmitting lines where solitons with minimal possible space width may be effective information carriers. The very sharp shock front in these strongly nonlinear dispersive systems may also be useful for the development of fast-rise-time electromagnetic nonlinear transmission lines. Another possible application may include a neurophysiology where the ultimate response at a given point is affected by the stimulation of other neighboring points, and where corresponding wave equations may be strongly nonlinear, at least for intensive excitations.

Research in this area was pioneered in my paper published in 1983. The subject is relatively new and is currently an active research area. This book is the first attempt to introduce it as a qualitatively new domain of wave dynamics of broad interest. The area is a step away from the weakly nonlinear case for discrete materials and naturally includes it.

Chapter 2 analyzes the micromechanical aspects of granular material deformation by strong shocks, which are not thoroughly analyzed in existing books. Single pore models are presented and analyzed, based on comparison of their results with experimental data. For these materials, it is very difficult to identify phenomena that can be modeled in numerical calculations. The highly heterogeneous state behind shock in granular materials is a well-established fact now. This means that the difference between averaged parameters like pressure, temperature, and their local values is larger than the averaged value itself.

A class of new experimentally observed micromechanical phenomena is discussed in this chapter. It includes the quasistatic–dynamic transition of particle deformation under shock loading, "cold" boundary layers on the solid/powder interface, shear localization on particle interfaces, microspall inside particles, and shock front dynamics. The results can be used for the validation of two- and three-dimensional numerical models. Shock-induced separation in porous mixtures and metallization of initially dielectric granular materials are also presented.

The interaction of an explosively induced shock wave with surface and bulk cavities is considered in Chapter 2. Shock-induced local tensile stresses can cause crack initiation and provide the basis for the discussed method of fracture applied for thick-walled structures. The increase of shock amplitude results in jet formation, which can be used for the experimental validation of computer codes.

In many practical cases, the pores are loaded under plane shock loading or under impact by projectile. The ballistic performance of Ti-6Al-4V-based homogeneous and "high-gradient" porous composite materials, obtained by hot isostatic pressing against long rod impact, is discussed in Chapter 2. Powder-filled voids and rods induce a volume distributed, highly heterogeneous pattern of damage initiated by cavities and their interactions, replacing a few dominant shear bands that cause plugging in homogeneous material. As a result, deflection of the rod penetrator is induced.

A nonsymmetrical pore collapse, "forced" shear localization creates a unique damage pattern in these "high-gradient" materials that can be interesting for a ballistic application. For example, such materials are able to deflect and to stop a long rod projectile within the target, reducing the absorbed linear momentum.

In Chapter 3, strong shock transformation by laminar materials and the comparison of laminar materials with their granular counterparts are considered, based on experimental results. The essentially nonlinear effect of anomalous strong shock attenuation in laminates depending on their cell size is analyzed. The effect of foams and powders on blast mitigation in air and in explosive chambers is analyzed utilizing different containers and perforated structures.

Chapter 4 introduces a Thick-Walled Cylinder (TWC) method for investigating shear instability in a broad class of materials, including monocrystals and polycrystals, fractured and densified ceramics, and energetic materials. The emphasis of this chapter is on a shear band patterning. This method allows for the observation of such new phenomena as the reaction initiation in energetic materials (titanium + silicon, niobium + silicon) in a single shear band, comminution or sintering of ceramics (silicon carbide, alumina) inside a shear localization region, dependence of shear instability on porosity, and grain size in granular materials. One of the very important features of this method is a relatively large sample size that allows support of uniform boundary conditions on the outside surface of the sample irrespective of the shear band patterning on the inside surface. The number of shear bands can be very large, up to a few hundred, and pattern development can be "frozen" on different stages of deformation. It provides a unique condition for the verification of new theories developed by D. Grady, M. Kipp, T. Wright, H. Ockendon, and A. Molinari for shear band self-organization resulting in periodic spacing.

Chapter 5 represents an analysis of nonequilibrium heat release in granular materials under strong shock. The proposed models are critically considered and analyzed based on experimental results, including the thermoelectric method proposed by the author.

In Chapter 6, practical attempts to use dynamic densification to fabricate new materials are considered, based on the results of the previous chapters. Only aspects straightforwardly connected with the mechanics and physics of shock loading are included. The criteria for strong bonding are presented, and their merits are analyzed. Examples including rapidly solidified and amorphous materials (including pressure-induced amorphization) are considered in detail.

The area of static and dynamic behavior of heterogeneous materials is very broad and developing very quickly. This present book is a monograph and mainly based on my own results, inevitably reflecting my tastes and preferences. A sincere attempt was made to include the results of others (approximately 750 references in total), in order to provide an objective background for my findings.

Nevertheless, many important approaches and results either are not included or are presented in a very short version. Many of the topics considered in this book represent active areas of research, with some uncertainty regarding the mechanisms of the phenomena and with the possibility of bringing about qualitatively new results. The reader may find some suggestions for future research herein, and open up questions that may be especially useful for young scientists. I also hope that the reader of this book will receive a more or less accurate picture of the current state of the art of the dynamic behavior of heterogeneous materials, and of its probable future developments.

The following books published recently, whose scope is connected with the nonlinear behavior of heterogeneous materials but whose approach is essentially different from this monograph, can be recommended to the reader:

Batsanov, S.S. (1994) *Effects of Explosion on Materials: Modification and Synthesis Under High-Pressure Shock Compression.* Springer-Verlag, New York.

Chéret, R. (1993) *Detonation of Condensed Explosives*, Springer-Verlag, New York.

Christensen, R.M. (1979) *Mechanics of Composite Materials*, Chichester, Wiley, UK.

Classic Papers in Shock Compression Science (1998) Johnson, J.N., Chéret, R., Eds. Springer-Verlag, New York.

Disorder and Granular Media (1994) Bideau, D., and Hansen, A., Eds. North-Holland, Amsterdam.

Duran, J. (2000) *Sands, Powders, and Grains.* Springer-Verlag, New York.

Gibson, L.J., and Ashby, M.A. (1988) *Cellular Solids, Structure and Properties.* Pergamon Press, Oxford.

Graham, R.A. (1993) *Solids Under High-Pressure Shock Compression:*

Mechanics, Physics, and Chemistry. Springer-Verlag, New York.

Granular Matter, An Interdisciplinary Approach (1994) Mehta, A., Ed. Springer-Verlag, New York.

High-Pressure Shock Compression of Solids (1993) Asay, J.R., and Shahinpoor, M., Eds. Springer-Verlag, New York.

High-Pressure Shock Compression of Solids II: Dynamic Fracture and Fragmentation (1996) Davison, L., Grady, D.E., and Shahinpoor, M., Eds. Springer-Verlag, New York.

High-Pressure Shock Compression of Solids III (1998) Davison, L., and Shahinpoor, M., Eds. Springer-Verlag, New York.

High-Pressure Shock Compression of Solids IV: Response of Highly Porous Solids to Shock Loading (1997) Davison, L., Horie, Y., and Shahinpoor, M., Eds. Springer-Verlag, New York.

Jones, N. (1989) *Structural Impact*, Cambridge University Press, Cambridge, UK.

Kunin, I.A. (1982) *Elastic Media with Microstructure I, One-Dimensional Models.* Springer-Verlag, Berlin.

Kunin, I.A. (1983) *Elastic Media with Microstructure II, Three-Dimensional Models.* Springer-Verlag, Berlin.

Meyers, M.A. (1994) *Dynamic Behavior of Materials.* Wiley, New York.

Mobile Particulate Systems (1995) In: *Proceedings of the NATO Advanced Study Institute on Mobile Particulate Systems*, Cargese, Corsica, France, July 4–15, 1994, Guazzelli, E., and Oger, L., Eds. Kluwer Academic, Dordrecht.

Murr, L.E. (1988) *Shock Waves for Industrial Applications.* Noyes, Park Ridge, New Jersey.

Nemat-Nasser, S., and Hori, M. (1993) *Micromechanics: Overall Properties of Heterogeneous Materials.* Elsevier Science, North-Holland, Amsterdam.

Prummer, R. (1987) *Explosivverdichtung Pulvriger Substanzen/Grundlagen, Verfahren, Ergebnisse.* Springer-Verlag, Berlin.

Shock Waves in Materials Science (1993) Sawaoka, A.B., Ed. Springer-Verlag, Tokyo.

Sobczyk, K. (1985) *Stochastic Wave Propagation.* Elsevier, Amsterdam.

Remoissenet, M. (1999) *Waves Called Solitons (Concepts and Experiments)*, 3rd revised and enlarged edition. Springer-Verlag, Berlin.

Trunin, R.F. (1998) *Shock Compression of Condensed Materials.* Cambridge University Press, Cambridge, UK.

La Jolla, California   Vitali F. Nesterenko

Acknowledgments

In 1989, shortly after the APS conference on shock waves in Albuquerque, New Mexico, Marc Meyers and I sat on the beach at La Jolla Shores and wrote a detailed plan for a joint book on the dynamic behavior of materials—the dream that never came true. Instead, my book *High-Rate Deformation of Heterogeneous Materials* was published in Russia in 1992, and a very popular book by Marc entitled *Dynamic Behavior of Materials* appeared in 1994. This new book includes the former but contains comprehensively more materials, including many results published by myself and others throughout the last 8 years. This book has grown to over 500 pages (instead of 200), with a different structure, and includes a new chapter on shear localization. Problem sections to each chapter are introduced, making it suitable for use in graduate teaching. It is essentially a new book. Marc Meyers provided an important support especially on the first stage of this project.

I am obliged to many people who have made a great impact on my life and research activity. Andrei Deribas and Anatoly Staver were the Ph.D. advisers who taught me the broad approach to the different phenomena in the explosive working of materials and its applications. Joint work with M. Bondar helped create an initial material emphasis in research on powder densification and the dynamic deformation of materials. Interactions during seminars, conferences, and meetings with C. Coste, A. Dremin, S. Godunov, Y. Horie, J. Gilman, J. Goddard, D. Grady, S. Grigoryan, A. Ivanov, P. Miller, V. Mineev, V. Mitrofanov, M. Mogilevskii, A. Molinari, V. Panin, A. Potapov, S. Psakhie, G. Ravichandran, A. Rosakis, Z. Rosenberg, A. Samsonov, L. Seaman, S. Sen, N. Thadhani, V. Titov, V. Vladimirov, T. Vreeland, and many others were an important part in promoting my better understanding of dynamic phenomena.

It would not be an exaggeration to say that my life and research would have been absolutely different without the titanic scientific activity of M.A. Lavrentyev, who tirelessly promoted the Special Physical-Mathematical School and Novosibirsk University, from which I graduated; he was the first director and founder of the Lavrentyev Institute of Hydrodynamics, the Special Design Office of High Rate Hydrodynamics, and the first president of the Siberian Branch of the USSR Academy of Sciences (now the Russian Academy of Sciences), where I started my scientific career and have spent a wonderful 23 years of my life in

Akademgorodok. The considerable amount of material included in this book was obtained during my association with the Siberian Branch of the Russian Academy of Sciences, and this book would never have been written without this.

It is equally true that this book would never have appeared without the remarkable opportunity to continue my research during the last seven years at the University of California at San Diego.

I was really lucky to work with, and be influenced by, my collaborators David Benson, Michael Carroll, Marc Meyers, Sia Nemat-Nasser, Marc Wilkins, and Tim Wright, with whom I discovered and published new phenomena and mechanisms, which are now part of this book. The expertise of the patriarch of the dynamic response of materials under impact, Werner Goldsmith, provided an invaluable source of knowledge. His remarkable life and energy were always a source of inspiration to me, and his support during the most difficult stages of my life will always be remembered.

Discussions with Tom Ahrens, Donald Curran, Jack Gilman, Bill Nellis, and Don Shockey, on many occasions, provided a great source of new and unexpected ideas. Judah Goldwasser was exceptionally kind and tolerant with me; despite that, however, this book took much more time to complete than originally expected. Without his support, this book would never have happened. His guidance in the research of the mechanics–chemistry linkage in energetic materials played a great part in the successful activity of our group at the University of California at San Diego. Andrew Crowson and John Bailey were also very helpful, and promoted a broad application of the Thick-Walled Cylinder (TWC) method for different materials. This book would not have been completed without the support provided by the U.S. Office of Naval Research, Contract N00014-94-1-1040 (Program Manager Dr. Judah Goldwasser), and by the U.S. Army Research Office, Contract ARO MURI DAAH 04-96-2-0376 (Program Manager Dr. David Stepp).

Bob Graham initiated this project and encouraged me to finish it. Lee Davison provided many insightful and helpful suggestions to improve the book, and, together with Yuki Horie, helped me to overcome the last few hurdles.

My coauthors and colleagues S. Afanasenko, N. Singh Brar, B. Chen, C. Cline, A.F. Demchuk, S.L. Gavrilyuk, S. Indrakanti, K.T. Kim, A. Kosenkov, A. Kusubov, Yu.G. Kuznetsov, J.C. LaSalvia, A. Lazaridi, V. Levitas, Yu. Mesheryakov, S. Pershin, J. Shih, V.A. Simonov, E. Sybiryakov, Q. Xue, Gu Yabei, and others contributed to our joint research to a great extent. The beautiful pictures presented in this book of SiC after the TWC test were prepared by James Shih. Martin Ostoja-Starzewski helped in preparing Section 1.6 on the phenomenological description of shock waves in nonlinear particulate systems. Andrea Brunson made great efforts to edit and improve the text.

My wife, Olga, was one of the most active promoters of this project, and her love was a vital support for me during this long journey. My son Igor helped me on the first stage of writing to type the English text.

I am extremely grateful to my beloved father and mother, Fedor Petrovich Nesterenko and Alexandra Andreevna Germanenko, who lived through difficult times but always provided support, love, and encouragement for their son.

La Jolla, California   Vitali F. Nesterenko

Contents

Preface   vii

Acknowledgments   xiii

1. Nonlinear Impulses in Particulate Materials   1

1.1 Introduction...   1
1.2 Long-Wave Equation for a "Strongly Compressed" Chain..............   3
1.3 Equation for a "Weakly Compressed" Chain...........................   7
1.4 Stationary Solutions for the "Sonic Vacuum" Equation.............. 13
1.5 Stability of Nonlinear Periodic Waves in "Sonic Vacuum"........... 22
1.6 Wave Dynamics of Unstressed One-Dimensional
     Granular Materials—Numerical Calculations......................... 26
1.7 Phenomenological Description of Waves in Particulate Systems..... 62
1.8 Experimental Observation of a New Type of Solitary Waves in a
     One-Dimensional Granular Medium................................... 65
1.9 Wave Dynamics in 2- and 3-D Granular Materials.................... 82
1.10 Nonlinear Waves in Discrete 1-D Power-Law Materials............... 90
1.11 General Case of Nonlinear Discrete Systems........................ 104
1.12 Conclusion.. 123
     Problems.. 124
     References.. 126

2. Mesomechanics of Porous Materials Under
Intense Dynamic Loading   137

2.1 Introduction.. 137
2.2 Dynamics of Pore Collapse: The Carroll–Holt Model................ 138
2.3 Modified Carroll–Holt Model....................................... 145
2.4 The Front Width of a Strong Shock................................. 156
2.5 Quasistatic and Dynamic Deformation of Powders Under Shock
     Loading... 161
2.6 Two-Dimensional Numerical Calculations of Shock
     Densification... 169

2.7   Metallization of Dielectric Powders Under Shock Loading.......... 205
2.8   Boundary Layers... 212
2.9   Separation of Components in Powder Mixtures...................... 222
2.10 Three-Dimensional Simulations... 223
2.11 Interaction of Plane Shock Waves with a Cavity on the Free
       Surface... 224
2.12 Interaction of a Long Rod Penetrator with a Porous Target......... 230
2.13 Validation of Numerical Modeling Against Experiments............ 232
2.14 Conclusion... 233
       Problems.. 234
       References.. 235

3. Transformation of Shocks in Laminated and Porous Materials   246

3.1   Introduction.. 246
3.2   Experiments on Shock Transformation in Solid Laminar
       Materials (LMs)... 248
3.3   Numerical Calculations of Shock Attenuation, the Mechanism
       of the Anomalous Effect of Cell Size................................. 251
3.4   Experimental Comparison of Shocks in Laminar Material and
       in a Random Heterogeneous Medium................................. 262
3.5   Damping of Shock Waves by Porous Materials...................... 274
3.6   Blast Mitigation by Porous Materials, Criteria for Damping....... 285
3.7   Conclusion.. 296
       Problems.. 296
       References.. 300

4. Shear Localization and Shear Bands Patterning in Heterogeneous Materials   307

4.1   Introduction.. 307
4.2   Thick-Walled Cylinder (TWC) Method 308
4.3   Experimental Observation of Shear Bands Patterning
       in Polycrystals... 320
4.4   Grady–Kipp Model of the Collective Behavior of Shear Bands..... 328
4.5   Wright–Ockendon Model for Shear Band Spacing 329
4.6   Molinari Model for Shear Band Spacing 332
4.7   Theoretical Predictions Versus Experiments......................... 333
4.8   Dependence of Shear Bands Patterning on Initial Structure in
       Inert (Fractured and Granular) Materials............................. 340
4.9   Behavior of Reactant Granular Materials Under Conditions
       of Controlled Localized Shear... 367
4.10 Conclusion.. 371
       Problems.. 372
       References.. 374

5. Nonequilibrium Heating of Powders Under Shock Loading   385

5.1	Introduction	385
5.2	Experimental Results	386
5.3	Thermodynamic Models for Heterogeneous Heating Under Shock-Wave Loading	398
5.4	"Skin" Model and Thermal Relaxation in Shocked Powders	415
5.5	Nonequilibrium Thermodynamics of Powder Mixtures	418
5.6	Shock and Chemical Reactions in Powders	427
5.7	Conclusion	431
	Problems	431
	References	432

6. Advanced Materials Treatment by Shock Waves   442

6.1	Introduction	442
6.2	Powder Densification and Consolidation by Shock Waves	443
6.3	Preservation of the Amorphous State Under Dynamic Loading	455
6.4	Obtaining of Supercooled States Under Shock-Wave Loading	464
6.5	Sintering, Hot Isostatic Pressing of Shocked Powders	472
6.6	Hot Explosive Compaction	483
6.7	Shock Densification of Composites	487
6.8	Particle Comminution and Activation by Dynamic Loading	488
6.9	Conclusion	490
	Problems	490
	References	492

Afterword   504

Index   507

1
Nonlinear Impulses in Particulate Materials

1.1 Introduction

Impulse propagation in granular (particular) materials due to the loading under impact or contact explosion is of practical interest for many applications. For example, granular bed from iron shot is used for the damping of contact explosions during technological operations in explosive chambers. It effectively prevents the chamber wall from the high-amplitude shock wave. The propagation and reflection of nonlinear waves with large amplitudes in sand or soil is important for the detection of foreign objects. The nature of waves in these materials is also of general interest because they represent the collective dynamic response strongly effected by mesostructure. At the same time, these materials pose some fundamental questions which demand reconsideration of the basic foundation of wave dynamics including shock-wave propagation and shock dynamics particularly.

Granular materials provide an example of media where linear and even weakly nonlinear descriptions fail. They are highly nonlinear materials with a qualitatively new behavior according to a few physically different reasons (Duffy and Mindlin [1957]; Nesterenko [1983a,b]; Herrmann, Stauffer, and Roux [1987]; Kuwabara and Kono [1987]; Goddard [1990]; Jaeger and Nagel [1992], [1996]; Falcon, Laroche, Fauve, and Coste [1998a,b] and others) including:

Nonlinearity of the interaction law: The Hertz law for force developing under the compression of two elastic granules (or more general power-law) has no linear part, even for relatively small displacements in the vicinity of a zero compression force.

Another type of nonlinearity is connected with no tensile strength for granular material. The interaction law in general consists of different parts: force under compression is combined with a zero interaction force under tension if there is no contact between particles.

Contact interaction under strong compression or for the displacement of grains smaller than the surface roughness can be essentially different from the Hertz law and can be responsible for hysteresis phenomena.

Structural rearrangements under applied load: Microscopically—the amount of neighbors in contact with given particles is dependent on the loading conditions.

Mesoscopically—dramatic changes in material fabric (force chains) result even from a small amount of disorder in the particle radii. Their lengths are of a few tens of particle diameter and they are extremely sensitive to the applied load, and can be essentially rearranged even under very small external action.

Dissipation: Friction and viscoelastic dissipation between particles can also be a nonlinear function of applied load, or relative displacements or displacement rates resulting in dependence of the restitution coefficient on velocity.

As was emphasized by Jaeger and Nagel [1996]: "a granular material behaves differently from any of the other familiar forms of matter—solids, liquids, or gases—and should therefore be considered an additional state of matter in its own right." Sinkovits and Sen [1996] pointed out that granular systems are excellent examples of many particle systems which are strongly affected by an external field.

Additional challenges with granular materials are connected to the problems of measurements of such parameters, which can be interpreted in theory or through computer calculations.

This book will primarily address nonlinearities connected with interaction law. Even this part of the general nonlinear behavior of granular materials is able to produce a qualitatively different mode of wave propagation. It places granular materials in a special class according to wave dynamics. For example, the analysis makes very clear that wave phenomena in granular materials for "weak" perturbations cannot be interpreted even qualitatively in terms of the "sound" propagation approach, if the amplitude of impulse is larger than the initial compression. This was the reason for the introduction of the concept of "sonic vacuum"—the medium where the traditional wave equation is not a basic equation for wave dynamics (Nesterenko [1983b]). The calculations of the effective elastic moduli for a granular material and sound propagation in initially compressed granular materials can be found in papers by Brandt [1955], Duffy and Mindlin [1957], Deresiewicz [1958], Digby [1981], Walton [1987], and others. The synthesis of all components of highly nonlinear behavior is definitely a very exciting area for future research.

As usually happens, the physically based approaches developed for one particular case can be applied for different nonlinear phenomena and materials later. This creates a broader interest in the qualitatively new wave phenomena, especially for such remarkably simple systems like granular materials. Wave propagation in this type of material can be important for geophysical applications and for the development of a qualitatively new delay, impulse transformation acoustic, and electrical lines. Granular materials afford a very simple example of where qualitatively new wave dynamics should be developed. Hopefully this development will also extend to other materials.

1.2   Long-Wave Equation for a "Strongly Compressed" Chain

A one-dimensional chain of particles is used to find solutions to important critical problems in wave dynamics. It is sufficient to recollect Newton's first derivation of the formula for sound speed, the Fermi–Pasta–Ulam problem (Fermi, Pasta, and Ulam [1965]), the exact solutions for the nonlinear Toda lattice (Toda [1981]), or the shock wave structure in a lattice (Duvall, Manvi, and Lowell [1969]). We will consider wave propagation in one-dimensional granular material, taking into account that particles interact with each other according to the Hertz law:

$$F = \frac{2E}{3(1-v^2)} \left(\frac{R_1 R_2}{R_1 + R_2} \right)^{1/2} \left\{ (R_1 + R_2) - (x_2 - x_1) \right\}^{3/2},$$

(1.1)

where F is the compression force between granules, E is Young's modulus of the material, R_1 and R_2 are the granule's radii, v is the Poisson coefficient, and x_1 and x_2 are the coordinates of the spherical granules centers $(x_2 > x_1)$.

This partial case of particle interaction is important not only as a representation of the interaction in real powder materials, but also as an example of strongly nonlinear media which cannot be represented in the frame of linear approximation for an initial displacement equal to zero (Nesterenko [1983b].

It is necessary to point out that the dependence of F in the form $\delta^{3/2}$, where δ is the current approach of particle centers, is valid not only for spheres, but also for contacts of other finite bodies (Landau and Lifshitz [1995]). Such unusually strong nonlinear behavior is caused by the finite particle sizes of a perfectly linear elastic material constituting the granular medium.

The static Hertz law for dynamic problems implies the following restrictions:

(1) the maximum shear stresses achieved in the vicinity of contact must be less than the elastic limit;
(2) the sizes of the contact surface are much smaller than the radii of curvature of each particle; and
(3) the characteristic times of the problem τ are much longer than the oscillation period of the basic shape for the elastic sphere T:

$$\tau \gg T \approx 2.5 \frac{R}{c_1},$$

(1.2)

where c_1 is the velocity of sound in the material of the interacting spheres.

Conditions (1) through (3) restrict the particle velocities to quantities of the order of several meters per second for metallic particles with radii in the interval 1 to 5 millimeters.

Friction between particles and the resulting dissipation processes are not taken into account in this consideration. The validity of the Hertz law for the

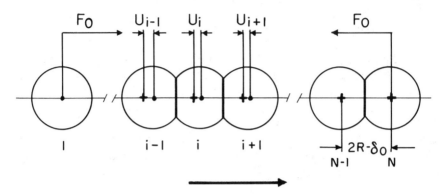

Figure 1.1. The one-dimensional chain of particles compressed by a large (in comparison with force changes in the impulse) static force F_0 causing an initial displacement δ_0 between neighboring centers. The crosses correspond to the initial positions of particle centers in a statically compressed chain. The black circles mark the current positions of the spheres centers. The right end of the chain is undisturbed, and the arrow at the bottom shows the direction of impulse propagation.

collision between particles of different materials is discussed in detail by Goldsmith [1960] and Johnson [1987].

In recent experiments and calculations, Falcon, Laroche, Fauve, and Coste [1998a,b] considered different types of interaction laws including those recently proposed (Goddard [1990], de Gennes [1996]) and confirmed the validity of the Hertz law for such different materials as tungsten carbide, steel, glass, nylon, and brass. As usual this law works well even outside its static limit, which may be the result of strain rate hardening for some materials.

Consider a one-dimensional chain of identical spherical granules (Fig. 1.1) with radius R. We assume that it is subjected to constant compression forces F_0 applied to both ends and securing the initial displacement δ_0 between the particle centers. As will be clear from the following, it is convenient to introduce δ_0 into the equation explicitly. For this purpose we replace the coordinate x_i by the displacement u_i of the given particle from its equilibrium position in the initially compressed chain (Fig. 1.1). Using the expression for the force from Eq. (1.1), the particle equation of motion becomes

$$\ddot{u}_i = A \left(\delta_0 - u_i + u_{i-1}\right)^{3/2} - A \left(\delta_0 - u_{i+1} + u_i\right)^{3/2}, \quad N-1 \geq i \geq 2, \tag{1.3}$$

$$m = \tfrac{4}{3}\pi R^3 \rho_0, \quad A = \frac{E(2R)^{1/2}}{3(1-v^2)m},$$

where m is the mass of the particle and ρ_0 is the density of particle material. It is assumed that the distance between the particle centers does not exceed $2R$ if the particles are spherical.

Equation (1.3) also describes the propagation of one-dimensional perturbations in a three-dimensional simple cubic packing of spheres, if the wavefront plane is parallel to the cube boundaries. For other packing of identical spheres, the interactions between neighbors will have the same character with differences only in the numerical coefficient A.

Equation (1.3) can be transformed to a well-investigated system of nonlinear oscillators under the assumption of small deformation in the medium by comparison with the initial one, i.e., considering

$$\frac{\left|u_{i-1} - u_i\right|}{\delta_0} \ll 1. \tag{1.4}$$

In the anharmonic approximation, Eq. (1.3) (using Eq. (1.4)) can be written (Nesterenko [1983b]) as follows:

$$\ddot{u}_i = \alpha\left(u_{i+1} - 2u_i + u_{i-1}\right) + \beta\left(u_{i+1} - 2u_i + u_{i-1}\right)\left(u_{i-1} - u_{i+1}\right), \quad N-1 > i \geq 2, \tag{1.5}$$

$$\alpha = \tfrac{3}{2} A\delta_0^{1/2}, \qquad \beta = \tfrac{3}{8} A\delta_0^{-1/2}.$$

Equations of this type with a quadratic nonlinearity were solved numerically in a number of studies (Anderson [1964]; Tsai and Beckett [1966]; Duvall, Manvi, and Lowell [1969]; Manvi, Duvall, and Lowell [1969]; Strenzwik [1979]) and a brief review can be found in Hill and Knopoff [1980]. It was shown that in the problem of a piston, moving with a constant velocity v_0, there is an oscillating nonstationary structure, similar to a "shock" wave type. The leading part of this disturbance is formed by a soliton with the velocity in the maximum equal to $2v_0$. Oscillations are "damped" behind the front even in the absence of dissipation. Dissipative dashpots were introduced by Anderson [1964] for a description of shock waves in the lattice (also see Duvall, Manvi, and Lowell [1969]). Equations valid for finite but small perturbations and similar to Eq. (1.5) can be obtained for different choices of interaction potentials. Undamped in time oscillations near the piston for sufficiently high values of its velocity were observed in numerical calculations of a Toda chain as well as with a particle interaction determined by the Morse potential (Strenzwik [1979]; Hill and Knopoff [1980]). The existence of solitons in a chain with a Morse potential was shown numerically by Strenzwik [1979].

In the long-wave approximation ($L > a = 2R$, where L is a characteristic spatial size of the wave perturbation), using the replacement (Kunin [1975]):

$$u_i = u(x), \quad u_{i-1} = e^{-aD}u(x), \quad u_{i+1} = e^{aD}u(x), \quad D \equiv \frac{\partial}{\partial x}, \tag{1.6}$$

one can obtain the nonlinear wave equation from Eq. (1.5):

$$u_{tt} = c_0^2 u_{xx} + 2c_0 \gamma u_{xxxx} - \sigma u_x u_{xx} , \tag{1.7}$$

$$c_0^2 = A\delta_0^{1/2} 6R^2 , \quad \gamma = \frac{c_0 R^2}{6} , \quad \sigma = \frac{c_0^2 R}{\delta_0} .$$

All terms of the order

$$\frac{c_0^2 u}{L^2}\left[\left(\frac{a}{L}\right)^4 + \left(\frac{a}{L}\right)^3 \left(\frac{u}{\delta_0}\right)\right] \tag{1.8}$$

and higher were omitted in deriving Eq. (1.7). In addition, the nonlinear term which is connected with the convective derivative (the main nonlinear one in hydrodynamics) was neglected. This approximation is true if the particle velocity in the wave is much less than its phase velocity, which is expected to be close to the sound speed c_0 despite nonlinearity. The ratio of this omitted nonlinear term (Eq. (1.8)) to the others taken into account in Eq. (1.7), is an order of magnitude of δ_0/R.

It was demonstrated by Zakharov [1974] for the equation similar to Eq. (1.4), that the inverse problem technique can be applied and it has an infinite set of commuting integrals of motion.

Equation (1.7) could be obtained from Eq. (1.3) through two steps only because we have two small parameters (as a result of weak nonlinearity and long-wave approximation). Eq. (1.7) can be transformed (Kunin [1975]) inside the same approximation into the Korteweg–de Vries (KdV) equation (Korteweg and de Vries [1895]) which describes propagation of disturbances in one direction:

$$\xi_t + c_0\xi_x + \gamma\xi_{xxx} + \frac{\sigma}{2c_0}\xi\xi_x = 0, \qquad \xi = -u_x. \tag{1.9}$$

The solutions of Eq. (1.9) are well known and their exciting history can be found in many papers and books (e.g., Sander and Hutter [1991]). The authors cited the clear definition of wave motion made by Russell [1845]: "the wave motion is therefore a transcendental motion; ... the motion of motion—the transference of motion without the transference of the matter, of form without the substance, of force without the agent."

From the properties of Eq. (1.9) it can be concluded that in the granular material, initially compressed by a static load, weakly nonlinear periodic, solitary and shock wave (the latter if dissipation is included) solutions can exist. The qualitative peculiarities of propagated impulses are determined by the initial characteristics of disturbances. For an equation similar to Eq. (1.7), Samsonov [1987] obtained supersonic and subsonic stationary solutions that were different from the corresponding solutions of Eq. (1.9).

A remarkable exact soliton solution of Eq. (1.9) is

$$\xi - \xi_0 = \Delta\xi = \Delta\xi_m \operatorname{sech}^2\left\{\left(\frac{\sigma\Delta\xi_m}{24c_0\gamma}\right)^{1/2}(x - Vt)\right\}, \qquad (1.10)$$

$$V = c_0 + \frac{\sigma}{6c_0}\Delta\xi_m, \qquad L = \left(\frac{24c_0\gamma}{\sigma\Delta\xi_m}\right)^{1/2},$$

where V is the soliton phase velocity and L the characteristic width. The soliton was discovered first experimentally by Russell [1838] in connection with solving the applied problem of ship motion in channels. Its main properties (Russell [1838], [1845]) were described based on the experiments. It took practically 60 years to derive a wave equation which has Eq. (1.10) as a stationary solution (Korteweg and de Vries [1895]).

It is easy to see from the expression for initial long-wave sound speed c_0 (Eq. (1.7)) and from the expressions for soliton parameters (Eq. (1.10)) that if $c_0 \to 0$ ($\delta_0 \to 0$), then the soliton velocity $V \to \infty$ and $L \to 0$, if the value $\Delta\xi_m$ is the constant. It is clear that approximation which includes only weak nonlinearity fails and the anharmonic approximation (Eq. (1.5)) is no more valid.

A qualitatively different approach should be developed if deformation connected with the wave is much larger than the initial one. The propagation of compression pulses in the materials where the long-wave sound velocity c_0 is equal to zero or small compared to the phase velocity of disturbances was theoretically considered by Nesterenko [1983b] and corresponding solitons of new type were experimentally discovered by Lazaridi and Nesterenko [1985]. For such media the concept of a "sonic vacuum" was introduced by Nesterenko [1992a], implying the impossibility of the propagation of linear waves described by a standard wave equation, or their insignificance for the description of strongly nonlinear disturbances. Finally, it is worth mentioning that wave propagation in uncompressed granular media poses a new fundamental problem of wave dynamics.

1.3   Equation for a "Weakly Compressed" Chain

A very interesting wave behavior appears if granular material is weakly compressed—the change of the neighboring particle displacements in a wave is much larger than the initial one δ_0 resulting from the static compression (see Fig. 1.2). The principal difference between Figures 1.1 and 1.2 is due to the lack of a small parameter with respect to the wave amplitude in the latter case. The anharmonic approximation (Eq. (1.5)) is not valid any more. For materials with zero (or very small) sound speed we cannot use the standard linear wave equation (Eq. 1.7 without two last terms) as a starting point for the next approximations, as has been done before. For the same reason there is no point in considering the KdV equation as the basic one for this case. Of course the equations for a discrete chain of particles interacting according to the Hertz law (Eq. (1.3)) can

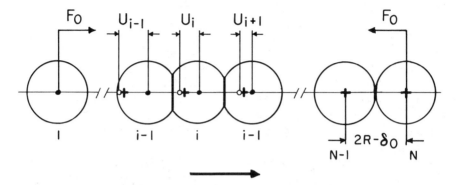

Figure 1.2. The one-dimensional chain compressed by a small (in comparison with forces in the impulse) static force F_0 causing an initial displacement δ_0 between neighboring centers. The crosses represent the initial positions of the particle centers in the statically compressed chain, the black circles correspond to the current positions of the spheres centers, and the initial positions of the spheres in the uncompressed chain are shown by open circles. The right end of the chain is undisturbed, and the arrow at the bottom shows the direction of impulse propagation.

still be valid, if the particle velocity is small enough to satisfy the restrictions, ensuring the elastic behavior and characteristic time for the propagating wave satisfies Eq. (1.2).

In the case of Figure 1.2, when the chain is initially "weakly" compressed by the static force F_0

$$\frac{|u_{i+1} - u_i|}{\delta_0} \underset{\sim}{>} 1 , \tag{1.11}$$

only an equation corresponding to the long-wave approximation ($L >> a$) can be obtained from Eq. (1.3) for the discrete chain. It can be done using a replacement similar to the Eq. (1.6) being used during the derivation of Eq. (1.7). We continue to suppose that despite Eq. (1.11), the differences in the displacements $(u_{i-1} - u_i)$, $(u_{i+1} - u_i)$, of the particles neighboring ith, are still very small in comparison with the particle diameter $a = 2R$ (but not with δ_0!).

Let us start with the equations for the discrete system of particles where the initial relative displacement δ_0 is included in displacement u_i calculated from the particle positions in the uncompressed system when the distance between the particles' centers equals $a = 2R$. This change of the starting points is not essential because the expressions for contact forces involve only $(u_{i-1} - u_i)$, $(u_{i+1} - u_i)$. In this case, the statically compressed chain at the moment $t = 0$ is described by the relation $u_{i-1} - u_i = \delta_0$ at $N > i > 1$ (Fig. 1.2), and Eq. (1.3) becomes (1.12)

$$\ddot{u}_i = A\left(u_{i-1} - u_i\right)^{3/2} - A\left(u_i - u_{i+1}\right)^{3/2}.$$

(1.12)

For long-wave disturbance the displacements u_{i-1}, u_{i+1} in Eq. (1.12) can be expanded in a power series according to the small parameter $\varepsilon = a/L$ up to the fourth order

$$u_i = u,$$

$$u_{i-1} = u - u_x a + \frac{1}{2} u_{xx} a^2 - \frac{1}{6} u_{xxx} a^3 + \frac{1}{24} u_{xxxx} a^4 + \cdots,$$

$$u_{i+1} = u + u_x a + \frac{1}{2} u_{xx} a^2 + \frac{1}{6} u_{xxx} a^3 + \frac{1}{24} u_{xxxx} a^4 + \cdots.$$

(1.13)

Subtracting one of the Eqs. (1.13) from another leads to the following expressions:

$$u_{i-1} - u_i = - u_x a + \frac{1}{2} u_{xx} a^2 - \frac{1}{6} u_{xxx} a^3 + \frac{1}{24} u_{xxxx} a^4 + \cdots = N(x) + \varphi(x),$$

$$u_i - u_{i+1} = - u_x a - \frac{1}{2} u_{xx} a^2 - \frac{1}{6} u_{xxx} a^3 - \frac{1}{24} u_{xxxx} a^4 + \cdots = N(x) + \psi(x),$$

(1.14)

$$N(x) = - u_x a = a\xi, \qquad \xi = - u_x,$$

$$\varphi(x) = + \frac{1}{2} u_{xx} a^2 - \frac{1}{6} u_{xxx} a^3 + \frac{1}{24} u_{xxxx} a^4 \cdots,$$

$$\psi(x) = - \frac{1}{2} u_{xx} a^2 - \frac{1}{6} u_{xxx} a^3 - \frac{1}{24} u_{xxxx} a^4 \cdots.$$

Now we can write expressions for two force terms in the right part of Eq. (1.12) using Eq. (1.14):

$$A\left(u_{i-1} - u_i\right)^{3/2} \approx A\left(N(x) + \varphi(x)\right)^{3/2},$$

$$A\left(u_i - u_{i+1}\right)^{3/2} \approx A\left(N(x) + \psi(x)\right)^{3/2}.$$

(1.15)

One can see that $N(x)$ and $\varphi(x)$, $\psi(x)$ have different orders of magnitude with respect to the small parameter ε:

$$\frac{\varphi(x)}{N(x)} \sim \frac{\psi(x)}{N(x)} \sim \varepsilon << 1.$$

(1.16)

Taking into account Eq. (1.16) we can expand the power functions from Eqs. (1.15) into series

$$\left(N(x) + \varphi(x)\right)^{3/2}$$
$$\approx \left(N(x)\right)^{3/2} + \frac{3}{2}\left(N(x)\right)^{1/2}\varphi(x) + \frac{3}{8}\left(N(x)\right)^{-1/2}\left(\varphi(x)\right)^2 - \frac{1}{16}\left(N(x)\right)^{-3/2}\left(\varphi(x)\right)^3,$$

$$\left(N(x) + \psi(x)\right)^{\frac{3}{2}}$$
$$\approx \left(N(x)\right)^{3/2} + \frac{3}{2}\left(N(x)\right)^{1/2}\psi(x) + \frac{3}{8}\left(N(x)\right)^{-1/2}\left(\psi(x)\right)^2 - \frac{1}{16}\left(N(x)\right)^{-3/2}\left(\psi(x)\right)^3. \tag{1.17}$$

The acceleration of the ith particle can be obtained from Eqs. (1.12), (1.15), and (1.17) as follows:

$$\ddot{u} \approx \frac{3}{2}\left(N(x)\right)^{1/2}\left(\psi(x) - \varphi(x)\right) + \frac{3}{8}\left(N(x)\right)^{-1/2}\left(\psi(x)^2 - \varphi(x)^2\right) \tag{1.18}$$
$$- \frac{1}{16}\left(N(x)\right)^{-3/2}\left(\psi(x)^3 - \varphi(x)^3\right).$$

Equations (1.14) are used to replace functions φ and ψ in terms of spatial derivatives of the displacement field u:

$$\varphi - \psi \approx a^2 u_{xx} + \frac{a^4}{12}u_{xxxx} + O(\varepsilon^6),$$
$$\varphi^2 - \psi^2 \approx -\frac{a^5}{3}u_{xx}u_{xxx} + O(\varepsilon^7), \tag{1.19}$$
$$\varphi^3 - \psi^3 \approx \frac{a^6}{4}u_{xx}^3 + O(\varepsilon^7).$$

The procedure of keeping the corresponding terms and neglecting others in Eq. (1.19) is based on the fact that different powers of $N(x)$ have different orders of magnitude with respect to the small parameter ε. The orders of magnitude for the corresponding terms in Eqs. (1.18) ensure that in the final equation all terms, including the order of magnitude ε^2 according to the main term, are kept. If the convective derivative in acceleration is ignored (that may be a reasonable approximation in this case as will be shown later), the wave equation for the simplest example of "sonic vacuum," chain of spherical granules (Fig. 1.2), is obtained from Eqs. (1.18) and (1.19) (Nesterenko [1983b], [1988]):

$$u_{tt} = c^2 \left\{ \frac{3}{2}(-u_x)^{1/2}u_{xx} + \frac{a^2}{8}(-u_x)^{1/2}u_{xxxx} - \frac{a^2}{8}\frac{u_{xx}u_{xxx}}{(-u_x)^{1/2}} - \frac{a^2}{64}\frac{(u_{xx})^3}{(-u_x)^{3/2}} \right\}, \tag{1.20}$$

$$-u_x > 0, \qquad\qquad c^2 = \frac{2E}{\pi\rho_0\left(1 - v^2\right)} = Aa^{5/2}.$$

It is necessary to mention that neglecting the nonlinear convective derivative term (made at this stage in writing acceleration as the partial derivative with respect to time) has not been proved. Equation (1.20) doesn't contain a priori characteristic phase speeds of the wave in comparison with Eqs. (1.1) and (1.9) that would describe a weakly nonlinear acoustic. The proof for this step can be made only later, when we obtain the characteristic phase velocity of a stationary wave V, dependent on the particle velocity v. It will be shown that at some physically meaningful range of v (where the Hertz law holds) $V >> v$.

Additionally, the order of magnitude of the neglected nonlinear convective derivative terms is much less than the order of the kept nonlinear dispersive terms (the last three terms in the right part of Eq. (1.19). They are of the order of magnitude ε^2 with respect to the main term.

The qualitative difference of Eq. (1.20) from the weakly nonlinear case (Eqs. (1.7) and (1.9)) is that this equation is "one small parameter" equation (only a long wave approximation is supposed to be valid). This allows for the extension of the weakly nonlinear case and results in qualitatively new solutions as will be demonstrated later.

Equation (1.14) can be rewritten in divergent form (also in strains $\xi = -u_x > 0$) (Nesterenko [1992c], [1994]). These forms provide straightforward derivation of the momentum and energy conservation laws (Whitham [1974]) and are useful for further transformations and obtaining the solutions

$$u_{tt} = -c^2 \left\{ (-u_x)^{3/2} + \frac{a^2}{12} \left[\left((-u_x)^{3/2} \right)_{xx} - \frac{3}{8} \left((-u_x)^{-1/2} \right) u_{xx}^2 \right] \right\}_x , \tag{1.21a}$$

$$\xi_{tt} = c^2 \left\{ \xi^{3/2} + \frac{a^2}{12} \left[\left(\xi^{3/2} \right)_{xx} - \frac{3}{8} \xi^{-1/2} \xi_x^2 \right] \right\}_{xx} . \tag{1.21b}$$

Additional rearrangements result in the following equations:

$$u_{tt} = -c^2 \left\{ (-u_x)^{3/2} + \frac{a^2}{10} \left[(-u_x)^{1/4} \left((-u_x)^{5/4} \right)_{xx} \right] \right\}_x , \tag{1.22a}$$

$$\xi_{tt} = c^2 \left\{ \xi^{3/2} + \frac{a^2}{10} \left[\xi^{1/4} \left(\xi^{5/4} \right)_{xx} \right] \right\}_{xx} . \tag{1.22b}$$

An equation written in the form of Eq. (1.22) is appropriate for finding the stationary solutions due to evident variable replacement. The Lagrangian density L for Eqs. (1.21) to (1.22), which will be used further for stability analysis, is equal to

$$L = \frac{u_t^2}{2} - c^2 \left\{ \frac{2}{5}(-u_x)^{5/2} + \frac{a^2}{2} \left[\frac{3}{8}(-u_x)^{1/2} u_{xx}^2 - \frac{1}{3}(-u_x)^{3/2} u_{xxx} \right] \right\}. \tag{1.23}$$

Derivation of Eq. (1.21) or (1.22) from the Lagrangian L is simpler than the used method based on the expansion in a power series of displacements and interaction law, but the latter is more physically clear.

Partial differential equations with the fourth spatial derivative such as Eqs. (1.7), (1.21), and (1.22) in the linear approximation have some undesirable properties, which have inspired interesting and emotional discussions (Benjamin, Bona, and Mahony [1972]; Kruskal [1975]) on the shortcomings of the KdV Equation (1.9), which inherited these properties from the Boussinesq Equation (1.7). One part of this problem can be very easily clarified if we take a look at the dispersion relation for the linearized Eq. (1.7),

$$\omega^2 = c_0^2 k^2 \left(1 - \frac{2\gamma}{c_0} k^2 \right), \tag{1.24}$$

where ω is the frequency and k is the wave vector of the linear periodic wave solution. It is easy to see from Eq. (1.24) that at $k > 2 \cdot 3^{1/2}/a$, or wavelength $\lambda < \lambda_{min} = \pi a/3^{1/2} \approx 1.8a$, where a is a distance between particles (in our case $2R$), the ω becomes imaginary. As a result, the perturbations with small wavelength (less than two-particle diameter in our case) are amplifying arbitrarily fast. This means ill-posedness of the initial-value problem for this equation. This is the same case for the KdV equation. One can overcome this problem by the regularization procedure which is based on replacement of the spatial derivative on the time derivative using the first-order approximation (Whitham [1974]). The analogous regularization can easily be made for the "sonic vacuum" Eq. (1.21). From the first approximation, we have

$$\left((-u_x)^{3/2} \right)_{xx} = -\frac{u_{tt}}{c^2}. \tag{1.25}$$

Substituting Eq. (1.25) rather than the first term in quadrant parentheses in Eq. (1.21), we obtain the regularized wave equations (Nesterenko [1992c], [1994]):

$$u_{tt} = -\left\{ c^2(-u_x)^{3/2} - \frac{a^2}{12} \left[u_{ttx} + \frac{3}{8} c^2 (-u_x)^{-1/2} u_{xx}^2 \right] \right\}_x, \tag{1.26}$$

and

$$\xi_{tt} = \left\{ c^2\, \xi^{3/2} + \beta\, \xi_{tt} - b\, \xi^{-1/2}\, \xi_x^2 \right\}_{xx},$$ (1.26b)

$$\beta = \frac{a^2}{12}, \qquad b = \frac{a^2 c^2}{32}.$$

Equations (1.26) and (1.21), (1.22) are formally equivalent according to the order of the error terms as approximations of the same discrete system.

It is necessary to emphasize that the continuum approach on the scale of the distance between particles in discrete systems cannot be considered as being validated by the regularization procedure. This helps to avoid the problems for short wavelength where the equation by itself is not supposed to be applicable just by derivation. The presented nonlinear wave equations are, by determination, the long-wave equations and are based on the assumption $L \gg a$ and should be used under this assumption. Any use of these equations for L comparable with a (which nevertheless may produce meaningful results) should be verified by computer calculations of a corresponding discrete system, or by experiment.

In papers by Rosenau [1986a,b], [1988] and by Rosenau and Hyman [1993], an equation for discrete systems with an arbitrary interaction law between particles was derived. It is partially analogous to Eq. (1.26b), but does not contain the last nonlinear dispersive term. This term for strongly nonlinear solutions has the same order of magnitude as the kept traditional dispersive term (the latter is exactly the same as in the linear case), and that's why it cannot be dropped. Later Rosenau [1994], [1997] formally introduced essentially a nonlinear wave equation $K(m,n)$ in partial cases similar to Eq. (1.21).

1.4 Stationary Solutions for the "Sonic Vacuum" Equation

1.4.1 General Stationary Solution

The derived wave equations are different from traditional weakly nonlinear equations and can demonstrate unique properties. Let us look at the stationary solutions of Eq. (1.21) which can be found in the form $u(x - Vt)$. Introducing the strain as a new variable $\xi = -u_x$, we obtain from Eq. (1.21a):

$$\frac{V^2}{c^2}\, \xi_x = \frac{3}{2}\xi^{1/2}\, \xi_x + \frac{a^2}{8}\frac{(\xi\, \xi_{xx})_x}{\xi^{1/2}} - \frac{a^2}{64}\frac{\xi_x^3}{\xi^{3/2}}.$$ (1.27)

Equation (1.27) can be integrated, performing the replacement of variable $\xi = z^{4/5}$ and supposing that $\xi(x = +\infty) = \xi_0$, $\xi_x(x = +\infty) = 0$, $\xi_{xx}(x = +\infty) = 0$:

$$\frac{V^2}{c^2}z^{4/5} = z^{6/5} + \frac{a^2}{10}z^{1/5}z_{xx} + C_1,$$ (1.28)

where C_1 is an integration constant determined by ξ_0 and V (Nesterenko [1983b]). Equation (1.28) can be conveniently written in dimensionless form, with additional variable replacements:

$$y^{4/5} = y^{6/5} + y^{1/5}y_{\eta\eta} + C_2, \tag{1.29}$$
$$z = \left(\frac{V}{c}\right)^5 y, \qquad x = \frac{a}{\sqrt{10}}\eta,$$

where C_2 is a constant.

It is interesting to study the qualitative behavior of the solution of Eq. (1.29) for different values of constant C_2. For this purpose we rewrite Eq. (1.29) as follows with a new constant C_3:

$$y_{\eta\eta} = -\frac{\partial}{\partial y}W(y), \tag{1.30}$$
$$W(y) = -\frac{5}{8}y^{8/5} + \frac{1}{2}y^2 + C_3 y^{4/5}.$$

This form prompts use of the analogy with particle motion in the "potential field" $W(y)$ (Kunin [1975]) where η is the "time" and y is the "coordinate."

The "potential energy" $W(y)$ is an effective quantity and of course is different from the real potential energy of the particle's interaction.

From the expression for $W(y)$ one can find that for the values of constant C_3 in the interval $0 < C_3 < 5/27$, the function $W(y)$ has form of a curve with two extremes like $y_{1, II}$ and $y_{2, II}$ as shown in Figure 1.3 for the intermediate value $C_3 = 5/81$ (curve II). In Figure 1.3 various $W(y)$ curves correspond to cases $C_3 = 10/81$ (curve III) and $C_3 = 5/27$ (curve IV without extremes). Deviation of C_3 from zero, and being in the interval $0 < C_3 < 5/27$, is very important for the behavior of the function $W(y)$ and the nature of the solution. The solitary waves are possible only under this condition.

Indeed, near the maximum y_1, $W(y)$ can be expanded in powers of $(y - y_1)$:

$$W(y) \approx W(y_1) - \alpha(y - y_1)^2. \tag{1.31}$$

This form of potential energy leads, within the mechanical analogy, to an infinite "time" of particle fall with total energy $W(y_1)$ at the point y_1, which corresponds to the formation of a solitary wave (Kunin [1975]).

The zero value of constant C_3 (curve I in Fig. 1.3) corresponds to a special case when only periodic waves are allowed where the strain approaches a zero value. At the same time, $W(y)$ curves may approach curve I as close as desired through selection of the appropriate positive value of C_3. Physically it means that one hump of periodic solution corresponding to $C_3 = 0$ may represent solitary solutions for infinitesimally small initial prestrains.

The relation between the extremes of functions $W(y)$—y_1 and y_2 do not depend on C_3 and can be obtained from the extreme conditions $W_y(y_1) = W_y(y_2) = 0$ (Coste, Falcon, and Fauve [1997]):

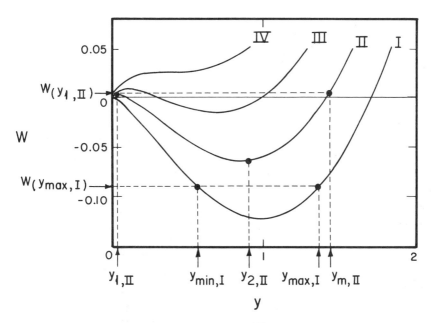

Figure 1.3. Effective "potential energy" $W(y)$ at different values of constant C_3: curve I, $C_3 = 0$; II, $C_3 = 5/81$; III, $C_3 = 10/81$; and IV, $C_3 = 5/27$.

$$y_2 = \left\{ \frac{1}{2} \left[1 - y_1^{2/5} + \sqrt{\left(1 - y_1^{2/5}\right)\left(1 + 3y_1^{2/5}\right)} \right] \right\}^{5/2} . \tag{1.32}$$

The general solution of Eq. (1.30) for periodical motions is well known (Landau and Lifshitz [1976]):

$$\eta = \eta_0 + \int_{y_0}^{y} \frac{dy}{\sqrt{-2\left[W(y) - W(y_{max})\right]}}, \tag{1.33}$$

where y_{max} is connected with the maximum strain in the periodic wave (in Fig. 1.3 shown for curve I, $y_{max,I}$).

Dependence of the phase speed of the periodic wave V_p on minimal and maximum strains ξ_{min} and ξ_{max} can be obtained from equations $W(y_{min}) = W(y_{max})$ and $W_y(y_2) = 0$:

$$V_p = c \left\{ \frac{2}{5\left(\xi_{max} - \xi_{min}\right)} \left[\frac{2\left(\xi_{max}^{5/2} - \xi_{min}^{5/2}\right) - 5\xi_2^{3/2}\left(\xi_{max} - \xi_{min}\right)}{\xi_{max} + \xi_{min} - 2\xi_2} \right] \right\}^{1/2} , \tag{1.34}$$

where ξ_2 is connected with the nondimensional variable y_2—the coordinate of the minimum of function $W(y)$ (Fig. 1.3): $\xi_2 = (V/c)^4 y_2^{4/5}$.

One can demonstrate that V_p approaches $c_0(\xi_2)$ in the case that $\xi_{min} \sim \xi_{max} \sim \xi_2$. It is necessary to be careful in opening the uncertainty in Eq. (1.34). For example, the introduction of natural (from the first look) relations $\xi_{min} = \xi_2 - \delta\xi$ and $\xi_{max} = \xi_2 + \delta\xi$, and expansion on the powers of $\delta\xi$, will not result in the correct answers because deviations of ξ_{min} and ξ_{max} from ξ_2 are not independent due to the equation $W(y_{min}) = W(y_{max})$. The correct way would be to introduce a small deviation in the first step ($\delta\xi_1 = \xi_{max} - \xi_{min}$) with expansion on its powers and then introduce another small quantity $\delta\xi_2 = \xi_2 - \xi_{min}$ with a subsequent secondary expansion.

The solitary phase speed V_s is a function of the minimum ξ_0 and maximum ξ_m strains corresponding to y_1 and y_m and can be determined from the equation $W(y_1) = W(y_m)$, taking into account that $W_y(y_1) = 0$. These two equations result in the equality

$$\frac{5}{8} y_1^{8/5} - \frac{3}{4} y_1^2 = -\frac{5}{8} y_m^{8/5} + \frac{1}{2} y_m^2 + C_3 y_m^{4/5}. \tag{1.35}$$

The physical meaning of the constant C_3 can be determined for a solitary type of the disturbance with the final state similar to the initial one using the condition $y_{\eta\eta} = 0$ at infinity. In this case,

$$C_3 = \frac{5}{4} \left(\frac{c}{V}\right)^4 \xi_0 \left\{ 1 - \xi_0^{1/2} \left(\frac{c}{V}\right)^2 \right\}, \tag{1.36}$$

where ξ_0 is the initial strain.

From Eqs. (1.35) and (1.36) (using relations between y and ξ) the dependence of the phase speed V_s of the solitary wave on the strain at infinity (ξ_0) and the strain at amplitude ξ_m may be found:

$$V_s = \frac{c}{\left(\xi_m - \xi_0\right)} \left\{ \frac{2}{5} \left[3\xi_0^{5/2} + 2\xi_m^{5/2} - 5\xi_0^{3/2}\xi_m \right] \right\}^{1/2}. \tag{1.37}$$

Eq. (1.37) for V_s can be obtained from Eq. (1.34) for V_p in this case if $\xi_{min} \rightarrow \xi_0$.

It is a useful exercise to show that the soliton phase speed V_s from Eq. (1.37) approaches the sound speed c_0 in the case that $\xi_m \rightarrow \xi_0$. In the next approximation the relation between V_s and the strain amplitude (Eq. (1.10)) typical for the KdV soliton can be recovered. The strain ξ_0 in Eq. (1.37) is a strain at infinity and at this point of analysis cannot be associated with the initial strain because it can be done only when the initial value-problem is solved. The initial-value problem will be solved through computer calculations for a discrete chain and it will be clear that ξ_0 in Eq. (1.37) and in others can be associated with the initial strain in the system.

It is important that the "potential energy" $W(y)$ at $0 < C_3 < 5/27$ ensures the existence of stationary solutions like only the compression solitary waves, as is clear from the relative positions of the maximum at y_1 (the "initial" point) and the minimum at y_2 ($y_2 > y_1$, Fig. 1.3). At $C_3 < 0$ or at $C_3 > 5/27$ these solutions, as well as solutions in the form of rarefaction waves, do not exist.

Thus, the strongly nonlinear periodic and solitary waves are the stationary solutions of Eqs. (1.20) to (1.22). It is important to mention that a solitary wave represents the extreme case of a periodic stationary solution (at $C_3 > 0$). The humps of the latter are separated under certain conditions by infinite space and for this reason they are considered representative of a single solitary wave. Actually, a single hump is possible as a stationary solution for a continuum approach only as being an extreme case of some periodic solution. This is physically clear from analogy with particle motion in the potential field, which allows only periodic motions including one with an infinite time of return—a solitary wave.

In the case when $C_3 = 0$, the periodic solution does not correspond to the sequence of infinitely remote humps, because sound speed in the space between them would be equal to zero and they cannot be interacting parts of one periodic solution. Instead these humps are connected to each other at zero strains. Each such hump at the same time may be considered as a representative of a solitary wave of Eq. (1.22) under infinitely small prestrain in a sense that the difference between them can be made as small as desired by managing the initial strain.

1.4.2   Properties of a Solitary Wave in Sonic Vacuum

The existence of a solitary wave as a stationary solution of the strongly nonlinear wave Eqs. (1.20) or (1.21) is a very interesting fact, because these equations are more general than the KdV wave equation. The latter is established in many publications as a basic wave equation for nonlinear problems practically in all types of materials. Let us consider the properties of this solitary wave in comparison with the KdV soliton described by Eq. (1.10).

First of all, this solitary wave is a supersonic one as well as the KdV soliton. In fact, restriction of $C_3 < 5/27$ guarantees a value of the phase velocity of a solitary compression wave larger than the initial velocity of sound, this can be easily verified from Eq. (1.31).

We can use another approach by simply comparing the phase velocity of a solitary wave V_s and the velocity of a long-wave sound c_0 which are equal, correspondingly,

$$V_s = c\xi_0^{1/4} y_1^{-1/5},$$

(1.38a)

$$c_0 = c\xi_0^{1/4}\left(\frac{3}{2}\right)^{1/2},$$

(1.38b)

where ξ_0 is the "initial" deformation caused by static compression. The expression for V_s is obtained from previous variable transformations, supposing that y_1 corresponds to the initial strain ξ_0, and the second one for c_0 can be obtained from the determination of the constants in Eqs. (1.3) and (1.20).

By comparing the above expressions, it is seen that the inequality $V_s > c_0$ is satisfied for $y_1 < (2/3)^{5/2}$. The given condition is valid for $C_3 < 5/27$, as it can be seen from the shape of $W(y)$. For $C_3 = 5/27$ we have $y_1 = (2/3)^{5/2}$. The constant C_3 determines the ratio of the soliton amplitude to the initial strain ξ_0.

Let us find the amplitude and spatial size of the solitary wave corresponding to the extreme case with infinitesimally small values of "initial" strain ξ_0. This can be expected to be mostly different from the weakly nonlinear case.

For $C_3 = 0$ a solution of Eq. (1.30) in the form of periodic waves, being a sequence of positive humps connected at the points with zero strains, can be obtained from the general solution (Eq. (1.33)). In a system moving with velocity V_p, this solution, written for the strain and the particle velocity, is

$$\xi = \left(\frac{5V_p^2}{4c^2}\right)^2 \cos^4\left(\frac{\sqrt{10}}{5a}x\right),$$

(1.39a)

$$\upsilon = V_p\left(\frac{5V_p^2}{4c^2}\right)^2 \cos^4\left(\frac{\sqrt{10}}{5a}x\right).$$

(1.39b)

At the local point

$$x = \frac{5a}{\sqrt{10}}\left(\frac{\pi}{2} \pm \pi n\right), \qquad n = 0, 1, ...,$$

(1.40)

ξ vanishes. This contradicts the condition $\xi > 0$, under which Eq. (1.20) was derived. Nevertheless, it is easy to verify that the solution described by Eq. (1.39) satisfies Eq. (1.20) at these points (Eq. (1.40)) as well.

For $C_3 \to 0$ the value of the maximum $W(y_1)$ tends to zero, as does the value of its coordinate y_1. The important point here is that the maximum at y_1 and a corresponding solitary solution exist at any indefinitely small value of constant C_3 which corresponds to the "initial" (the strain at infinity is also indefinitely small) compression strain of the chain ξ_0 (Eq. (1.36), Fig. 1.4).

If the value of the total "energy" in this case is $W(y_1)$, the behavior of the solution for $y \gg y_1$ coincides with the behavior of one hump described by Eq. (1.39), corresponding to the case $C_3 = 0$. Physically it means that the solitary

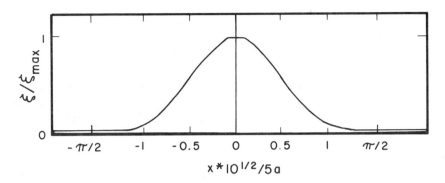

Figure 1.4. The shape of a solitary wave for very small initial compression.

shape can be taken as one hump of a periodic solution (Eq. (1.39)), with finite wave-length everywhere except in the vicinity of infinitesimally small ξ_0 (Fig. 1.4). Consequently, in this case, the maximum strain value ξ_m and the corresponding particle velocity υ_m in a solitary wave are close to the amplitude values of the periodic wave (Eq. (1.39)):

$$\xi_m = \frac{\upsilon_m}{V} = \left(\frac{5V_s^2}{4c^2}\right)^2 = \left(\frac{25}{16}\right)^{1/5} c^{-4/5} \upsilon_m^{4/5}. \tag{1.41}$$

This relation between V_s and ξ_m can also be obtained from the more general Eq. (1.37), supposing $\xi_0 = 0$. In this case, the characteristic spatial size of a soliton L_s is determined by the period of the solution described by Eq. (1.39), which equals

$$L_s = \left(\frac{5a}{\sqrt{10}}\right)\pi \approx 5a \tag{1.42}$$

with an accuracy of 0.65 %.

The phase speed of soliton V_s has a nonlinear dependence on maximum strain in the wave

$$V_s = \frac{2}{\sqrt{5}} c\xi_m^{1/4} = \left(\frac{16}{25}\right)^{1/5} c^{4/5} \upsilon_m^{1/5}. \tag{1.43}$$

Equation (1.43) can be used to prove that $V_s \ll c_0$ at some range of particle velocity and it makes the use of the Hertz law physically justified.

It can be concluded that the one hump described by Eq. (1. 39) is able to represent a solitary solution, which can be excited in an initially compressed chain with infinitesimally small initial strain. More exactly, the difference between one hump of a stationary solution in the case of $\xi_0 > 0$ and one hump of a periodic solution described by Eq. (1.39) can be made as small as desired by selecting an appropriately small ξ_0. Physically it means that this solution can be expected to describe the solitary impulse in an uncompressed chain, as was verified by computer calculations and in experiments (Nesterenko [1992d]; Sinkovits and Sen [1995]; Coste, Falcon, and Fauve [1997]; Chatterjee [1999]).

It should be emphasized that strongly nonlinear solitary waves with properties similar to those described in this section can be supported by the general interaction law between elements in discrete systems (see Sections 1.10 and 1.11) and are not only specific for the Hertz type of interaction.

1.4.3   Soliton with Compact Support

The finite spatial size of the soliton in a "sonic vacuum" is a unique feature in comparison with the KdV soliton (Nesterenko [1983b]; Lazaridi and Nesterenko [1985]). Chatterjee [1999] pointed out that one hump of periodic solution described by example Eq. (1.39a) for the interval $x \in (-\sqrt{10}\, \pi a/4, \sqrt{10}\, \pi a/4)$, in combination with setting $\xi(x) \equiv 0$ outside that interval, provides a function $\xi(x)$ that is three times differentiable as required, satisfies Eq. (1.21b) (provided that $\xi(x) \equiv 0$ is additionally accepted as a valid solution), and satisfies the basic conditions on the traveling wave solution. This approach only violates the condition under which Eq. (1.20) was derived $\xi(x) > 0$.

The most important properties of this soliton are independence of spatial size on amplitude and strongly nonlinear dependence of speed on particle velocity (or strain) in the crest (Nesterenko [1983b], Remoissenet [1999]).

Wright [1984], [1985] found analytically the finite length solitary wave (solitary wave with compact support) in a nonlinear elastic rod and pointed out the analogy between steady waves in elastic rods and granular materials.

There is also another example of the soliton with compact support (Scott and Stevenson [1984]; Takahashi and Satsuma [1988]; Takahashi, Sachs, and Satsuma [1990]) in very different media. This soliton is a solution of the so-called "magma" equation describing the flow of two kinds of fluid with different densities; for example, when low-density fluid is injected continuously from a tube at the bottom of a tank filled with a high-density fluid or when melt in the Earth's mantle migrates through the compacting media of partially molten rock by the buoyant force caused by the density difference between melt and matrix.

Recent results on solitons with compact support can be found in the book by Remoissenet [1999] and in papers by Rosenau and Hyman [1993], Kivshar [1993], Dusuel, Michaux, and Remoissenet [1998], Dinda and Remoissenet [1999], and others.

1.4.4   Convective Term Can Be Small for the Stationary Wave

The convective derivative was dropped during the derivation of Eq. (1.20) without any justification. The physical reason for this could only be the small particle velocity in comparison with the phase speed of the wave. It cannot be done a priori because there is no characteristic wave speed in Eq. (1.20) for a strongly nonlinear case. Having introduced the phase speed of stationary wave V, it can be proved that the previously dropped nonlinear term, connected with the convective derivative, is really small in comparison with the kept terms. It has the order of magnitude

$$\upsilon \frac{\partial \upsilon}{\partial x} \sim \xi^2 \frac{V^2}{L}. \tag{1.44}$$

Here υ, ξ, V, L are typical parameters in the wave for particle velocity, strain, phase velocity, and wave width. The orders of magnitude of the corresponding nonlinear terms at the right part of Eq. (1.22) are

$$c^2\left(\left(-u_x\right)^{3/2}\right)_x \sim \frac{c^2 u^{3/2}}{L^{5/2}} \sim \frac{c^2 \xi^{3/2}}{L}, \tag{1.45a}$$

$$c^2 \frac{a^2}{10}\left(\left(-u_x\right)^{1/4}\left(\left(-u_x\right)^{5/4}\right)_{xx}\right)_x \sim \frac{c^2 \xi^{3/2} a^2}{L^3}. \tag{1.45b}$$

The ratio of last terms in Eqs. (1.45) and Eq. (1.44) have the orders of magnitude $\sim \xi$ and $\xi L/a$, correspondingly, taking into account Eq. (1.43) for V_s and supposing $\xi \sim \xi_{max}$. That is why the relatively small strain ξ (but not in connection with the initial one ξ_0!), ensures the neglect of the convective derivative in comparison with other nonlinear terms. In a specific case of granular material with steel particles, strain of the order of magnitude $\xi \sim 10^{-3}$ and $L/a \sim 5$ at particle velocity ~ 1 m/s, it makes this hypothesis suitable (also see Problem 23).

The considered solitary waves cannot be associated with true solitons without investigation of their behavior in mutual interactions. For example, equation "φ^4" has stationary solutions in the form of a solitary wave, kink, or antikink, which are not true solitons. However, the comparison of solitary waves for Eq. (1.20) to (1.22) with the soliton solution of a discrete system of particles (Eq. (1.3)), which will be done in Section 1.4, provides some hope that this new type of solitary wave may be associated with a true soliton. To make a final decision, the analysis of nonstationary regimes for Eq. (1.20) is necessary.

1.4.5 Stationary Shock Waves

In the presence of dissipation in the medium, viscous-like terms should be added to Eq. (1.20). The exact type of dissipation does not effect the equilibrium state behind shock. For $C_3 \to 0$ the strain and particle velocity behind the front of the stationary shock correspond to the minimum $W(y_2)$ and equal the values ξ_c and υ_c which depend on the shock speed V_{sh}:

$$V_{sh} = c\xi_c^{1/4},$$

(1.46a)

$$V_{sh} = c^{4/5}v_c^{1/5}.$$

(1.46b)

This type of "Hugoniot" is different in comparison with traditional ones ($V_{sh} = c_0 + \text{const } v_c$). Comparison of Eqs. (1.46a,b) with corresponding equations for a soliton (Eq. (1.43)) supports the conclusion that at the same phase speeds the final equilibrium strain behind the shock is always smaller than in the soliton maximum.

If $\xi_{m,sh}$ represents the maximum possible amplitude of oscillations in the case of shock wave (for very weak dissipation), then (according to the potential energy analogy) the leading hump with this strain must be close to a soliton. From Eq. (1.41) the speed of this soliton V_{sh}, which is equal to the speed of a stationary shock, can be connected with $\xi_{m,sh}$. By comparing the quantities $\xi_{m,sh}$ and ξ_c from Eq. (1.46a) one can see that the ratio of maximum strains in the front of the stationary shock wave to its established value behind the front (and the corresponding values of particle velocities) reaches the value

$$\frac{\xi_{m,sh}}{\xi_c} = \frac{v_{m,sh}}{v_c} = \left(\frac{5}{4}\right)^2 \approx 1.56.$$

(1.47)

It is important to emphasize that on the stage of establishment of the steady shock the amplitude of the leading hump may exceed this value.

The difference found in the maximum and final equilibrium strains leads to the ratio of the maximum pressure at the front of the stationary shock wave P_m to the established pressure behind P_c:

$$\frac{P_m}{P_c} = \left(\frac{\xi_m}{\xi_c}\right)^{3/2} = \left(\frac{5}{4}\right)^3 \approx 1.95.$$

(1.48)

Further investigation of the strongly nonlinear one-dimensional chain of oscillators, interacting at compressive loading according to the Hertz law, is carried out numerically and may be found in Section 1.5.

1.5   Stability of Nonlinear Periodic Waves in "Sonic Vacuum"

In Section 1.3, it was shown that the essentially nonlinear wave Eq. (1.20) for the "sonic vacuum" has qualitatively new periodic and solitary solutions. A partial exact periodic solution of this equation was found in explicit form (Eq. (1.39)). It allows straightforward application of the Whitham method (Whitham [1970]) for the investigation of the modulation stability of this solution.

"Sonic vacuum" Eqs. (1.20) to (1.22) are the Euler equations for the variational principle

$$\delta \iint L \, dt \, dx = 0 , \tag{1.49}$$

where the Lagrangian density L is described by Eq. (1.23).

Following Whitham, we will seek the solution of Eq. (1.20) or (1.21) and (1.22) in the form

$$u = \frac{\Psi(T, X)}{\varepsilon} + \varphi(\theta, T, X) + \cdots . \tag{1.50}$$

This form represents the basic hypothesis that the solution contains fast and slowly varying parts in time and space. Here, points denote neglected terms of higher order in ε; $T = \varepsilon t$, $X = \varepsilon x$ are the slow variables; $\theta = \theta(T, X)/\varepsilon$ is the phase; ε is a small parameter, which is the ratio of the length of the periodic wave to that of the modulation wave. The functions Ψ, φ, and θ are to be determined. The quantities $k = \theta_X$ and $\omega = \theta_T$ are the local wave number and frequency.

Substituting Eq. (1.50) into Eq. (1.23), we can obtain L_0 in the zeroth approximation (Gavrilyuk and Nesterenko [1993]):

$$L_0 = \frac{\left(\Psi_T - \omega\varphi_\theta\right)^2}{2} - \frac{2c^2\left(-\Psi_X - k\varphi_\theta\right)^{5/2}}{5} \tag{1.51}$$

$$- \frac{c^2 a^2}{2}\left\{\frac{3}{8}\left(-\Psi_X - k\varphi_\theta\right)^{1/2} k^4 \varphi_{\theta\theta}^2 - \frac{1}{3}\left(-\Psi_X - k\varphi_\theta\right)^{3/2} k^3 \varphi_{\theta\theta\theta}\right\}.$$

It is useful to introduce the following notation:

$$\xi = -\Psi_X - k\varphi_\theta , \qquad V = \frac{\omega}{k} . \tag{1.52}$$

Using Eq. (1.52), the Lagrangian represented by Eq. (1.51) can be transformed into the form

$$L_0 = \frac{\left(\Psi_T + V\left(\Psi_X + \xi\right)\right)^2}{2} - \frac{2c^2 \xi^{5/2}}{5} - \frac{c^2 a^2 k^2}{2}\left\{\frac{3}{8}\xi^{1/2}\xi_\theta^2 + \frac{1}{3}\xi^{3/2}\xi_{\theta\theta\theta}\right\}. \tag{1.53}$$

The dependence of ξ on θ is determined by varying Eq. (1.50) with respect to φ. It results in the equation

$$\frac{V^2}{c^2}\xi_\theta = \frac{3}{2}\xi^{1/2}\xi_\theta + \frac{a^2 k^2}{8}\left\{\xi^{-1/2}\left(\xi\xi_{\theta\theta}\right)_\theta - \frac{\xi_\theta^3 \xi^{-3/2}}{8}\right\} . \tag{1.54}$$

As is shown in Section 1.3, Eq. (1.54) has the exact partial solution

$$\xi = \left(\frac{5V^2}{4c^2}\right)^2 \cos^4\left(\frac{\sqrt{10}}{5ak}\theta\right),$$

(1.55)

which vanishes to zero at discrete values of θ. Formally, Eq. (1.54) has a singularity at these points. Its derivation was based on the inequality $\xi > 0$. However, direct substitution proves that in fact there is no singularity for this specific solution. It is the result of a very smooth behavior of solution in the vicinity of these points. The stability of the solution described by Eq. (1.55) will be studied.

Let us introduce the averaging operator

$$\langle\cdot\rangle = \frac{1}{\tau}\oint\cdot d\theta,$$

(1.56)

where integration by period τ is done. From Eqs. (1.51) and (1.52), it follows that

$$\langle\xi\rangle = -\Psi_X,$$

(1.57)

$$\langle L_0\rangle = \frac{\Psi_T^2 - V^2\Psi_X^2 + V^2\langle\xi^2\rangle}{2} - \frac{2c^2\langle\xi^{5/2}\rangle}{5} - \frac{c^2 a^2 k^2}{2}\left\{\frac{3}{8}\langle\xi^{1/2}\xi_\theta^2\rangle + \frac{1}{3}\langle\xi^{3/2}\xi_{\theta\theta\theta}\rangle\right\}.$$

It is useful for further transformation to get an expression for the integral of the power function of cosine (Gradshteyn and Ryzhik [1980]):

$$\frac{1}{\tau}\oint\cos^{2n}\left(\frac{\sqrt{10}}{5ak}\theta\right)d\theta = \frac{1}{\pi}\oint\cos^{2n}(\varpi)\,d\varpi = \frac{(2n-1)!!}{2^n n!}.$$

(1.58)

Using this expression we can get values of the averaged functions in Eqs. (1.57) based on the solution of Eq. (1.55):

$$\langle\xi\rangle = \left(\frac{5}{4}\frac{V^2}{c^2}\right)^2\frac{3}{8} = -\Psi_X, \quad \langle\xi^2\rangle = \left(\frac{5}{4}\frac{V^2}{c^2}\right)^4\frac{7!!}{2^4 4!}, \quad \langle\xi^{5/2}\rangle = \left(\frac{5}{4}\frac{V^2}{c^2}\right)^5\frac{9!!}{2^5 5!},$$

(1.59)

$$\langle\xi^{1/2}\xi_\theta^2\rangle = \left(\frac{5}{4}\frac{V^2}{c^2}\right)^5\frac{32}{5a^2k^2}\langle\cos^8\varpi - \cos^{10}\varpi\rangle = \left(\frac{5}{4}\frac{V^2}{c^2}\right)^5\frac{32}{5a^2k^2}\left(\frac{7!!}{2^4 4!} - \frac{9!!}{2^5 5!}\right),$$

$$\langle\xi^{3/2}\xi_{\theta\theta\theta}\rangle = \left(\frac{5}{4}\frac{V^2}{c^2}\right)^5\left(\frac{\sqrt{10}}{5ak}\right)^3 8\langle(3\cos^7\varpi - 8\cos^9\varpi)\sin\varpi\rangle = 0.$$

Since

$$\frac{5}{4}\frac{V^2}{c^2} = \sqrt{-\frac{8}{3}\Psi_X},$$

(1.60)

then

$$\langle L_0\rangle = \frac{\Psi_T^2}{2} - \frac{10\sqrt{2}}{9\sqrt{3}}c^2(-\Psi_X)^{5/2}.$$

(1.61)

The Lagrangian (Eq. (1.61)) averaged on an exact partial solution (Eq. (1.55)) is analogous to the first two terms of the Lagrangian L for the investigated system (Eqs. (1.19) to (1.21)). The difference is only in coefficients, with

$$\frac{10\sqrt{2}}{9\sqrt{3}} > \frac{2}{5} . \tag{1.62}$$

Thus, for Ψ we have the equation

$$\Psi_{TT} + \frac{25\sqrt{2}}{9\sqrt{3}}c^2\left(\left(-\Psi_x\right)^{3/2}\right)_x = 0 , \tag{1.63}$$

which is equivalent to the equations of the one-dimensional theory of nonlinear elasticity or to the equations of motion of a barotropic gas written in Lagrangian mass coordinates.

In a system described by Eq. (1.63), a "gradient catastrophe" is probable if we have an increasing profile of $-\Psi_x$, resulting in shock waves. It may be overcome if the corrections of higher order in ε are allowed. The corresponding equation is likely to be analogous to Eq. (1.19) to (1.21). In this case, one can expect the envelope solitons to appear.

In Eq. (1.63) the following parameter c_1 plays the role of the local phase velocity of the envelope wave with deformation $-\Psi_x$:

$$c_1\left[\left(-\Psi_x\right)\right] > c_0\left[\left(-\Psi_x\right)\right]. \tag{1.64}$$

It is interesting to compare the ratio of c_1 with other characteristic velocity values in the present system, for example, with long-wave sound speed c_0 at the same deformation $-\Psi_x$. For c_0 we have the expression (Eq. (1.38))

$$c_0^2 = \frac{3}{2}\left(-\Psi_x\right)^{1/2}c^2. \tag{1.65}$$

From Eqs. (1.64) and (1.65), it is easily seen that by virtue of Eq. (1.59) the following relationship is valid:

$$c_1^2 = \frac{25}{3\sqrt{6}}\left(-\Psi_x\right)^{1/2}c^2 . \tag{1.66}$$

The hyperbolic character of Eq. (1.63) implies the stability of the solution described by Eq. (1.55). Note that the corresponding gas dynamics analogy was also found in describing rapidly oscillating solutions of the equations of a bubbly liquid with an incompressible liquid phase at oscillations close to resonance ones, as was shown by Gavrilyuk [1989].

1.6 Wave Dynamics of Unstressed One-Dimensional Granular Materials—Numerical Calculations

1.6.1 Equation of Motion and Control Procedures

The analytical treatment of an essentially nonlinear system of particles is very restricted and, actually, all known results for particles interacting by the Hertz law are described in previous sections. The numerical solution of this system allows investigation of a nonstationary regimes of wave motions and comparison with analytical results for stationary cases and with experimental data.

For the numerical study of wave processes in the one-dimensional chain of particles with arbitrary radii R_i the second-order equations of motion were reduced to a first-order system of equations (Nesterenko [1983b]):

$$\dot{x}_i = F_i(x), \qquad x = (x_1, x_2, ..., x_{2N}), \qquad i = 1, ..., 2N;$$

$$F_i(x) = \varphi_i(x) - \psi_i(x), \qquad i = 1, ..., N;$$

$$F_i(x) = x_{i-N}, \qquad i = N+1, ..., 2N;$$

$$\varphi_i(x) = \left(\frac{R_i R_{i-1}}{R_i + R_{i-1}} \right)^{1/2} \frac{E\delta_{i-1}^{3/2}}{2\pi\rho_0(1-\nu^2)R_i^3}, \qquad \text{if } \delta_{i-1} > 0, i = 2, ..., N;$$

$$\varphi_i(x) = 0, \qquad \text{if } \delta_{i-1} \leq 0, i = 2, ..., N;$$

$$\psi_i(x) = \left(\frac{R_i R_{i+1}}{R_i + R_{i+1}} \right)^{1/2} \frac{E\delta_i^{3/2}}{2\pi\rho_0(1-\nu^2)R_i^3}, \qquad \text{if } \delta_i > 0, i = 1, ..., N-1;$$

$$\psi_i(x) = 0, \qquad \text{if } \delta_i \leq 0, i = 1, ..., N-1;$$

$$\delta_{i-1} = R_i + R_{i-1} - (x_{N+i} - x_{N+i-1}), \qquad i = 2, ..., N;$$

$$x_{N+i}(t=0) = x_{N+i-1} + R_i + R_{i-1}, \qquad i = 2, ..., N;$$

$$x_{N+1}(t=0) = 0.$$

(1.67)

The quantities x_i for $i = 1, ..., N$ are the velocity of the ith particle, while for $i = N+1, ..., 2N$ they are the coordinates values. The initial conditions for the velocities and forces at the chain ends are specified below for each separate case. A similar system of equations for identical particles was used by Herrmann and Seitz [1982] to describe the multiple collisions in "Newton's Cradle." It will be discussed later in this chapter.

For the numerical solution of the system of nonlinear Eqs. (1.67), the Hamming and fourth-order Runge–Kutta methods were used. For convenience of comparison with experiment the problem was solved in real time. The physically based choice of the initial time step Δt was made by starting from the natural physical condition

$$T_1 = 2.94 \left(\frac{5}{8A} \right)^{2/5} \frac{1}{v_0^{1/5}} \gg \Delta t, \tag{1.68}$$

where T_1 is the impact time of two spheres with relative initial velocity v_0 (Landau and Lifshitz [1995]). In the following the size of the time step was varied in order to reach a compromise between computational accuracy and computer time. In most calculations the time step was equal to $0.25 \cdot 10^{-5}$ s.

Control of the solution was managed by the momentum and energy conservation laws and through comparison of the results obtained by various methods. During the calculation the momentum conservation law is satisfied by an accuracy no less than $10^{-5}\%$, and the total energy conservation law was satisfied within 10^{-3} to $10^{-2}\%$.

To verify the correctness of the calculation we also compared the numerical solution of the problem of the impact of a single particle with a chain of 100 identical particles, arranged with relatively large initial gaps with its obvious exact solution—the consecutive movement of a single particle. This comparison showed coincidence of the numerical and exact solutions to within $10^{-2}\%$.

Chatterjee [1999] presented results of computer calculations with much higher accuracy (the energy error is $2 \cdot 10^{-8}\%$) but with the same qualitative results, the difference in soliton parameters not exceeding a few percent.

1.6.2 Dimensionless Analysis of a Discrete Chain

Very important conclusions can be drawn using dimensionless analysis of the "exact" equations for a discrete chain. Nesterenko [1983b] pointed out that for $\delta_0 = 0$ and identical R_i Eqs. (1.3), (1.67), with variable replacement,

$$\zeta_i = \frac{u_i}{v_0}, \qquad \tau = t \left(A^2 v_0 \right)^{1/5}, \tag{1.69}$$

are transformed into dimensionless form

$$\ddot{\zeta}_i = \left[\int_0^\tau (\zeta_{i-1} - \zeta_i) \, d\tau \right]^{3/2} - \left[\int_0^\tau (\zeta_i - \zeta_{i+1}) \, d\tau \right]^{3/2}. \tag{1.70}$$

In Eq. (1.70) the dot means derivative with respect to the dimensionless time τ. Clearly, for a chain with free ends ($\varphi_1 = \psi_N = 0$), identical τ values correspond to identical ζ_i values if the problem is solved with the initial conditions

$$\zeta_i(t=0) = 1, \quad i = 1, ..., k; \quad \zeta_l(t=0) = 0, \quad k < l \le N. \tag{1.71}$$

The same conclusion is true for a piston problem or for a stationary pulse. This comment is very important when finding the temporal width and wave speed dependencies on the amplitude of the stationary solitary wave as soon as its existence is established. It clearly predicts the independence of normalized particle velocity profiles in solitary waves (and of course their space width) on amplitude: according to Eqs. (1.69), (1.70) the dimensionless particle's velocities, normalized by velocity amplitude in a solitary wave, are just repeated in the corresponding dimensionless times (Nesterenko [1983b]).

A similar conclusion is also valid for the problem of a piston moving one of the ends of the chain with constant velocity when the overall motion is not a stationary one. Therefore, an additional control method was verification of the observed equality of the dimensionless particle velocity ζ_i at identical moments τ for problems with different values of initial velocities which included soliton formation and nonstationary "shocks" (Nesterenko [1983b], [1992d], [2000b]).

1.6.3   Solitary Waves Formation in a Chain of Granules with Free Ends and Chain Fragmentation

1.6.3.1   Solitons Formation

A chain with free ends has zero sound speed. This case is probably the most interesting one and it is possible to expect qualitative differences with systems having nonzero sound speed, like a chain which is precompressed. In this subsection, all particle radii R_i are identical and equal to $R_i = 3 \cdot 10^{-3}$ m. The following values were chosen for the quantities A and ρ_0, typical for steel particles: $A = 5.6 \cdot 10^{12}$ N/(m$^{3/2}$ kg), $E = 200$ GPa, $\rho_0 = 7.8 \cdot 10^3$ kg/m^3.

Let us consider the case where the initial conditions for particle velocities correspond to the impact of two particles on the left end of the chain:

$$x_1(t=0) = x_2(t=0) = \upsilon_0 = 5 \text{ m/s}; \quad x_i(t=0) = 0, \quad N \ge i > 2, \quad N = 100. \tag{1.72}$$

Boundary conditions correspond to free ends of the chain $\varphi_1 = \psi_N = 0$. The time dependencies obtained for x_1/υ_0 are shown in Figure 1.5. It is seen that for these initial data two solitary waves are formed with material being practically at rest between them.

The process of an initial perturbation decomposition into solitary waves is very quick and already completed on the twentieth particle, so the typical length of impulse travel resulting in its decomposition (40R) is comparable with the width of the solitary wave (10R). After that, each of the solitary waves propagates as a stationary one. As a consequence of formation of solitary waves the final velocity of the last particle after its detachment from the chain at the

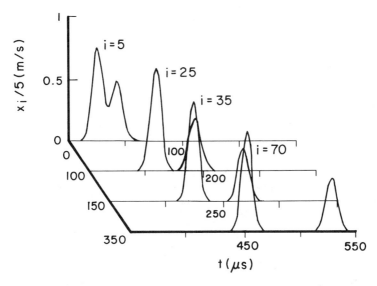

Figure 1.5. Generation of two solitary waves by the impact of the first two particles (moving at the same velocity 5 m/s) upon the left end of the chain.

right free end (x_N) differs from v_0. For example, for different length of the chain we obtain $x_{20}/v_0 = 1.2338$, $x_{50}/v_0 = 1.2342$, $x_{100}/v_0 = 1.2342$. These values are in perfect agreement with results by Hinch and Saint-Jean [1999]. The equality of the final velocities $x_{50} = x_{100}$ serves as further verification of the stationary nature of a large-amplitude solitary wave at least after traveling the distance equal to $100R$ (50 particles).

Since the solitary wave amplitudes are different, the distance between them will continuously increase. A completely similar pattern is also observed for other values of the initial velocities of the first two particles with varying time scale according to Eq. (1.69). It verifies the numerical results.

The impact caused by the first four, five, and six particles with the same initial velocity was additionally investigated. In this case, the perturbation was decomposed correspondingly into four, five, and six solitary waves (Nesterenko [1983b], [1992a]; Hinch and Saint-Jean [1999]).

If only one first particle impacts a system, a single soliton is formed in the system (Nesterenko [1983b]; Chatterjee [1999]; Hinch and Saint-Jean [1999]) in contrast to the famous demonstration of momentum conservation where momentum is supposed to be concentrated in one single particle moving through a system ("Newton's Cradle").

Numerical results demonstrating a solitary wave in this chain are consistent with the existence theorem for solitary waves on lattices by Friesecke and Wattis [1994]. This is because the Hertz interaction force evidently results in potential energy being superquadratic with the other properties required being satisfied for the existence of a solitary wave, as was remarked by MacKay [1999].

1.6.3.2   Chain Fragmentation

It is necessary to mention that due to the free ends of the chain there occurs a rebound of the first particles and a "destruction" of the initial part of the system behind traveling stationary compression waves. So the behavior of the whole system is nonstationary, despite the formation of stationary solitary waves.

For example, at the moment $t = 2 \cdot 10^{-4}$ s, when the second wave of smaller amplitude (Fig. 1.5, impact by two particles) is located near the twenty-seventh particle, the particles from first to seventh bounced back and shows a negative velocity. Those from first to eighth are separated from each other by a distance between their centers larger than $2R$. The velocity values of the first seven particles are equal to -0.11, -0.048, -0.013, -0.0068, -0.0029, and -0.0004 m/s; destruction of the chain is restricted by the first 10 particles. The rebounded part removed 2% of the initial value of momentum and 0.04% of energy. So, the "destruction" of the system from a certain moment in time does not affect the soliton shape because the velocity of the leading part of the "fracture" wave is much smaller than the velocity of the solitary wave (Nesterenko [1983b]).

It is seen that the velocity of the rebound particles are decreasing very fast with particles number. Hinch and Saint-Jean [1999] found that in the case of impact by one ball, its velocity after rebound is equal to $-0.07v_0$ and the velocity of the nth ball from the end is equal to $-0.084v_0 e^{-0.55n}$. It allows the propagation of a strongly nonlinear steady compression impulse even in the chain with zero tensile strength.

At the same time, the very fact of destruction renders that it is impossible to accurately determine the wave amplitude from the initial conditions in any continuum description of this chain with zero tensile strength.

The high accuracy of numerical calculations by Chatterjee [1999] provide a detailed picture of soliton propagation and system "fracture" due to the impact by one ball. For example, he found a very fast drop of rebounding velocity magnitudes with a ball number starting from the impacted end. The practically zero velocity of balls is maintained behind the propagated soliton, and is 12 orders of magnitude smaller than the velocity in the soliton maximum.

Hascoet, Herrmann, and Loreto [1999] investigated the impulse behavior in another example of a strongly nonlinear chain where the contact law is described by a step function with a linear part for compression and no force when two beads are not in contact.

This example clarifies the influence of one type of high nonlinearity, broken contacts under tension and represents another system, which is qualitatively modified by the propagating wave itself. The sequence of peaks with decreasing amplitudes and velocities were observed and the scaling law for the decay of the main peak was found. This type of strong nonlinearity was not able to ensure the steady propagation of pulses contrary to the case with the Hertz type of nonlinearity. They observed that for the appropriate boundary conditions (one end of the chain is fixed) and for the initial condition corresponding to the prescribed initial compression of the two neighboring end particles, the behavior of the propagated impulse is qualitatively different in comparison with the classical harmonic chain with linear interaction laws for both compression and tension. The difference includes the amplitudes of the displacements in the leading peaks as well as the behavior of impulse profiles.

The qualitative sensitivity of the fragmentation behavior of a cohesionless chain of balls on the interaction law (linear and Hertz law) under impact by one particle was demonstrated by Hinch and Saint-Jean [1999]. In the first case no stationary solitary wave was formed and the amount of balls flying off from the far end increased with the number of balls in chain (N) as $1.5N^{1/3}$, the furthest at a velocity $1.4v_0N^{-1/6}$, and the others at similar but slower speeds. The majority of balls rebound, the impacting ball at $-0.13v_0$ and the nth ball from the end at a velocity of $-0.16v_0n^{-5/6}$ at large n.

This behavior is in striking contrast to the Hertz law where only two balls fly off from the far end with velocities of $0.986v_0$ and $0.149v_0$, that do not depend on the number of balls, when the majority hardly move. This is apparently the result of stationary wave formation. The velocity of the impacting particle after rebound is two times smaller and equal to $-0.07v_0$, and the velocity of the nth ball from the end is equal to $-0.084v_0e^{-0.55n}$.

The fragmentation behavior of chains is a fascinating subject because it combines features that can be explained in the continuum approach with others that may be interpreted only based on system discreteness. The important point is that the momentum and energy of the solitary waves (which can be described in continuum approximation) propagating into the chain together with initial conditions predetermine the total momentum and energy of particles in the fragmented end. But the velocity distribution of the separated particles probably cannot be described, using the results of continuum approximation.

1.6.3.3   Solitons and "Newton's Cradle"

"Newton's Cradle" is a chain of five or six steel balls suspended on ropes and arranged in a straight line, and which is frequently used in a lecture demonstration of impulse and energy conservation (see, e.g., web page http://www.physics.gla.ac.uk/~kskeldon/PubSci/exhibits/D10/). The impact at one end of this chain by a few (say one, two or three) balls, which were simultaneously pulled to one side and released, results in an equal number of balls that break away from the other end having the same velocity. This motion is repeated without noticiable changes through many cycles, and is called "dispersion free" behavior by Herrmann and Schmalzle [1981].

Thus, momentum and energy are apparently conserved as evidenced by the fact that velocities are transferred without significant losses manifesting by the same angle of deflection of breaking balls. In some textbooks (see, e.g., citations in Herrmann and Schmalzle [1981]) this behavior is considered as the only possible consequence of the conservation laws of energy and momentum—a thought which is common even these days among students and even faculty.

At the same time, it was recognized relatively long ago (Chapman [1960]; Herrmann and Schmalzle [1981]; Herrmann and Seitz [1982], and others) that when collisions among a group of more than two spheres are involved, the infinite number of velocity combinations will satisfy two conservation laws. In experiments about 10% of the initial momentum is left in the impacting particle and the end particle on the opposite side of the chain does not have all the energy, as shown by Chapman [1960].

The mainly "dispersion-free" behavior according to Herrmann and Seitz [1982] is due to the fact that after the first collision sequence the chain of balls with five particles is in a different state that it was initially. They emphasized that "as a consequence of the slight dispersion within the initial chain, the balls no longer touch after the first collision sequence." Because each ball is separated from its neighbors by a small distance, it results in a "dispersion-free" propagation through the subsequent sequence of binary impacts.

Despite that, these authors did not identify the new type of stationary wave emerging through the transition period from impact by one particle, they used a system of equations similar to Eqs. (1.67) and made very interesting observations of the process including the difference between linear and nonlinear chains. They also noticed that the propagation time of perturbation measured from impact by the first ball to the interruption of contact between the last balls was approximately proportional to the length of the system varying from 2 to 15 balls. The effective velocity of perturbation found from this data was about 585 m/s—much less than the speed of sound in steel. This speed was not and cannot be identified with some stationary wave process because of a relatively long transition period probably when the solitary wave was established. Nevertheless, it gives a right order of magnitude for the possible value of perturbation speed under given conditions of impact.

As is known now the chain of contacting balls is not "dispersion free," on the contrary it is characterized by a strongly nonlinear dispersion resulting in a unique solitary wave. As was demonstrated in computer calculations and in experiments, momentum and energy conservation can be equally satisfied by the motion of a sequence of solitons, with each of them having five balls. Impact by one particle results in one soliton, and impact by two particles in two solitons after the initial impulse travels a distance equal approximately to 10 to 20 particles diameters (Nesterenko [1983b]). It is true that impact by one particle is followed mainly by the break of one particle from the other end, however, approximately 10% of momentum is carried by the second particle also (Chatterjee [1999]). The difference is even more striking with impact by two balls resulting in the creation of two solitons with the separation distance between them being a function of the length of "Newton's Cradle."

So this toy (if long enough, say, 30 balls) can be a simple and excellent example of complex highly nonlinear wave dynamics, in addition to the demonstration of energy and momentum conservation for small numbers of balls. In the latter case there is too little time for solitons to be formed simply due to a small propagation distance and not due to the absence of dispersion.

1.6.4 Interaction of the Solitary Waves

1.6.4.1 Collision of Solitary Waves Moving in Opposite Direction

To create solitary waves with different amplitudes moving in opposite directions the following initial velocity conditions were used:

$$x_1(t = 0) = x_2(t = 0) = 5 \text{ m/s}, \qquad x_{100}(t = 0) = x_{99}(t = 0) = -5 \text{ m/s}, \qquad (1.73)$$
$$x_i(t = 0) = 0, \qquad i \neq 1, 2, 100, 99.$$

These conditions correspond to the impact of two particles onto the opposite ends of the chain. The ends of the chain are assumed to be free ($\varphi_I = \psi_N = 0$), and the total number of particles $N = 100$.

The result of the collision is shown in Figure 1.6, where variation of the particle velocities for the twenty-fifth and seventy-fifth particles with time are depicted before and after collision. It is evident that, first, there is a decomposition of initial disturbances mainly into four solitons—two pairs moving toward each other in opposite directions. Until the collision, each of the pairs moves similarly to the case described in the previous subsection (Fig. 1.5). Following the interaction in the middle of the chain, the solitary waves

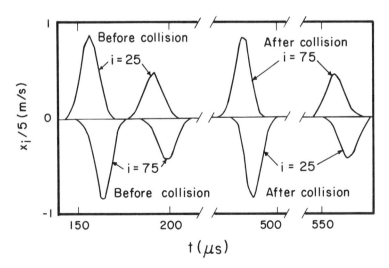

Figure 1.6. Interaction of two identical pairs of solitary waves moving in the opposite direction.

propagate without changing their shapes (Fig. 1.6). Only phase changes occur. The difference in the arrival time of the first velocity maximum at the seventy-fifth particle between this case and the case corresponding to Figure 1.5 is around 10 μs. A similar solitons interaction without a change in their shape was observed during reflection from a rigid wall of a sequence of six solitary waves, generated by the impact of a piston with mass $5m$ ($N = 80$) (Nesterenko [1983b]; Lazaridi and Nesterenko [1985]).

The secondary spawned solitons after collision of the two original solitons were observed by Manciu, Sen, and Hurd [1999a]. Energy contained in these secondary solitons is less than 0.5% of the energy contained in the original solitons. It means that solitary waves in a discrete chain may be considered as true solitons only within some approximation.

Calculations with a larger number of particles and with strict control of the stationarity of pulses before collision are desirable to clarify the details of interaction and the final conclusion about the nature of solitary waves.

1.6.4.2   Overtaking of Solitary Waves by One with a Larger Amplitude

Solitary waves with different amplitudes moving in the same direction were created using the following initial conditions for velocities:

$$x_1(t = 0) = x_2(t = 0) = 10 \text{ m/s}, \qquad x_{18}(t = 0) = x_{19}(t = 0) = 5 \text{ m/s}, \qquad (1.74)$$
$$x_i(t = 0) = 0, \qquad i \neq 1, 2, 18, 19.$$

The ends of the chain are assumed to be free ($\varphi_1 = \psi_N = 0$), the total number of particles $N = 150$. These conditions correspond to the simultaneous impact of two particles onto the left end of the chain and the "inner" impact by a pair of particles inside the chain with smaller velocity.

These initial conditions resulted in formation of the leading group composed from three solitary waves with different amplitudes moving in the same direction (Fig. 1.7). The time interval $2 \cdot 10^{-4} < t < 3 \cdot 10^{-4}$ s corresponds to the velocity profile of the 50th particle, the interval $4 \cdot 10^{-4} < t < 5.5 \cdot 10^{-4}$ s to the velocity profile of the 85th particle, and the interval $t > 8 \cdot 10^{-4}$ s to the velocity of the 145th particle. In the latter case soliton 2 is not shown, since for $t > 5 \cdot 10^{-4}$ s it does not participate in the interaction between solitary waves. As seen from Figure 1.7 the solitary wave 3 subsequently has passed waves 2 and 1.

The amplitude and shape of the solitary waves did not change during these interactions. It is interesting that for the used initial conditions a total of six solitary waves were formed in the system, despite the fact that the initial perturbation, consisting only of two moving particles, decomposes into two solitary waves. The given fact is the consequence of the nonlinear nature of the particle interaction.

Thus, compression solitary waves originating at the decomposition of the initial disturbances in the discrete, strongly nonlinear (interaction law without the linear part under compression, zero tensile strength) chain of particles

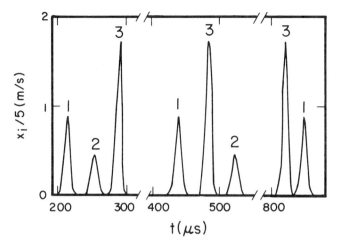

Figure 1.7. Catching up of two solitary waves with small amplitudes (1, 2) by a larger solitary wave (3) moving in the same direction.

possess the main properties of solitons. They are stationary waves, which do not change their properties on interaction with each other. The aforementioned similarity between these solitons and solitary solutions of the continuum Eq. (1.19) for the sonic vacuum can put forward a hypothesis that the latter are also solitons. Nevertheless, this statement was not yet proved analytically for nonstationary perturbations satisfying Eq. (1.19).

1.6.5   Solitons in Numerical Calculations for a Discrete Chain Versus Solitary Waves in Continuum

The numerical calculations clearly demonstrate the existence of stationary compressive solitons (Figs. 1.5 to 1.7) in a discrete chain of particles with zero tensile strength. If this is the case, then it follows from the "exact" equation (Eqs. (1.69), (1.70)) for a discrete chain that the temporal width of this solitary wave T_s and its phase velocity $V_{s,c}$ must depend on the particle velocity at the maximum of the solitary wave v_m, as follows:

$$T_s \sim \left(v_m\right)^{-1/5}, \quad V_{s,c} \sim \left(v_m\right)^{1/5}. \tag{1.75}$$

From Eq. (1.75), it is evident that the space width of a stationary solitary wave does not depend on wave amplitude. It is also easy to understand taking into account that solutions of Eqs. (1.69), (1.70) with the dimensionless velocites of given particles will be the same at corresponding dimensionless times for solitons with different amplitudes resulting in identical normalized soliton profiles (Nesterenko [1983b]).

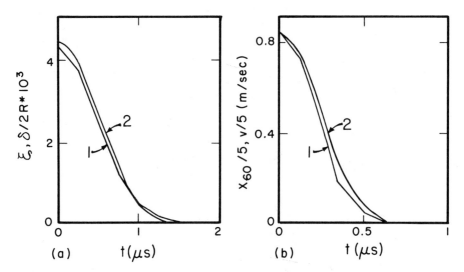

Figure 1.8. The (a) comparison of strains and (b) particle velocities profiles for a solitary solution of Eq. (1.20) (continuum) and Eq. (1.67) (discrete system): time dependence of $\delta/2R$ ((a), curve 1) is shown for particles with numbers 60 and 61, and $x_{60}/5$ ((b), curve 1) for the 60th particle.

The existence of solitary waves in discrete chains for "superquadratic" potentials was proved by Friesecke and Wattis [1994], and a partial case of Hertz interaction was considered by MacKay [1999]. It will be shown later (Section 1.11.3.1, Nesterenko [2000], [2001]) that conditions for soliton existence in continuum approximation for "normal" interaction law (like the Hertz law) are similar to the condition for soliton existence in a discrete chain obtained by Friesecke and Wattis [1994]. It explains why continuum approximation, being a truncated version of the equations for a discrete chain, is able to support the same solitary solutions.

Equations (1.75) valid for stationary waves in a discrete chain are similar to Eqs. (1.42) and (1.43) (note that $T_s \sim L/V$) describing the solitary solution of a sonic vacuum wave equation in a continuum limit (Eq (1.20)). So these properties of solitons in a discrete chain are also recovered in a continuum limit. It is interesting to compare in detail solitary waves obtained as stationary solutions of the "sonic vacuum" wave equation (Eq. (1.20)) and solitary waves found in computer calculations, which are stationary solutions for systems of discrete particles described by another equation (see Eq. (1.67)). Both cases are shown in Figure 1.8. The functions ξ and v described by Eqs. (1.39) are represented by the corresponding curves 2, in which the value of the phase velocity V was taken equal to its value in the numerical calculations.

There are three important reasons why these waves could be different in general. First, the solutions of the continuum Eq. (1.20) in the form of solitary waves (strictly speaking we did not prove that they are solitons!) exist only for any nonvanishing, though infinitesimally small, initial deformation ξ_0 and solitons in a discrete chain that exist at $\xi_0 = 0$. In the continuum approach the finite value ξ_0 guarantees the required asymptotic behavior of the solution when ξ is approaching ξ_0, leading to solitary wave formation.

Another reason is due to the fact that the spatial size of the solitary wave for a sonic vacuum equation is equal to only five particle sizes. That raises a general question of applicability of the continuum approach for waves with length comparable to the characteristic size of the system and reliability of the obtained results based on the wave equation Eq. (1.20)—long-wave limit of the system of equations for a discrete chain.

Finally, the system of Eqs. (1.67) and the continuum approximation presented by Eq. (1.20) differ because the nonlinear convective derivative of the particle velocity was dropped in the latter case. The relative value of this neglected term in comparison with others may be easily estimated for a solitary solution (see Section 1.3 and Problem 23).

The quick answer to these important arguments is that the solitons in a discrete chain and the solitary waves in an asymptotic case of very small ξ_0 for the sonic wave equation (Eq. (1.20)) are the same objects with very similar characteristics at some range of particle velocity (see Fig. 1.8).

To clarify this issue in detail, the specific numerical calculations demonstrated that the presence of the initial deformations $\xi_0 << \xi_m$, guaranteed by the application of constant external forces to the ends of the discrete chain (see also Section 1.6.9), essentially did not change the pattern of perturbation decomposition into solitons with respect to the case $\xi_0 = 0$. It was found that the only assignment of an initial deformation $\xi_0 >> \xi_m$ qualitatively changes the nature of the numerical solution (Nesterenko [1983b], [1992d]).

The solitary solutions obtained for the continuum Eq. (1.20) with those obtained in numerical calculations of the system of Eqs. (1.67) when $\xi_0 = 0$ are presented in Figure 1.8 for strain and particle velocity profiles. The solitary waves of Eq. (1.20) for strains in the wave much larger than the initial ξ_0 are very well described by one hump of the periodic solution from continuum approximation given by Eq. (1.34). The comparison of curves 1 and 2 clearly demonstrates that these profiles are very similar and "close enough" to each other. There are important differences only for small values of strains and velocities. In both cases the soliton is created by collective motion mainly of five particles. In the numerical calculations velocities of the extreme left and right particles in the soliton (third from the crest) are about 1% of the velocity of the central particle (Nesterenko and Lazaridi [1987]). Numerical calculations by Chatterjee [1999] with a much higher accuracy demonstrated the result similar to curve 1, in Figure 1.8(b), except the neighborhood of the wave crest where the maximum velocity was 4% higher than the maximum velocity for the analytically derived solitary wave. In his calculations it was possible to find the velocities of the next particles: the fourth particle from the crest has a velocity about $10^{-3}\%$ with respect to the velocity of the central particle and the fifth particle has $10^{-8}\%$.

From Eqs. (1.34) the relation between the phase speed V_s and maximum velocity v_m in the solitary wave for the continuum limit can be obtained:

$$V_s = c^{4/5}\left(\frac{16}{25}v_m\right)^{1/5} = \left(\frac{16}{25}\right)^{1/5}\left(\frac{2E}{\left(1-v^2\right)\pi\rho_0}\right)^{2/5}\left(v_m\right)^{1/5}. \tag{1.76}$$

Based on the analysis of results from computer calculations the phase speed of the stationary solitary wave $V_{s,c}$ is described by the following dependence:

$$V_{s,c} = B\left(\frac{2E}{\left(1-v^2\right)\pi\rho_0}\right)^{2/5}\left(v_m\right)^{1/5}. \tag{1.77}$$

Equation (1.77) was obtained from the calculated phase speed of a soliton with larger amplitude (Fig. 1.5) from particle velocity histories approximately for the sixtieth particle and the eightieth particle for the problem with impact by two balls. The value of constant B was found to be 0.915 from these calculations (Nesterenko [1983b]). The specific particles were selected in such way that the numerical data for each particle were symmetric with respect to the soliton maximum. This helped to maintain a reasonable accuracy in identification of the place of soliton maximum. The stationarity of this soliton was also checked according to the equality of velocities of the detached fiftieth and hundredhth particles in corresponding chains (Nesterenko [1983b]).

Strictly speaking, it is impossible to prove that a soliton in numerical calculations is exactly stationary. It may be slowly evolving with some an unknown large time scale from an initial disturbance into an asymptotically stationary solution. Unfortunately, even perfect conservation of momentum and energy does not guarantee that a numerical soliton is really stationary and we do not know how many particles are necessary to guarantee this. The comparison between Eq. (1.76) and Eq. (1.77) ($(16/25)^{1/5} = 0.91461$) demonstrates that they are very close despite this remark.

There is a basic problem with comparison of solitons in a discrete chain and a stationary solution of some type of wave equation like Eq. (1.20). It is connected with the fact that numerical results are always taken for the finite length of the chain from the initial-value problem, and an analytical stationary solution exists (so to speak) by itself as a property of a differential equation. Nevertheless, their comparison is very useful and can be trusted in some limits. An interesting discussion of solitons in discrete systems and in continuum limits is presented in a paper by Druzhinin and Ostrovskii [1991].

Sen and Manciu [1999] found an elegant analytical solution of the form A $tanh(f_n)$ for a stationary wave in a discrete chain, where f_n is represented by series.

For the Hertz chain first four terms provide very good accuracy and the result is very close to a soliton for long-wave approximation (Eq. (1.39)).

Numerical calculations made by Chatterjee [1999] with essentially higher accuracy (the energy error is only $2 \cdot 10^{-8}\%$), confirmed the existence of a solitary wave in this chain. Its phase speed was connected with particle velocity in maximum also by Eq. (1.77) with only a slightly different coefficient $B = 0.951$ instead of 0.915 obtained by Nesterenko [1983b].

Chatterjee [1999] used an asymptotic description of the tail of the soliton in a discrete chain and developed a new asymptotic solution for the full solitary wave. It is closer to the results of numerical calculations for a discrete chain than the result of long-wave approximation, presented by Eq. (1.76).

In conclusion, the solitary waves found in computer calculations ($\xi_0 = 0$) and the stationary solitary solutions of the continuum sonic vacuum equation (if $\xi_0 \ll \xi_m$) are the same objects ("close enough") within an accuracy of continuum approximation and calculations. They have the same space and temporal width, a similar dependence of phase speed on maximum velocity and strain. It is really remarkable that presumably the existing stationary solutions of a discrete system described by Eq. (1.3) are recovered as exact analytical solutions of a strongly nonlinear long wave (Eq. (1.20)) which is a heavily truncated version of Eq. (1.3) with a dropped convective derivative of particle velocity.

1.6.6   KdV Soliton Versus a Strongly Nonlinear One

The soliton solution of the KdV equation is a weakly nonlinear one. The obvious difference between the solitary waves of Eq. (1.20), the solitons for a discrete system described by Eq. (1.67) and the soliton solutions of the KdV equation, is the qualitatively different dependence of wave speed and of the typical width on the velocity (strain) amplitude (compare Eqs. (1.10) and (1.42), (1.43)).

Another interesting feature is that the spatial size of the solitary wave in the case of a sonic vacuum is finite ("compact support") and independent of the velocity and strain amplitudes. Its size equals five particle diameters (see Figs. 1.5 to 1.7) and Eq. (1.42) (Nesterenko [1983b]).

There is the drastic contrast with the KdV soliton (Eq. (1.10)), where the soliton width is equal to infinity when the strain amplitude is approaching zero. The initial long-wave sound speed c_0 is the very important parameter of the weakly nonlinear KdV soliton and does not affect the property of the strongly nonlinear soliton in sonic vacuum. At $\xi_{max} \gg \xi_0$ the sound speed c_0 does not influence the properties of this new type of solitary wave—the space width and dependence of the phase speed on the strain and particle velocity in maximum.

Finally, KdV solitons are two small parameter solitons in comparison with one small parameter solitary wave in the strongly nonlinear case. For a small strain in the wave, in comparison with ξ_0, the solution of Eqs. (1.20) to (1.22) in the form of a solitary wave degenerates into the weakly nonlinear soliton of the KdV equation.

1.6.7   Fast Decomposition of an External Impulse into Solitons

Impulsive forces in granular materials can be created by contact explosion. Actually, properties of granular materials are ideal for damping shock waves created by explosions with charge mass ~1 to 100 kg in real devices like explosive chambers and for the explosive welding of items with complex shapes (Apalikov et al., [1983]; Tsemahovich [1988]). The typical time scale of explosive loading is 10 to 100 μs. A microexplosive type of loading was also used in one-dimensional experiments in granular materials, by Rossmanith and Shukla [1982], with total pulse time being approximately 0.4 to 0.5 μs.

In computer calculations the boundary conditions on the left end of the chain with zero velocity of particles are given as follows:

$$\varphi_1 = 2 \cdot 10^6 (1 - 10^4 t) \text{ m/s}^2, \quad 0 \le t \le 10^{-4} \text{ s}; \quad \varphi_1 = 0, \quad t > 10^{-4} \text{ s}. \tag{1.78}$$

The right end of the chain is assumed free ($\psi_N = 0$), $N = 100$.

As seen from Figure 1.9, where the velocity–time profiles of the 35th and 60th particles are shown (x_{35} and x_{60} correspondingly), the perturbation decomposed into seven solitons. When the duration of force was decreased down from 100 μs to 10 μs the number of solitons decreased to one.

From general considration, the number of solitons n_s should be scaled as

$$n_s \sim T_i \, c^{4/5} \, v_{mi}^{1/5} \, a^{-1}, \tag{1.79}$$

where T_i is the duration of external impulse, v_{mi} is the maximum velocity of the boundary, a is a particle diameter, and c is a constant (see problem 27).

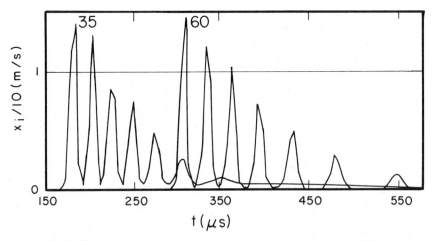

Figure 1.9. Decomposition of the initial triangular impulse in the initially uncompressed chain of particles: velocity profiles for the 35th and 60th particles.

Impulses different from the linear decay behind the front, with a typical duration of 200 μs, can be created by a piston impact with a velocity of 1 m/s and masses equal 5 to 10 mass of particle in the chain. An increase in piston mass results in a proportional increase in the number of solitons at the same distance from the impact end and during the same time interval (Nesterenko and Lazaridi [1987]).

The pulse duration of 10 to 100 μs is characteristic for explosive loading experiments. Therefore, the discrete structure of the granular medium can have a substantial effect on the transformation of compression pulses in this case. Again, it is worth mentioning that decomposition of the initial impulse into the soliton train goes very quickly, being one of the main properties of a given strongly nonlinear system with nonlinear dispersion. The short pulses with duration less than 1 μs should degenerate into one soliton as was observed in experiments by Zhu, Shukla, and Sadd [1996].

1.6.8   The Piston Moving with Constant Velocity into an Uncompressed Chain of Identical Particles

1.6.8.1   Waves in a Nondissipative Chain

The system of Eqs. (1.67) for a discrete chain does not contain dissipative terms. Physically, it means that piston impact should not result in the stationary shock wave with properties similar to that described by Eqs. (1.46) to (1.48). At the same time the absence of dissipation produces extreme conditions for development of the wave properties determined by inner material structure.

The initial conditions for the velocities to simulate impact by a very long piston are

$$x_1(t = 0) = 5 \text{ m/s}, \qquad x_i(t = 0) = 0, \qquad N \geq i \geq 1. \tag{1.80}$$

The right end of a chain is assumed free ($\psi_N = 0$, $N = 200$), while the left end is subject to the condition $\varphi_1 = \psi_2$, guaranteeing constant velocity of the first particle. The values of all particle velocities at the moment $t = 5.45 \cdot 10^{-4}$ s are given in Figure 1.10. For convenience of perception the velocity values of neighboring particles are connected by the straight lines. The real velocity values correspond only to integers of the particle number i. A few important conclusions based on this calculation are presented below.

1.6.8.2   Formation of a Leading Soliton

The amplitude of the first maximum v_m approaches with time twice the value of the piston velocity v_0. This result is similar to those obtained for a Toda chain (Hill and Knopoff [1980]), for particles interacting according to the Morse potential (Strenzwik [1979]), and for the nonlinear KdV equation (Kunin [1975]). This property seems to be a very common property of any discrete system and does not depend on the interaction law between particles. It is also similar to the

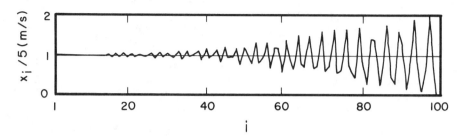

Figure 1.10. Particle velocity profile when the left end of a chain is moving with constant velocity 5 m/s, system of identical particles.

situation in the collision of a rigid wall moving with a constant speed v_0 and a particle at rest. It is easy to see from the system connected with the wall, that after the collision, a particle will move with velocity $2v_0$ for any interaction law if collision is perfectly elastic.

It is also seen from Figure 1.10 that a stationary soliton with the velocity amplitude equal to $2v_0$ is formed at the head of the globally nonstationary wave. It is an interesting feature because both the leading wave and particles in the piston vicinity approach the stationary states very quickly resembling a situation in the stationary shock wave. For example, starting with particle number $i = 20$, the velocity of the leading front is practically equal to the soliton velocity with an amplitude of $2v_0$ within an accuracy of 1%.

In this nondissipative case, one can find speed of the leading front of a nonstationary wave V_{ns} (V_{ns} is practically constant on small distances from the left end of the system). Using the numerical results for soliton speed $V_{s,c}$ (Eq. (1.77) with $B=0.951$) and supposing that $v_m = 2v_0$, we obtain

$$V_{ns} \approx B\,(2)^{1/5}\left(\frac{2E}{\left(1-v^2\right)\pi\rho_0}\right)^{2/5}(v_0)^{1/5}. \tag{1.81}$$

The result of the calculation shows that in the problem under consideration the maximum pressure in the first oscillation P_m substantially exceeds the pressure established near the piston P_c. For example, if the piston velocity is $v_0 = 5$ m/s (at the moment where the wave front is at the 50th particle), $P_m/P_c = 2.24$ (Nesterenko [1983b], [1992d]).

One more fact should be mentioned in the numerical calculations. Since the velocity amplitude of the leading soliton is quickly approaching $2v_0$ (Fig. 1.10), the value of the maximum velocity of the last particle on the free end after "spall" and separation from the chain, will exceed value $2v_0$ for relatively long chains. The values of v_N/v_0 for a different number of particles in the system $N = 20$, 50, 100, and 200 are equal, respectively, 2.708, 2.836, 2.876, and 2.895. And as already noted, the ratio v_N/v_0 is independent of v_0.

1.6.8.3   Separation of Particles into Two Groups

It should be clear that the global nonstationarity of motion consists of a continuous increase in the number of particles participating in the oscillatory motion in the area between the leading wave and the piston.

This allows for the separation of particles into two groups: the leading group has relative velocities comparable to the velocity of the piston, and the tail is composed of particles with practically no relative motion. Both numbers are increasing with the propagation distance (later we will see that this property is not preserved in the case of a random chain of particles). The number of particles participating in the oscillatory motion (N_h) is increasing faster than the number of particles being practically at rest (N_e). The latter number describes particles with velocities different from the piston velocity of 10% or less in the area adjacent to the piston. This separation is already clear when the leading wave is only at the 20th particle with $N_h = 14$ and $N_e = 6$. When the leading wave is at the 60th particle $N_h/N_e = 1.5$ and at a later time when it approaches the 100th particle (Fig. 1.10) $N_h/N_e = 2.4$.

The particle's separation will result in a different behavior when this wave approaches the free end of the chain: the leading group produces multiple "spall" and particles separate each from other and the tail moves as a solid body.

This example clearly demonstrates that a very "weak" nonstationarity of the leading front could be accompanied by "strong" nonstationarity of the material motion behind it, in the absence of dissipation or at its small value. Therefore, fixing only the velocity of the leading front, one can reach erroneous conclusions concerning the stationarity of the process based only on the steady motion of the front and the constant piston velocity.

In nonlinear heterogeneous systems with dispersion, the measurements of total wave profiles at different times should be used to control the stationarity conditions. This comment is especially important for granular, reactive materials where stationarity conditions are very often inexplicitly postulated to justify use of the Hugoniot relations.

Strenzwik [1979] and Hill and Knopoff [1980] have found that for a weakly nonlinear case (Eqs. (1.5)) with a sufficiently large coefficient of nonlinearity the particles even near the piston carry out undamped oscillations in the nondissipative case. Unlike these results, it was found in the numerical experiment for strongly nonlinear systems (Eq. (1.12)) that the velocity profiles in the coordinates $\zeta_i - i$ are identical at values v_0/V_{ns} equal to 0.95 and $5 \cdot 10^{-3}$ at the corresponding moments of times. It is in agreement with the dimensionless scaling of a discrete chain represented by Eqs. (1.69) to (1.71).

The piston problem has some analogy with calculations and experiments by Falcon, Laroche, Fauve, and Coste [1998b], where the 1-D chain collides with a rigid wall. It was shown that the maximum force felt by the rigid wall was independent of the number of beads—the natural result of a wave process in the chain. The different behavior was observed in experiments of chain collisions with a soft wall where the maximum force depends on the particles number if

$N < 10$ and saturates on a level smaller than for rigid wall.

It can be expected from computer results with a piston moving into the chain with constant velocity (see Fig. 1.10), that the beads at the top of the column separate with a velocity greater than the initial velocity of the chain (analog velocity of the piston). Falcon, Laroche, Fauve, and Coste [1998b] demonstrated that after wave interaction with the free end, particles are separated into two groups: with velocities larger and smaller than the velocity of the piston. It is consistent with the existing two groups of particles under piston impact discussed earlier. This resulted in different types of particle detachments: the upper particles separate one after the other and the particles at the bottom move as a cluster. The velocity of the separated end particle was equal to $1.5v_0$ for the number of particles 7.

1.6.8.4 Stationary Wave in Dissipative Chain Versus Nonstationary Wave in Nondissipative Chain Under Piston Impact

If a particle chain exhibits dissipative behavior then the stationary regime of the wave motion may be established. Velocity D_1 of a stationary shock wave may be found based on the equation of state of the medium and the mass and momentum conservation equations (Nesterenko [1983b]):

$$D_1 = \left(\frac{2E}{\left(1-v^2\right)\pi\rho_0} \right)^{2/5} v_0^{1/5} = c^{4/5} v_0^{1/5} . \tag{1.82}$$

This speed does not depend on the specific type of dissipation. Using the previous result for a speed of V_{ns} (Eq. (1.81), $B = 0.951$) the obtained ratio of speeds of two principally different regimes of motions in a discrete lattice is $V_{ns}/D_1 \approx 1.09$. This leads to another practically important conclusion: the wave speed for steady shock is close to the speed of a leading pulse of a completely different nonsteady wave! It makes it very difficult to justify experimentally which type of wave is propagated in the system based on the speed of the leading front.

From the equation for the shock wave speed D_1 (Eq. (1.82)) neither D_1 nor the speed V_{ns}, (Eq. (1.81)) for nonstationary wave motion without dissipation, depend on the size of the particles.

The stationary shock wave can be monotonous or oscillatory depending on the dissipation level. The simplest way to understand it is to look on the stationary shock wave from the same point of view as we did for solitary motion using the effective "potential energy" approach (with an added dissipative term) in the continuum limit. If the dissipation is very weak, the shape of the leading front of the stationary wave can be expected to be near a soliton shape as is evident from the "potential energy" curve $W(y)$. The amplitude of this leading soliton v_{m1} must guarantee the value of the stationary front velocity $D_1(V_s(v_{m1})$ $= D_1)$. Using the expression for the soliton velocity (Eq. (1.43)) and Eq. (1.82) we obtain $v_{m1} \approx 25/16v_0 \approx 1.56v_0$.

Thus, for weak dissipation, the particle velocity at the stationary front can exceed the piston velocity by 1.56 times. A comparison with the previous result for maximum particle velocity in the leading hump for the nondissipative nonstationary wave ($2v_0$) shows that the stationary wave is characterized by a smaller particle velocity in the leading hump. For strong dissipation, the shock front will be monotonic and the final value of the particle velocity is approached without intermediate maximum.

We recollect that on the basis of continuum approximation (Eq. (1.48)) for the stationary shock wave, the pressure ratio for weak dissipation may reach the value $P_m/P_c = 1.95$. We saw that the corresponding value in numerical calculations for the nonstationary wave is $P_m/P_c \approx 2.24$. Thus, the pressure in the oscillatory wave at the nonstationary regime can exceed the pressure at the front of the oscillatory stationary shock wave. From a practical point of view this means that under impact, particles in granular material adjacent to the shock entrance may be subjected to larger stress than particles in the depth.

1.6.8.5   Kinetic Versus Potential Energy

As a result of piston impact the system is accumulating kinetic energy E_k and potential energy E_p. It is interesting to investigate the relation between these energies (see Fig. 1.11) during the propagation of nonstationary waves which nevertheless carry some features of steady process as discussed in the previous section.

It is evident from Figure 1.11 that the practically steady ratio $E_k/E_p = 1.249$ is established by the time $t \approx 3 \cdot 10^{-5}$ s. At this instant only six particles participate in the motion. It follows that E_k/E_p is practically constant after this moment despite the nonstationarity of particle motion and increasing the number of moving particles. The deviation from this value does not exceed 0.8%.

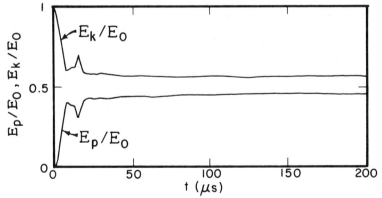

Figure 1.11. Establishment of the stationary ratio E_k/E_m and E_k/E_p for the nonstationary piston problem.

The given value of $E_k/E_p = 1.249$ is close to the expected one by the virial theorem for particles interacting according to the hertz law and performing finite oscillatory motion

$$\langle E_k \rangle = -\frac{1}{2}\left\langle \sum_{i=1}^{N} \vec{r}_i \vec{F}_i \right\rangle \implies \frac{\langle E_k \rangle}{\langle E_p \rangle} = 1.25. \tag{1.83}$$

It does not follow from the virial theorem, however, that deviations from the mean value will be small, as is observed in the numerical calculation. The small values of deviations from the mean values are obviously explained by the large number of particles in the system. A similar energy distribution between E_k and E_p was characteristic for an impact by two particles and soliton interaction as well as for the soliton itself. So, the steady ratio of kinetic and potential energy is a distinguished feature of the collective motion of the strongly nonlinear system of particles which may also be the case for three-dimensional systems.

1.6.8.6   Nonlinear Chains in Liquid

If the nonlinear chain of particles is placed into a liquid with relatively small viscosity, then it is possible to expect soliton-like slightly decaying pulses, as well as the stationary oscillatory shock waves propagate there, similar to the previously considered chain. It is interesting to find critical values of the liquid viscosity that will allow propagation of the oscillatory shock waves.

Sen, M. Manciu, and F. Manciu [1999] made a very interesting proposal to use soliton-like pulses in a nonlinear magnetically ordered chain of grains suspended in water to eject the very small ink droplets comparable with the size of the ferrofluid grains. If this idea results in practical devices, it may help to design inkjet printers of unparalleled resolution.

1.6.9   Influence of Initial Compression on Wave Transformation in Strongly Nonlinear Chains

In previous sections we considered the properties of waves in uncompressed chains, where the amplitude of waves was much larger than the initial strain in a compressed system, or even at zero initial compression. In an analytical approach, the initial prestrain (though it can be indefinitely small) is of fundamental importance, providing the necessary asymptotic for stationary solution, resulting in a soliton. The initial prestrain may also be very important for nonstationary wave behavior which we did not consider analytically.

1.6.9.1   Soliton in a Precompressed Chain

The dependence of the phase speed of a soliton on initial compression (ξ_0 is a strain in infinity) is given by Eq. (1.37). Using Eq. (1.38) for sound speed in a chain, the following dependence between sound speed in the initial state c_0 and the phase speed of soliton V_s with strain amplitude ξ_m can be obtained:

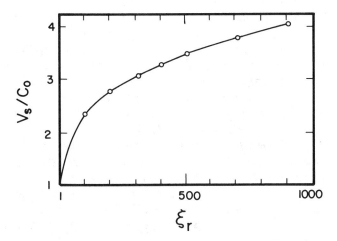

Figure 1.12. Ratio of the solitary wave speed V_s to the sound speed c_0 in the initially compressed chain versus the relative amplitude ξ_r, the parameter of nonlinearity.

$$\frac{V_s}{c_0} = \frac{1}{(\xi_r - 1)} \left\{ \frac{4}{15} \left[3 + 2\xi_r^{5/2} - 5\xi_r \right] \right\}^{1/2}, \tag{1.84}$$

where $\xi_r = \xi_m/\xi_0$ is the relative strain amplitude of the soliton (see Fig. 1.12). The distinguished feature of the soliton speed is a very weak dependence on relative amplitude ξ_r. The excess of the soliton speed over initial sound speed, only in 10% is provided by a relative strain equal to 3.

We can see that to obtain a few times difference between soliton speed and sound speed, the relative strains above 100 are necessary. It is easy to achieve it in slightly compressed granular materials even under conditions of low velocity impact (~1 m/s) which makes them quite different from other substances.

It is clear from Figure 1.12 that the rate of change of colitary speed is decreasing with with an increase of relative strain amplitude.

1.6.9.2 Nonstationary Waves in a Precompressed Chain

There are no analytical results on nonstationary wave behavior in a strongly nonlinear case which are similar to results for a weakly nonlinear case (Whitham [1974a]). For example, the relation between the number of emerging solitons and initial impulse duration is not established. Nevertheless, the general pictures in these two cases are qualitatively similar—the longer pulse creates a larger number of solitons. The number of solitons in numerical calculations is proportional to the mass of the striker, reflecting the tendency toward longer impulse duration with the striker's mass increase.

To investigate a nonstationary wave behavior with a different strain amplitude the piston impact (piston mass M_p and velocity v_s) on one end was used, and the wall supported the other end. The static force F_0 was applied to the end of the chain to ensure initial uniform precompression. The system of equations (Eq. (1.67)) modified to include the static force F_0, resulting in initial displacement δ_0 and strain ξ_0, was used (Nesterenko and Lazaridi [1987]; Nesterenko [1992d]). The forces between the flat surfaces of the piston, wall, and corresponding particles were prescribed by the Hertz law.

Four characteristic cases were considered: (1) $\xi_0 = 0$; (2) $\Delta\xi_m = \xi_m - \xi_0 \gg \xi_0$; (3) $\Delta\xi_m \sim \xi_0$; and (4) $\Delta\xi_m \ll \xi_0$ (Fig. 1.13). The selection of each specific case was ensured by the different constant forces F_0 and piston velocity. The particles, wall, and piston for further comparison with experiments had properties of steel (Young's modulus $E = 2.1 \cdot 10^{11}$ N/m^2, Poisson coefficient $v_1 = 0.29$), the particle radius is 2.375 mm, and the piston mass is equal to five particles mass.

Results of the calculation are presented in Figure 1.13. Moment $t = 0$ corresponds to the initial instant of impact. From Figure 1.13 the following conclusions can be made (Nesterenko and Lazaridi [1987]).

If the wave amplitude is relatively large ($\Delta\xi_m = \xi_m - \xi_0 \gg \xi_0$), then the impulse transformation is very similar to the case when $\xi_0 = 0$ at the same type of loading (compare Figs. 1.13(a) and (b)). The soliton trains were created in both cases at short distances from the impulse entrance comparable with soliton width (five diameters of particles)—another distinguished feature of a strongly nonlinear system.

It is necessary to mention that, in general, initial prestress will change the entering loading impulse, even at the same impact conditions (mass and velocity of piston). In this case, the time dependencies of force acting between piston and end particle were very similar. The piston rebounded from the chain at $t = 50 \cdot 10^{-5}$ s for the case of $F_0 = 0$ and at $t = 49 \cdot 10^{-5}$ s for $F_0 = 2$ N. Only in the former case an additional collision of a particle, with rebounded piston at $t = 75 \cdot 10^{-5}$ s, was observed. The velocity history of particles demonstrates the slightly faster decomposition of the initial impulse into the soliton train for $F_0 = 0$ (the process was accomplished for the twentieth particle). The same qualitative tendencies were observed at the initial precompression with $F_0 = 20$ N, the piston rebounded from the chain at $t = 25 \cdot 10^{-5}$ s. The velocity maximum in the first impulse exceeds (as in the case $F_0 = 0$) the impact velocity of the piston, decreasing with an F_0 increase: 1.2 m/s for $F_0 = 2$ N and 1.05 m/s for $F_0 = 20$ N.

When the wave amplitude is comparable with initial prestrain ($\Delta\xi_m \sim \xi_0$), the process of impulse transformation at the same length of the chain is qualitatively different in the case of: (1) $\xi_0 = 0$; and (2) $\Delta\xi_m = \xi_m - \xi_0 \gg \xi_0$ (compare Figs. 1.13(a), (b), and (c)). Two new features can be mentioned; the process of impulse decomposition into soliton train is retarded, and the periodic wave tail was created following the main impulse. Also the dispersion process results in transformation of the leading impulse and its periodic tail, which is evident after

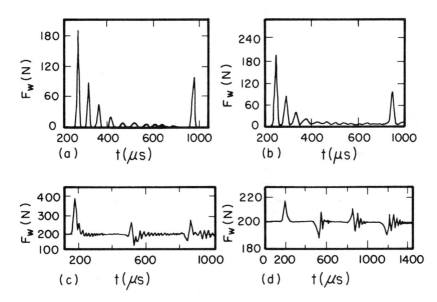

Figure 1.13. Dependence of the forces on the wall F_w on time demonstrating qualitative change in the system behavior at increasing initial precompression and decreasing amplitude of impulse: (a) case (1), $F_0 = 0$, $v_s = 1$ m/sec; (b) case (2), $F_0 = 2$ N, $v_s = 1$ m/s; (c) case (3), $F_0 = 200$ N, $v_s = 1$ m/s; (d) case (3), $F_0 = 200$ N, $v_s = 0.1$ m/sec. Number of particles $N = 40$, piston mass $M_p = 5m$.

multiple reflections from the wall and from the impacted end of the chain. The interaction time of the piston with the impacted end, in this case, is decreased by 2.66 times in comparison with $F_0 = 0$, mainly due to the stage of interaction with reduced velocities of the piston. In contradiction with case $F_0 = 0$, the velocity maximum in the first impulse (0.85 m/s) does not exceed the impact velocity of the piston. It is interesting that, despite a qualitative change of impulse shape, the maximum dynamic increase of force exerted on the wall does not depend on F_0 at the same impact conditions (see Figs. 1.13(a), (b), and (c)).

In the case of a weak wave ($\Delta\xi_m \ll \xi_0$), a quasistationary single leading impulse (later modified by dispersion) accompanied by a periodic tail, with very small amplitude, propagates in the system. The increase in the number of particles from 40 to 80 did not change the amplitude of force (15 N) in the reflected wave, only a slight change on the right part of the impulse was observed (Nesterenko and Lazaridi [1987]). This leading impulse is created by the motion of 17 particles (with velocities $v > 0.01v_{max}$) and the phase speed close to the initial sound speed, $c_0 = 1151$ m/s. This space width is larger than for the KdV soliton with a corresponding amplitude (Eq. (1.10)) suggesting that even the length of the system, equal to 80 particles diameters, was not enough to form a KdV soliton—a typical feature of a weakly nonlinear system, as opposed to a strongly nonlinear one. Again we can conclude that relatively weaker waves have a less pronounced tendency for decomposition into a soliton train. In Figures 1.13 (c), (d), it is clear that the linear dispersion modifies the wave reflected from the impacted end due to the absence of stationary solitary rarefaction waves in this system. It is interesting that the speed of the first strongly nonlinear impulse in case (3) is close to the speed of the leading weakly nonlinear impulse in case (4). This behavior is in agreement with a very weak soliton speed dependence on the relative strain (Fig. 1.12).

Finally, we can conclude that, in addition to the differences in the properties of stationary waves propagating in strongly and weakly nonlinear systems, these materials are also different qualitatively in respect to the transformation of nonstationary impulses. A strongly nonlinear system has a distinguished feature revealed in a very fast impulse decomposition on the distances comparable to the width of a stationary soliton. A very weak initial compression does not result in qualitatively different behavior in comparison with $F_0 = 0$, except in the appearance of a periodic tail with small amplitude. This means that the prediction of the existence of a new type of solitary wave, as a stationary solution of Eq. (1.20) in the limit $\xi_0 \to 0$, is physically justified.

1.6.9.3   Effect of Gravity

If the chain of particles is under gravitational load (has a vertical position), then the initial precompression is not uniform along the chain. In this case, when the impulse amplitude is much higher than the static strain at the bottom of the system, then it should propagate mainly in the same manner as in the unstressed chain. When the initial strain becomes comparable to the strain increase in the wave, then this disturbance will approach the weakly nonlinear case.

Sen and Sinkovits [1995], [1996] numerically investigated 1-D and 2-D gravitationally loaded granular systems. They found that the impulse shape is essentially modified during the propagation toward the more loaded part of the system for the modest pulse amplitudes. This means that gravitational loading may be very important for the dispersion of weakly nonlinear pulses and must be incorporated into the media description. Power laws for the dependence of grain displacement and velocity on the depth for a weak signal were obtained analytically and in numerical simulation by Hong, Ji, and Kim [1999].

1.6.10   Soliton Interaction with the Boundary
of Two "Sonic Vacuums"

The interaction of compression pulses with the contact of two "sonic vacuums" is of particular interest. Indeed, in this case, there is no concept of sonic impedance determining the reflected and passed impulse amplitudes, as in the traditional acoustic approach.

The behavior of compression pulses when passing through the contact of two "sonic vacuums"—the adjacent chains of steel spherical granules with diameters

4.75 mm (SV-1) and 7.9 mm (SV-2), the number of particles equals 20 in each system—was studied in numerical calculation (Nesterenko, Lazaridi, and Sibiryakov [1995]). The system of the equations for identical particles (Eq. (1.67)) for each of the "sonic vacuums," with the same Hertz law for particle interaction with different radii on the interface, was used. The initial impulses in each of these "sonic vacuums" were excited by the striker (steel ball) impact with its mass equal to the mass of the particle in impacted chain and velocity 1 m/s.

In both cases, impact resulted in single soliton excitation, which was close to stationary in the middle of the corresponding system, then later interacted with the interface. The compression forces (F_{av}) were calculated both in incident and reflected waves (averaging between the 14th and 15th and the 15th and 16th particle contact forces in the corresponding system) and in passed impulses (averaging between the 5th and 6th and the 6th and 7th particle contacts from the interface). The results of the calculations are presented in Figures 1.14 and 1.15 with geometry of impact and points of calculations shown schematically at the top of each figure.

The resulting wave pictures were quite different in these two cases. The first case situation is qualitatively close to the expected from a traditional acoustic approach. When the incident soliton (I) hits the contact from the side of SV-1 ("light") a reflected soliton (R) appears in it (Fig. 1.14(a)). The transient soliton (T) propagates in SV-2 and later, after reflection from the wall, arrives at the points of observation in SV-2 (TW) (Fig. 1.14(b)).

In the second case when the incident soliton (I) falls on the interface from the side of SV-2 ("heavy" system), then the reflected compression pulse is not observed (Fig. 1.15(a)). While in SV-1, instead of a single soliton, as in the

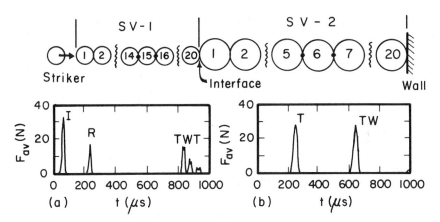

Figure 1.14. Numerical calculations of the soliton interaction with the contact of two "sonic vacuums": (a) corresponds to the averaged forces at the point before interface (dots on the contacts of the 14th and 15th and the 15th and 16th particles in SV-1); (b) averaged forces behind the interface (dots on the contacts of the 5th and 6th and the 6th and 7th particles in SV-2). A soliton was created by striker impact ($M_s = m_1$, $v_s = 1$ m/s) on the SV-1, see picture at the top.

previous case, an asymmetrical nonstationary triangular pulse, with a steep front (T) arises near the contact, which later decomposes into a series of seven solitons (TW) as it propagates forward and later reflects from the rigid wall (Fig. 1.15(b)). This train of solitons later interacts with the interface resulting in a soliton train in SV-2 at the corresponding points of observation in SV-2 (TWT). It is apparent that such a pattern of pulse interaction with the interface of two "sonic vacuums" qualitatively differs from expected traditional acoustics.

The reflected and transient soliton parameters for the first case, corresponding to Figure 1.14, may be found from the laws of momentum and energy conservation, supposing that all pulses are solitons. The total momentum and energy should be preserved in the system if dissipation is absent:

$$p_I = p_T - p_R, \quad E_I = E_T + E_R, \tag{1.85}$$

$$E_I = \frac{p_I^2}{2m_{1ef}}, \quad E_T = \frac{p_T^2}{2m_{2ef}}, \quad E_T = \frac{p_R^2}{2m_{1ef}},$$

where p_I, p_T, and p_R; E_I, E_T, and E_R are the momenta and energies of the incident (I), transient (T), and reflected (R) solitons, respectively; and m_{1ef} and m_{2ef} are the effective soliton masses in each system. An expression for the effective masses of the solitons ($m_{ief} = 0.27m_i$) can be obtained on the basis of Eq. (1.41). From Eq. (1.85) the momenta p_1, and p_2 may be obtained as follows:

$$p_T = \frac{2\,p_I}{1+k}, \quad p_R = \frac{1-k}{1+k}\,p_I, \quad k = \frac{m_1}{m_2}. \tag{1.86}$$

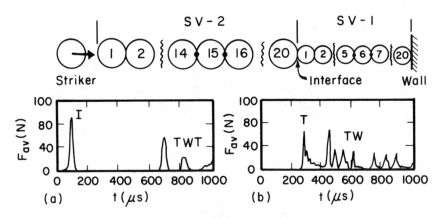

Figure 1.15. Numerical calculations of the soliton interaction in contact with two "sonic vacuums": (a) the averaged forces at the points before the interface (dots on the contacts between the 14th and 15th and the 15th and 16th particles in SV-2); (b) averaged forces behind the interface (dots on the contacts between the 5th and 6th and the 6th and 7th particles in SV-1). The soliton was created by striker impact ($M_s = m_2$, $v_s = 1$ m/s) on the SV-2, see picture at the top.

These relations between p_I, p_T, and p_R are in agreement with computer calculations. It is apparent that such an approach is possible only when the pulse falls from "light" SV into "heavy" SV. It should be mentioned that the conservation laws support the realization of only one reflection scenario, whose uniqueness is not proved by these laws.

The observed transformation of the impulses at the contact with the subsequent rapid decomposition of the transient impulse into the sequence of solitons allows the systems of "sonic vacuum" type to transform external actions on the system into the pulse sequence required.

1.6.11   Waves in Two-Particle Periodical Chains

1.6.11.1   Analytical Approach

The behavior of a two-particle 1-D periodical chain of particles is qualitatively different from the behavior of the chain with equal particle masses, even in the case of the linear interaction law (Kunin [1975]). For example, in the former system, for every wave number there are two characteristic frequencies corresponding to the two branches of vibration specter—acoustical and optical. Even long waves have qualitatively different dispersion relations. Another important difference is the existence of the forbidden frequencies. The vibrations with these frequencies can-not propagate without decay. It is interesting to investigate the behavior of a strongly nonlinear 1-D periodical chain of particles with alternating masses.

The long-wave approximation analogous to Eq. (1.20) can be derived (Nesterenko and Lazaridi [1990]) using the approach proposed by Dash and Patnaik [1981] (see also papers by Mokross and Buttner [1981] and by Mertens and Buttner [1986]) for a weakly nonlinear case. In an extreme case, when the mass of one particle is much larger than the mass of another ($k = m_1/m_2 \gg 1$), and both have the same diameter a, this equation has the solitary solution

$$\xi = \left(\frac{5V^2}{8c^2}\right)^2 \cos^4\left(\frac{x}{a\sqrt{10}}\right), \tag{1.87}$$

with a characteristic space scale of solitary wave $L \approx 10a$, and the width of the solitary disturbance is twice that in the case of particles with the same masses. So we can see that the mere redistribution of mass between neighboring particles can result in a wider soliton.

Bogdanov and Skvortsov [1992] considered a weakly nonlinear case for a diatomic lattice and proposed wave equations for low-frequency perturbation and for the envelope of high-frequency oscillations similar to Zakharov's equations (Zakharov [1972]) in plasma.

1.6.11.2   Numerical Simulation

Computer calculations were used to investigate the behavior of periodical uncompressed diaparticle chains with mass ratio $k = 2, 4, 16, 24, 64$ (Nesterenko and Lazaridi [1990]). Only cases with $k = 2$, 4, and 64 will be considered here. The chain's macroproperties (linear density and elastic properties, resulting in the same long-wave sound speed in the case of $\xi_0 \neq 0$) were kept the same, and equal to the properties of the chain with equal mass m. It was ensured by selecting the condition $2m = m_1 + m_2$ (redistribution of masses between neighboring particles) and choosing the same distances between particle centers and interaction constants for both types of chains.

The dynamic loading of these systems was performed by the impact of a striker with mass $M_s = 5m$ on the left end of the chains. The other end contacted with the steel plate. The interactions between particles in the chains, particles, plain wall, and piston obeyed the Hertz law. The distance between particle centers was 4.75 mm, the curvature radii in the contact points 2.375 mm, and Young's modulus of the particles material and wall $E = 2.1 \cdot 10^{11}$ N/m^2. The striker velocity was equal to $v_s = 1$ m/s, and the first particles adjacent to the piston were the lighter ones, particles are numerated starting from the wall.

1.6.11.3   Diatomic Chain with a Large Mass Difference

The chain with $k = 64$ was considered as representative for a system with large mass differences of neighboring particles. The results of computer calculations for the force acting on the wall (F_w), and the velocity history of the 15th heavy particle (u_{15}) are shown in Figures 1.16(a) and (b) for the chain with $k = 64$ and particle number $N = 40$ (zero time corresponds to the impact instant). Young's modulus of particles are the same and equal to $E = 2.1 \cdot 10^{11}$ N/m^2, radii of contacts 2.375 mm, distance between centers 4.75 mm, striker mass $M_s = 5m$, velocity $v_s = 1$ m/s, impact on light end particle.

For comparison in Figure 1.17(a) and (b), the corresponding force on the wall (F_w) and velocity history of the 15th particle (v_{15}) for the chain with equal particles ($k = 1$, $m_1 = m_2$, $N = 40$) at the same method of loading are depicted. Young's modulus of particles $E = 2.1 \cdot 10^{11}$ N/m^2, radius 2.375 mm, striker mass $M_s = 5m$, velocity $v_s = 1$ m/s.

From the comparison of Figures 1.16 and 1.17 we can see that the phase speeds of leading solitons in the two systems are practically identical (≈ 772 m/s). The subsequent leading soliton amplitudes are slightly different with forces on the wall and particle velocities being larger in the chain with identical particles (see Figs. 1.16 and 1.17). At the same time, there is a qualitative difference in the wave dynamics of these two systems having the same macroscopic properties under the same type of loading.

In the chain with equal particles, six clearly detectable solitons were created (Fig. 1.17, a soliton train with negative velocities in Fig. 1.17(b) is connected with reflections from the wall).

In the diatomic chain, their number was three; the last peak in Figure 1.16(a)

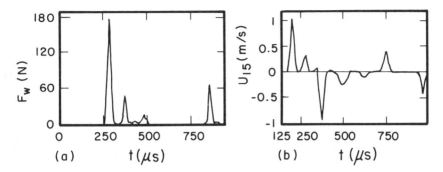

Figure 1.16. Impulses in a diatomic chain ($m_1 + m_2 = 2m$) with large mass difference ($k = 64$): (a) the dependence of force F_w acting on the wall; (b) the velocity history of the 15th heavy particle.

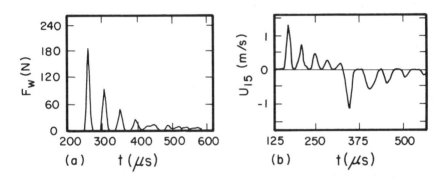

Figure 1.17. Impulses in the chain with identical masses of particles (m, $k = 1$): (a) the dependence of force F_w acting on the wall; (b) velocity history of the 15th particle.

and the last two peaks in Fig. 1.16(b) are connected with the solitons reflection from the piston. The time history of the neighboring light 14th particle was similar to the velocity profile for the heavy 15th particle (Fig. 1.16(b)) only with smaller amplitude (0.75 m/s). This is a general relation between velocities of neighboring light and heavy particles. Also the motion of light particles exhibits a qualitatively new feature: high-frequency small-amplitude modulation of the velocity profile, more pronounced with the decrease of wave amplitude (Nesterenko and Lazaridi [1990]). The solitons spatial size and time width in this case is approximately two times larger than in the former example, which agrees with the discussed analytical prediction for $k \gg 1$.

1.6.11.4   Diatomic Chain with Comparable Masses of Particles

The new type of behavior was observed at relatively smaller mass differences (k = 2 and 4). At $k = 2$ (see Fig. 1.18, Young's modulus of particles are the same E = 2.1·10^{11} N/m², radii of contacts is 2.375 mm, distance between centers is 4.75 mm, striker mass $M_s = 5m$, velocity $v_s = 1$ m/s, impact on light end particle) the behavior of the system is qualitatively different from the previous cases when $k = 64 >> 1$ (Fig. 1.16) or when $k = 1$ (Fig. 1.17).

Despite that, in this case the system was macroscopically the same as in previous chains with $k = 64$ and 1, the amplitude of the force on the wall (F_w = 144 N) became smaller at the same type of loading for $k = 1$ ($F_w = 200$ N) and k = 64 ($F_w = 177$ N) (see Figs. 1.18(a), 1.16(a), and 1.17(a)). The speed of the first nonsymmetrical impulse was approximately the same as in the former two cases, its temporary width became smaller than for $k = 64$. The history of the force acting on the wall does not demonstrate any periodic sequence of solitons.

In this case the velocity histories for light and small particles are not similar (compare Figs. 1.18(b) and (c)). For example, the light 14th particle velocity profile can be characterized by the detachment from the following oscillating tail of the leading wave with two maximums (Fig. 1.18(b)). Time interval between maximums in the first peak was 0.15·10^{-4} s, being different from intervals between corresponding peaks in the oscillating tail: $\Delta t_{4-5} = \Delta t_{5-6} = \Delta t_{6-7} = \Delta t_{7-8}$ = 0.275·10^{-4} s. The leading two-peak wave and time intervals between the following peaks, as well as their amplitudes, were constant (with accuracy 5%) during wave propagation inside the chain. This was verified by increasing the number of particles up to 80. This behavior suggests that the first double peak impulse, and the corresponding periodic tail in the velocity profile of the light particle, may be described as quasistationary waves. The new feature for this case is the negative value of light particle velocity in the propagating profile.

The velocity profile for the heavy 15th particle, has a nonsymmetrical leading peak (1), followed by another peak (2), with smaller amplitude and space width close to 10 particles diameters as a soliton in the system with $k >> 1$ (Fig. 1.18(c)). While the wave propagates, these two peaks preserve their amplitude and shape with an increase of time interval between them.

So the velocity profile of heavy particles has two quasistationary soliton-like waves with phase speed depending on the amplitude. The wave profile of heavy particles also has a characteristic plateau corresponding to approximately constant velocity. The train of negative impulses arriving at the moment $t \approx 350$ µs (as for light particles, Fig. 1.18(b)) corresponds to reflection waves from the lower end of the chain contacting with the wall.

It is interesting that in macroscopically identical systems, the force acting on the wall has a minimal value observed at $k = 4$ (90 N), at the same loading conditions as in the investigated diapason of $k = 2, 4, 16, 24, 64$ (Nesterenko and Lazaridi [1990]). This is more than two times less than the force on the wall for a system with identical particles! This example demonstrates the possibility of the microscopical structural optimization of macroscopically identical

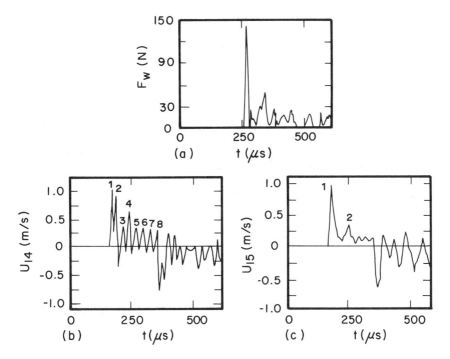

Figure 1.18. Impulses in a diatomic chain $(m_1 + m_2 = 2m)$ with relatively small mass difference $(k = 2)$: (a) the dependence of force F_w acting on the wall; (b) velocity history of the 14th light particle u_{14}; (c) velocity history of the 15th heavy particle u_{15}. Number of particles $N = 40$.

nonlinear systems for dispersional wave damping. The speed of the leading impulse practically does not depend on k.

For comparison with experimental results, a diatomic chain composed from steel spheres with a radius of 4.75 mm, and tungsten cylinders with a diameter of 4.75 mm equal to its height, was considered.

The results of calculations are presented in Figure 1.19, Young's modulus of steel spheres $E = 2.1 \cdot 10^{11}$ N/m^2, the Poisson coefficient is $v_1 = 0.29$, the radius is 2.375 mm. Young's modulus of tungsten cylinders $E = 3.9 \cdot 10^{11}$ N/m^2, Poisson coefficient $v_2 = 0.34$, the cylinders height is equal its diameter 4.75 mm; the striker mass $M_s = 10m$, velocity $v_s = 1$ m/s, impact on light end particle (steel sphere), another steel sphere is in contact with the wall.

In this case, the wave behavior is qualitatively similar to the case where $k = 2$ (Fig. 1.18) and $k = 4$, for a system with the same elastic property of particles. The velocity profile of light particles has a double peak, followed by a quasistationary periodic tail, and the velocity of heavy particle has three soliton-

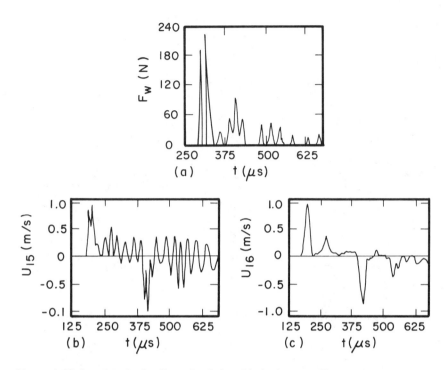

Figure 1.19. Impulses in the diatomic chain with the larger cell mass ($m_1 + m_2 = 5m$, $k = 4$): (a) the force F_w acting on the wall; (b) velocity history of the 15th light (steel sphere) particle; (c) velocity history of the 16th heavy (tungsten cylinder) particle. Number of particles $N = 41$.

like impulses. Only the force on the wall has a double peak connected with the contact of light particles, with the wall in contrast with the contact of heavy particles in previous cases.

1.6.12 Randomized One-Dimensional Granular Materials

The random strongly nonlinear chain of particles can have properties quite different from a periodically ordered system. one of the reasons for this expectation is that the existence of periodic nonlinear solitary waves is associated with the accumulation of nonlinear effects over the distance ~10 particles diameters. This process can be essentially effected by strong chaotic changes in the neighboring particles radii (Nesterenko [1983b], [1992d]).

The random one-dimensional system was obtained by assigning to each particle radius R_i, equal to some number in a sequence of random numbers in the interval (0, 1) found by means of the standard computer program *RAND*:

$$R_i = \{R_0 + RAND(1, 0)\} \times 10^{-3} \text{ m}, \tag{1.88}$$

where R_0 is reference radius. Its variations allow change to the degree of randomness. The values of the constants in governing Eqs. (1.67) are equal: $E = 2 \cdot 10^{11}$ N/m², $\rho_0 = 7.8 \cdot 10^3$ kg/m³, $v = 0.29$. Radius R_0 was equal to 0.5, 2, and 3 mm ensuring a different level of randomness varying from 3 to 1.16 if measured as a ratio of the maximum and minimum diameters of particles. The number of particles in this system is $N = 200$. Two problems were considered.

1.6.12.1 Piston Impact on a Random Uncompressed Chain

The initial and boundary conditions for the system of equations (1.67), are similar to the case with the ordered system. The velocity profile for the most disordered chain with $R_0 = 0.5$ (at the moment of time $t = 3.8 \cdot 10^{-4}$ s), at the constant velocity $v_0 = 5$ m/s at the left end particle, is depicted in Figure 1.20(a). Figure 1.20(b) shows the velocity of the 25th particle as a function of time for the same R_0 and piston velocity. The radius of this particle is $0.782465 \cdot 10^{-3}$ m.

The important differences between this system with chaotic particle sizes and the case where their sizes are identical (see Fig. 1.10) are the following:

(1) In the system with chaotic particle sizes (Fig. 1.20) there is no tendency toward the uniform steady state of the velocities profile, even near the piston, unlike in the case of identical sizes (see Fig. 1.10). It was also characteristic for the cases $R_0 = 2$ and 3 at the same piston velocity 5 m/s.

(2) In disordered nondissipative chains there is no possibility of identifying two groups of particles with different values of relative velocities, as with the case for a uniform chain. A disordered chain should result in a completely different behavior of the system after the shock wave emerges from material on the free surface. In the case of random particles, complete disintegration of granular media is expected with multiple collisions in comparison with the case of a uniform chain, where a cluster of particles with close velocity may be formed from a "shock" tail for a relatively large number of particles.

(3) The velocity amplitude at the front does not represent the maximum velocity in the system any more. It can be lower than its value behind the front (Fig. 1.20), and does not tend monotonically to increase to the doubled value of the piston velocity during wave propagation. The value of the particle velocity in the first leading oscillation varies with the propagation into the material in a nonmonotonic way, and can be smaller than v_0 or substantially exceed this value. In the chaotic chain, the smallest particle mass does not necessarily correspond to the highest velocity in the first maximum, and vice versa.

(4) Behind the wave front, different particles can reach velocity values even slightly higher than $2v_0$ (see Fig. 1.20). Particle velocities larger than $2v_0$ were determined in all three investigated cases with $R_0 = 0.5, 2, 3$ for a piston

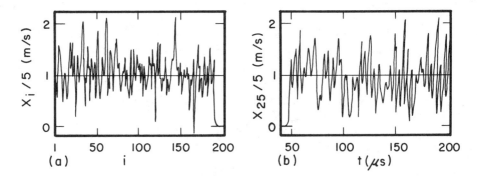

Figure 1.20. "Shock" waves in random chains, left end moving with constant velocity $\upsilon_0 = 5$ m/s: (a) particle velocity profile for a random system with $R_0 = 0.5$, (b) dependence of the particle velocity for the 25th particle on time.

velocity $\upsilon_0 = 5$ m/s. The excess of individual particle velocities over $2\upsilon_0$ increased with decreasing R_0, i.e., with the increase in the chaotic spread of particle sizes and masses. The maximum velocity value behind the front (11.76 m/s) was fixed at the 25th particle with $R_0 = 0.5$, $\upsilon_0 = 5$ m/s. The wave front at this moment is at the 170th particle.

(5) Thus, despite the chaotically varying particle sizes causing disordered oscillations of their velocities behind the front, the nonlinearity of the interaction and mass differences lead to substantial excess in the velocity amplitude at the front and behind in comparison with the piston velocity. It means that after chain disintegration, due to wave interaction with the free end of the chain, particles with velocities a few times higher than the velocity of piston may be found.

(6) The appearance of negative velocities (absent in uniform chains) was a typical phenomenon for a chaotic chain. For example, in the problem with $R_0 = 0.5$, $\upsilon_0 = 5$ m/s the same 25th particle, as in the previous case, had a velocity of -2.142 m/s. At this time the wave front was at the 148th particle.

(7) When the perturbation emerges from the free end of the chain the final velocity for the last particle after the "spall" may be smaller (or larger) than in the case of identical particles. For example, when $R_0 = 2$, after separation of the last 100th particle from chain, its relative velocity is $x_{100}/\upsilon_0 = 2.366$, distinctly smaller than the value 2.876 for the case of identical particles. But for a more chaotic chain ($R_0 = 0.5$), the corresponding velocity value after "spall" at the free chain end is equal to $x_{100}/\upsilon_0 = 2.969$, which is larger than the similar velocity ratio in the case of identical particles.

(8) Despite the highly disordered oscillations of particle velocities, the ratio of kinetic to potential energy in a system with random particle sizes also

oscillates near the mean value, as in the case for a system with identical particle sizes (Fig. 1.11). It is an interesting illustration that nonstationary behavior of the mechanical system, according to one set of microvariables, can coexist with steady behavior with respect to macroparameters. The differences between these two cases are:

—the maximum deviations from the mean value are substantially larger in the random case; and

—the mean values of E_k/E_p deviate more from the value 1.25, following from the virial theorem. For example, the values E_k/E_p, averaged over the time interval during which the leading front of the wave traverses the distance from the 70th to the 90th particle for the cases $R_0 = 0.5$, 2, and 3, are equal, respectively, to 1.178, 1.236, and 1.232. In this case the maximum relative deviations from these mean values are equal, respectively, to 9, 7, and 6%.

1.6.12.2 Decomposition of Initial Perturbation

Short impulse loading of a random chain was created by the impact of two particles with a velocity of 5 m/s similar to the uniform chain (Fig. 1.5). The time dependencies of the velocities for the 61st particle and the first oscillation for the 96th particle are given in Figure 1.21 ($R_0 = 2$). In comparison with the case of identical particles (Fig. 1.5), the perturbation does not decompose into two solitons, but has a significantly random character.

At the same time, it is characteristic that the leading perturbation still resembles the soliton shape, demonstrating the robustness of this phenomena even in highly randomized chains (see also Manciu, Sen, and Hurd [2000]). Of course in this case the leading pulse is not a stationary one.

The important feature is the decay in amplitude, even in the absence of dissipative losses. It is evident from Figure 1.21 that this damping is related to the scattering of energy and its redistribution among a large number of particles. The damping of the velocity amplitude is conveniently traced by the final

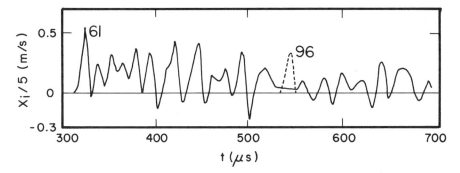

Figure 1.21. Decay of the quasisolitary impulse in the random chain: particle velocity for the 61th particle and first oscillation (broken line) for the 96th particle.

velocity υ_N of the last particle at the free end of the chain after "spall." Thus, for $R_0 = 0.5$, $\upsilon_{100}/\upsilon_0 = 0.692$, and for less random chains with $R_0 = 2$, $\upsilon_{100}/\upsilon_0 = 0.507$, and for $R_0 = 3$, $\upsilon_{50}/\upsilon_0 = 1.003$. It is seen that these values are smaller than 1.2342, which is characteristic for a system of identical particles in the similar case of impact by two particles with the same velocity.

A significant increase in the amount of particle chaotization in sizes with R_0 decreasing from 3 to 0.5 does not lead to an enhanced damping of the velocity amplitude. On the contrary, damping is smaller in the most random case.

1.7   Phenomenological Description of Waves in Particulate Systems

There are other approaches to analyze wave propagation in random particulate systems. The generalization of the classical method of characteristics to materials with granular-type microstructures, with linear and nonlinear elastic, and with an elastic/dissipative response of its elements, was made by Ostoja–Starzewski [1984], [1991a,b], [1994]. The basic idea is to represent the one-dimensional linear elastic medium in the x-direction by a vector random process $\{l, \rho, E\}_x$ with piecewize continuous realizations, whereby l denotes the element lengths, ρ their density, and E their Young's modulus. In the case of a bilinear elastic medium, a process $\{l, \rho, E_0, E_1\}_x$ is employed, where E_0 stands for the first elastic modules, while E_1 stands for the second elastic modules.

In this case, any given element (layer) was considered as a separate continuum for which classical nonlinear wave equations were applied combined with conditions of continuity of tractions and displacements on plane interfaces. This approach is different from the case of a granular system, considered at the beginning of this section, where the wave processes in the granules were neglected and the response of individual grains was considered as a quasistatic one with interface interaction according to the Hertz law.

In the above-mentioned papers by Ostoja–Starzewski, a stochastic method has been developed to assess the effects of material randomness, usually of high signal-to-noise ratio, on wave propagation taking into account the structure of discrete material. This method takes advantage of the Markov property of forward propagating disturbances—any single point of the pulse $f(t)$ applied at $x = 0$. The random variation of the elements properties has two effects on a traveling disturbance: random zig-zaging of characteristics and amplitude attenuation caused by the generation of backscattered waves (e.g., energy loss) on element boundaries. Consequently, the evolution of any disturbance is described by a vector Markov diffusion process, and the entire pulse in space–time can be assessed. It was found that due to the random scatter of disturbances and their attenuation the character of some pulse propagation problems, as compared to the deterministic homogeneous media cases, may be altered qualitatively. The limiting case of zero noise in these parameters corresponds to a homogeneous medium case described simply by the averages of constitutive coefficients, for which the transient wave solutions are well known (e.g., Cristescu [1967]; Nowacki [1978]). While studies on this subject were mainly focused on one-dimensional media problems, this approach offers a way of analyzing two- and three-dimensional phenomena, especially under conditions of spherical and cylindrical symmetry.

At the same time, it is necessary to mention that in an essentially nonlinear case, or in a case of weak dissipation and a large number of elements, the backscattered waves (disregarded in these papers) can produce qualitatively new results. This was experimentally and theoretically demonstrated by Nesterenko, Fomin, and Cheskidov [1983a,b] for strong shock waves in laminar materials (see Chapter 3). For relatively short sequences of elements, or for weak nonlinearity and for relatively short impulse duration, the influence of these backscattered waves can be neglected.

A stochastic generalization of a deterministic problem of acceleration-to-shock wave transition was considered by Ostoja–Starzewski [1993], [1995]. This problem for a uniform continuum was originally studied in the 1960s through the 1980s (e.g., Bland [1969]; Chen [1973]; McCarthy [1975]; Menon, Sharma, and Jeffrey [1983]). The problem is described by a Bernoulli equation that governs a competition of two effects: elastic nonlinearity and dissipation on the evolution of wave amplitude a:

$$\frac{da}{dx} = \beta(x)a^2 - \mu(x)a. \tag{1.89}$$

The first term in the right-hand side of Eq. (1.89) is dependent on a material coefficient β, and tends to make the acceleration wave with initial amplitude a_0 degenerate to a shock, while the second one, depending on a material coefficient μ, tends to attenuate it. For material with $\mu = \text{const} > 0$, and $\beta = \text{const}$ there is a critical amplitude a_c, for the wave to change into a shock wave

$$a_c = \mu/\beta, \tag{1.90}$$

and the distance ξ_c to form a shock

$$\xi_c = -1/\mu \, \ln[1 - \mu/\beta a_0]. \tag{1.91}$$

Determination of the random, rather than deterministic, critical amplitude a_c and distance ξ_c to form a shock, was first considered by Ostoja–Starzewski [1991a,b]. Four cases of the vector random process $\{\mu, \beta\}_x$, driving the evolution of an acceleration wave, may be distinguished depending on whether both μ_x and/or β_x are white or nonwhite (of Markov type) processes. Analysis presented by Ostoja–Starzewski [1993] was focused on the white noise case. The principal conclusion is that the average critical amplitude of the random medium problem is greater than the critical amplitude of the corresponding deterministic homogeneous medium problem. Also, effects of the positive, zero, and negative correlatedness of μ_x and β_x on the second, third, and fourth moments of the critical amplitude were investigated. A discussion of the other nonwhite noise cases is given by Ostoja–Starzewski [1994], which includes a comparative review of the governing equations, available solution methods (analytical versus numerical), and some results on the problem of "distance to form a shock," which is cast into a classical stochastic problem of "random exit times."

The phenomenological description of granular materials, with an additional degree of kinematic freedom reflecting the existence of local kinetic energy in elastic material around pores, was introduced by Goodman and Cowin [1972]. This approach was successfully applied to the evolution of acceleration waves in granular materials (Nunziato and Walsh [1977]; Nunziato and Cowin [1979]; Parker and Seymour [1980]; and others) including the solid explosive PBX-9404 (Nunziato, Kennedy, and Walsh [1978]). This approach can help to predict the generation of shock wave from acceleration wave which can result in detonation initiation in porous explosives.

The one-dimensional chain of spherical grains with random radii corresponds to the chain of particles with random masses, and force interaction constants. Continuum approximation for a one-dimensional weakly nonlinear lattice, with spatially varying distribution of mass and interaction constants, was considered by Ono [1972]. He proposed the following nonlinear equation:

$$\psi_\tau + \psi^p \, \psi_\xi + \psi_{\xi\xi\xi} + \nu(\tau) \, \psi = 0 \,, \tag{1.92}$$

$$\nu(\tau) = \frac{1}{4p} \frac{d}{d\tau} \ln\left(m^{4-p} \, k^{3p-4} \, \alpha^{-4}\right) + 12c^4\lambda \,,$$

where ψ is the velocity, ξ and τ the stretched time and coordinates, c the local sound speed, α and λ stand for the nonlinearity and the dissipation coefficients correspondingly, p is the positive constant, describing the power-law nonlinear term in interaction force, m is the particle mass, and k is the linear spring constant. Equation (1.92) was obtained by expanding the displacements y_{i+1} and y_{i-1}, in the Taylor series, as well as by expanding neighboring masses, spring constants, nonlinear and dissipation parameters into series. The terms of the orders h^5, αh^{p+1}, σh, and $\sigma\alpha h^p$ were neglected (σ is an additional small parameter, measuring the strength of heterogeneity). It was shown that initial disturbance in the form of the KdV soliton may be disintegrated into a number of solitons and/or oscillatory waves, depending upon the property of the inhomogeneity function $\nu(\tau)$. The equations similar to Eq. (1.92), can be obtained for low-dispersive, long nonlinear waves in weakly inhomogeneous media (e.g., for waves in shallow water with randomly varying depth, ion–acoustic waves in inhomogeneous plasma) and were analyzed by Asano [1974], Madsen and Mei [1969], Sobczyk [1992].

This approach evidently considers only the weak inhomogeneity, with small and gradual changes of parameters between adjacent particles. In the case of granular materials considered at the beginning of this section, the masses and interaction parameters are changing abruptly, from one grain to another, and can not be represented by expansion of parameters of the closest particle. Also, Eq. (1.92) represents only a weakly nonlinear case, which cannot be considered as proved for conditions of sonic vacuum. That is why it is unclear how the Eq. (1.92) can be used to describe the waves in the chain with random particles. At the same time, we can observe some qualitatively similar features in the behavior of these two systems: solitons in weakly nonlinear and strongly nonlinear random particulate systems decay by disintegration into other disturbances even without dissipation.

1.8   Experimental Observation of a New Type of Solitary Waves in a One-Dimensional Granular Medium

1.8.1   What Can Be Expected from Experiments?

The analysis of the differential equation, representing the long-wave approximation for the system of particles, interacting according to the Hertz law, demonstrated the existence of steady solitary waves of a qualitatively new type. They are consistent with the numerical calculations of a corresponding discrete string of particles. The nonlinear Eq. (1.20) is more general than the KdV equation, which includes weakly nonlinear and dispersion effects in the first approximation for a broad class of physical systems.

Wave propagation with characteristic space dimensions the same as the typical space scale of the system, is a very interesting and yet unresolved problem. For example, there is a dispersion relation problem for short waves for the Bussinesq equation, which can be formally eliminated in its regularized form. It is also a general problem for any continuum description in the behavior of a nonlinear discrete system of particles. Recent discussions on this topic can be found in the paper by Albert and Bona [1991]. In our case, where the space scale of solitary waves is equal only to five particle diameters, they can be nevertheless successfully described as stationary solutions of the "zonic vacuum" Eq. (1.20), in accord with computer calculations for a discrete chain.

From this point of view, it is very interesting to compare results of the theoretical analysis and computer calculations for strongly nonlinear particulate systems with the experiment. The investigated system has this distinctive feature—the law of interaction between particles does not contain a linear component, even in the zeroth approximation. The given interaction for a one-dimensional chain (or for 3-D simple cubic packing) corresponds to an equation state of the medium, for uniaxial static compression in the form

$$\sigma = \frac{E\xi^{3/2}}{3(1-\nu^2)},$$
(1.93)

where σ is the stress in equivalent continuum, ξ is the strain, and E and ν are the Young's modulus and the Poisson ratio of the particle's material. The long-wave sound speed in a medium, described by this equation of state, is

$$c_0^2 = \frac{1}{\rho}\frac{\partial\sigma}{\partial\xi} = \frac{E}{2\rho(1-\nu^2)}\,\xi^{1/2},\tag{1.94}$$

where ρ is the density of the medium. For $\xi = 0$, i.e., in a chain of particles not subjected to an external force, $c_0 = 0$. Consequently, the standard wave equation cannot be used to describe perturbations of arbitrary amplitude in the case of zero initial strain.

On the basis of the common approach in this case, it is reasonable to accept that the impact by a piston, with any velocity, will result in a shock wave if dissipative processes are present. The numerical calculations (see Figs. 1.9 and 1.10, Nesterenko [1983b]) have shown that the propagating wave is represented by a train of solitary pulses in the absence of dissipation, under different types of impulse loading.

1.8.2   Solitons Excited in "Sonic Vacuum" with Identical Particles

From an experimental point of view, it is convenient to excite the given system by the impact of a finite mass piston. The parameter determined in the experiment, and then compared with numerical analysis, in this case may be the reaction of a rigid wall contacting with the end of the chain opposite to the impact-driven end (Fig. 1.14) (Lazaridi and Nesterenko [1985]), or the compressive stress inside the particle (Nesterenko and Lazaridi [1990]).

A one-dimensional experimental system of particles (Fig. 1.22) was created with the help of a quartz tube having an inside diameter of 5 mm. A chain of chromium steel balls (bearing type), with a diameter of 4.75 mm, was placed inside this tube. It was vertically rested on a disk of hardened (HRC = 40) chromium–manganese–silicon steel, with a thickness of 2 mm. The chain was prevented from contact with the quartz tube, the latter being rested in the holder.

A lead–zirconate–titanate PZT wafer was bonded to the underside of the disk with epoxy resin, serving as a piezoelectric transducer of the force acting on the steel disk. The wafer, in turn, was soldered to a brass rod, with a length of 200 mm, which was coated with epoxy resin and inserted in a copper case. This configuration of the force sensor makes it possible to eliminate the influence of reflected waves in the rod; the latter condition was tested separately.

The signal from the piezoelectric transducer was transmitted by means of a copper electrode, soldered into the upper part of the PZT wafer, to the oscilloscope. A long enough characteristic time of electrical circuit, electrical resistance multiplied by electric capacity = 0.25 s, was ensured by a special electrical scheme. The gauge was calibrated by impact of a single ball, and then, comparing the measured symmetrical time dependence of the stress, with the dependence calculated according to Hertz's law. The oscilloscope operated in the

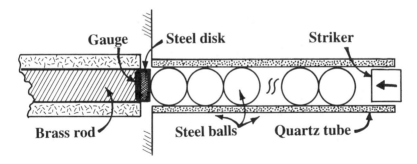

Figure 1.22. Experimental set-up for observation of a soliton generated by striker impact upon an one-dimensional granular chain.

internal-trigger mode, using the working signal. The impact on the system was provided by the drop of the piston, with mass $M = 5m$ (m is equal to the particle mass) and velocity 0.5 m/s.

Figure 1.23(a),(b) shows the experimental time dependence of the force F_w, acting on the lower disk for a number of particles, $N = 20$ and 40 particles, respectively. It clearly demonstrates the existence of the train of solitary waves and its evolution.

The amplitudes and time parameters of the pulses, averaged over four independent experiments, are shown in Table 1.1. The quantities τ_{12}, τ_{13}, and τ_{14} are the time intervals between the maximum signal corresponding to the first soliton and the subsequent (as numbered by the second index) solitons. The quantities F_1–F_4 (in N) are equal to the amplitudes of the pulses for the first four solitons. The relative measurement error does not exceed 17% for the values of F_1–F_4 and 7% for the time intervals τ_{12}, τ_{13}, and τ_{14}.

Coste, Falcon, and Fauve [1997] used a physically similar geometry of experimental set-up shown in Figure 1.22, with larger particle diameters. They applied vibration exciter for impulse loading.

For a comparison with the experimental results, the computer calculations of the discrete system (see Eq. (1.67)) were carried out according to a fourth-order Runge–Kutta scheme, with a time step of $0.25 \cdot 10^{-5}$ s. The interaction between the piston, the first particle, and between the last particle and the rested wall was chosen on the basis of Hertz's law. The energy and momentum in the numerical calculations were monitored within the respective error limits of 10^{-2} and $10^{-5}\%$. The mass of the piston was equal to five masses of one particle, as it was in the experiments. The constants in the interaction law (Eq. (1.1)) were taken equal to $E = 2 \cdot 10^{11}$ N/m^2 and $\nu = 0.29$, corresponding to the properties of the steel. The velocity of the piston in calculations was 0.5 m/s, the particle diameter was $2R = 4.75 \cdot 10^{-3}$ m, and the density of the particle material was $\rho_0 = 7.8 \cdot 10^3$ kg/m^3.

Figure 1.24(a), (b) shows the variation of the force F_w found in the numerical

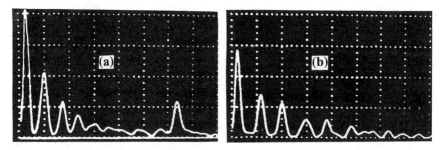

Figure 1.23. Evolution of the soliton train excited in an experiment by striker impact with the distance from the entrance: (a) $N = 20$, (b) $N = 40$. The vertical scale is 18.3 N between large divisions, the large horizontal division is equal to 50 µs.

Table 1.1. Experimental and calculated parameters of a solitary wave.

Impulses parameters	Experiment ($N = 20$)	Calculations ($N = 20$)	Experiment ($N = 40$)	Calculations ($N = 40$)
τ_{12} (µs)	35	37.5	48	55
τ_{13} (µs)	72	72.5	91	110
τ_{14} (µs)	105	110	135	170
F_1 (N)	71	83	52	85
F_2 (N)	40	39	24	39
F_3 (N)	21	19	18	20
F_4 (N)	10	11	11	11

calculations between a plate, and the end of the chain for $N = 20$ and 40 particles, respectively. The beginning of the piston interaction with a chain was taken as a zero moment. It is evident from Figures 1.23 and 1.24 that the initial disturbance decomposes into a train of solitary waves. The occurrence of an isolated pulse at $t = 430$ µs in Figure 1.24(a) corresponds to the arrival of the reflected wave from the piston at the lower disk. A similar wave in Figure 1.24(b) is observed at 860 µs because of the longer chain. The numerical calculations exhibited a steady-state behavior in the propagation of generated solitary waves, which are similar to those previously obtained in computer calculations for different conditions of loading (Fig. 1.7). An increase in mass of

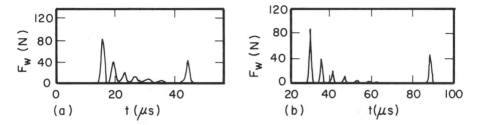

Figure 1.24. Evolution of the soliton train with the distance from computer calculations: (a) $N = 20$, (b) $N = 40$. Initial impulse is excited by striker impact ($M_s = 5m$, $v_s = 0.5$ m/s).

the piston causes the number of the waves to increase. The calculated values for τ_{12}, τ_{13}, and τ_{14} and F_1–F_4 are shown in the corresponding columns of Table 1.1.

It is evident from a comparison of Figures 1.23(a) and 1.24(a), and the data of Table 1.1, that there is qualitative and quantitative agreement between the amplitudes of the solitary waves, their number, and characteristic time parameters for a short system with $N = 20$. For a longer chain with $N = 40$ the experimentally observed qualitative pattern is also completely consistent with the numerical calculations (see Figs. 1.23(b) and 1.24(b)). But the amplitudes of the first solitary waves (F_1 and F_2) in the experiments are essentially smaller than the theoretical values (see Table 1.1), owing to the presence of dissipation, which is ignored in calculations. As a result, a larger disagreement between experiments, and computed data is observed in the time intervals τ_{12}–τ_{14}.

Figure 1.25 shows the sequential decomposition of an initial impulse in a given one-dimensional system, depending on the time of propagation (N is the number of particles between the piston and lower disk) in the experimental set-up analogous to Figure 1.22. It should be compared with the experimental results from Figure 1.23 for impact by a smaller striker mass in the same experimental set-up. From Figure 1.25 one of the remarkable features of the "sonic vacuum" is evident—very rapid decomposition of the initial impulse, on the distances comparable with the soliton width. In fact, under the given conditions, the impulse is split after traveling through only 10 (!) first particles. This peculiarity can be used for controlled impulse transformation in very short transmission lines of a different nature. We notice that this property cannot be obtained through stationary analysis of Eq. (1.20), which makes a very desirable nonstationary analysis of this equation. The role of dissipation is also clear from Figure 1.25, but the existence of essential dissipation does not change the process of initial impulse splitting into solitons and their propagation.

In experiments by Coste, Falcon, and Fauve [1997], Figure 1.26, and Falcon [1997], a systematic and quantitative study of the velocity and shape of solitary waves was performed. They demonstrated the negligible decay of the

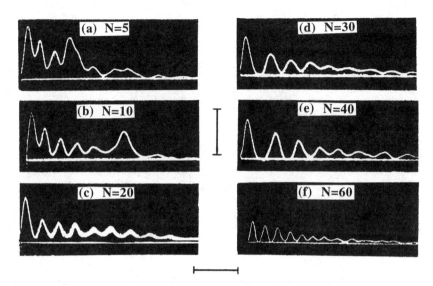

Figure 1.25. Evolution of the soliton train excited in experiments by striker impact (M_s = 10 m, υ_s = 0.5 m/s) with the propagation distance: (a) $N = 5$, (b) $N = 10$, (c) $N =$ 20, (d) $N = 30$, (e) $N = 40$, (f) $N = 60$. Vertical scale corresponds to 80 N, and the horizontal scale to 50 μs (a),(b),(c),(d),(e) and 100 μs (f).

soliton after traveling a distance equal to 50 particle diameters—one order of magnitude longer than soliton width. They concluded that experimentally observed solitons are in very good agreement with the solitary solutions of strongly nonlinear Eq. (1.20).

Fig. 1.26 represents their experimental results in comparison with theoretical prediction for soliton shape (Eq. (1.39)), when no static force is applied to the chain. The chain is composed of steel beads, each 8 mm in diameter, and the number of beads is equal to 51. Each graph displays the evolution of force with time. Dots are experimental values and solid lines are the theoretical predictions based on Eq. (1.39). Their agreement with the theory is very satisfactory. At lower impulse amplitudes (60 N) the agreement was not as good.

Comparison of experimental data for soliton speed, depending on the force amplitude with theoretical predictions for no applied static force (Eq. (1.37)), demonstrates an excellent agreement between them (Fig. 1.27). Experiments conducted by Coste, Falcon, and Fauve [1997] did not find essential dissipation, in comparison with results by Lazaridi and Nesterenko [1985]. In their chain containing 51 particles they found in general excellent agreement with the theory for either the velocity of the pulses, or their complete shape. The reason for stronger dissipation in experiments by Lazaridi and Nesterenko [1985] could be connected with smaller radii of the particles, resulting in their worse alignment, and with higher velocity of impact.

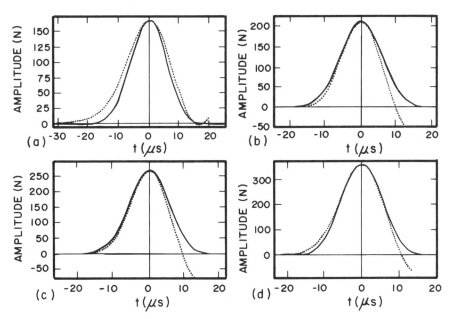

Figure 1.26. Shape of the solitary waves when no static force is applied to the chain: (a) $V_s = 811$ m/s, $F_m = 168.5$ N; (b) $V_s = 845$ m/s, $F_m = 216.2$ N; (c) $V_s = 877$ m/s, $F_m = 269.7$ N; (d) $V_s = 919$ m/s, $F_m = 356.4$ N. (Reprinted from *Phys. Rev., E*, **56**, no. 5, Coste, C., Falcon, E., and Fauve, S. Solitary Waves in a Chain of Beads Under Hertz Contact, pp. 6104–6117, 1997, with permission from author).

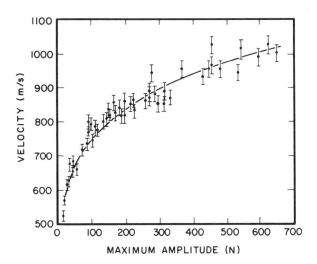

Figure 1.27. The wave speed versus the amplitude of the wave for no applied static force. The solid line is the theoretical prediction based on Eq. (1.43). (Reprinted from *Phys. Rev., E*, **56**, no. 5, Coste, C., Falcon, E., and Fauve, S. Solitary Waves in a Chain of Beads Under Hertz Contact, pp. 6104–6117, 1997, with permission from author).

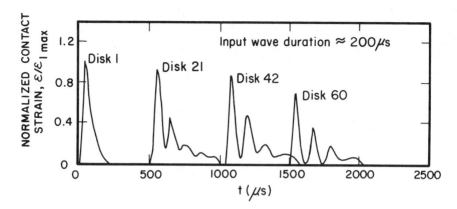

Figure 1.28. Impulse decomposition and solitons generation in the system of Homalite-100 disks, input wave duration is 200 µs. (Reprinted from *J. Mech. Phys. Solids*, **41**, no. 11, Shukla, A., Sadd, M.H., Xu, Y., and Tai, Q.M. Influence of Loading Pulse Duration on Dynamic Load Transfer in a Simulated Granular Medium, pp. 1795–1808, 1993, with permission from Elsevier Science).

One of the interesting features in the experiments by Coste, Falcon, and Fauve [1997] is connected with the relative width of experimental profiles, and results of long-wave approximation (Eq. (1.39)) for relatively small impulse amplitudes (see Fig. 13 in their paper). In their experiments, solitons were essentially wider than predicted by continuum approximation. The natural explanation that it may be connected with the role of discreteness does not help, because solitons in discrete chains are more narrow than solitons described by Eq. (1.39) in the frame of long-wave approximation (Nesterenko [1983b]; Chatterjee [1999]). More experiments with precise measurements of soliton profiles are necessary to make final judgments on the mechanism of this discrepancy.

Coste and Gilles [1999] experimentally demonstrated that the nonlinear solitons in noncompressed chains, composed from very different materials like steel, glass, nylon, and brass beads, follow the predictions of long-wave theory based on the Hertz law (Eqs. (1.39), (1.43)) with remarkable accuracy, except the solitons speed for brass, where agreement is not as good as for other materials.

Shukla, Sadd, Xu, and Tai [1993] experimentally investigated the dynamic response of a one-dimensional chain of 100 circular disks (diameter 25.4 mm and thickness 6.35 mm) fabricated from Homalite 100 (Fig. 1.28). They found the qualitatively different behavior of short (90 µs) and longer waves (duration > 200 µs). The latter undergoing separation into a series of short waves (Fig. 1.28). The authors explain this behavior with the approach based on Eq. (1.20) for "sonic vacuum."

Experimental investigation of impulses in the one-dimensional chain of particles from optically active material, was carried out by Zhu, Shukla, and

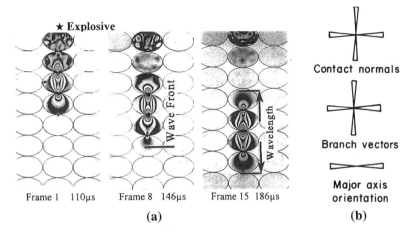

Figure 1.29. Generation of a short impulse in the two-dimensional body-centered assembly of elliptical particles under concentrated explosive loading: (a) typical isochromatic fringes; (b) fabric parameter distribution. (Reprinted from *J. Mech. Phys. Solids*, **44**, no. 8, Zhu, Y., Shukla, A., and Sadd, M.H. Effect of Microstructural Fabric on Dynamic Load Transfer in Two Dimensional Assemblies of Elliptical Particles, pp. 1283–1303, 1996, with permission from Elsevier Science).

Sadd [1996]. The most interesting observation is the development of impulse composed from four particles (Fig. 1.29) in just the first few (six) particles. This phenomen is specific for strongly nonlinear behavior (Nesterenko [1983b]; Lazaridi and Nesterenko [1985]), and the impulse length of four particle diameter can be explained by the power-type interaction law with exponent $n = 1.68$. At the same time, too small a distance traveled by the impulse, combined with extremely short and intensive loading conditions, can strongly influence the shape of the optically observed wave and provide the difference between its profile and a possible stationary soliton in the experiments. Actually no definite conclusions about the existence of a stationary solitary wave (soliton) can be made based on experiments with six particles only. We have seen (Fig. 1.25) that, despite a fast impulse splitting into solitons, at least 10 particles are still necessary to separate a stationary soliton (see also Fig. 1.28).

Thus, a new type of solitary wave was observed in the independent experiments. These waves agree with the theoretical calculations of a discrete chain of particles, and with the stationary solutions of the "sonic vacuum" Eq. (1.20). The qualitative difference of this "one small parameter" soliton from "two small parameter" KdV soliton is considered in detail in Section 1.3. The solitons in this case act as an elementary steady-state excitations of the system replacing sound waves.

1.8.3   Collision of a 1-D Chain with a Wall

The piston problem considered earlier in this chapter has some analogy with experiments by Falcon, Laroche, Fauve, and Coste [1998b], where a 1-D chain collides with a rigid wall. In their experiments, it was shown that the maximum force felt by the rigid wall was independent of the number of beads—the natural result of a wave process in the chain. It is interesting that the beginning of the force histories are almost identical for collision of one or four particles with the same velocity. The duration of impact increases linearly with the number of particles in the chain. The authors observed that the column of particles does not bounce as a whole from the rigid wall, at a number of particles equal to 5.

The formation of clusters bouncing from the wall may be expected for a larger number of particles, based on the numerical calculations of the piston problem for uniform chains (see Fig. 1.10). At the same time, formation of clusters even for long random chains is improbable, if dissipation is weak (see Fig. 1.20).

The different behavior has been observed in experiments of a chain collision with a soft wall, where the maximum force depends on the particles number if N < 15 to 20 and saturates on the level smaller than for the rigid wall.

Falcon, Laroche, Fauve, and Coste [1998b] emphasized that the phenomena observed for the 1-D experiments of chain collision with the wall are not specific only for this geometry, but also may be typical for two or three dimensions. The bead detachment effect which is a purely dispersive effect is probably a precursor mechanism for the fluidization of the vibrated granular media. The dissipative losses may result in the enhancing of clustering tendencies.

1.8.4   Soliton Excitation in an Uncompressed Two-Particle Chain

An experimental set-up for a soliton generation in a two-particle system is presented in Figure 1.30. The system is made of alternating steel spheres with diameter 4.75 mm and tungsten cylinders with height and diameter equal to 4.75 mm. This system has a particle's mass ratio of $k = 4$ and is similar to the system considered in computer calculations with the results presented in Figure 1.19. A steel striker, with a mass equal to $10m$ (m is the mass of the sphere) and velocity 1 m/s, impacts the two-particle chain on the right end sphere.

The force acting on the wall, and the compression force inside the 15th particle was measured in experiments. The gauge in the wall was made as in previously described experiments (Fig. 1.22). The transducer inside the steel particle was also made from a lead–zirconate–titanate PZT wafer (diameter 3.8 mm, thickness 0.5 mm) soldered from both sides to copper foils and bonded with epoxy resin to steel hemispheres from the same material and radius as the spherical particles. This particle with the gauge inside, has the same mass as the other particles. The signal from this gauge was transmitted through thin wires with a cross-section surface 0.1 mm^2. A characteristic time of electrical circuit (0.01 s) was enough to prevent signal distortion.

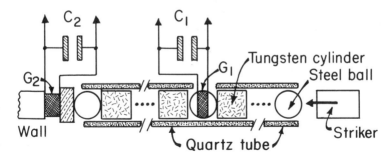

Figure 1.30. Experimental set-up for observation of a soliton generated by striker impact in a 1-D diatomic chain, the piezoelectric gauge G_1 records the compression force inside steel ball, and the G_2 records the contact force F_w.

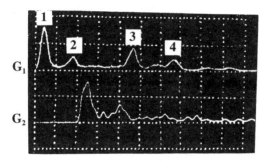

Figure 1.31. Impulses in a 1-D diatomic granular chain generated in experiments by striker impact ($M_s = 10$ m, $v_s = 1$ m/s): (a) compression forces inside the 15th particle, (G_1); (b) contact force $F_w(G_2)$. Number of particles $N = 41$. Vertical scale, 95 N/div (G_1) and 92 N/div (G_2); horizontal scale, 50 μs/div.

Experimental results are presented in Figure 1.31. The oscilloscope was triggered by a signal from the gauge inside the 15th particle. In experiments were observed two seperated incident impulses, recorded by gauge G_1 inside the 15th particle (Fig. 1.31, upper beam, impulses 1 and 2).

Also, two reflected impulses (3 and 4) from the wall are present that are similar to those found in computer calculations of the velocity profile of the neighboring heavy 16th particle (Fig. 1.19(c)). The time interval (70 μs) between these impulses in experiments is consistent with the time interval in computer calculations. The similarity is also evident for the time interval between impulse arrivals on the 15th particle and on the wall (gauge G_2, Fig. 1.31, low beam) and between widths of the impulses. A characteristic double hit of the wall (compare Fig. 1.31, low beam, with Fig. 1.19(a)) was not observed in experiments, and the force amplitude (150 N) was essentially lower than in computer calculations (216 N) (Nesterenko and Lazaridi [1990]). The former is probably due to the overlapping of impulses in the experiment, and the latter can be explained by dissipation. For a smaller number of particles (21), the agreement between experimental amplitude (230 N) and calculations (258 N) was very good. It is possible to conclude that the main peculiarities of the two-particle chain observed in experiments are consistent with computer calculations and that soliton-like objects also exist in chains with periodically alternating masses.

1.8.5   Soliton Interaction with the Boundary of Two "Sonic Vacuums" and Impurities

1.8.5.1   Soliton Interaction with the Interface

The behavior of compression pulses when passing through the contact of two "sonic vacuums,"—chains of steel spherical granules with diameters 4.75 mm (SV-1) and 7.9 mm (SV-2)—were studied through experiments and numerical calculations (Nesterenko, Lazaridi, and Sibiryakov [1995]). The compressive stresses were measured both in incident and reflected waves (transducer G_1 in the 15th particle from the impacted end) and in passed impulses (transducer G_2 in the 6th particle from the interface) (see Figs. 1.32 and 1.33).

A single incident soliton was created by striker impact ($M_s = m_1$, $v_s = 1$ m/s) on the SV-1 (Fig. 1.32) and by striker impact ($M_s = m_2$, $v_s = 1$ m/s) on the SV-2 (Fig. 1.33). Pictures at the top of Figures 1.32 and 1.33 represent the

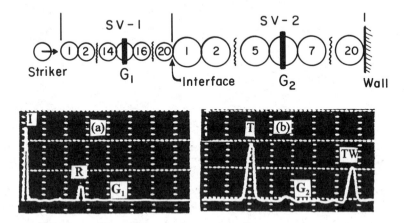

Figure 1.32. Soliton interaction with contact of two "sonic vacuums": (a) waves recorded by gauge (G_1) inside the 15th particle before the interface; (b) waves recorded by gauge (G_2) inside the 6th particle behind the interface. Vertical scales are 10 N/div (G_1) and 7 N/div (G_2), and horizontal scales are 100 µs/div.

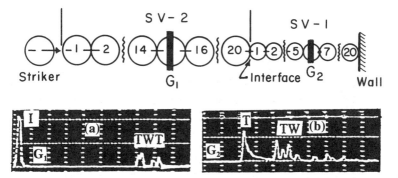

Figure 1.33. Soliton interaction with contact of two "sonic vacuums": (a) waves recorded by gauge (G_1) inside the 15th particle before the interface; (b) waves recorded by gauge (G_2) inside the 6th particle behind the interface. Vertical scales are 31.5 N/div (G_1) and 50 N/div (G_2), horizontal scales are the same, 100 µs/div.

corresponding geometry of experiments. These experimental results should be compared with computer calculations (Figs. 1.14 and 1.15).

When the incident soliton (I) hits the contact from the side of SV-1 ("light"), a reflected soliton (R) appears in it, and the transient soliton (T) propagates in SV-2. The latter reflects from the wall and arrives at the same point of measurement marked TW (Fig. 1.32(a),(b)). This is in complete agreement with the results of computer calculations (Fig. 1.14), except for the decrease in amplitude of the R and TW impulses in experiments due to dissipation. It is apparent that this pattern of pulse interaction with the interface of two "sonic vacuums" qualitatively differs from the expected in traditional acoustics.

An even more unusual picture is observed when the incident soliton (I) strikes from the side of SV-2 ("heavy" system) (Fig. 1.33). Then the reflected compression pulse is not observed (Fig. 1.33(a)). While in SV-1, an asymmetrical triangular transmitted pulse (T) with a steep front arises near the contact, which later decomposes into a series of solitons (TW) as it propagates and later reflects from the rigid wall. The remnants of this impulse (TWRT) being transmitted into SV-2 are also noticeable in Figure 1.33(a). This behavior is observed in computer calculations as well (Fig. 1.15) except for the visible dissipation effect in experiments.

1.8.5.2   Soliton Interaction with Impurities

Soliton interaction in a strongly nonlinear system with impurities is another subject of interest. Even in linear chains, the impurities are responsible for qualitatively new effects, such as localization of perturbation in its vicinity (see, e.g., Lee, Florencio, and Hong [1989]). Sen, Manciu, and Wright [1998] proposed "soliton pulse spectroscopy" as a potential method for probing buried objects in granular beds (see also Naughton et al. [1998]). They were able to detect a minor, but regular and noticeable, difference in the backscattered pulses for the light and heavy mass of impurities (Manciu, Sen, and Hurd [1999b]; Hascoet and Herrmann [2000]). Their idea is based on the fact that sand grains are rather heavy objects in comparison with objects composed from such materials as rubber, plastic, or wood. At the same time, the reflected impulse can be observed from the contact of granular media with the same overall density (see previous section). This area is completely unexplored and in need of intensive attention. In the case of successful development, it can also provide a very useful tool to probe the local irregularities in the granular materials packing, before hot isostatic pressing, and other densification techniques.

Singh, Shukla, and Zervas [1996] conducted a study of impulse propagation with a shape close to a soliton, through a 1-D chain of Homalite-100 particles (disks) with one disk being damaged by a crack of a different orientation. It did not effect the group velocity with comparison of the undamaged chain of disks, but resulted in the localized decrease of peak contact load after the stress wave propagated through the damaged disk.

1.8.6 Soliton Excitation in a Precompressed Chain

Experimental data on impulse behavior in a precompressed chain of particles were obtained by Nesterenko and Lazaridi [1987] and by Coste, Falcon, and Fauve [1997]. The latter authors used an experimental setup which allowed them to explore a wide range of the amplitude of stationary pulses, and to vary the static force applied to the chain from 9.8 N to 167 N. The experimental pulses for a static force of 167 N for different wave speeds (maximum force in the pulse), together with theoretical predictions based on Eq. (1.33) for the soliton, are presented in Figure 1.34. The chain is composed from steel beads each 8 mm in diameter, and the number of beads is equal to 51. Each graph displays the evolution of the force with time. Dots are experimental values and solid lines are the theoretical predictions based on the stationary solutions of Eq. (1.20).

The dependence of soliton speed on force amplitude is presented in Figure 1.35, as well as theoretical dependence similar to the function $V_s/c_0(\xi_r)$ from Eq. (1.84) (only ξ_r is expressed through the relative force amplitude F_m/F_0). All experimental data lie on the single curve coinciding with the theoretical prediction for soliton speed in the long-wave approximation (Eq. (1.84)).

The results of computer calculations and experiments, with the impact of the piston on the prestressed chain of the same particles resulting in nonstationary wave propagation (static forces $F_0 = 2$ and 20 N), also agree satisfactorily with computer calculations (Nesterenko and Lazaridi [1987]). A small initial prestrain of 2 N did not change the wave characteristics in comparison with zero prestrain. This observation proves the previous statement that solitons in the chain with small prestrain are similar to the solitons with zero prestrain.

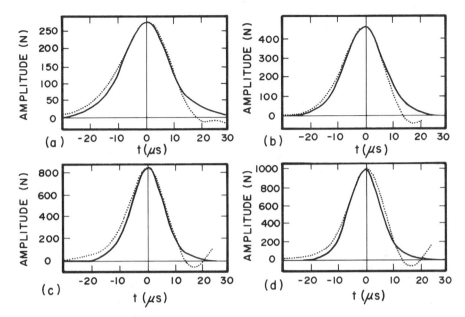

Figure 1.34. Soliton waves for a static force of 167 N: (a) $V_s = 1131$ m/s, $F_m = 274$ N; (b) $V_s = 1178$ m/s, $F_m = 474$ N; (c) $V_s = 1239$ m/s, $F_m = 842$ N; (d) $V_s = 1260$ m/s, $F_m = 1005$ N. (Reprinted from *Phys. Rev., E*, **56**, no. 5, Coste, C., Falcon, E., and Fauve, S. Solitary Waves in a Chain of Beads Under Hertz Contact, pp. 6104–6117, 1997, with permission from author).

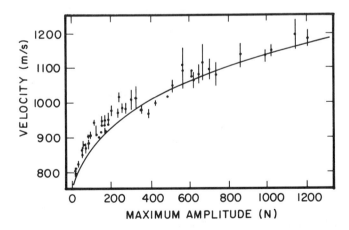

Figure 1.35. Soliton wave speed versus amplitude of the wave for a static force of 29.7 N. (Reprinted from *Phys. Rev., E*, **56**, no. 5, Coste, C., Falcon, E., and Fauve, S. Solitary Waves in a Chain of Beads Under Hertz Contact, pp. 6104–6117, 1997, with permission from author).

1.8.7   Effect of Plastic Deformation of Particles Contacts on a Wave Structure

The elastic nonlinear interaction, described by the Hertz law and discreteness of the chain, are the main reasons for the observed wave phenomena. Nevertheless, this law has a natural limit resulting from the possible plastic flow of material in the vicinity of the contact. It is possible to derive the limit for particle velocity of the soliton $\upsilon_{m,l}$ based on the equation for maximal allowable shear stresses for the contact of two spherical particles (Timoshenko and Goodier [1987]), and using the dependence of strain amplitude in the soliton on the particle velocity amplitude (Eq. (1.41)):

$$\upsilon_{m,l} \approx 34\, c_1 \left(\frac{\sigma_s}{E}\right)^{5/2}, \qquad c_1^2 = \frac{E}{\rho_0}, \tag{1.95}$$

where σ_s is the shear yield strength and the Poisson coefficient ν is equal to 0.3. For quenched and tempered steel: $\sigma_s = 380$ MPa, $E = 200$ GPa, and $c_1 = 5 \cdot 10^3$ m/s. For these parameters, the maximal velocity in the soliton, ensuring the elastic behavior, should be less than 3 cm/s.

In reality, the solitons in the chain of steel particles are reproduced at least at a velocity of about 1 m/s, despite the probable local plastic flow in the contact area that can cause additional attenuation. Strain rate hardening can increase the allowable level of shear stresses in the particles. For particles from brittle materials, the maximal tensile stresses at the circular boundary of the contact surface will restrict the range of the particle velocities for elastic behavior during soliton propagation.

Nesterenko and Lazaridi [1987] and Nesterenko [1994] experimentally investigated the differences between geometrically identical systems of spherical particles (diameter $4.75 \cdot 10^{-3}$ m) made from steel (mainly elastic behavior), and lead (mainly plastic behavior). The same method of loading, piston impact with velocity 1 m/s and mass equal to 30 masses of steel particles, was applied to both chains. The experimental setup is similar to Figure 1.22. The geometry of experiments was identical in both cases: the number of particles $N = 20$, particle diameter $4.75 \cdot 10^{-3}$ m, impulses were created by the steel striker with $M_s = 30m$, $\upsilon_s = 1$ m/s, m is the mass of steel sphere in the system. Experimental results, corresponding to the reaction of the wall, are shown in Figure 1.36 for steel (Fig. 1.36(a)) and lead particles (Fig. 1.36(b)) correspondingly.

It is evident that the essential plastic deformation of particle contacts for lead particles resulted in the qualitatively different shock structures: monotonic for lead particles and oscillating for steel particles. Unlike the chain with steel particles, it also provided the essential decrease of force amplitude acting on the wall (17 times), and an increase of shock rise time (40 times). Comparison of Figures 1.36(a) and (b) demonstrates that there are no two distinguished groups of particles behind the shock front according to their relative velocities in a chain

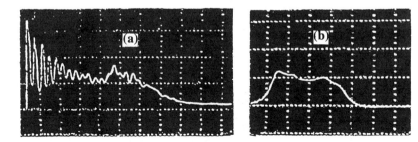

Figure 1.36. Dependence of the force F_w acting on the lower steel disk of the particle type: (a) steel particles; (b) lead particles. Vertical scales are (a) 92 N/div and (b) 18.5 N/div, horizontal scales are (a) 50 μs/div and (b) 200 μs/div.

composed of lead spheres. This will result in the cluster-type behavior of these particles after shock interaction with free surface.

Negreskul [1993] studied the system of lead spheres in one-dimensional computer calculations with the following interaction law between particles:

$$F_{i,i+1} = \begin{cases} \pi d BHN\, h_{i,i+1}, & h_{i,i+1}(t) > h_{i,i+1}(t-dt), \\ 0, & h_{i,i+1}(t) < h_{i,i+1}(t-dt), \end{cases} \tag{1.96}$$

$$h_{i,i+1} = d - r_{i,i+1},$$

where d is particle diameter, BHN is the Brinel hardness number, dt is the time increment, and $r_{i,i+1}$ is the distance between particle centers. The calculations with this "history-dependent" interaction law demonstrated the decaying monotonic profile of shock impulse with a wide shock front in perfect agreement with experimental results (Fig. 1.36(b)).

Manciu, Sen, and Hurd [2000] numerically observed that a soliton-like decaying impulse propagates in the Hertzian chain with dissipation introduced via a restitution coefficient.

Sadd, Tai, and Shukla [1993] emphasized the effects of various normal contact laws on the wave propagational characteristics in 1-D granular materials including wave attenuation. They found that a hysteretic law, where loading and unloading contact responses were taken to be different, provided the best fit to their experimental data for wave speed, amplitude attenuation, and dispersion characteristics of granular media composed from disks of Homalite 100 material.

The chains of plastically deformed objects (disks, hollow and filled rings) can be a good model of impact arresting systems (Reddy, Reid, and Barr [1991]).

1.9   Wave Dynamics in 2- and 3-D Granular Materials

1.9.1   Experiments and Numerical Simulation with 2-D Models

If granular material is composed from identical spherical particles packed in a "crystal" lattice, the propagation of disturbances in some cases may be considered exactly in the same manner as for a 1-D lattice, with only the change of numerical coefficient A in Eq. (1.3). A 1-D case is equal to simple cubic packing with a wave front parallel to the plane of the maximal coordination number.

For the 1-D Hertz chain, macroscopic behavior reflects a nonlinearity on a microscopic scale. It may not be the case for the 3-D granular materials, where a nonlinearity on a macroscopic scale is not solely caused by contact law. It may be a result of increasing the number of force chains supporting the applied compressive load even for linear elastic forces between grains (Herrmann et al. [1987]). The power-law exponent is sensitive to the sample size and to applied compressive loads, and may be as high as 10! (Travers et al. [1988]).

In real granular material there is a natural "free" volume resulting from non-ideal packing, even for identical particles. Granular materials have a mesostructure which is characterized by a so-called fabric tensor or "force chains" (Mehrabadi and Nemat-Nasser [1983]), where granules are placed practically one-dimensionally. The typical length of these "force chains" is much larger than the particle size and may exceed 20 particles diameter (Travers et al. [1987]; see also Liu et al. [1995], Radjai et al. [1996]). At the same time even essentially disordered 1-D chains still preserve some qualitative properties of ordered systems, as discussed previously.

It suggests that real granular material may resemble, in its behavior, some qualitative features of 1-D chains. For example, it is reasonable to expect that the characteristic speed of perturbation in a 3-D case will have a similar dependence on average particle velocity (the same power law). The value of the exponent may depend on the velocity (stress) amplitude in impulse, reflecting the effect of the particles rearrangement and the increase in coordination number under compression. Nevertheless, nonideal packing, a possibility for particle rotations and friction, can produce a new qualitative feature for 3-D granular assemblies in comparison with the 1-D chains.

Liu and Nagel [1992], [1993a,b] experimentally demonstrated that microscopic changes, in the dimensions of one particle (~3000 A) with a radius of 5 mm, resulted in qualitative differences of the overall wave behavior of a granular sample. The tiny change of single particle temperature by 0.8 K caused a signal change by as much as 25% in a reproducible manner.

They emphasized that granular media have some features that distinguish them from other solids and liquids: even very small vibrations resulting in structure evolution could lead to the wave localization. The authors use the term "sound" which can hardly be used for this material, as is clearly shown by one-dimensional analysis. This approach practically avoids the important amplitude parameter, which will produce different "sound" speeds, even in the arbitrary small range of wave amplitudes for uncompressed granular assemblies. The results of 1-D analysis demonstrate that granular materials require the development of principally new wave dynamics by reconsidering the main terms of the traditional one, such as sound speed and wave impedance, if amplitude of deformation in the wave is much larger than that caused by the initial prestrain.

A 2-D model system of balls connected by linear springs, with a random distribution of the spring constants, was used by Leibig [1994] to reproduce the complex frequency response seen in granular materials under forced vibrations. A velocity-dependent dissipation with a uniform damping constant is included in the model. The main result of the paper is that experimentally measured (Liu and Nagel [1992]) vibrational properties of initially compressed granular material can be reproduced with a simple linear spring and ball model, where the disorder is introduced only through random distribution of spring constants. A power law in the behavior of the power spectrum ($S(f) \sim f^{\alpha}$ with $\alpha \sim 2$, Liu and Nagel [1992]) for granular media composed from spherical glass beads with 0.5 cm diameter is reproduced by the model.

The effect of fabric in granular materials on dynamic load transfer in two-dimensional assemblies of elliptical particles was investigated by Zhu, Shukla, and Sadd [1996]. The loading was performed with a small amount of explosive on the top of one of the granules, and the results are shown in Figure 1.37. The experimental results obtained for different geometries of packing clearly demonstrate the importance of specific fabric such as the major axis orientation, contact normal, and branch vectors on the wave propagation in 2-D particulate materials. The small amount of particles (4 to 5) involved in the interaction does not provide clear evidence for soliton-type behavior.

Different contact laws for particle interaction in one- and 2-D packing, including deviation from elasticity through incorporating a hysteretic law, were considered by Sadd, Tai, and Shukla [1993]. They were able to fit their experimental data for granular media prepared from disks of Homalite 100 with reasonable accuracy for hysterasis normal contact law. They also investigated the possibility of introducing velocity-dependent damping in combination with non-linear contact law, which resulted in unacceptable high dispersion being in contradiction with experiments.

Boutrex, Raphael, and de Gennes [1997] investigated a propagation of a pressure step in granular material, including wall friction with linear interaction law for contact forces between particles.

Melin [1994] studied in computer simulation a two-dimensional packing of monodisperse and polydisperse grains under gravity. It was found that dependence of the wave speed V on the depth h excited by the step-shaped excitation is very weak and does not obey the relation $V \sim h^{1/6}$ following the Hertz law of particles interaction, and also has weak dependence on the gravitational constant. This can be easily understood taking into account that for this material the sound speed is not a characteristic speed. The soliton speed (Eq. (1.84)), with weak dependence on initial compression (for small ξ_r) should be used instead.

Numerical molecular-type dynamic simulation of the vertical propagation of

★ Explosive

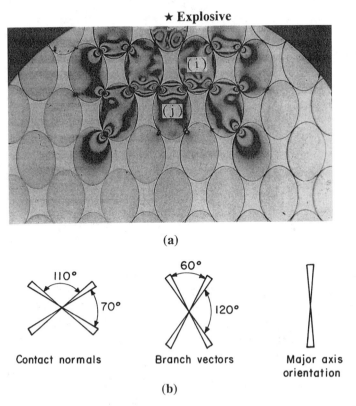

(a)

110°   60°

70°   120°

Contact normals   Branch vectors   Major axis
orientation

(b)

Figure 1.37. Wave propagation in the two-dimensional HCPS3 assembly of elliptical particles under concentrate explosive loading: (a) typical isochromatic fringes (time 129 μs); (b) fabric parameter distribution. (Reprinted from *J. Mech. Phys. Solids*, **44**, no. 8, Zhu, Y., Shukla, A., and Sadd, M.H. Effect of Microstructural Fabric on Dynamic Load Transfer in Two Dimensional Assemblies of Elliptical particles, pp. 1283–1303, 1996, with permission from Elsevier Science.)

weak and strong impulses in deep gravitationally compacted 1- and 2-D granular columns was performed by Sen and Sinkovits [1995], [1996]. They obtained the scaling law $c_0 \sim z^{1/6}$, where z is the depth, following from Hertzian contact theory in continuum approach for vertical sound propagation at large depths (see Eqs. (1.93) and (1.94)). They also found a profound effect of voids in the 2-D column on the wave propagation, due to the fact that strong inhomogeneities are introduced in the force network. The presence of voids resulted in random motion of the grains in the bulk of the column (see a similar effect for the disordered 1-D case, Fig. 1.21) and decreased the sound speed value.

Substantial differences were observed between the behavior of the columns containing mass defects and those containing voids. A void cannot be thought of as a defect of zero mass, as reflected by the fact that removing a grain from the column leads to the loss of the corresponding intergrain contacts, resulting in a sound speed decrease despite the decrease of the density. This type of study is very important in developing acoustic probes for studying the position distribution of buried inclusions in dry granular soil.

Gilles and Coste [2001] experimentally demonstrated that a uniaxial and isotropic compression of the 2-D packing of nylon and steel beads leads to very different acoustical behavior. They attributed this to a collective effect caused by the different reaction (due to these loadings) of random lattice of contacts (fabric).

Dynamic stress bridging in granular materials under weak shock compression by a piston traveling at 33 m/s was considered in numerical calculations by Bardenhagen and Brackbill [1998]. A two-dimensional array (ten rows of ten particles) of cylinders with properties appropriate to Plexiglass provided a geometry with well-distinguished force chains under quasistatic loading. It was observed that the stress wave is propagated in this system with "stress fingers" preceding the planar compaction wave. The most pronounced "stress finger" in a dynamic simulation coincided with the same path of force chain in the quasistatic simulation. The authors concluded that "stress fingers" steadily outpaced the compaction wave, making the latter one intrinsically dispersive. It was demonstrated that addition of a binder substantially reduced the force fluctuations under both static and dynamic loading. The relationship between the compaction wave and wave propagation along force chains will be considered separately.

1.9.2   Nonlinear Acoustoelastic Properties

The acoustoelastic approach can be applied to granular materials if they are under static compression and if the stress amplitude of the wave disturbance is small in comparison with initial compression.

Hara [1935], Iida [1939], Takahashi and Sato [1949], [1950], and Gassmann [1951] pioneered the research on granular acoustics based on the Hertz law between particles. Duffy and Mindlin [1957] were the first who considered both normal and tangential components of contact forces for elastic spheres arranged in ordered fcc-type packing. Digby [1981] and Walton [1987] derived the effective elastic modules of the random packing of identical elastic spheres. The calculations are made relative to two initial deformed states: hydrostatic compression and uniaxial compression. The effect of friction is taken into account. Computer simulated deformation of statically loaded granular materials was investigated by Thornton and Barnes [1986]. Goddard [1990] introduced contact number variation due to particle-chain buckling and modified the power law for conical contacts.

Recent experimental results by Jia, Caroli, and Velicky [1999] are in agreement with data by Duffy and Mindlin [1957] on the dependence of sound speed on static compression which follows quasipower law with a decreasing exponent from 1/4 at low applied stresses to 1/6 at high load. They observed the coexistence of a coherent ballistic pulse (E signal) traveling through an "effective contact medium" and a speck-like multiple scattered signal (S signal). Only the

S signal was configurationally sensitive. Also, the relative amplitudes of these signals strongly depend on the ratios of the bead size to the wavelength and may be considered as another structure sensitive parameter.

It was emphasized by Zaikin [1990] that long-wave longitudinal (V_P) and transverse (V_S) velocities in consolidated granular media depend on two structural parameters of the pore space: porosity (f) and the product (η) of the specific surface (per unit of volume) and the mean grain size (as the diameter of a sphere with the same volume to surface ratio as in real granular material). An increase in the latter leads to a decrease in the wave velocities even if the porosity is the same. The corresponding dependencies of V_P and V_S on f and η were obtained; the longitudinal waves are more sensitive to the change in both structure parameters compared to the transverse waves.

Liu and Nagel [1993a] addressed the question: "What is the velocity of sound in granular medium?" They discovered a very interesting and unexpected behavior of granular media—the velocity of "sound" varied by a factor of 5 (280 m/s and 57 m/s), depending upon whether one measured either the arrival time of the rising edge of the pulse or the quantity analogous to the group velocity. They concluded that this discrepancy is not due to a nonlinear response of the medium. At the same time, as demonstrated by the 1-D model, the concept of sound speed is not adequate for wave dynamics in granular material, due to the strong nonlinearity and coupling of nonlinearity with dispersion. It should be replaced with a soliton-type approach, which can also be strongly affected by disorder. Results of the experimental study of the nonlinear elastic properties of granular media with nonideal packing are presented in papers by Belyeva, Zaitsev, and Ostrovskii [1993] and Belyeva, Zaitsev, and Timanin [1994]. They obtained the following σ–ε relation (instead of Eq. (1.93) for ideal simple cubic packing):

$$\sigma = \frac{n(1-\alpha)E\xi^{3/2}}{3\pi(1-v^2)}, \tag{1.97}$$

where n is the average number of contacts per spherical particle and α is the porosity coefficient ($\pi/6$ for simple cubic packing). This equation results in the sound speed (instead of Eq. (1.94)):

$$c_0^2 = \frac{n(1-\alpha)E}{2\rho(1-v^2)}\,\xi^{1/2}. \tag{1.98}$$

From these equations, Belyeva, Zaitsev, and Ostrovskii [1993] obtained in the weakly nonlinear case the second-order ($\Gamma_s^{(2)}$) and cubic ($\Gamma_s^{(3)}$) nonlinearity parameters correspondingly (which are the same as for Eq. (1.93)):

$$\Gamma_s^{(2)} = \frac{1}{2\xi_0}, \tag{1.99}$$

$$\Gamma_s^{(3)} = \frac{1}{6\xi_0^2}, \tag{1.100}$$

where ξ_0 is the initial static compressive strain. From Eqs. (1.99) and (1.100), it follows that nonlinearity parameters of the granular medium are influenced only by the initial static compression. These parameters were measured by the method of the second- and third-harmonic generation in lead shot and tufa with different grain sizes. The experiments confirmed that the nonlinearity parameters of the granular medium do not depend on the material or size of the grains, but are solely determined by ξ_0. The measured values for $\Gamma_s^{(2)}$ fall into the interval from $1.3 \cdot 10^2$ to $1.5 \cdot 10^3$ and for $\Gamma_s^{(3)}$ from $4 \cdot 10^4$ to $5.5 \cdot 10^6$ for initial strains ξ_0 from $3 \cdot 10^{-4}$ to $3.4 \cdot 10^{-3}$ being in good agreement with Eqs. (1.99) and (1.100).

This approach is also in qualitative agreement with data obtained by Groshkov et al. [1990] for loamy soil at a depth of 0.5 m: $\Gamma_s^{(2)} = (0.7-1.0) \cdot 10^3$ and $\Gamma_s^{(3)} = (1.6-4.1) \cdot 10^7$, the calculations based on Eqs. (1.99) and (1.100) give $\Gamma_s^{(2)} = 10^3$ and $\Gamma_s^{(3)} = 10^7$.

From experimental data, Belyeva et al. [1994] estimated that (based on Eq. (1.97)) for a coarse lead shot (particle sizes 1.65–1.8 mm) and for polymer particles (sizes 0.3–0.35 mm) the average number of contacts per spherical particle is equal to 3.5–4. This is essentially smaller than the coordination numbers for the random ($n = 8.84$) or closest ($n = 12$) packing of spherical particles. This confirms that a considerable number of potential intergranular contacts in real 3-D systems are not involved in sound propagation.

1.9.3   Impact Loading of Real Granular Materials and 1-D Model

1.9.3.1   Three-Dimensional and 1-D Granular Assemblies Under Equivalent Impact

Many important properties of real granular materials are not included in the 1-D model as was discussed at the beginning of this chapter. Nevertheless, it is interesting to compare the behavior of real granular materials with 1-D modeling representing the simple cubic packing.

The experimental pressure profiles on the rigid wall for real 3-D granular material (cast iron shot with particle sizes 3–4 mm) and its 1-D model at analogous conditions of loading (the same velocity of piston and its mass per unit surface) are compared by Nesterenko [1988]. Pressure amplitude in the former case was $\sim 10^5$ Pa. That is close to the maximum normalized pressure P in the corresponding computer calculations of the 1-D model ($M = 5m$, $v_0 = 0.1$ m/s). This was calculated from force F_{max}, according to formula $P = F_{max}/4R^2$ for simple cubic packing of spherical particles. The time of force increase in 1-D computer calculations was ~ 20 μs, that is, six to ten times less than obtained in experiments for real granular material at the same conditions of impact loading.

This order of magnitude difference in front width between 1- and 3-D systems under equivalent load cannot be explained in the frame of the 1-D approach. We saw that at initially unstressed 1-D chains, with ordered particle packing and typical nonlinearity from the Hertz law, nonlinear dispersion can result only in front width comparable with the particle size. This conclusion does not depend on the particle packing, because this parameter can change only the propagation velocity, but not the order of the space scale of the signal.

Randomization of particle sizes, which to some extent may represent packing defects in 3-D material, also did not result in the spreading of the leading front (Fig. 1.21). In this case, random oscillations are excited behind the shock front and their energy is not transformed into kinetic energy of movement in the shock direction, or into potential energy. On a macroscale, this energy can be considered as lost energy. But this type of "dissipation," connected with the collective particle oscillations on the scale of the particle sizes, does not provide the spreading of the leading front (Fig. 1.21), resulting only in an oscillating tail and spreading of its back part. Finally, initial prestress of a 1-D chain by the force ·200 N (corresponding to the average pressure in experiments 90 ·10^5 Pa) increases the leading front thickness, which in this case is close to eigth particles diameter with typical ramp time 30 μs. But this increase cannot explain the observed time rise of ~100–200 μs at experiments with 3-D granular materials.

We can conclude that the real 3-D granular material has essentially different dynamic behavior in comparison with the 1-D model, even if the σ–ξ relations are close. The characteristic order of time and space parameters, determining the typical front width in a 3-D system, are connected with processes which are physically different than in the 1-D model. It can be associated with the transition time for configuration change from one state to another, depending on the applied pressure and its time dependence. The more dispersed character of shock impulses in real 3-D material can be explained by additional dispersion due to friction processes between particles. It is also possible that energy is dissipated as a result of excitation of the rotational degree of particle movement. The simplest way to describe such behavior is to introduce some "viscosity," which can provide reasonable results (Nesterenko and Lazaridi [1987]).

1.9.3.2   Compaction Wave and Elastic Waves Along Force Chains in 3-D Granular Assemblies

Despite the essential difference between three- and 1-D granular assemblies the former contains a mesoscopic element with 1-D geometry—force chains the length of a few tens of particle diameters, as previously discussed. We saw that this length is enough to form a soliton under some conditions of loading (see Figs. 1.23 and 1.25) or to establish a quasistationary speed of "shock" wave. In the 1-D case the speed of the stationary shock is always larger than the speed of the soliton with equal corresponding particle velocities. Three- and 2-D granular assemblies may carry other types of waves which are absent in the 1-D case.

For example, a stationary compaction wave induced by piston moving with velocity υ and transforming material from the initial state with density ρ_{00} to the final state with increased density ρ_1 may be considered. The speed of this wave D_c may be easily found from the mass conservation law

$$D_c = \frac{\upsilon}{(1-r)}, \qquad r = \frac{\rho_{00}}{\rho_1}. \tag{1.101}$$

This speed D_c does not depend on the acoustic properties of the material.

From another point of view, the soliton (shock) can propagate along presumably 1-D force chains. For example, the speed of soliton V_s may be presented in convenient form (see Problem 18)

$$V_s = \left(\frac{16}{25}\right)^{1/5}\left(\frac{2(1-2\nu)}{\pi(1-\nu)^2}\right)^{2/5} c_L \left(\frac{\upsilon_m}{c_L}\right)^{1/5},$$
(1.102)

where υ_m is the particle velocity in the soliton crest and c_L is a longitudinal sound speed in the bulk of the particle.

Comparison between Eqs. (1.101) and (1.102) results in the conclusion that at relatively small υ, υ_m and appropriate parameter r, soliton speed will always exceed the speed of a compaction wave (Nesterenko [1999], [2000]). It is true even if $\upsilon_m \ll \upsilon$ because of the very weak dependence of the soliton speed on particle velocity. For example, soliton speed in the chain of steel balls is about 1000 m/s at the particle velocity of a few m/s (Lazaridi and Nesterenko [1985]; Coste, Falcon, and Fauve [1997]). The same speed of compaction front in 3-D granular material with $r = 0.5$ corresponds to the particle velocity $\upsilon = 500$ m/s and the shock pressure $P = 2$ GPa ($P = \rho_{00} D_c \upsilon$).

So it is possible to expect a nonlinear "heterogeneous" elastic precursor in similar 3-D granular materials, under shock loading generating compaction waves with particle velocities up to 500 m/s. The speed of the shock wave (stationary or quasistationary) in a 1-D chain (Eqs. (1.81), (1.82)) has the same order of magnitude as V_s. Thus the same conclusion is valid for propagation of 1-D shocks in force chains ahead of the front of the compaction wave.

The difference between V_s and D_c results in the propagation of elastic disturbances in some parts of the granular material: along previously existing force chains that will not essentially effect material density ahead of the compaction front. They propagate due to the specific deformation mechanism —contact interaction of the particles with quasistatic loading of their bulks. An increase of particle velocity will suppress this mechanism of heterogeneous elastic precursor that may be important only for weak shocks. Evaluation of the critical velocity of the piston υ_{cr}, above which an elastic precursor along force chains will not propagate, can be made based on equation $D_c = V_s \cos \Theta$:

$$\upsilon_{cr} = \left(\frac{16}{25}\right)^{1/5}(1-r)\left(\frac{2(1-2\nu)}{\pi(1-\nu)^2}\right)^{2/5}\left(\frac{\upsilon_{m,1}}{c_L}\right)^{5/4} c_L,$$
(1.103)

where $\upsilon_{m,1}$ is a maximal allowable particle velocity in the soliton, and Θ is the angle between the direction of the force chain and the speed of the compaction wave. The low boundary for this velocity may be calculated based on Eq. (1.95). This consideration, despite evident and crude approximations, may be useful for the evaluation of different regimes of weak shock propagation in granular materials. The additional coefficient should be added to Eq. (1.103) due to the relative position of the force chain, with respect to the direction of shock speed.

The dynamic "stress fingers" propagated ahead of the compaction wave were observed numerically by Bardenhagen and Brackbill [1998]. Using data for their material (Poisson coefficient $\nu = 1/3$ and $c_L = 2000$ m/s) and selecting $r = 0.9$ and $\upsilon_{m,l} = 1$ m/s in Eq. (1.103) we obtain a velocity of $\upsilon_{cr} = 24$ m/s which is of the same order of magnitude as in the numerical calculations (33 m/s). This means that soliton (nonstationary or stationary shocks) propagating along force chains may cause the heterogeneous elastic precursor ("stress fingering") in front of a the compaction wave. Nesterenko [1992d] (see also Figs. 2.9(a) to (c)) observed an irregular precursor wave in front of the compaction shock in granular Teflon with large particle sizes like 1.5–2 mm. A different mechanism of shock dispersion is connected with strong compression waves propagating in the body of particles and is numerically observed by Benson [1994].

1.9.4   Granular Materials as Shock Absorbers

Despite the lack of understanding in nonlinear behavior of granular materials, they are effectively applied in practice. Granular materials such as cast iron shots are very effectively used in the damping of explosive loading, and support for the foundation plate at explosive welding process, and in explosive chambers. A group from Barnaul Politechnical Institute (Apalikov, Bronovskii, Konon et al. [1983]; Tsemahovich [1988]) introduced this media for damping purposes and widely applied it for explosive manufacturing. The granular bed from iron shot is a traditional shock absorber in the design of explosive chambers.

Granular material of the same type can effectively prevent spall in reinforced concrete constructions at contact explosive loading (Nesterenko et al. [1986]). In another example, the layer from the cast iron shot with the thickness of only 20 mm (particle size 3–5 mm) placed between the explosive charge and a steel plate with a thickness of 300 mm can prevent the fracture developing from a surface defect (Ipatiev, Kosenkov, and Nesterenko et al. [1986]). The research on shock-wave decay in granular assemblies is essential for the development of strong shock absorbers for violent dynamic loading from contact explosion.

1.10   Nonlinear Waves in Discrete 1-D Power-Law Materials

1.10.1   Wave Equation for Power-Law Particulate Materials

The propagation of compression pulses in the materials, where the long-wave sound speed c_0 is equal to zero or small in comparison with the speed of disturbances, is considered in Sections 1.1 to 1.8. For such media, the concept of "sonic vacuum" is introduced to emphasize the fact that the linear waves described by a standard wave equation cannot propagate in it, or they are insignificant for the description of strongly nonlinear disturbances. The general example of such elastic nonlinear media are systems with power-law dependence of force F on the displacement $\delta(F \sim \delta^n)$, like the Hertz law for contact interaction of two particles. From a physical point of view, it is sensible to consider only the case when the exponent in the power interaction law $n > 0$. If $n = 3/2$, solitary waves of a new type are found analytically within the long-wave approximation, by numerical calculations, and as well experimentally. Goddard [1990] demonstrated that, for the conical geometry of particle contact, n should be taken as 2.

In this section we will address the question: Are the solitary waves found for power-law interaction with $n = 3/2$ also typical for the different exponent n? Nesterenko [1992a–d] demonstrated that similar waves are the basic excitations for power-law materials, if $n > 1$ and rarefaction waves exist if $n < 1$.

In a general case of power-law material, the equation of motion is:

$$\ddot{u}_i = A_n\left(u_{i-1} - u_i\right)^n - A_n\left(u_i - u_{i+1}\right)^n , \qquad\qquad N \geq i \geq 2 . \tag{1.104}$$

In long-wave approximation $(L >> a)$ Eq. (1.104) can be transformed to a strongly nonlinear wave equation similar to Eq. (1.19):

$$u_{tt} = -c_n^2\left\{\left(-u_x\right)^n + \frac{na^2}{24}\left[(n-1)\left(-u_x\right)^{n-2}u_{xx}^2 - 2\left(-u_x\right)^{n-1}u_{xxx}\right]\right\}_x , \tag{1.105}$$

$$c_n^2 = A_n\, a^{n+1}, \qquad \xi = -u_x > 0 ,$$

where u is the displacement, a is the initial distance between the particles of the undisturbed system, and c_n is the constant with speed dimensionality. The requirement $\xi > 0$ is necessary for the case of discrete systems with zero tensile strength and for certain values of n. Equation (1.105) can be written in more compact form (Nesterenko [1992a], [1993b], [1994]):

$$u_{tt} = -c_n^2\left\{\left(-u_x\right)^n + \frac{na^2}{6(n+1)}\left[\left(-u_x\right)^{\frac{n-1}{2}}\left(\left(-u_x\right)^{\frac{n+1}{2}}\right)_{xx}\right]\right\}_x . \tag{1.106}$$

Gavrilyuk and Shugrin [1996] demonstrated that Eq. (1.106) can be written as a system of quasilinear equations after the introduction of the new variables $u_t \to \chi$, $u_x \to \sigma$:

$$\sigma_t - \chi_x = 0, \qquad \chi_t + p_x = 0, \tag{1.107}$$

$$p = -c_n^2\left\{\left(-\sigma\right)^n + \frac{na^2}{6(n+1)}\left[\left(-\sigma\right)^{\frac{n-1}{2}}\left(\left(-\sigma\right)^{\frac{n+1}{2}}\right)_{xx}\right]\right\}.$$

This system belongs to a wide class of nonlinear equations with partial derivatives describing wave propagation in dispersive media, which was introduced by Gavrilyuk and Shugrin [1996] in variational form

$$\frac{\partial}{\partial x^\alpha}\left(\frac{\delta L^\alpha}{\delta q^\beta}\right) = 0, \tag{1.108}$$

where x^α are independent variables (t, x in the case of Eq. (1.106)), q^β are dependent variables (χ, σ in the case of Eq. (1.106)), and L^α are, in general, the nonlinear functions of q^β and their derivatives with respect to x^α. The system of Eqs. (1.108) admits the conservation law (Gavrilyuk and Shugrin [1996]):

$$\frac{\partial \Phi^\alpha}{\partial x^\alpha} = 0, \quad \Phi^\alpha \equiv q^\beta \frac{\delta L^\alpha}{\delta q^\beta} - L^\alpha + q_\gamma^\beta \frac{\delta L^\gamma}{\delta q_\alpha^\beta}. \tag{1.109}$$

Equation (1.108) is the natural generalization for dispersing media of Godunov's system of quasilinear equations (Godunov [1961]). Substitution of a term with higher derivatives with respect to x in Eq. (1.105), by the term with the mixed derivatives with respect to x, t, is considered to be more preferable for wave equations with spatial dispersion (Whitham [1974a]; Albert and Bona [1991]). This preserves the accuracy of the equations in the considered long-wave limit, but in the linear case eliminates imaginary frequencies in the dispersion relation when $\lambda < \lambda_{\min} = \pi a/3^{1/2}$. Although, for this short space scale, neither the former nor the latter equation can be used correctly without independent proof by numerical calculations of the corresponding genuine discrete system or by experiment. Substituting the lower approximation for the term with the higher derivative in square brackets, the more "preferable" equation is obtained

$$u_{tt} = -\left\{c_n^2(-u_x)^n - \frac{a^2}{12}\left[u_{ttx} + \frac{c_n^2 n(n-1)}{2}(-u_x)^{n-2}u_{xx}^2\right]\right\}_x. \tag{1.110}$$

Equations (1.105) and (1.110) may be obtained using the variational approach with the appropriate Lagrangians L_1 and L_2 (Nesterenko [1992c]):

$$L_1 = \frac{u_t^2}{2} - c_n^2\left\{\frac{(-u_x)^{n+1}}{n+1} + \frac{a^2}{2}\left[\frac{n}{4}(-u_x)^{n-1}u_{xx}^2 - \frac{1}{3}(-u_x)^n u_{xxx}\right]\right\}, \tag{1.111}$$

$$L_2 = \frac{u_t^2}{2} + \frac{a^2}{24}u_{xt}^2 - c_n^2\left[\frac{(-u_x)^{n+1}}{n+1} - \frac{(-u_x)^n}{24}u_{xxx}\right]. \tag{1.112}$$

Lagrangian formalism was used in Section 1.4 to prove the stability of the periodical solutions of Eq. (1.20) in the partial case $n = 3/2$. No attempts to apply the same approach for general exponent n have been made. It is also useful to write Eq. (1.106) in strains $\xi = -u_x$:

$$\xi_{tt} = \left\{ c_n^2 \xi^n + \beta \xi_{tt} - \gamma \xi^{n-2} \xi_x^2 \right\}_{xx}, \tag{1.113}$$

$$\beta = \frac{a^2}{12}, \qquad \gamma = \frac{a^2 c_n^2}{24} n(n-1).$$

It should be noted that c_n is not a long-wave sound speed c_0. The expression for c_0 may be obtained from the linearized Eq. (1.113):

$$c_0 = c_n \sqrt{n} \, \xi_0^{\frac{(n-1)}{2}}. \tag{1.114}$$

For power-law materials when $\xi_0 \to 0$, then $c_0 \to 0$ if $n > 1$.

1.10.2 Stationary Solutions

Stationary solutions $\xi(x - Vt)$ of Eq. (1.105) may be obtained. In this case, upon integration with respect to x with boundary conditions $\xi(x = +\infty) = \xi_0$, $\xi_x(x = +\infty) = 0$, $\xi_{xx}(x = +\infty) = 0$ and after the change of variables $(\xi, x) \to (y, \eta)$.

Equation (1.105) is reduced to the equation of motion of the nonlinear oscillator in the potential field $W(y)$, similar to Eq. (1.30):

$$y_{\eta\eta} = -\frac{\partial W}{\partial y}, \tag{1.115}$$

$$W(y) = \frac{y^2}{2} - \frac{(n+1)}{4} y^{\frac{4}{(n+1)}} + C y^{\frac{2}{(n+1)}},$$

$$\eta = \frac{x}{a} \sqrt{\frac{6(n+1)}{n}}, \qquad y = \xi^{\frac{(n+1)}{2}} \left(\frac{c_n}{V} \right)^{\frac{(n+1)}{(n-1)}},$$

where C is the constant. The existence of the maximum and minimum effective potential energy $W(y)$ determines the possibility of solitary solution. Using the condition for the existence of extremes $dW/dy = 0$, one may obtain the following equation:

$$\frac{(n+1)}{2} \Phi(\alpha) = C, \tag{1.116}$$

$$\Phi(\alpha) = \alpha - \alpha^n,$$

$$\alpha = y^{\frac{2}{(n+1)}}.$$

Function $\Phi(\alpha)$ is shown in Figure 1.38 for two specific cases corresponding to $n < 1$ and $n < 1$. Using necessary condition for the extreme $\Phi'(\alpha) = 0$, find that this function has one extreme at α_{ext}:

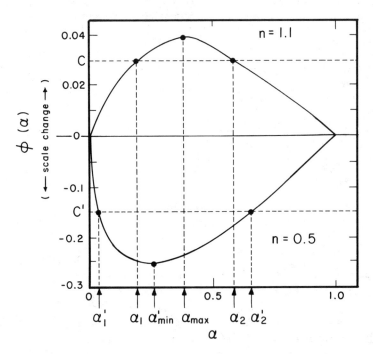

Figure 1.38. Graphical illustration of equation for $W(y)$ extremes: (a) normal compressibility, $n = 1.1$; (b) abnormal compressibility, $n = 0.5$.

$$\alpha_{ext} = (n)^{1/(1-n)}. \tag{1.117}$$

It is important that for the exponent n being in different regions, the behavior of function $\Phi(\alpha)$ is qualitatively different: for $n > 1$ it has a maximum at α_{max} and for $n < 1$ a minimum at α'_{min} ($\alpha_{ext} = \alpha_{max}$ if $n > 1$ and $\alpha_{ext} = \alpha'_{min}$ if $n < 1$, Fig. 1.38). Using Eq. (1.117) also find the value of the maximum (minimum) of the function $\Phi(\alpha)$ for the given exponent n:

$$\Phi(\alpha_{ext}) = (n - 1)n^{n/(1-n)}. \tag{1.118}$$

It is evident from Eqs. (1.116) and (1.118) (see also Fig. 1.38) that the effective potential energy $W(y)$ will have two extremes only if:

for $n > 1$:

$$0 < C < \frac{(n^2 - 1)}{2}n^{n/(1-n)} = N_1, \tag{1.119}$$

and for $n < 1$:

$$0 > C' > \frac{(n^2 - 1)}{2} n^{n/(1-n)} = N'_1. \tag{1.120}$$

The points shown in Figure 1.38 with coordinates $\alpha_1 < \alpha_{max}$ and $\alpha_2 > \alpha_{max}$, and $\alpha'_1 < \alpha'_{min}$ and $\alpha'_2 > \alpha'_{min}$ (solutions of Eq. (1.116) for different n and constants C, C') naturally represent the extreme points of function $W(y)$, correspondingly, for $n > 1$ and $n < 1$. For the relative positions of the extremes of function $W(y)$, determine what kind of soliton may propagate in the media. To identify them, find the second derivatives of $W(y)$ in points with coordinates α_1 and α_2 ($n > 1$) and in points α'_1 and α'_2 ($n < 1$):

$$\frac{d^2 W}{dy^2} = \frac{2y^{\frac{(3-n)}{(n+1)}}}{(n+1)} (n\alpha^{n-1} - 1). \tag{1.121}$$

The signs of the second derivative of $W(y)$ in points y_1 and y_2 ($y_1 < y_2$) corresponding to α_1 and α_2, and in points y'_1 and y'_2 ($y'_1 < y'_2$) corresponding to α'_1 and α'_2 can be found from the previous equation.

If $n > 1$ ($\alpha_1 < \alpha_{max}$ and $\alpha_2 > \alpha_{max}$) we obtain at y_1:

$$\alpha_1 < \alpha_{max} \; \rightarrow \; n\alpha_1^{n-1} < n\alpha_{max}^{n-1} = 1 \; \rightarrow \; \frac{d^2 W(y = y_1)}{dy^2} < 0, \tag{1.122}$$

and at y_2:

$$\alpha_2 > \alpha_{max} \; \rightarrow \; n\alpha_2^{n-1} > n\alpha_{max}^{n-1} = 1 \; \rightarrow \; \frac{d^2 W(y = y_2)}{dy^2} > 0. \tag{1.123}$$

From the signs of the second derivative of $W(y)$, conclude that for $n > 1$ the relative positions of the extreme are as shown in Figure 1.39(a), $y_2 > y_1$. One of the interesting properties of $W(y)$ at $n > 1$ is that if $C = 0$, it has only one extreme—minimum at the coordinate independent of n ($y^0_{min} = 1$) with a value of $W^0_{min} = (1-n)/4$.

If $n < 1$ ($\alpha'_1 < \alpha'_{min}$ and $\alpha'_2 > \alpha'_{min}$) we obtain at y'_1:

$$\alpha'_1 < \alpha'_{min} \; \rightarrow \; n\alpha'^{n-1} > n\alpha'^{n-1}_{min} = 1 \; \rightarrow \; \frac{d^2 W(y = y'_1)}{dy^2} > 0, \tag{1.124}$$

and at y'_2:

$$\alpha'_2 > \alpha'_{min} \; \rightarrow \; n\alpha'^{n-1}_2 < n\alpha'^{n-1}_{min} = 1 \; \rightarrow \; \frac{d^2 W(y = y'_2)}{dy^2} < 0. \tag{1.125}$$

The signs of the second derivative of $W(y)$ allow for the conclusion that for $n < 1$ the relative positions of the extreme are opposite to the case $n > 1$, as shown in Figure 1.39(b) for $n = 0.5$ and $C' = 0.5 \cdot (N'_1 + N_2)$, where N_2 is specified later.

These two cases represent a qualitatively different behavior of materials. The former ($n > 1$) represents a "normal" behavior—hardening under increasing of the

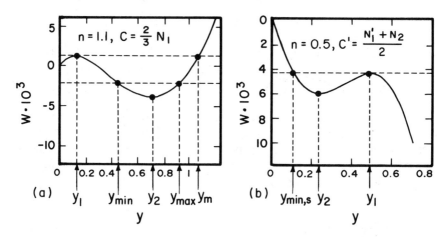

Figure 1.39. Effective potential energy $W(y)$ for power-law materials at the corresponding values of the constants C and C': (a) "normal" compressibility, $n = 1.1$; (b) "abnormal" compressibility, $n = 0.5$.

load and the latter ($n < 1$) represents "abnormal" softening under the load. For this reason the properties of stationary waves in these two classes of materials will be considered separately. It is possible that in the different ranges of the load, the behavior of real material can be essentially different, switching from hardening under the load ($n > 1$) to softening ($n < 1$).

1.10.2.1   Compression Solitary Waves in a "Normal" Material

This material ($n > 1$) represents a general example of a "sonic vacuum" as is evident from Eq. (1.114) for the sound speed c_0. There is a solution of Eqs. (1.105), (1.115) in the form of a compression (the strain amplitude is larger than the strain in infinity (or "initial"), $y_m > y_1$) solitary wave when the integration constant C is within the limits described by Eq. (1.119).

The physical sense of these restrictions is that infinitesimally small initial deformation ξ_0 (or sound velocity c_0) is necessary for the existence of the solitary compression wave, which is a supersonic ($V > c_0$) one. This is similar to the partial case $n = 3/2$ (Section 1.3). The general stationary solution of Eq. (1.115) has a form presented by Eq. (1.33) with a corresponding replacement of $W(y)$. It is possible to obtain a general equation for the speed of periodic wave V_p using equations $W(y_{min}) = W(y_{max})$ and $W_y(y_2) = 0$ (see Fig. 1.39(a)):

$$V_p = c_n \left\{ \frac{2}{(n+1)\left(\xi_{max} - \xi_{min}\right)} \left[\frac{\left(\xi_{max}^{n+1} - \xi_{min}^{n+1}\right) - (n+1)\xi_{2}^{n}\left(\xi_{max} - \xi_{min}\right)}{\xi_{max} + \xi_{min} - 2\xi_2} \right] \right\}^{1/2} . \qquad (1.126)$$

In this equation, ξ_2, ξ_{max}, and ξ_{min} are connected with y_2, y_{max}, and y_{min} through the variable transformation (Eq. (1.115)). One can demonstrate that $V_p \to c_0(\xi_2)$ with c_0 from Eq. (1.114) in the case if $\xi_{min} \sim \xi_{max} \sim \xi_2$.

The expression for solitary phase speed can be obtained from equations $W(y_1) = W(y_m)$ and $W_y(y_1) = 0$ (see Fig. 1.39(a)):

$$V_s = \frac{c_n}{\left(\xi_m - \xi_0\right)} \left\{ \frac{2\left[n\xi_0^{n+1} + \xi_m^{n+1} - (n+1)\xi_0^n\xi_m \right]}{(n+1)} \right\}^{1/2}. \tag{1.127}$$

The soliton phase speed V_s from Eq. (1.123) approaches the sound speed c_0 (Eq. (1.114)), if $\xi_m \to \xi_0$ and in the next approximation, the relation for the KdV soliton between V_s and the strain amplitude (Eq. (1.10)) can be recovered. The KdV soliton also corresponds to the limit

$$C \lesssim N_1. \tag{1.128}$$

If the constant C is close to zero (which means $\xi_m \gg \xi_0$) the compression solitary waves qualitatively differ from the KdV solitons. In this case, they are the main "carrying tones" of this strongly nonlinear system characterized by the following dependence of the phase velocity V on ξ_m, or on the maximum particle velocity υ_{max}, and the characteristic spatial length L_n (Nesterenko [1992 a–d], [1994]):

$$V_s = c_n\sqrt{\frac{2}{n+1}}(\xi_m)^{\frac{(n-1)}{2}} = \left(\frac{2c_n^2}{n+1}\right)^{\frac{1}{(n+1)}}(\upsilon_{max})^{\frac{(n-1)}{(n+1)}}, \tag{1.129}$$

$$L_n = \frac{\pi a}{n-1}\sqrt{\frac{n(n+1)}{6}}. \tag{1.130}$$

The phase velocity and the width of these solitons are independent of the sound velocity c_0, opposite to the KdV equation solutions, where the medium properties in linear description are inherited. A similar equation for L_n for the power-law 1-D materials was also presented by Shukla et al. [1993].

These new solitary solutions exist with $L_n \gg a$ even if the power-law is approaching the linear interaction law ($n \to 1$, see Problem 20). This means that detection of this stationary wave with the corresponding L_n and the dependence of phase speed V_s on the maximum particle velocity υ_{max} may be used as the proof of the slightly nonlinear behavior of materials with $n = 1 + \psi$, $\psi \ll 1$.

At large values of exponent $n \gg 1$, L_n is approaching the distance between particles in a discrete system ($L_n \to \pi a/6^{1/2}$). The numerical verification of the results of long-wave approximation for a power-law discrete chain is especially desirable and was provided by Sen [1998b], Manciu and Sen [1999], and Manciu, Sen, and Hurd [1999a], and Hascoet and Herrmann [2000].

A periodical solution of Eq. (1.105) can be obtained at $C = 0$ (Nesterenko [1992b,c], [1994]):

$$\xi = \left\{ \frac{V^2 (n+1)}{c_n^2} \frac{}{2} \right\}^{\frac{1}{n-1}} \left| \sin \left(\frac{(n-1)}{(n+1)} \sqrt{\frac{6(n+1)}{n}} \frac{(x-Vt)}{a} \right) \right|^{\frac{2}{n-1}} . \tag{1.131}$$

Equation (1.131) has the form of a sequence of positive humps joined at $\xi_0 = 0$. One such hump may represent a solitary solution at $\xi_m \gg \xi_0$. For $n = 3/2$ (a chain of elastic spheres) Eq. (1.39) can be recovered. A similar solution for 1-D power-law granular material was presented by Shukla et al. [1993] and the $K(m,n)$ equation was presented by Rosenau [1996] in the case of $m = n$.

In a "sonic vacuum," (assuming that the soliton is well described at $\xi_0 = 0$ by one hump of the periodical solution (Eq. (1.131)) the following connection between the momentum and total energy of the soliton can be derived

$$E_0 = \frac{p^2}{2m_{ef}}, \qquad m_{ef} = \alpha m, \tag{1.132}$$

where m is the mass of the particles of the discrete system, m_{ef} is the effective mass of a soliton, and α is the coefficient depending on n.

Modulation stability of the periodical solution for $n = 3/2$ (Eq. (1.39)) is proved in Section 1.4. There is no proof of modulation stability for the general power-law materials, despite the existence of an analytical function as a solution.

As was already discussed, the granular materials with the classical ellipsoid contacts are the simple natural example of power-law materials with $n = 3/2$. For the conical contact of particles Goddard [1990] demonstrated that $n = 2$. This results in the replacement of the $P^{1/3}$ law, due to the Hertzian contact theory, by the $P^{1/2}$ law for the dependence of elastic stiffness of the compressed granular material. This change of exponent will result in stronger dependence of the soliton speed on the maximum particle velocity and in the shorter soliton width L_n $(n = 2) = \pi a$.

We can see that strongly nonlinear solitary waves, with properties similar to those described in Section 1.4.3 for the Hertz law, are supported by the general power-law for the interaction between elements in the discrete system. This means that they are not specific for only Hertz interaction (see also Sections 1.10.2.3 and 1.10.2.4).

The compression shock waves can exist in power-law materials if dissipation is included, similar to the case with $n = 3/2$ (Eq. (1.82), see also Problem 10).

1.10.2.2   Discrete Chain Versus Continuum Approximation

The appearance of compression solitary waves in power-law materials in long-wave approximation is in complete agreement with the existence theorem for solitary waves on discrete lattices by Friesecke and Wattis [1994]. Really, the power-law interaction force with $n > 1$ evidently results in potential energy being superquadratic, with other properties required for the existence of the solitary wave being satisfied also. So, for this case, the solutions of long-wave approximation and second-order forward–backward differential–difference equations for discrete systems are consistent.

For a strongly nonlinear case, Eq. (1.104) for the discrete chain may be rewritten similarly to Eq. (1.69), with a corresponding variable replacement (see Problem 8). This results in a similar conclusion, as for the partial case for a discrete lattice with $n = 3/2$: the width of the strongly nonlinear impulse does not depend on its amplitude. Also for a disrete chain the dependence of a phase speed on a particle velocity in the maximum follows Eq. (1.129) derived in continuum limit. Additionally, long-wave approximation provides coefficients for the dependence of phase speed on initial conditions (ξ_0) and strain (or particle velocity) in maximum (Eq. (1.127)) and the length of the solitary pulse. The accuracy of this approximation should be established through the analysis of neglected terms or in comparison with numerical results for a discrete chain.

Sen and Sinkovits [1995], [1996] investigated the propagation of weak and strong perturbations in gravitationally compacted granular columns characterized by the power law ($n = 3/2$ and 5) intergrain potential. They observed a difference between sound and soliton speeds for relatively high impact velocity and small depth which they attributed to the soliton-like pulses similar to the considered case of "sonic vacuum" (see Section 1.5) for the pristine 1- and 2-D granular systems.

In the case of a strong impulse, its speed increases more slowly with depth in comparison with sound speed. They also investigated the effect of additional packing defects. The main difference between the perfectly packed system, and the system with voids, is manifested by the energy transfer from soliton-like impulses into the random motion of particles similar to the result for randomized 1-D system (Fig. 1.21). The existence of voids also decreased the sound speed without changing its dependence on depth (initial compression).

It is important to mention that analyzed stationary solutions correspond to long-wave approximation, and should not be expected to coincide with stationary solutions in a discrete system of particles. Sen [1998] and Manciu and Sen [1999] demonstrated, in computer calculations, that such stationary solutions also exist for discrete systems with power-law interaction and exponent in the interval $n = 1.2–19$. The width of these solitary waves was close to that predicted by Eq. (1.130) in a whole investigated range of exponent n. Better agreement was obtained for smaller n.

At the same time, the deviations in the functional dependence of L_n on n between Eq. (1.130) and the results for the discrete chain were noticeable at $n > 3$. It is especially natural for large n, because long-wave approximation cannot be expected to work perfectly when the space scale of a solitary solution is close to the particle diameter (Hascoet and Herrmann [2000]). It is amazing that long-wave approximation resulted in the compact analytical solution and provided such reasonable results when the soliton width was equal to five particle diameters or less. Chatterjee [1999] found an asymptotic solution for the solitary wave when n was slightly greater than 1. It is interesting that this solution describes results of computer simulation for $n = 3/2$ with better accuracy than Eq. (1.39), even though 1.5 is not "slightly larger" than 1.

1.10.2.3   Transverse Vibrations of a Fiber with Discrete Particles

There is another example of a "sonic vacuum" which can be realized in simple experiments. An exponent $n = 3$ corresponds to the strictly transverse vibrations of linearly elastic unstressed fiber with discrete masses (Fig. 1.40, Nesterenko [1992c], [1993a]). The mass of fiber between neighboring particles is negligible in comparison with the mass of the particle. For this case there is a remarkable simple nonlinear partial periodical solution of Eq. (1.105) (no restriction of $\xi > 0$ is required):

$$\xi = \sqrt{2}\frac{V}{c_3}\sin\frac{\sqrt{2}}{a_0}(x - Vt). \tag{1.133}$$

This function is a harmonic one and nonlinearity is manifested by linear dependence of the wave amplitude on speed V. The corresponding solitary solution is represented by one hump of periodic solution; in this case Equation (1.105) also supports the "breather" solution $\xi = Q(t)Z(x)$ (Rosenau [1994]).

The existence of regular, and in this case even, harmonic solutions for a strongly nonlinear motion of elastic materials is completely counterintuitive. The incorrect intuitive guess is reflected in D. Bernoulli's [1741] remark: "for the elongation will not be proportional to the extending force ... and everything must be irregular." It is instructive to compare this statement with his perfect motto especially valid in a highly nonlinear world: "To find the solution I used new principles, and therefore I wanted to confirm the theorems with experiments, so ... people ... would have no doubt about their truth" (D. Bernoulli [1738]).

It will be very interesting to get experimental confirmation of the existence of solitary waves in the strictly transverse vibrations of linearly elastic unstressed fibers with discrete masses.

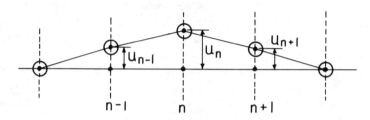

Figure 1.40. Transverse wave in a system of particles with vertical displacements only.

1.10.2.4   Plug-Chain Gas–Liquid System

A compression wave in a plug-chain periodic gas–liquid system, with the gas being near the thermodynamic critical point, is obeying the equation similar to Eqs. (1.104) and (1.105) for the power-law material with an exponent $n = 3$ (Gavrilyuk and Serre [1995]). In this case, a gas being in thermodynamic equilibrium with a liquid is responsible for strongly nonlinear "springs" connecting the liquid plugs. The nonlinearity is due to the fact that the first and second derivatives of the gas pressure are equal to zero (isentropic case) at the critical point.

Gavrilyuk and Serre [1995] found explicit solutions of Eq. (1.104) for a discrete chain by variable separation. They also considered the modulation equations of this system in the long-wave approximation and established connection with the Korteweg theory of capillarity.

The material similar to the strongly nonlinear periodic gas-liquid system may be solid/foam laminate considered in Section 3.5.1.2.

1.10.2.5   Rarefaction Solitary Waves in "Abnormal" Material

Materials with $n < 1$ represent different, "abnormal"-type behavior in comparison with $n > 1$. In the former case, the sound speed is equal to infinity if $\xi_0 = 0$ (Eq. (1.114)). The value $n < 1$ stipulates anomalous behavior under compression, i.e., the decrease of its elasticity modules as the deformation grows. This is typical for media subjected to phase transformation or failure.

Nevertheless, it is interesting to investigate the properties of strongly nonlinear stationary solutions in a discrete medium with a power interaction law, if the exponent n is less than 1 $(0 < n < 1)$, even without dissipation taken into account. The reason for this is connected to the possibility of the existence of quasisolitary waves resembling a nondissipative case for weak dissipation as was the case in experiments with a chain of spherical particles (Figs. 1.23 and 1.25).

Long-wave approximation is represented by Eq. (1.102) and for stationary solutions Eq. (1.115) holds. For constant C' being in the interval described by Eq. (1.120) the effective potential $W(y)$ has two extremes y_1 and y_2 with $y_1 > y_2$, where y_1, y_2 correspond to the maximum and minimum, respectively (Figures 1.39(b) and 1.41). The $W(y)$ curves for different values of constant C' and their extremes are shown in Fig. 1.41 for $n = 0.1$. The relative positions of extremes, when $0 < n < 1$, are reversed to their locations for $n > 1$ (compare Figs. 1.39(a), 1.39(b), and 1.41).

If an additional condition applies

$$\frac{n-1}{2}\left(\frac{n+1}{2n}\right)^{n/(n-1)} = N_2 > C' > N_1. \tag{1.134}$$

then the values of the maximum of curves $W(y)$ are negative (see Fig. 1.41, curve III). At $C' = N_2$ the curve $W(y)$ touches the abscissa axis y at the maximum point (see Fig. 1.41, curve II):

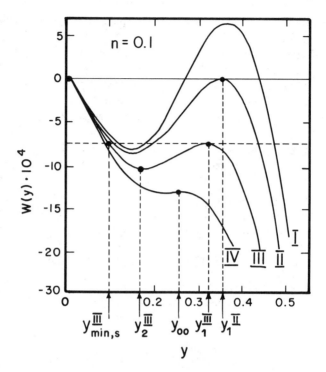

Figure 1.41. Effective potential energy $W(y)$ for power-law materials with abnormal compressibility ($n = 0.1$) at corresponding values of constant C': curve I, $C' = 0.99$ N_2; II, $C' = N_2$; III, $C' = (N_1 + N_2)/2$; IV, $C' = N_1$.

$$y_1^{II} = \left(\frac{2n}{n+1}\right)^{(1+n)/2(1-n)}.$$ (1.135)

The important conclusion follows from the property of effective potential energy with restrictions for C' from Eq. (1.120): if C' is negative but outside the region specified by Eq. (1.130), then $W(y)$ has a positive maximum (see Fig. 1.41, curve I). In this case a solitary solution is not possible for a system which has zero strength under tension. Such a system cannot support this amplitude of a solitary wave.

In the case when $C' \to N'_1$ both extremes are combined in one point y_{00} (see Fig. 1.41, curve IV):

$$y_{00} = n^{(1+n)/2(1-n)}.$$ (1.136)

According to the physical sense y_1 corresponds to the strain in the infinity which may also be associated with the initial state of the system. The relative arrangement of y_1 and y_2 (Figs. 1.39(b) and 1.41) allows the conclusion that, in this case, periodical and solitary rarefaction waves exist.

Solitary compression waves in the system are absent in contrast to the exponent $n > 1$ (compare Figs. 1.3, 1.39(b), and 1.41). If the physical sense allows negative values of ξ (the system has a finite tensile strength), then solitary waves are possible at $0 > C' > N_2$. A solitary solution corresponding to the case $C' = N_2$ represents an important partial case, when the system attains zero strain in the soliton.

Using the above-mentioned properties of the function $W(y)$, one can derive the dependence of phase velocity V_p of the periodic wave on the maximum (ξ_{max}) and minimum (ξ_{min}) strains and ξ_2 similar to Eq. (1.126). The speed of the solitary rarefaction wave $V_{s,r}$, depending on the strain in its minimum ($\xi_{min,s}$) at total "energy" of the nonlinear "oscillator" $W_0 = W(y_1)$, can also be obtained in a manner similar to the derivation of Eqs. (1.37) and (1.127):

$$V_{s,r} = \frac{c_n}{\left(\xi_0 - \xi_{min,s}\right)} \left\{ \frac{2\left[n\xi_0^{n+1} + \xi_{min,s}^{n+1} - (n+1)\xi_0^n\xi_{min,s}\right]}{(n+1)} \right\}^{1/2}. \tag{1.137}$$

If $\xi_{min} \to 0$ (that corresponds to a solitary solution at $C' \to N_2$), obtain from Eq. (1.137) the phase speed for this soliton

$$V_{s,r} \Rightarrow c_n\sqrt{n}\, \xi_0^{(n-1)/2} \left(\frac{2}{n+1}\right)^{1/2}. \tag{1.138}$$

From the comparison of the long-wave sound speed at the initial state c_0 (Eq. (1.114)) and $V_{s,r}$ from Eq. (1.138), we can conclude that this solitary rarefaction wave is supersonic according to the initial state. The unique feature of this solution with zero strain in the minimum, is the infinite sound speed c_0 at this point. The maximum possible ratio of phase speed of this strongly nonlinear rarefaction soliton, and sound speed in infinity, is close to 1.4 for $n \ll 1$.

In the opposite case at $C' \to N_1$ the soliton speed $V_{s,r}$ approaches the initial sound speed c_0

$$V_{s,r} \Rightarrow c_n\sqrt{n}\xi_0^{n-1/2}. \tag{1.139}$$

The supersonic property of the rarefaction solitary wave, in a general case, can be seen from the relation between $V_{s,r}$ and c_0 following from variable transformation (Eq. (1.111)):

$$\frac{V_{s,r}}{c_0} = \frac{y_1^{(1-n)/(1+n)}}{\sqrt{n}}. \tag{1.140}$$

Due to the relative positions of function $W(y)$ extremes in the case $n < 1$, and when Eq. (1.117) holds, obtain $y_1 > y_{00}$. Using Eq. (1.133) for y_{00} obtain the general property of rarefaction solitary waves

$$V_{s,r} > c_0. \qquad (1.141)$$

Thus, in discrete systems with abnormal compressibility, the compression solitary waves are prohibited and instead the rarefaction solitary waves are allowed. The existence of expansion solitary waves in power-law materials at $n < 1$ in long-wave approximation is in complete agreement with the existence theorem, for solitary waves on lattices, by Friesecke and Wattis [1994].

The stationary rarefaction shock waves can exist in power-law materials at $n < 1$ if dissipation is included into consideration similar to the case of materials with abnormal compressibility considered in the following section.

1.11   General Case of Nonlinear Discrete Systems

1.11.1   Wave Equation for 1-D Particulate Materials

In Sections 1.1 to 1.10 different concrete cases of interaction laws between particles, including a relatively general case of power-law material, were considered. It was shown that for these materials, there are two qualitatively different cases: $n > 1$ (normal compressibility) and $n < 1$ (abnormal compressibility). If $n > 1$ ("sonic vacuum"), a new solitary compression wave with a maximum strain ξ_{max} much greater than the initial ξ_0 can exist, being qualitatively different from the soliton of the KdV equation. The main differences are the following: the initial sound speed has no effect on the soliton's characteristics, its finite size does not depend on the amplitude, and they emerge from initial perturbation very fast, or in short distances from the entrance.

If $n < 1$, compression solitons do not exist, but solitary rarefaction waves can propagate in a system. As $n \to 1$, the smooth transition between the corresponding stationary solutions for $n > 1$ and $n < 1$ does not exist.

The wave equations (1.20), (1.102) are more general than the KdV Eq. (1.9), and can be transformed into it by making additional approximation, and introducing a second small parameter according to the amplitude of disturbance.

One-dimensional chains of discrete particles are traditional systems (Fermi, Pasta, and Ulam [1965]; Kruskal [1975]; Kunin [1975]) to obtain physically based partial differential wave equations, including the cases with spatial derivatives of order higher than the second, for essentially nonlinear systems (Nesterenko [1983b]; Shukla, Sadd, Xu, and Tai [1993]; Triantafyllidis and Bardenhagen [1993], [1994]). This type of equation is especially interesting for the localized nonlinear disturbances in structured media when their space scale is approaching the typical scale of the system (shear bands, solitons, shock waves, necking). Therefore, it is vital to obtain a continuous approximation (without assuming that the amplitude of perturbation is small) for a discrete system of particles in which the interaction law is more general than the power law. The latter one is already able to grasp an essentially different type of material behavior (depending on $n > 1$ or $n < 1$).

Its worth mentioning here that, as was demonstrated by the wonderful and sometimes dramatic histories of the soliton and Kortweg–deVries or Sine–Gordon equations or others (Sander and Hutter [1991], Filippov [2000]), the equation obtained for some concrete physical systems, as usual, has many general applications, sometimes in absolutely different areas.

Now let us examine the system of equations for a discrete chain of particles interacting according to general law only with its closest neighbors (Nesterenko [1995]):

$$m\ddot{u}_i = f(u_{i-1} - u_i) - f(u_i - u_{i+1}), \qquad N \geq i \geq 2, \tag{1.142}$$

where m is the mass of the particles, $f(u_i - u_i)$ is the interactive force between them with numbers i, $i - 1$, u_i is the displacement of the ith particle from the initial equilibrium position, which may include some initial displacements causing static deformation of the system, N is the number of particles in the system.

As was done in Sections 1.1 to 1.3, we examined a continuous approximation of the 1-D system described by Eq. (1.142), supposing that the spatial dimension of the disturbance $L >> a$, a is the distance between particles in the equilibrium state. It is assumed that the displacements of neighboring particles can be obtained by expanding the displacement of the ith particle in a fourth-order series (which will provide the simplest different result from the traditional continuum approach) in terms of a small parameter $\varepsilon = a/L$. Then,

$$u_i = u, \tag{1.143}$$

$$u_{i-1} = u - u_x a + \frac{1}{2}u_{xx}a^2 - \frac{1}{6}u_{xxx}a^3 + \frac{1}{24}u_{xxxx}a^4 + \cdots ,$$

$$u_{i+1} = u + u_x a + \frac{1}{2}u_{xx}a^2 + \frac{1}{6}u_{xxx}a^3 + \frac{1}{24}u_{xxxx}a^4 + \cdots .$$

The higher-order terms can be considered as well. The term of the fourth order gives the first term in a differential equation, that is, qualitatively different from the classical wave equation. It is amazing, as demonstrated by the comparison of the analytical solutions of the corresponding high-gradient, essentially nonlinear wave Eq. (1.20) with computer calculations and experiments (Nesterenko [1983b]; Lazaridi and Nesterenko [1985]; Coste, Falcon, and Fauve [1997]) that cutting off the expanding procedure on this step can be very successful for the description of system behavior with large gradients. This holds true even if the characteristic scale of disturbances is comparable with the characteristic space scale of the system. It gives some hope that this type of effort can provide the next step in describing material behavior at large gradients in more complex systems.

The following derivation is simple, but is presented in detail because, in some papers, the resulting equations contain inaccuracies.

From Eq. (1.143) we can obtain

$$
\begin{aligned}
&u_{i-1} - u_i = -u_x a + \tfrac{1}{2} u_{xx} a^2 - \tfrac{1}{6} u_{xxx} a^3 + \tfrac{1}{24} u_{xxxx} a^4 + \cdots = N(x) + \varphi(x), \\
&u_i - u_{i+1} = -u_x a - \tfrac{1}{2} u_{xx} a^2 - \tfrac{1}{6} u_{xxx} a^3 - \tfrac{1}{24} u_{xxxx} a^4 + \cdots = N(x) + \psi(x), \\
&N(x) = -u_x a = a\xi, \qquad \xi = -u_x, \\
&\varphi(x) = + \tfrac{1}{2} u_{xx} a^2 - \tfrac{1}{6} u_{xxx} a^3 + \tfrac{1}{24} u_{xxxx} a^4 \cdots , \\
&\psi(x) = - \tfrac{1}{2} u_{xx} a^2 - \tfrac{1}{6} u_{xxx} a^3 - \tfrac{1}{24} u_{xxxx} a^4 \cdots .
\end{aligned}
\tag{1.144}
$$

Equations (1.144) can be used to write the following relations for interaction forces between neighboring particles

$$
\begin{aligned}
&f(u_{i-1} - u_i) \approx f\{N(x) + \varphi(x)\}, \\
&f(u_i - u_{i+1}) \approx f\{N(x) + \psi(x)\}.
\end{aligned}
\tag{1.145}
$$

It is easy to see that $N(x)$ has a different order in ε than $\varphi(x)$ and $\psi(x)$:

$$
\frac{\varphi(x)}{N(x)} \sim \frac{\psi(x)}{N(x)} \sim \varepsilon \ll 1.
\tag{1.146}
$$

We may use an expansion procedure once more. By using Eq. (1.142), we can expand the function f in a series

$$
\begin{aligned}
&f\{N(x) + \varphi(x)\} \approx f\{N(x)\} + \frac{f'}{a}\varphi + \frac{f''}{2a^2}\varphi^2 + \frac{f'''}{6a^3}\varphi^3 + \cdots , \\
&f\{N(x) + \psi(x)\} \approx f\{N(x)\} + \frac{f'}{a}\psi + \frac{f''}{2a^2}\psi^2 + \frac{f'''}{6a^3}\psi^3 + \cdots .
\end{aligned}
\tag{1.147}
$$

Here, the prime denotes differentiation of the function $f(a\xi)$ with respect to ξ.

An expression for the force acting on the ith particle can be obtained from Eqs. (1.145) and (1.147)

$$
\begin{aligned}
&f(u_{i-1} - u_i) - f(u_i - u_{i+1}) \\
&\approx \frac{f'}{a}(\varphi - \psi) + \frac{f''}{2a^2}(\varphi^2 - \psi^2) + \frac{f'''}{6a^3}(\varphi^3 - \psi^3) + \cdots .
\end{aligned}
\tag{1.148}
$$

Using Eq. (1.144), we have

$$
\varphi - \psi \approx a^2 u_{xx} + \frac{a^4}{12} u_{xxxx} + O(\varepsilon^6),
\tag{1.149a}
$$

$$
\varphi^2 - \psi^2 \approx -\frac{a^5}{3} u_{xx} u_{xxx} + O(\varepsilon^7),
\tag{1.149b}
$$

$$\varphi^3 - \psi^3 \approx \frac{a^6}{4} u_{xx}^3 + O(\varepsilon^7). \tag{1.149c}$$

The necessity of the retention of terms of the different order in ε in Eqs. (1.149a) and (1.149b,c) results from the fact that the ratio of multipliers of the corresponding differences in Eq. (1.148) are of the following order in ε:

$$a\frac{f''}{f'} \sim \varepsilon, \qquad a^2\frac{f'''}{f'} \sim \varepsilon^2. \tag{1.150}$$

Due to the expressions represented by Eq. (1.150), the retained terms in Eq. (1.148) lead to a final equation including all terms up to order ε^2 as compared with the main term.

The convective derivative can be neglected if the phase velocity is much larger than the particle velocity, which was the case for power-law granular materials at $n = 3/2$ (see Section 1.3) and can be expected to be valid at least at some range of disturbance parameters for the general interaction law. This approximation cannot be justified a priori for essentially nonlinear materials without the solution of the corresponding wave equation. This is due to the fact that the characteristic wave phase speed as the material parameter (well determined in linear or weakly nonlinear cases) doesn't exist in a strongly nonlinear case for zero compression.

The resulting wave equation, after substitution of Eqs. (1.148) and (1.149) into Eq. (1.142), has the form

$$\rho u_{tt} = f' u_{xx} + a^2 \left[\frac{f'}{12} u_{xxxx} - \frac{f''}{6} u_{xx} u_{xxx} + \frac{f'''}{24} u_{xx}^3 \right], \tag{1.151}$$

where $\rho = m/a$ is the linear density. The equation for the strain is found by differentiating both sides of Eq. 1.151 with respect to x and setting $\xi = -u_x$:

$$\rho \xi_{tt} = \left\{ f' \xi_x + a^2 \left[\frac{f'}{12} \xi_{xxx} + \frac{f''}{6} \xi_x \xi_{xx} + \frac{f'''}{24} \xi_x^3 \right] \right\}_x. \tag{1.152}$$

The expression in square brackets in Eq. (1.152) can be transformed into a divergent form

$$\frac{f'}{12} \xi_{xxx} + \frac{f''}{6} \xi_x \xi_{xx} + \frac{f'''}{24} \xi_x^3 = \frac{1}{24} \left[2f' \xi_{xx} + f'' \xi_x^2 \right]_x. \tag{1.153}$$

From Eqs. (1.152) and (1.153), we obtain the final long-wavelength approximation for the discrete system Eq. (1.142):

$$\rho\xi_{tt} = \left\{ f + \frac{a^2}{24}\left[2f'\xi_{xx} + f''\xi_x^2\right]\right\}_{xx}.$$ (1.154)

Equation (1.149) can easily be rewritten in the general form of the system of two equations similar to equations (1.107) and (1.108).

It is interesting that the specific structure of the right term is common for very different phenomenon, like the response of sensory receptors in neurophysiology, which do not operate independently (Coleman [1971a,b]). For example, in vision, the magnitude of the signal reported to the central nervous system by a photoreceptor at a point is determined by the intensity of light received at that point and by the signals from other points, as concluded by Ernst Mach [1868] based on his own experiments.

Another example may be repesented by periodic structured metamaterials (Yablonovitch [1993]; Pendry et al., [1999]; Smith et al. [1999]). If the product of effective permittivity and magnetic permeability in such material has a large value then nonlinear electromagnetic properties of matrix material may result in strongly nonlinear mode of wave propagation described by Eq. (1.154) and may be in certain conditions similar to the considered wave properties of "sonic vacuum."

Dynamics of the strongly nonlinear discrete chains with nonmonotone dependence of the springe force on spring elongation is considered by Balk, Cherkov, and Slepyan [2001].

1.11.2 Regularized Wave Equation

The corresponding regularized equation with mixed derivatives can be obtained from Eq. (1.154). To do this, we replace the term with a fourth-order space derivative by the term with a mixed derivative, using the zero approximation (without the term in square brackets in Eq. (1.154))

$$f'\xi_{xx} = \rho\xi_{tt} - f''\xi_x^2.$$ (1.155)

By substituting Eq. (1.155) into Eq. (1.154), we obtain

$$\rho\xi_{tt} = \left\{ f + \frac{a^2}{24}\left[2\rho\xi_{tt} - f''\xi_x^2\right]\right\}_{xx}.$$ (1.156)

The term in Eq. (1.156) with mixed derivatives is a dispersion term, as in the classical linear approximation, and the additional term in square brackets has the same order as the former, and represents the contribution of nonlinear dispersion. Eqs. (1.154) and (1.156) are equivalent, according to the order of neglected error terms.

Rosenau [1986a,b], [1988] and Rosenau and Hyman [1993] proposed a long-wave equation for nonlinear discrete systems, with an arbitrary interaction law between particles. It is partially analogous to Eq. (1.156), but does not contain the last essentially nonlinear dispersive term. This term for strongly nonlinear solutions has the same order of magnitude as the kept traditional dispersive term (the latter is exactly the same as in the linear case) and cannot be ignored under consistent consideration.

There are no theorems on the relative merits of Eq. (1.156) in comparison with Eq. (1.154). At the same time, it is interesting and instructive to follow the discussion on the transformation of the equations with high space derivatives to the regularized form in the example of the KdV equation which can be found in papers by Benjamin, Bona, and Mahony [1972]; Kruskal [1975]; Bona, Pritchard, and Scott [1983]; and Albert and Bona [1991]. They studied the relative merits of the weakly nonlinear KdV equation (the weakly nonlinear case which can be obtained from the more general Eq. (1.156):

$$\eta_t + \eta_x + \eta\,\eta_x + \eta_{xxx} = 0, \tag{1.157}$$

and its regularized form

$$\zeta_t + \zeta_x + \zeta\,\zeta_x - \zeta_{xxt} = 0. \tag{1.158}$$

Here η, ζ represent some material parameters (strain, velocity) and are functions of two dimensionless variables x and t. These papers are very informative for the comparison of these equations and are highly recommended for detailed study. Here we will summarize their main results:

(1) On a time scale where weakly nonlinear and dispersive effects are expected to provide qualitatively new effects in comparison with linear theory, the solutions of the initial-value problems for Eqs. (1.157) and (1.158) agree within the accuracy of either model.

(2) Both equations appear to be at least qualitatively good models in comparison with experiments with surface water waves at relatively small Stokes numbers. Zabusky and Galvin [1971], Hammack and Segur [1974], Sander and Hutter [1992], demonstrated it for the KdV equation and Hammack [1973] obtained experimental confirmation of the validity of the regularized equation.

There are also some essential differences between the equations (1.157) and (1.158) (Benjamin, Bona, and Mahony [1972]; Kruskal [1975]; Bona, Pritchard, and Scott [1983]; Albert and Bona [1991]):

(1) Eq. (1.157) is asymptotically pure, which is not the case for the regularized Eq. (1.158).

(2) The initial-value problem for Eq. (1.158) is well posed in contrast with Eq. (1.157).

(3) For two-point boundary-value problems, well-posedness results are straightforward for the regularized Eq. (1.158), no such results are known for the KdV equation.

(4) There is a problem with the linearized dispersion relation for the KdV equation for short wavelengths resulting in its dispersive blow-up. At the same time, we should not forget that any differential equation has definite limits if it is seriously considered to describe real discrete systems. From this point of view the natural step to improve the situation with dispersion relations is to include the higher space derivatives which will remove dispersive blow-up.

(5) One of the problems with the regularized equation is caused by the possibility of introducing different types of regularized equations within the same accuracy, and there is no selection criterion for which one is real.

(6) In computer simulation, the regularized Eq. (1.158) is far easier to be approximated by high-order, unconditionally stable numerical schemes.

The aforementioned relations between the KdV equation and its regularized form, are also valid for a more general class of nonlinear models, as shown by Albert and Bona [1991].

From a physical point of view, it is evident that any continuum approach fails on a scale less than the distance between particles in the discrete system, even if this approach has a good dispersion relation for any short wavelength. Computer calculations based on the continuum approximations, should not include disturbances with the wavelength shorter than some critical wavelength, because this is out of the validity range of continuum approximation itself. It can be done, for example, using a frequency-dependent dissipation process. In this case, these models will be equivalent.

In any case, the validity of one of these two continuum approaches, especially the case when the typical space scale of disturbances L is close to the distances between particles a, can be justified only by comparison with experiment or numerical calculation for genuine discrete models. This step cannot be replaced by artificially sustaining proper behavior of the wave equation at any small length scale.

As demonstrated in this chapter, the essentially nonlinear equations with high space derivatives could predict qualitatively new stationary solitary solutions with a length comparable to the distance between particles in agreement with experiment and computer calculations. Certainly it is not clear which equation with restricted wavelength can be more convenient for computer codes, especially concerning the boundary conditions problem.

Equations (1.154) and (1.156) can be considered general wave equations for various physical phenomena at large gradients in media with structure, where strong nonlinearity results in nonlinear dispersion.

1.11.3   Stationary Solutions of the General Wave Equation

We recall that in the case of a power-law interaction, the new type of stationary solitary waves was easy to find, because of a change of variables

$$z = \xi^{n + 1/2}. \tag{1.159}$$

The structure of the expression in square brackets in Eq. (1.154) allows for the use of a variable transformation that solves the same problem as it does in Eq. (1.159) for a power-law interaction. In this case, the following variable transformation

$$z = B \int_{\xi_0}^{\xi} \left(f' \right)^{1/2} d\xi, \tag{1.160}$$

where B is a constant $(f' > 0)$, helps to reduce Eq. (1.154) to the form which is easy to analyze at least for the stationary case

$$\rho[M(z)]_{tt} = \left[K(z) + \frac{a^2}{12} P(z) z_{xx} \right]_{xx}. \tag{1.161}$$

Here the functions $M(z) = \xi(z)$, $K(z) = f\{aM(z)\}$, and $P(z) = [f'\{aM(z)\}]^{1/2}$ are found from the inverse transformation to Eq. (1.160). It is supposed that $f(\xi)$ is always the increasing function of its argument.

In the case of stationary waves $z(x - Vt)$, Eq. (1.161) can be reduced to the ordinary differential equation of a nonlinear oscillator in a potential field $W(z)$. It can be done by integrating Eq. (1.161) for the stationary case twice from $+\infty$ to x under the boundary conditions for strain and their derivatives: $\xi(x = +\infty) = \xi_0$, $\xi_x(x = +\infty) = 0$, $\xi_{xx}(x = +\infty) = 0$, $\xi_{xxx}(x = +\infty) = 0$:

$$\frac{a^2}{12} z_{xx} = -\frac{\partial W(z)}{\partial z}, \tag{1.162}$$

$$-\frac{\partial W(z)}{\partial z} = \rho V^2 \frac{M(z)}{P(z)} - \frac{K(z)}{P(z)} - \frac{C}{P(z)}, \qquad P(z) \neq 0,$$

where C is the constant which can be determined by the value of ξ_0.

In the case of a power-law interaction $(f(a\xi) = mA_n(a\xi)^n)$, the characteristic shape of $W(z)$, with two maximums, returns us to a compression solitary wave at $n > 1$ or to a rarefaction solitary wave if $n < 1$. The conditions for the existence of solitons, periodic waves, and their parameters, in the continuous approximation for an arbitrary interaction law between the particles, can thus be determined by analyzing the extremes of the function $W(z)$ and their relative positions. If the function $W(z)$ can be obtained explicitly, using the inverse transformation $\xi(z)$, then the form of stationary perturbations is obtained from Eq. (1.162) at least in a quadrature.

It is important to note that the structure of Eq. (1.154) changes qualitatively if the discrete system significantly differs from a periodic one. In that case the terms of lower space derivatives will not be canceled as they were for a symmetric periodic structure. At the same time, as shown previously in Section 1.5, the behavior of single pulses can be qualitatively close to a periodic system if the particles are only slightly chaotisized compared to their average size; they can be significantly different if the "chaos" is strong enough. The effects of nonlinear dispersion can form solitons on the early stages of propagation when the effects of randomness need a large number of particles to be involved to result in essential wave changes.

It is interesting that in the stationary case ($\xi = \xi(x - Vt)$) Eq. (1.154) corresponds to the equation obtained by Coleman [1983], [1985] for the slowly deformed thin polymeric fiber

$$T^0 = \tau(\lambda) + \beta(\lambda)\lambda_x^2 + \gamma(\lambda)\lambda_{xx}, \tag{1.163}$$

where λ is the local stretch ratio and T^0 is the tensile force. It is easy to see this upon making the replacements

$$\lambda \to \xi, \quad \tau(\lambda) \to f(\xi) - \rho V^2\xi, \tag{1.164}$$

$$\beta(\lambda) \to \beta(\xi) = \frac{a^2}{24}f'', \quad \gamma(\lambda) \to \gamma(\xi) = \frac{a^2}{12}f',$$

which represents the complete analogy between theories for nonlinear rods and granular materials first proved by Wright [1984], [1985]. It demonstrates that often equations and phenomena established for one system are analogous to very different areas.

It is possible to make some general conclusions about the properties of stationary solutions (if they exist) without finding their concrete form based on Eq. (1.154) (Nesterenko [1999], [2000]).

According to Eq. (1.162) the extreme of the effective "potential energy" $W(z)$ must satisfy the equation similar to Eq. (1.116):

$$\Phi(z) = C, \tag{1.165a}$$

where

$$\Phi(z) = \rho V^2 M(z) - K(z), \tag{1.165b}$$

To investigate the behavior of function $\Phi(z)$ find its first and second derivatives expressing functions in Eq. (1.165b) through ξ and f:

$$\frac{d\Phi(z)}{dz} = (\rho V^2 - f_\xi)\xi_z, \tag{1.166}$$

$$\frac{d^2\Phi(z)}{dz^2} = (\rho V^2 - f_\xi)\xi_{zz} - f_{\xi\xi}\xi_z^2. \tag{1.167}$$

From Eq. (1.166), find that function $\Phi(\zeta)$ has one extreme at z_{ext} (ξ_{ext}), where a physically clear condition must be satisfied according to the determination of the sound speed c_0:

$$\rho V^2 - f_\xi(z_{ext}) = 0 \quad \to \quad V = c_0(\xi_{ext}). \tag{1.168}$$

Equation (1.168) means that at point z_{ext} (ξ_{ext}) the local sound speed c_0 is equal to the phase speed of the solitary solution V. At the same point (where $\Phi_z = 0$) the sign of a second derivative Φ_{zz} is determined by the shape of function $f(a\xi)$. As follows from Eqs. (1.166) and (1.167), for materials with normal behavior ($f_{\xi\xi} > 0$), function $\Phi(z)$ has maximum Φ_{max} at $z_{ext}(\xi_{ext})$, similar to the function $\Phi(\alpha)$ depicted in Figure 1.38 for $n = 1.1$ where ξ_{ext} corresponds to α_{max}. It is supposed that $f = 0$ at $\xi = 0$, so zero strain corresponds to the system without external loading resulting in $\Phi(z) = 0$ at $z = 0$.

If $0 < C < \Phi_{max}$ then Eq. (1.165a) has two roots: one at $z_1 < z_{ext}$ and the other at $z_2 > z_{ext}$. They represent the extremes of the effective "potential energy" $W(z)$. Evidently, the case where $C \to \Phi_{max}$ corresponds to a weakly nonlinear wave and small $C \to 0$ describes a strongly nonlinear one. To determine the nature of these extremes find the second derivative of $W(z)$ at the points of $z_1 < z_{ext}$ and $z_2 > z_{ext}$:

$$\frac{d^2 W}{dz^2} = -\frac{\Phi'(z)}{P(z)}. \tag{1.169}$$

The sign of the second derivative $W(z)$ at $z_1 < z_{ext}$ is apparently negative because z_{ext} corresponds to the maximum of $\Phi(z)$. This means that function $W(z)$ also has a maximum at z_1. The sign of W_{zz} at $z_2 > z_{ext}$ is apparently positive and corresponds to the minimum of $W(z)$ at z_2. Finally, function $W(z)$ for materials with normal behavior has a shape similar to the function $W(y)$ in Figures 1.3 ($n = 3/2$) and 1.38 ($n = 1.1$) which result in a compression solitary wave.

For a chain with "normal" interaction, this result is similar to the condition for soliton existence in the discrete chain obtained by Friesecke and Wattis [1994]. It explains why continuum approximation, being the truncated version of the equations for a discrete chain, is able to support the same solitary solutions.

The convex behavior of function $f(a\xi)$ is a sufficient condition for the existence of the rarefaction-type solitary solution of Eq. (1.154). The proof is similar to the case of materials with normal compressibility.

Really the function $\Phi(z)$, in the case of material with abnormal behavior, also has one extreme at z_{ext} (ξ_{ext}) where the local sound speed c_0 is equal to the phase speed of the solitary solution $V_{s.r.}$. At this extreme point, the sign of the second derivative Φ_{zz} is determined by the shape of function $f(a\xi)$. As follows from Eqs. (1.166) and (1.167), for materials with abnormal behavior ($f'(a\xi) < 0$), function $\Phi(z)$ has minimum Φ_{min} at $z_{ext}(\xi_{ext})$, similar to the function $\Phi(\alpha)$ depicted in Figure 1.38 for $n = 0.5$.

If $0 > C > \Phi_{max}$, then Eq. (1.165a) has two roots: one at $z_1 > z_{ext}$ and the other at $z_2 < z_{ext}$ corresponding to the extremes of the effective "potential energy" $W(z)$. The type of these extremes is determined by the second derivative of $W(z)$ represented by Eq. (1.169) (or the first derivative of $\Phi(z)$).

It is evident from the described property of function $\Phi(z)$ that the sign of the second derivative $W(z)$ at $z_1 > z_{ext}$ evidently is negative, because z_{ext} corresponds to the minimum of $\Phi(z)$. This means that function $W(z)$ has a maximum at z_1.

The sign of W_{zz} at $z_2 < z_{ext}$ is positive corresponding to the minimum of $W(z)$ at z_2. Finally, function $W(z)$ has a shape similar to the function $W(y)$ in Figures 1.39(b) ($n = 0.5$) and 1.41 ($n = 0.1$) which results in a rarefaction solitary wave.

Let us suppose that a steady solution $\xi(x - Vt)$ of Eq. (1.154) exists. It means that the effective potential energy $W(z)$ from Eq. (1.162) has a shape similar to the shape of functions $W(y)$ for normal or abnormal behavior correspondingly (see examples for power-law materials, Fig. 1.39). We will try to find the equations for the phase speed V of these solutions. The relation $\xi_{tt} = V^2\xi_{xx}$ holds for any steady wave. Find solutions that approach a constant value in the infinity and that have properties

$$x(\xi = +\infty) = \xi_0, \quad \xi_x(x = +\infty) = 0, \quad \xi_{xx}(x = +\infty) = 0, \quad \xi_{xxx}(x = +\infty) = 0. \quad (1.170)$$

Then, after the two integrations of Eq. (1.154) from x to $+\infty$ and using Eqs. (1.170) one may obtain the relation valid at any point of the solution in the system moving with velocity V:

$$V^2 = \frac{f(a\xi) - f(a\xi_0)}{\rho(\xi - \xi_0)} + \frac{a^2\left[2f'(a\xi\,\xi_{xx}(\xi) - f''(a\xi)\xi_x^2(\xi)\right]}{24\rho(\xi - \xi_0)}. \quad (1.171)$$

Functions f, ξ and their derivatives are taken at the points x, ξ.

We would like to find the relationship between the deformation maximum ξ_m (or minimum ξ_{min}) and phase speed V for material with normal ($f''(a\xi) > 0$) and abnormal ($f''(a\xi) < 0$) behaviors. Let us make a natural hypothesis that the solutions have a symmetrical bell-shape profile with a central maximum (minimum). In the maximum (minimum), $\xi_x(\xi = \xi_m) = 0$ ($\xi_x(\xi = \xi_{min}) = 0$). Using these necessary conditions for the extremes obtain from Eq. (1.171) for compression V_s and rarefaction $V_{s,r}$ solitons, correspondingly,

$$V_s^2 = \frac{f(a\xi_m) - f(a\xi_0)}{\rho(\xi_m - \xi_0)} + \frac{a^2 f'(a\xi_m)\xi_{xx}(\xi_m)}{12\rho(\xi_m - \xi_0)}, \quad (1.172a)$$

$$V_{s,r}^2 = \frac{f(a\xi_{min}) - f(a\xi_0)}{\rho(\xi_{min} - \xi_0)} + \frac{a^2 f'(a\xi_{min})\xi_{xx}(\xi_{min})}{12\rho(\xi_{min} - \xi_0)}. \quad (1.172b)$$

The additional equation for solitary waves can be obtained, based on the fact that they exist, only if the "potential energy" $W(z)$ in Eq. (1.162) has at least two extremes z_1 and z_2. Their relative order can be similar to y_1 and y_2 as shown for function $W(y)$ in Figure 1.39 for the case of power-law materials with $n > 1$ and $n < 1$.

A solitary wave solution corresponds to the total "energy" of the system equal to $W(z_1)$. The quantities z_1 and $z_m(z_{min})$ are connected with the strains in infinity ξ_0 and with the maximal ξ_m (or minimum ξ_{min}) strain in the soliton through the relation presented by Eq. (1.160).

There is another important strain ξ_1 determining the soliton phase speed. It corresponds to the extreme of $W(z)$ positioned between the initial and maximal (minimal) strains which is shown at point y_2 in Figure 1.39 for power-law materials. The necessary condition ($W_z = 0$) for the extreme of $W(z)$ to be in $z_1(\xi_0)$ and z_2 (ξ_1) follows straightforwardly from Eq. (1.162):

$$C = \rho V^2 \xi_0 - f\left(a\xi_0\right),$$
(1.173)

$$C = \rho V^2 \xi_1 - f\left(a\xi_1\right).$$
(1.174)

From Eqs. (1.173) and (1.174) we can get the additional to Eqs. (1.172) expression for solitary wave phase speed depending on the initial strain ξ_0 and the intermediate strain ξ_1:

$$V^2 = \frac{f(a\xi_1) - f(a\xi_0)}{\rho\left(\xi_1 - \xi_0\right)} = \tan \gamma.$$
(1.175)

Physically, strain ξ_1 corresponds to the maximal strain rate in the solitary wave, and Eq. (1.175) is valid for both compression and rarefaction solitary waves. The difference between them is in the relative positions of ξ_0, ξ_1 and ξ_m or ξ_{min}. For the compression solitary wave $\xi_0 < \xi_1 < \xi_m$ and for rarefaction solitary wave $\xi_0 > \xi_1 > \xi_{min}$. It is useful to emphasize that ξ_1 corresponds to the extreme of $W(z)$ intermediate between ξ_0 and ξ_m or between ξ_0 and ξ_{min}. Eq. (1.175) is compatible with one of the Eqs. (1.172) only for materials with some specific properties. For example, in the case of power-law materials, compression solitary waves are possible for normal behavior and rarefaction solitary waves for abnormal behavior. It will be shown that this is also valid for the general case.

1.11.3.1   Compression Solitary Waves in "Normal" Materials

Materials with normal behavior ($f'(a\xi) > 0$) represent the elastic hardening due to the strain increase: the more material deformed, the more it resists the strain increase. This is represented in Figure 1.42.

The angles γ and β are related, correspondingly, to the speed of a solitary wave with strain in the crest ξ_m (Eq. (1.175)) and to the first term in the left part of Eq. (1.172a) where

$$\frac{f\left(a\xi_m\right) - f\left(a\xi_0\right)}{\rho\left(\xi_m - \xi_0\right)} = \tan \beta.$$
(1.176)

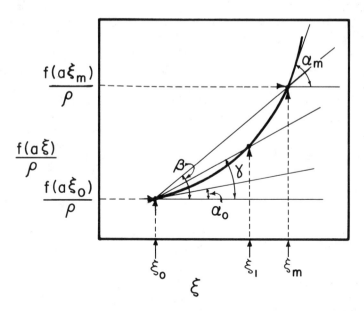

Figure 1.42. Graphical illustration of the relative speeds of sound, phase speeds of the compression solitary wave, and compression shock for materials with normal compressibility.

From the fact that $f''(a\xi) > 0$, obtain that for any intermediate point ξ_1 ($\xi_0 < \xi_1 < \xi_m$), $\tan \beta > \tan \gamma$. This simply reflects the general property of $f(\xi)$ for normal materials.

On the other hand, for the compression solitary wave we have $\xi_{xx}(\xi = \xi_m) < 0$, $\xi_0 < \xi_1 < \xi_m$. From Eqs. (1.172a) and (1.175) we obtain, in this case,

$$\rho V_s^2 < \tan \beta, \tag{1.177a}$$

$$\rho V_s^2 = \tan \gamma. \tag{1.177b}$$

The sound speeds corresponding to the state at infinity $c_0(\xi = \xi_0)$ and at the maximum $c_m(\xi = \xi_m)$ are represented by tangents of the angles α_0 and α_m in Figure 1.42:

$$c_0^2(\xi_0) = \frac{f'(a\xi_0)}{\rho} = \tan\alpha_0, \tag{1.178}$$

$$c_{0,m}^2(\xi_m) = \frac{f'(a\xi_m)}{\rho} = \tan\alpha_m. \tag{1.179}$$

The relations between V_s, c_0 and c_m are presented in graphical form as relations between γ, α_0, and α_m Figure 1.42. From Figure 1.42 and Eqs. (1.177) and (1.178), we obtain

$$V_s^2 < \tan \alpha_m = c_m^2. \tag{1.180}$$

Thus, phase speed V_s of the steady compression solitary wave for material with normal compressibility, if it exists, is always less than the sound speed $c_{0,m}(\xi_m)$ corresponding to maximal deformation in the wave. It is worth mentioning that sound speeds c_0 and $c_{0,m}$ are long wave characteristics, meanwhile, the essentially nonlinear wave Eq. (1.154), as we have seen for the Hertz chain (Section 1.3), can have soliton solutions with the width comparable to the distance between particles.

It is interesting to find the relation between the initial sound speed c_0 and phase velocity for the compression solitary wave V_s. From the comparison of the angles α_0 and γ representing the corresponding speeds, it is evident that

$$V_s > c_0. \tag{1.181}$$

For strong compression solitary waves ($\xi_m >> \xi_0$) from balancing main nonlinear terms with nonlinear dispersive terms in the right part (term in square brackets) of Eq. (1.154) we obtain that

$$L_s = B_1 a, \tag{1.182}$$

where constant B_1 does not depend on solitary wave amplitude, but is influenced by the shape of function $f(a\xi)$ (see, e.g., Eq. (1.130) for power-law materials with $n > 1$). This means that the spatial size of the strongly nonlinear solitary wave does not depend on its amplitude and is proportional to the characteristic size of the system in the considered continuum limit.

For the phase speed of the strong solitary wave (ξ_m, $\xi_1 >> \xi_0$) the following dependence of phase speed on amplitude may be obtained from Eq. (1.175):

$$V_s = \left(\frac{f(a\xi_1)}{\rho \xi_1} \right)^{1/2}, \tag{1.183}$$

where the intermediate strain $\xi_1 < \xi_m$ (Fig. 1.42) corresponds to the extreme of $W(y)$ positioned between maximal strain and strain at infinity. This is shown as point y_2 in Figure 1.39(a) for power-law material with normal compressibility. For quick order of magnitude estimations of the soliton speed V_s, which may be useful for the planning of experiments, the rough relation between ξ_1 and ξ_m ($\xi_1 \sim \xi_m/2$) may be used due to the approximate symmetry of potentials $W(y)$ in many cases (see Figures 1.3 and 1.39(a)). For example, Eq. (1.183) may be used to check that the neglected nonlinear term, corresponding to the convective derivative, is much smaller than kept terms. For comparison, see exact Eq. (1.129) for power-law material.

It is important to mention that conditions for the existence of strongly nonlinear solitary waves in continuum approximation, expressed by Eq. (1.154), are identical to the conditions of solitary wave existence in discrete chains described by Eq. (1.142) (Friesecke and Wattis [1994]). This helps to understand why continuum approximation, being a trancated version of equations for a discrete chain, still supports the same solitary wave as was shown in the example of the Hertz interaction law.

These strongly nonlinear solitary waves are probably the shortest possible stationary waves in discrete systems and represent the basic carrying tones for wave disturbances, just like sound waves in linear systems.

This may be appropriate based on their unique properties and taking into account that they are supported by a quite general interaction law between discrete elements and, therefore, can be expected in many physically different systems to separate them into a special class and provide the corresponding name. This will emphasize the necessity to develop a new wave dynamics for these systems when the amplitudes of wave disturbances are in a strongly nonlinear range. One suggestion can be based on the short version of words "carrying tone," in Russian "*nesushii ton*"–*neston*.

1.11.3.2   Compression Shock Waves in "Normal" Materials

A wave equation should include some additional, for example, viscous term, if material exhibits dissipative behavior. In this case, Eq. (1.154) is modified:

$$\rho\,\xi_{tt} = \left\{ f + \frac{a^2}{24}\left[2\rho\xi_{tt} + f''\xi_x^2\right] - \eta\xi_t \right\}_{xx}. \tag{1.184}$$

The double integration of this equation for the steady wave $\xi(x - Dt)$ with the boundary conditions (Eq. (1.170)), results in the expression

$$D_s^2\rho\big(\xi - \xi_0\big) = f(\xi) - f(\xi_0) + \frac{a^2\left[2f'(\xi)\,\xi_{xx}(\xi) + f''(\xi)\,\xi_x^2(\xi)\right]}{24} + \eta V\xi_x. \tag{1.185}$$

The change of variables (Eq. (1.160)) give the equation for the nonlinear oscillator with damping term

$$\frac{a^2}{12}z_{xx} = -\frac{\partial W(z)}{\partial z} - \eta D_s\Psi(z)z_x, \tag{1.186}$$

$$-\frac{\partial W(z)}{\partial z} = \rho D_s^2\frac{M(z)}{P(z)} - \frac{K(z)}{P(z)} - \frac{C}{P(z)}, \qquad P(z) \neq 0,$$

$$\Psi(z) = \frac{1}{Bf'\{aM(z)\}}.$$

We can use the method of "potential energy" $W(z)$, as in the case of solitary solution. The movement of the mass with initial coordinate z_1 and total energy $W(z_1)$ will result after the few damped oscillations (or monotonically, depending on the viscosity coefficient) in the final position z_2 (ξ_1). At that point, z_x, $z_{xx} = 0$, and, correspondingly, $\xi_x = \xi_{xx} = 0$. From Eq. (1.185) we can get

$$D_s^2 = \frac{f(a\xi_1) - f(a\xi_0)}{\rho(\xi_1 - \xi_0)} = \tan \gamma. \tag{1.187}$$

The graphical representation of this situation is also depicted in Figure 1.42. It is clear that the value of $\tan \gamma = D_s$ in Figure 1.42 is equal to the speed of the shock wave with final strain ξ_1.

According to the "potential energy" analogy, the steady strain ξ_1 behind the shock wave is smaller than the maximum amplitude in the head of the wave, if dissipation cannot prevent oscillating behavior of the shock profile. This is evident from the "potential energy" method. Also, the amplitude of the strain in the leading oscillation in the stationary shock wave $z_{sh,m}$ ($\xi_{sh,m}$) is smaller than the soliton amplitude z_m (ξ_m) at the same phase speed.

Let us consider the relation between soliton and shock parameters from another point of view, if the steady strains behind the shock wave and in the soliton crest are equal to, say, ξ_m. Then the corresponding phase speeds D_s and V_s are different. By comparing $\tan \beta (D_s)$ and $\tan \gamma (V_s)$ (the latter corresponding, in this case, to the intermediate state in the solitary wave ξ_1 between ξ_0 and ξ_m) it is easy to see from the plot in Figure 1.42 that

$$D_s(\xi_0, \xi_m) > V_s(\xi_0, \xi_m). \tag{1.188}$$

The state behind the shock wave can be considered as a structurally insensitive parameter and can be obtained by a continuum approach. Thus we can conclude that material in a solitary wave (structurally sensitive phenomenon) behaves as "softer" material in comparison with its behavior in a shock wave.

If dissipation is very weak, the shock profile can approach the oscillatory form, as is evident from Eq. (1.186). Comparison of Eq. (1.162) for a solitary wave with Eq. (1.186) for a shock wave, will help to understand that, in this case, the first oscillation can be close to the soliton type. For the shock to be able to have the speed $D_s(\xi_m)$, the leading soliton amplitude should exceed the final value ξ_m, as is clear from the inequality represented by Eq. (1.188).

It should be emphasized that the properties of stationary solutions of the strongly nonlinear wave equation, corresponding to a discrete system with a general interaction law between particles, are based on many assumptions incorporated into the derivation of this equation. They include the replacement of real particles in the system by rigid masses and neglecting the convective derivatives. For a specific system these assumptions should be checked as was done for the case of elastic particles interacting with the Hertz law.

1.11.3.3   Solitary Waves in "Abnormal" Materials

Materials with abnormal behavior $(f''(a\xi) < 0)$ represent the elastic softening due to the strain increase: the more material deformed, the less it resisted the strain increase. This is represented in Fig. 1.43 where ξ_0 is the strain at infinity for the rarefaction wave. In this case, the angles γ and β, representing the speed of the solitary wave with minimum strain ξ_{min} (Eq. (1.175)) and the first term in the left part of Eq. (1.172b) are shown in Figure 1.43:

$$\frac{f(a\xi_{min}) - f(a\xi_0)}{\rho(\xi_{min} - \xi_0)} = \tan \beta. \tag{1.189}$$

From the fact that $f''(a\xi) < 0$ we obtain that for any intermediate point ξ_1 $(\xi_0 > \xi_1 > \xi_{min})$ $\tan \beta > \tan \gamma$. It is the consequence of the convex shape of $f(\xi)$ for materials with abnormal behavior. This is opposite in the case of a compression solitary wave in materials with normal behavior $(f''(a\xi) > 0)$.

On the other hand, for the rarefaction solitary wave with deformation minimum at the soliton crest, we have $\xi_{xx}(\xi = \xi_{min}) > 0$, $\xi_0 > \xi_1 > \xi_{min}$. From Eqs. (1.172b) and (1.189), we obtain, in this case,

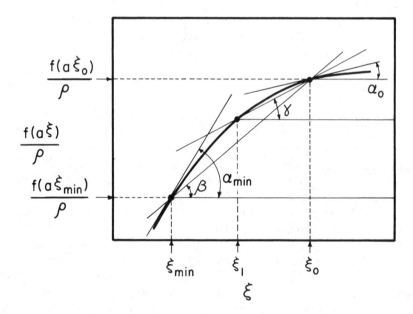

Figure 1.43. Graphical illustration of the relative speeds of sound, phase speeds of the rarefaction solitary wave, and rarefaction shock for abnormal materials.

$$\rho V_{s,r}^2 < \tan \beta, \tag{1.190a}$$

$$\rho V_{s,r}^2 = \tan \gamma. \tag{1.190b}$$

Equations (1.190) are compatible only with the convex shape of function $f(a\xi)$ ($f''(a\xi) < 0$). It is clear that the concave shape of function $f(a\xi)$ (normal behavior, $f''(a\xi) > 0$) is not compatible with the conditions for the rarefaction solitary wave represented by Eqs. (1.190).

Tangents of angles α_0 and α_{min} representing the sound speeds, correspondingly, in the state in infinity c_0 ($\xi = \xi_0$) and in the solitary wave crest c_{min} ($\xi = \xi_{min}$), are shown in Figure 1.43:

$$c_0^2(\xi_0) = \frac{f'(a\xi_0)}{\rho} = \tan \alpha_0, \tag{1.191}$$

$$c_{0,min}^2(\xi_{min}) = \frac{f'(a\xi_{min})}{\rho} = \tan \alpha_{min}. \tag{1.192}$$

The relations between $V_{s,r}$, c_0, and $c_{0,min}$ are presented in graphical form as the relations between γ, α_0, and α_{min} in Figure 1.43. From Figure 1.43 and Eqs. (1.190), (1.191), and (1.192), we obtain

$$V_{s,r}^2 < \tan \alpha_{min} = c_{min}^2. \tag{1.193}$$

Thus, the phase speed $V_{s,r}$ of the rarefaction solitary wave for material with abnormal compressibility, if it exists, is always less than the sound speed corresponding to minimum deformation in the wave crest. This is similar to the case of the compression solitary wave.

From the comparison of the angles α_0 and γ, representing the corresponding sound speed c_0 at strain ξ_0 and the solitary wave speed $V_{s,r}$ with strain in the minimum ξ_{min} ($\xi_0 > \xi_1 > \xi_{min}$), it is evident (Fig. 1.43) that

$$V_{s,r} > c_0. \tag{1.194}$$

This condition is similar to the supersonic nature of the compression solitary wave in materials with normal compressibility (Eq. (1.181)).

For a strong rarefaction solitary wave ($\xi_{min} \ll \xi_0$) from a balancing main nonlinear term with nonlinear dispersive terms in the right part (term in square brackets) of Eq. (1.154), we obtain that

$$L_{s,r} = B_3 a, \tag{1.195}$$

where the constant B_3 does not depend on the solitary wave amplitude, but is influenced by the shape of function $f(a\xi)$. So the spatial size of a strongly nonlinear rarefaction solitary wave does not depend on its amplitude and is proportional to the characteristic size of the system.

For the phase speed of the nonlinear solitary wave, the following dependence of the phase speed on amplitude may be found from Eq. (1.175):

$$V_{s,r} = \left(\frac{f(a\xi_1) - f(a\xi_0)}{\rho(\xi_1 - \xi_0)} \right)^{1/2}. \tag{1.196}$$

The intermediate strain ξ_1 in Eq. (1.196) ($\xi_1 < \xi_0$) corresponds to the extreme of $W(y)$ positioned between the initial and minimal strains, which is shown as point y_2 in Figures 1.39(b) and 1.41 for power-law materials with abnormal compressibility. For a quick order of magnitude estimation of the soliton speed $V_{s,r}$ for a strongly nonlinear regime, which may be useful for experiment planning, the rough relation between ξ_1 and ξ_0 ($\xi_1 \sim \xi_0/2$) may be used, due to the approximate symmetry of potentials $W(y)$ in many cases (see Figs. 1.39(b) and 1.41). This estimation may be useful to check that the nonlinear term corresponding to the convective derivative could be neglected at a given range of ξ_{min}, ξ_0 or particle velocity. For comparison, see the exact equation (Eq. (1.138)) for power-law material with $0 < n < 1$ when $\xi_{min} \ll \xi_0, \xi_1$.

1.11.3.4   Rarefaction Shocks: Comparison with Solitons

Stationary compression shock waves in materials with abnormal behavior do not exist. Instead, rarefaction shock waves may propagate if the dissipation term is included. Shock speed $D_{s,r} = \tan \gamma$ (Fig. 1.43) satisfies the equation similar to Eq. (1.187) where ξ_1 is the equilibrium strain behind the shock.

The steady strain ξ_1 behind the rarefaction shock wave, is larger than the minimum amplitude in the head of the wave, if dissipation cannot prevent oscillating behavior of the shock profile. This is evident from the shape of the effective potential energy for materials with abnormal behavior (see, e.g., Fig. 1.41, ξ_1 corresponds to the point y_2 for curve III).

It is interesting to note from Figure 1.41 that in materials with zero tensile strength, stationary rarefaction shock waves with some amplitude can only exist at a high enough dissipation rate, preventing the system from negative strains and destruction.

Also, the minimum of the strain, in the leading oscillation in the stationary rarefaction shock wave $z_{s,r,min}$ ($\xi_{sh,m}$) is larger than the minimum strain in the rarefaction soliton z_{min} (ξ_{min}) at the same phase speed as is evident from the shape of the effective potential energy (see, e.g., Fig. 1.41).

On the other hand, if the steady strains behind the shock wave and of the soliton maximum are equal, then the corresponding phase speeds $D_{s,r}$ and $V_{s,r}$ are different. It is possible to see this from the plot in Figure 1.43 by comparing angle $\beta(D_{s,r})$ corresponding to the steady strain ξ_{min} behind the shock and angle γ, in this case, corresponding to the intermediate state in the soliton between ξ_0

and ξ_{min}:

$$D_{s,r}(\xi_0, \xi_{min}) > V_{s,r}(\xi_0, \xi_{min}) \tag{1.197}$$

The state behind the rarefaction shock wave can be considered as a structurally insensitive parameter, and can be obtained in the continuum approach. That is why the material with abnormal compressibility in the soliton (structurally sensitive phenomenon) is softer in comparison with its behavior in the rarefaction shock wave.

1.12   Conclusion

Granular materials, even in the simplest one-dimensional form, as a chain of particles interacting according to a strongly nonlinear law (like the Hertz law for elastic spheres) represent a class of medium where new wave dynamics should be developed. Materials that may have the same type of wave behavior include unstressed chains of particles or molecules in transverse motion, systems such as jointed rocks, gas–liquid mixtures, and other examples.

The unusual feature of these materials is the negligible linear range of the interaction force resulting in zero or very small sound speed ("sonic vacuum"). In this case a new wave equation is based on only one small parameter—ratio of the particle size to the wave length. Such equation is more general than weakly nonlinear Korteweg–de Vries equation which is based on two small parameters (for a 1-D chain they are the ratio of wave amplitude to the initial stress and the ratio of the particle sizes to the wavelegth). This more general wave equation has stationary solutions with unique properties. The similar situation can be expected in other strongly nonlinear cases of wave motion, for example, like waves in water channels, electrical lines, plasma.

Periodic waves, compression solitary, and shock waves for strongly nonlinear behavior are qualitatively different from the weakly nonlinear case, and cannot be obtained in the frame of the latter approach. They have such unique features, i.e., the spatial extent of compression solitons and their shape do not depend on amplitude, initial sound speed does not determine the soliton parameters, speed has weak dependence on amplitude and the initial impulse is split into a soliton train quickly on very short distances from the entrance.

The existence of these new types of wave disturbances is confirmed by different authors through an analytical approach, numerical calculations, and experiments in 1-D granular materials.

It was shown that changes in the microstructural parameters of these materials, without changing their quasistatic macroscopic behavior, may result in a modified dynamic behavior, and may be used for structural probing by measuring the properties of nonlinear waves.

In the case opposite the "sonic vacuum," strongly nonlinear materials with abnormal behavior, there are periodic waves, rarefaction shock waves, and solitons. They are supersonic, relative to the initial state. No stationary compression waves are allowed in such materials.

The strongly nonliner mode of wave propagation may find applications not only in areas traditional for granular materials, but also in other areas like a delay lines, or signal processing (generating) devices which allow a strong modification of properties by external electrical fields or in an other areas. These waves also represent a very compact way of signal propagation in transmitting lines where solitons, with minimal possible space width, may be effective information carriers. The very sharp shock front in these strongly nonlinear dispersive systems may also be useful for the development of fast-rise-time electromagnetic nonlinear transmission lines. Structured metamaterials in the form of multidimensional periodic structures with strongly nonlinear response (for example, acoustic or electromagnetic) may be designed to use unique properties of waves described in this chapter.

Another possible application may include a neurophysiology where the ultimate response in a given point is affected by the simulation of other neighboring points, and where the corresponding wave equations may be strongly nonlinear, at least for intensive excitations.

Problems

1. Find the expression for the momentum and energy of a "sonic vacuum" soliton as a function of phase speed V_s using the dependence of particle velocity on a coordinate (Eq. (1.39b)).

2. Prove that the relation between the phase speed of a soliton V_s and the strain in its maximum ξ_m (Eq. (1.37)), can be reduced to Eq. (1.43) for a "sonic vacuum."

3. Find the phase speed of a "sonic vacuum" soliton V_s for the system of ceramic spheres with the density $\rho_0 = 3.94$ g/cm^3, Young's modulus of the ceramic material $E = 370$ GPa, Poisson's ratio $v = 0.26$, and particle velocity in the maximum $v_m = 1$ m/s.

4. Write in quadrature the shape of the soliton-type stationary solutions of Eq. (1.20) for positive values of constant C.

5. Obtain the KdV equation from the essentially nonlinear wave equation (1.20) in the case when maximum strain in the wave ξ_m is close to the initial strain ξ_0.

6. Demonstrate that in Eq. (1.34) phase speed $V_p \to c_0(\xi_2)$ is the case if ξ_{min} ~ ξ_{max} ~ ξ_2.

7. Show that the soliton speed V_s from Eq. (1.37) approaches the sound speed c_0 in the case of $\xi_m \to \xi_0$, and in the next approximation the relation for the KdV soliton between V_s and the strain amplitude (Eq. (1.10)) can be recovered.

8. Make a variable replacement in Eq. (1.104) for the strongly nonlinear case of the power-law materials similar to the partial case for the Hertz law (Eqs.

(1.69), (1.70)):

$$\zeta_i = \frac{\dot{u}_i}{v_0}, \qquad \tau = t\left(A_n v_0^{n-1}\right)^{1/(n+1)}.$$

Prove that if stationary solitary waves exist, then their profiles with different amplitudes are identical in the corresponding dimensionless coordinates. Explain why this results in independence of their space width on amplitude and in the dependence of phase speed on particle velocity described by Eq. (1.129).

9. Find the breathers of the form $\xi = Q(t)Z(x)$ for Eq. (1.106) using the separation of the variables in the case of $n = 3$ (Rosenau [1994]).

10. Find the following relation between the final strain and speed in a stationary compression shock for the uncompressed power-law ($n > 1$) 1-D granular material:

$$V_{sh} = c_n \xi_c^{(n-1)/2}.$$

Use the fact that the $W(y)$ has a minimum at $y = 1$ if $C = 0$.

11. Obtain the first integrals of momentum and energy conservation from the general nonlinear Eq. (1.154).

12. Find the sound speed $c_0(\xi_m)$ corresponding to a maximum of the strain (ξ_m) in the "sonic vacuum" soliton. Compare with the soliton phase speed V_s. Explain why solitons exist despite that $c_0(\xi_m) > V_s$.

13. Find the equation for plane nonlinear waves propagating in a 2-D sphere arrangement. Consider the equation of motion for particles in a plane, with closest packing density in the direction perpendicular to this plane.

14. Evaluate the long-wave sound speed for the granular material composed from glass beads with diameter 1 mm, density 3.94 g/cc, Young's modulus $E = 370$ GPa, Poisson's ratio $\nu = 0.26$, and pressure $= 10$ atm. Use Eqs. (1.97) and (1.98) for sound speed, calculate the typical forces between beads in a pressure pulse supposing dense random packing of particles.

15. Prove that Eqs. (1.20) and (1.26a) are formally equivalent within the same order of the error terms.

16. Obtain the inequality $V_{s,r} < c_{min}(\xi_{min})$ for "abnormal" power-law materials analytically using Eqs. (1.134) and (1.136).

17. Derive the variational form of Eq. (1.154) for the general interaction law between the particles, and obtain the corresponding conservation law using Eq. (1.106).

18. Find the following relation between longitudinal sound speed in the material of granules c_L, particle velocity in the soliton crest v_m, and the soliton speed in the 1-D chain V_s:

$$V_s = \left(\frac{16}{25}\right)^{1/5} \left(\frac{2(1-2\nu)}{\pi(1-\nu)^2}\right)^{2/5} c_L \left(\frac{v_m}{c_L}\right)^{1/5}.$$

19. Prove that Eq. (1. 104) with $n = 3$ has a partial solution in the form $u_i = g(i)f(t)$ with $f(t) = A + Bt$, $g(i) = \alpha + i\beta$, where A, B, α, and β are constants (Gavrilyuk and Serre [1995]). Find the equation for function $g(i)$.

20. Derive an equation for a strongly nonlinear soliton in the power-law material (Eq. (1.129)) based on Eq. (1.183) for the general interaction law and the relation between ξ_1 and ξ_{max} from the effective "potential energy" $W(y)$ at $C = 0$ (Eq. (1.115)).

21. Derive the following equations for sound speed c_0, speed V_s, and width $L_{1+\psi}$ for compression solitons in power-law materials with exponent $n = 1 + \psi$, where $\psi \ll 1$ (Nesterenko 1992c):

$$c_0 = c_1 \, \xi_0^{\psi/2}, \quad V_s = c_1 \, \xi_{max}^{\psi/2}, \quad L_{1+\psi} = \frac{\pi a}{\sqrt{3}\,\psi}.$$

22. Find a criterion for the existence of an elastic precursor along force chains ahead of a compaction wave in uncompressed power-law granular materials (similar to Eq. (1.103)).

23. Find the ratio of a dropped convective derivative and a left partial derivative of particle velocity with respect to time in Eq. (1.20) for the solitary solution described by one hump of Eq. (1. 39b).

24. Find the equation for the net displacement of particles caused by a passage of a solitary wave decribed by Eq. (1.39a).

25. Find the effective mass for soliton (Eq. (1.132)) in the power-law chain with $n = 3/2$, qualitatively explain the dependence of the effective mass on exponent n.

26. Explain the origin of highly nonlinear behavior of transverse waves in the system of heavy particles attached to the linear elastic fiber with negligible mass (see Fig. 1.40).

27. Show that the the number of solitons n_s should be scaled as

$$n_s \sim T_i c^{4/5} \upsilon_{mi}^{1/5} a^{-1},$$

where T_i is the duration of external impulse, υ_{mi} is the maximal velocity of the boundary, a is a particle diameter and c is constant (Eq. (1.20)). Use the equation for soliton duration.

References

Albert, J.P., and Bona, J.L. (1991) Comparisons Between Model Equations for Long Waves. *Journal of Nonlinear Science*, **1**, pp. 345374.

Anderson, G.D. (1964) Ph.D. thesis, Washington State University, Pullman.

Apalikov, Yu.I., Bronovskii G.A., Konon Yu.A., et al. (1983) Support for Explosive Cladding of Metallic Surfaces. Patent USSR (A.C. 330702A SSSR, MKI B23K20/08), *Discoveries, Inventions*, no. 24, p. 200 (in Russian).

Asano, N. (1974) Wave Propagation in Non-Uniform Media. *Prog. Theor. Phys.*, Supplement No. 55.

Balk, A.M., Cherkaev, A.V., and Slepyan, L.I. (2001) Dynamics of Chains with Non-Monotone Stress-Strain Relations. II. Nonlinear Waves and Waves of Phase Transition. *J. Mech. Phys. Solids*, **49**, pp. 149–171.

Bardenhagen, S., and Triantafyllidis, N. (1994) Derivation of Higher Order Gradient Continuum Theories in 2,3-D Non-Linear Elasticity from Periodic Lattice Models. *J. Mech. Phys. Solids*, **42**, no. 1, pp. 111–139.

Bardenhagen, S.G., and Brackbill, J.U. (1998) Dynamic Stress Bridging in Granular Material. *J. Applied Physics*, **83**, no. 11, pp. 5732-5740.

Belyaeva, I.Yu., Zaitsev, V.Yu., and Ostrovskii, L.A. (1993) Nonlinear Acoustoelastic Properties of Granular Media. *Acoustical Physics,* **39**, no. 1, pp. 11–15.

Belyaeva, I.Yu., Zaitsev, V.Yu., and Timanin, E.M. (1994) Experimental Study of Nonlinear Elastic Properties of Granular Media with Nonideal Packing. *Acoustical Physics*, **40**, no. 6, pp. 789–793.

Benjamin, T.B., Bona, J.L., and Mahony, J.J. (1972) Model Equations for Long Waves in Nonlinear Dispersive Systems. *Philosophical Transactions of the Royal Society of London A. Mathematical and Physical Sciences*, **272**, no. 1220, pp. 47–78.

Benson, D.J. (1994) An Analysis by Direct Numerical Simulation of the Effects of Particle Morphology on the Shock Compression of Copper Powder. *Modelling Simul. Mater. Sci. Eng.* **2**, pp. 535–550.

Bernoulli, D. (1738) Theoremata de Oscillationibus Corporum Filo Flexili Connexorum et Catenae Verticaliter Suspensae. *Comm. Acad. Sci. Petrop.*, **6**, pp. 108–122.
English translation: in *The Evolution of Dynamics: Vibration Theory from 1687 to 1742.* 1981, Springer-Verlag, New York, pp. 156–176.

Bernoulli, D. (1741) Letter of 28 January 1741, from *Correspondance Mathematique et Physique de Quelques Celebres Geometres du XVIII eme Siecle,* vol. 2, St. Petersburg, 1843. Facs. rep. Johnson, New York, 1968.
English translation: In: Cannon, J.T., and Dostrovsky, S. (1981) *The Evolution of Dynamics: Vibration Theory from 1687 to 1742.* Springer-Verlag, New York, p. 46.

Bland, D.R. (1969) *Nonlinear Dynamic Elasticity.* Blaisdell, New York.

Bogdanov A.N., and Skvortsov A.T. (1992) Nonlinear Elastic Waves in a Granular Medium. *Journal de Physique*, Coll. C1, **2**, pp. C1-779–C1-782.

Bona, J.L., Pritchard, W.G., and Scott, L.R. (1983) A Comparison of Solutions of Two Model Equations for Long Waves. Proceedings of the AMS–SIAM Conference on Fluid Dynamical Problems in Astrophysics and Geophysics, Chicago, July 1981. *Lectures in Applied Mathematics,* **20**, (Edited by N. Lebovitz). American Mathematical Society, Providence, RI, pp. 235–267.

Boutreux, T., Raphael, E., and de Gennes, P.G. (1997) Propagation of a Pressure Step in a Granular Material; The Role of Wall Friction. *Phys. Rev. B,* **55**, no. 5, pp. 5759–5773.

Brandt, H. (1955) A Study of the Speed of Sound in Porous Granular Media. *ASME Journal of Applied Mechanics*, **22**, pp. 479–486.

Chapman, S. (1960) Misconception Concerning the Dynamics of the Impact Ball Apparatus. *Am. J. Phys.*, **28**, no. 8, pp. 705–711.

Chatterjee, A. (1999) An Asymptotic Solution for Solitary Waves in a Chain of Elastic Spheres. *Phys. Rev. B,* **59,** no. 5, pp. 5912–5919.

Chen, P.J. (1973) Growth and Decay of Waves in Solids. In: *Encyclopedia of Physics* (Edited by S. Fluegge and C. Truesdell), Vol. **VI a/3.** Springer-Verlag, Berlin.

Coleman, B.D. (1971a) On Retardation Theorems. *Archive for Rational Mechanics and Analysis,* **43,** pp. 1–23.

Coleman, B.D. (1971b) A Mathematical Theory of Lateral Sensory Inhibition. *Archive for Rational Mechanics and Analysis,* **43,** pp. 79–100.

Coleman, B.D. (1983) Necking and Drawing in Polymeric Fibers Under Tension. *Archive for Rational Mechanics and Analysis,* **83,** pp. 115–137.

Coleman, B.D. (1985) On the Cold Drawing of Polymers. *Comp. & Maths. With Appls.,* **11,** nos. 1–3, pp. 35–65.

Coste, C., Falcon, E., and Fauve, S. (1997) Solitary Waves in a Chain of Beads Under Hertz Contact. *Physical Review E,* **56,** no. 5, pp. 6104–6117.

Coste, C., and Gilles, B. (1999) On the Validity of Hertz Contact Law for Granular Material Acoustics. *The European Physical Journal B,* **7,** pp. 155-168.

Cristescu, N. (1967) *Dynamic Plasticity.* North-Holland, Amsterdam.

Dash, P.C., and Patnaik, K. (1981) Solitons in Nonlinear Diatomic Lattices. *Progress in Theoretical Physics,* **65,** no. 5, pp. 526–1541.

de Gennes, P.G. (1996) Static Compression of a Granular Medium: The 'Soft" Shell Model. *Europhys. Lett.* **5,** no. 2, pp. 145–149.

Deresiewicz, H. (1958) Stress–Strain Relations for a Simple Model of a Granular Medium. *ASME Journal of Applied Mechanics,* **25,** pp. 402–406.

Digby, P.J. (1981) The Effective Elastic Moduli of Porous Granular Rocks. *ASME Journal of Applied Mechanics,* **48,** pp. 803–808.

Dinda, P.T., and Remoissenet, M. (1999) Breather Compactons in Nonlinear Klein–Gordon Systems. *Phys. Rev. E,* **60,** no. 5, pp. 6218–6221.

Druzhinin, O.A., and Ostrovskii, L.A. (1991) Solitons in Discrete Lattices. *Physics Letters A,* **160,** pp. 357–362.

Duffy, J., and Mindlin, R.D. (1957) Stress–Strain Relations and Vibrations of a Granular Medium. *ASME Journal of Applied Mechanics,* **24,** pp. 585–593.

Dusuel, S., Michaux, P., and Remoissenet, M. (1998) From Kinks to Compactonlike Kinks. *Phys. Rev. E,* **57,** no. 2, pp. 2320–2326.

Duvall, G.E., Manvi, R., and Lowell, S.C. (1969) Steady Shock Profiles in a One-Dimensional Lattice. *Journal of Applied Physics,* **40,** no. 9, pp. 3771–3775.

Falcon, E. (1997) Comportements Dynamiques Associes au Contact de Hertz: Processus Collectifs de Collision et Propagation D'Ondes Solitaires Dans les Milieux Granulaires. *These 191–97, Diplome de Doctorat.* L'Ecole Normale Superieure de Lyon.

Falcon, E., Laroche, C., Fauve, S., and Coste, C. (1998a) Behavior of One Inelastic Ball Bouncing Repeatedly off the Ground. *The European Physical Journal B,* **3,** pp. 45–57.

Falcon, E., Laroche, C., Fauve, S., and Coste, C. (1998b) Collision of a 1-D Column of Beads with a Wall. *The European Physical Journal B,* **5,** pp. 111–131.

Fermi, E., Pasta, J.R., and Ulam, S.M. (1965) Studies of Non-Linear Problems. Technical Report. LA-1940, Los Alamos National Laboratory. Reprinted in *Collected Works of E. Fermi,* vol. **II.** University of Chicago Press, pp. 978–988.

Filippov, A.T. (2000) *The Versatile Soliton.* Birkhauser, Boston.

Friesecke, G., and Wattis, J.A.D. (1994) Existence Theorem for Solitary Waves on Lattices. *Communications in Mathematical Physics*, **161**, no. 2, pp. 391–418.

Gassmann, F. (1951) Elastic Waves Through a Packing of Spheres. *Geophysics*, **16**, pp. 673-685.

Gavrilyuk, S.L. (1989) Modulation Equations for a Mixture of Gas Bubbles in an Incompressible Liquid. *Prikl. Mekh. Tekh. Fiz.*, **30**, no. 2, pp. 86–92 (in Russian).
English translation: *Journal of Applied Mechanics and Technical Physics (JAM)* 1989, September, pp. 247–253.

Gavrilyuk, S.L., and Nesterenko, V.F. (1993) Stability of Periodic Excitations for One Model of "Sonic Vacuum." *Prikl. Mekh. Tekh. Fiz.*, **34**, no. 6, pp. 45–48 (in Russian).
English translation: *Journal of Applied Mechanics and Technical Physics(JAM)* no. 6, 1993, pp. 784-787.

Gavrilyuk, S.L., and Serre, D. (1995) A Model of a Plug-Chain System Near the Thermodynamic Critical Point: Connection with the Korteweg Theory of Capillarity and Modulation Equations. *Proceedings of IUTAM Symposium on Waves in Liquid/Gas and Liquid/Vapor Two-Phase Systems* (Edited S. Morioka and L.van Wijngaarden), Kyoto, Japan, 1994. Kluwer Academic, Dodrecht, pp. 419–428.

Gavrilyuk, S.L., and Shugrin, S.M. (1996) Media with Equations of State that Depend on Derivatives. *Prikl. Mekh. Tekh. Fiz.*, **37**, no. 2, pp. 35–49 (in Russian).
English translation: *Journal of Applied Mechanics and Technical Physics (JAM)* 1996, **37**, no. 2, pp. 177–189.

Gilles, B., and Coste, C. (2001) Nonlinear Elasticity of a 2-D Regular Array of Beads. *Powders and Grains* (submitted).

Goddard, J.D. (1990) Nonlinear Elasticity and Pressure-Dependent Wave Speeds in Granular Media. *Proc. R. Soc. London A*, **430**, pp. 105–131.

Godunov, S.K. (1961) An Interesting Class of Quasiliner Systems. *Dokl. Akad. Nauk SSSR*, **139**, no. 3, pp. 521–523 (in Russian).

Goldsmith, W. (1960) *Impact*. Arnold, London.

Goodman, M.A., and Cowin S.C. (1972) A Continuum Theory for Granular Materials. *Archive for Rational Mechanics and Analysis*, **44**, pp. 249–266.

Gradshteyn, I.S., and Ryzhik I.M. (1980) *Table of Integrals, Series, and Products*. Academic Press, San Diego–Toronto, Harcourt Brace Jovanovich, p. 131.

Groshkov, A.L., Kalimulin, R.R., Shalashov, G.M., and Shemagin, V.A. (1990) Nonlinear Interbore-Hole Probing with Modulation of Acoustic Wave by Seismic Fields. *Izvestiya AN SSSR*, **313**, no. 1, p. 63–65 (in Russian).

Hammack, J.L. (1973) A Note on Tsunamis: Their Generation and Propagation in an Ocean of Uniform Depth. *J. Fluid Mech.*, **60**, pp. 769799.

Hammack, J. L., and Segur, H. (1974) The Korteveg–de Vries Equation and Water Waves. Part 2. Comparison with Experiments. *J. Fluid Mech.*, **65**, pp. 289–314.

Hara, G. (1935) Theorie der Akustischen Schwingungsausbreitung in Gekornten Substanzen und Experimentelle Untersuchungen an Kohlepulver. *Elektrische Nachrichtentechnik*, **12**, pp. 191–200.

Hascoet, E, Herrmann, H.J., and Loreto, V. (1999) Shock Propagation in a Granular Chain. *Phys. Rev. E*, **59**, no. 3, pp. 3202–3206.

Hascoet, E., and Herrmann, H.J. (2000) Shocks in Non-Loaded Bead Chains With Impurities. *The European Physical Journal* B, **14**, pp.183-190.

Herrmann, F., and Schmalzle, P. (1981) Simple Explanation of a Well-Known Collision Experiment. *Am. J. Phys.,* **49**, no. 8, pp. 761–764.

Herrmann, F., and Seitz, P. (1982) How Does the Ball-Chain Work? *Am. J. Phys.,* **50**, no. 11, pp. 977–981.

Herrmann, H.J., Stauffer, D., and Roux, S. (1987) Violation of Linear Elasticity due to Randomness. *Europhys. Lett.* **3**, no. 3, pp. 265–267.

Hill, T.G., and Knopoff, L. (1980) Propagation of Shock Waves in One-Dimensional Crystal Lattices. *Journal of Geophysical Research,* **85**, no. B12, pp. 7025–7030.

Hinch E.J., and Saint-Jean S. (1999) The Fragmentation of a Line of Balls by an Impact. *Proc. R. Soc. Lond. A,* **455**, pp. 3201–3220.

Hong, J., Ji, J.-Y., and Kim, H., (1999) Power Laws in Nonlinear Chain under Gravity. *Phys. Rev. Lett.,* **82**, no. 15, pp. 3058–3061.

Iida, K. (1939) Velocity of Elastic Waves in a Granular Substance. *Bulletin of the Earthquake Research Institute,* Japan, **17**, pp. 783–808.

Ipatiev, A.S., Kosenkov, A.P., Nesterenko, V.F., Meshcheriakov, Y.P., and Afanasenko, S.I. (1986) Experimental Investigation of Fracture Characteristics of Metal Plates with Semicylindrical Grooves at Explosive Loading with Contact Surface Charges. In: *Proc. IX Intern. Conf. on High Energy Rate Fabrication* (Edited by V.F. Nesterenko and I.V. Yakovlev), USSR Academy of Sciences, Siberian Division, Novosibirsk, pp. 64–70.

Jaeger, H.M., and Nagel, S.R. (1992) Physics of the Granular State. *Science,* 20, **255**, pp. 1523–1531.

Jaeger, H.M., and Nagel, S.R. (1996) Granular Solids, Liquids, and Gases. *Reviews of Modern Physics,* **68**, no. 4, pp. 1259–1273.

Jia, X., Caroli, C., and Velicky, B. (1999) Ultrasound Propagation in Externally Stressed Granular Media. *Phys. Rev. Letters,* **82**, no. 9, pp. 1863–1866.

Johnson, K.L. (1987) *Contact Mechanics.* Cambridge University Press, New York.

Kivshar, Y.S. (1993) Intrinsic Localized Models as Solitons with a Compact Support. *Phys. Rev. E,* **48**, no. 1, pp. R43–R45.

Korteweg, D.J., and de Vries, G. (1895) On the Change of Form of Long Waves Advancing in a Rectangular Canal, and on an New Type of Long Stationary Waves. *London, Edinburgh and Dublin Philosophical Magazine and Journal of Science,* ser. 5, **39**, pp. 422–443.

Kruskal, M. (1975) Nonlinear Wave Equations. In: *Dynamical Systems, Theory and Applications, Lecture Notes in Physics,* vol. **38**. Springer-Verlag, Heidelberg, pp. 310–354.

Kunin, I.A. (1975) *Theory of Elastic Media with Microstructure.* Nauka, Moscow (in Russian).

Kuwabara, G., and Kono, K. (1987) Restitution Coefficient in a Collision between Two Spheres. *Jap. J. Appl. Phys.* **26**, no. 8, pp. 1230–1233.

Landau, L.D., and Lifshitz E.M. (1976) *Mechanics.* Translated from the Russian by J.B. Sykes and J.S. Bell, 3rd ed. Pergamon Press, Oxford.

Landau, L.D., and Lifshitz E.M. (1995) *Theory of Elasticity.* Translated from the Russian by J.B. Sykes and W.H. Reid, 3rd English ed. Revised and enlarged by E.M. Lifshitz, A.M. Kosevich, and L.P. Pitaevskii. Butterworth, Oxford; Heinemann, Boston.

Lazaridi, A.N., and Nesterenko, V.F. (1985) Observation of a New Type of Solitary Waves in a One-dimensional Granular Medium. *Prikl. Mekh. Tekh. Fiz.,* **26**, no. 3, pp. 115–118 (in Russian).

English translation: *Journal of Applied Mechanics and Technical Physics (JAM),* 1985, pp. 405–408.

Lee, M.H., Florencio, J. Jr., and Hong, J. (1989) Dynamic Equivalence of a Two-Dimensional Quantum Electron Gas and a Classical Harmonic Oscillator Chain with an Impurity Mass. *Journal of Physics A*, **22**, no. 8, pp. L331–L335.

Leibig, M. (1994) Model for the Propagation of Sound in Granular Materials. *Phys. Rev. B*, **49**, no. 2, pp. 1647–1656.

Liu, C.-H., and Nagel, S.R. (1992) Sound in Sand. *Phys. Rev. Letters,* **68**, no. 15, pp. 2301–2304.

Liu, C.-H., and Nagel S.R., (1993a) Sound in a Granular Material, Disorder and Nonlinearity. *Phys. Rev. B*, **48**, no. 21, pp. 15646–15650.

Liu, C.-H., and Nagel, S.R. (1993b) Sound and Vibration in Granular Material. *J. Phys.: Condens. Matter,* **6**, pp. A433–A436.

Liu, C.-H., Nagel, S.R., Schecter, D.A., Coppersmith, S.N., Majumdar, S., Narayan, O., and Witten, T.A. (1995) Force Fluctuations in Bead Packs. *Science*, **269**, pp. 513–515.

Mach, E. (1868) *Sitzungsber. math.-naturwiss. Cl. kaiserlich. Akad. Wissenschaften,* Vienna, **57**, no. 2, pp. 11–19.

MacKay, R.S. (1999) Solitary Waves in a Chain of Beads under Hertz Contact. *Physics Letters A*, **251**, no. 3, pp. 191–192.

Madsen, O.S., and Mei, C.C. (1969) The Transformation of a Solitary Wave over an Uneven Bottom. *J. Fluid Mech.,* **39**, pp. 781–791.

Manciu, F.S., Manciu, M., and Sen, S. (2000) Possibility of Controlled Ejection of Ferrofluid Grtains from a Magnetically Ordered Ferrofluid Using High Frequency Non-linear Acoustic Pulses – a Particle Dynamical Study. *J. Magnetism and Magnetic Materials*, **220**, no. 2-3, pp. 285-292.

Manciu, M., Sen, S., and Hurd, A.J. (1999a) The Propagation and Backscattering of Soliton-Like Pulses in a Chain of Quartz Beads and Related Problems (I). Propagation. *Physica A,* **274**, pp. 588–606.

Manciu, M., Sen, S., and Hurd, A.J. (1999b) The Propagation and Backscattering of Soliton-Like Pulses in a Chain of Quartz Beads and Related Problems. (II). Backscattering *Physica A,* **274**, pp. 607–618.

Manvi, R., Duvall, G.E., and Lowell, S.C. (1969) Finite Amplitude Longitudinal Waves in Lattice. *Int. J. Mech. Sci.,* **11**, pp. 1–9.

McCarthy, M.F. (1975) Singular Surfaces and Waves. in *Continuum Physics* (Edited by A.C. Eringen), vol. **II**. Academic Press, New York, pp. 450–521.

Mehrabadi, M.M., and Nemat-Nasser, S. (1983) Stress, Dilatancy and Fabric in Granular Materials. *Mechanics of Materials,* **2**, no. 2, pp. 155–61.

Melin, S. (1994) Wave Propagation in Granular Assemblies. *Phys. Rev. B,* **49**, no. 3, pp. 2353–2361.

Menon, V.V., Sharma, V.D., and Jeffrey, A. (1983) On the General Behavior of Acceleration Waves. *Applicable Analysis*, **16**, pp. 101–120.

Mertens, F.G., and Buttner, H. (1986) Solitons on the Toda Lattice: Thermodynamical and Quantum-Mechanical Aspects. In: *Solitons*, (Edited by S.E. Trullinger, V.E. Zakharov, and V.L. Pokrovsky). Elsevier Science, New York, pp. 723–781.

Mokross, F., and Buttner H. (1981) Comments on the Diatomic Toda Lattice. *Phys. Rev.* A**24**, no. 5, pp. 2826–2828.

Naughton, M.J., Shelton, R., Sen, S., Manciu, M. (1998) Detection of Non-Metallic Landmines Using Shock Impulses and MEMS Sensors. In: *Second International Conference on Detection of Abandoned Land Mines* (IEE Conf. Publ. No. 458), (*Second International Conference on Detection of Abandoned Land Mines*, Edinburgh, 12–14 October 1998). IEE, London, pp. 249–252.

Negreskul, S.I. (1993) Modeling of Granular Materials under Dynamic Loading. *Author's Abstract of PhD Thesis.* Institute of Physics of Strength and Material Science, Tomsk, p. 17.

Nesterenko, V.F. (1983a) Propagation of Nonlinear Compression Pulses in Granular Media. (Abstract of presentation on theoretical seminar of academician L.V. Ovsyannikov, 24 March, 1982). *Izvestia Akademii Nauk SSSR, Mekhanika Gzidkosti i Gasa,* no. 1, p. 191 (in Russian).

Nesterenko, V.F. (1983b) Propagation of Nonlinear Compression Pulses in Granular Media. *Prikl. Mekh. Tekh. Fiz.,* **24**, no. 5, pp. 136–148 (in Russian). English translation: *Journal of Applied Mechanics and Technical Physics (JAM),* 1983b, no. 5, pp. 733–743.

Nesterenko, V.F., Fomin V.M., and Cheskidov P.A. (1983a) Attenuation of Strong Shock Waves in Laminate Materials. In: *Nonlinear Deformation Waves* (Edited by U. Nigul and J. Engelbrecht). Springer-Verlag, Berlin, pp. 191–197.

Nesterenko, V.F., Fomin, V.M., and Cheskidov P.A. (1983b) Damping of Strong Shocks in Laminar Materials. *Zhurnal Prikladnoi Mekhaniki i Tehknicheskoi Fiziki,* **24**, no. 4, pp. 130–139 (in Russian). English translation: *Journal of Applied Mechanics and Technical Physics,* January 1984, pp. 567–575.

Nesterenko, V.F., et al. (1986) Development of Protection Method for Reinforced Concrete Plates Against Spall after Contact Explosion. *Technical Report,* Special Design Office of High-Rate Hydrodynamics, Novosibirsk, #GR01850066141 (#SKB02870009933), p. 61.

Nesterenko, V.F., and Lazaridi, A.N. (1987) Solitons and Shock Waves in One-Dimensional Granular Media. In: *Problems of Nonlinear Acoustics, Proceedings of 11 International Symposium on Nonlinear Acoustics,* Novosibirsk, 24–28 August, Part 1, pp. 309–313 (in Russian).

Nesterenko, V.F. (1988) Nonlinear Phenomena under Impulse Loading of Heterogeneous Condensed Media. Doctor in Physics and Mathematics Thesis, Academy of Sciences, Siberian Branch, Russia, Lavrentyev Institute of Hydrodynamics, Novosibirsk.

Nesterenko, V.F., and Lazaridi, A.N. (1990) The Peculiarities of Wave Processes in Periodical Systems of Particles with Different Masses. In: *Obrabotka materialov impulsnymi nagruzkami,* Novosibirsk, pp. 30–42 (in Russian).

Nesterenko, V.F. (1992a) Nonlinear Waves in "Sonic Vacuum". *Fizika Goreniya i Vzryva,* **28**, no. 3 pp. 121–123 (in Russian).

Nesterenko, V.F. (1992b) Pulse Compression Nature in Strongly Nonlinear Medium. In: *Proc. of Second Intern. Symp. on Intense Dynamic Loading and Its Effects,* Chengdu, China, pp. 236–240.

Nesterenko, V.F. (1992c) A New Type of Collective Excitations in a "Sonic Vacuum." *Akustika neodnorodnykh sred,* Novosibirsk, pp. 228–233 (in Russian).

Nesterenko, V.F. (1992d) *High-Rate Deformation of Heterogeneous Materials.* Nauka, Novosibirsk, Ch. 2, pp. 51–80 (in Russian).

Nesterenko, V.F. (1993a) Examples of "Sonic Vacuum," *Fizika Goreniya i Vzryva*, no. 2, pp. 132–134 (in Russian).

Nesterenko, V.F. (1993b) Solitary Waves in Discrete Medium with Anomalous Compressibility. *Fizika Goreniya i Vzryva*, **29**, no. 2, pp. 134–136 (in Russian).

Nesterenko, V.F., (1994) Solitary Waves in Discrete Media with Anomalous Compressibility and Similar to "Sonic Vacuum," *Journal De Physique IV*, Colloque C8, supplement au *Journal de Physique III*, **4**, pp. C8-72–C8-734.

Nesterenko, V.F. (1995) Continuous Approximation for Wave Perturbations in a Nonlinear Discrete Medium. *Fizika Goreniya i Vzyva*, **31**, no. 1, pp. 119–123 (in Russian).
English translation: *Explosion, Combustion and Shock Waves,* July, 1995, pp. 116–119.

Nesterenko, V.F., Lazaridi, A.N., and Sibiryakov, E.B. (1995) The Decay of Soliton at the Contact of Two "Acoustic Vacuums." *Prikl. Mekh. Tekh. Fiz.*, **36**, no. 2, pp. 19–22 (in Russian).
English translation: *Journal of Applied Mechanics and Technical Physics (JAM)*, September, 1995, pp. 166–168.

Nesterenko, V.F. (1999) Solitons, Shock Waves in Strongly Nonlinear Particulate Media. Presentation at *11th APS Topical Group Meeting on Shock Compression of Condensed Matter*, Snowbird, Utah, 27 June–2 July 1999. *Bulletin of the American Physical Society*, **44**, no. 2, p. 86.

Nesterenko, V.F. (2000) Solitons, Shock Waves in Strongly Nonlinear Particulate Media. In: *Shock Compression of Condensed Matter—1999* (Edited by M.D. Furnish, L.C. Chabildas, and R.S. Hixson). AIP, New York, pp. 177–180.

Nesterenko, V.F. (2001) New Wave Dynamics in Granular State. In: *Granular State, Materials Research Society Symposium Proceedings* (Edited by S.Sen and M.L. Hunt). Materials Research Society, Warrendale, Pennsylvania, pp. BB3.1.1–BB.3.1.12.

Nowacki, W.K. (1978) *Stress Waves in Non-Elastic Solids.* Pergamon Press. Oxford.

Nunziato, J., and Walsh, E. (1977) On the Influence of Void Compaction and Non-Uniformity on the Propagation of One-Dimensional Acceleration Waves in Granular Materials. *Arch. Rational Mech. Anal.*, **64**, pp. 299–316.

Nunziato, J., and Cowin, S.C. (1979) A Nonlinear Theory of Elastic Materials with Voids. *Arch. Rational Mech. Anal.*, **72**, pp. 175–201.

Nunziato, J., Kennedy, J.E., and Walsh, E. (1978) The Behavior of One-Dimensional Acceleration Waves in Inhomogeneous Granular Solids. *Int. J. Engng. Sci.*, **16**, pp. 637–648.

Ono, H. (1972) Wave Propagation in an Inhomogeneous Anharmonic Lattice. *Journal of the Physical Society of Japan*, **32**, no. 2, pp. 332–336.

Ostoja-Starzewski, M. (1984) Stress Wave Propagation in Discrete Random Solids. In: *Wave Phenomena: Modern Theory and Applications* (Edited by C. Rogers and T. Bryant). Elsevier Science, Amsterdam. North-Holland Mathematics Studies, vol. **97**, pp. 267–278.

Ostoja-Starzewski, M. (1991a) Wavefront Propagation in a Class of Random Microstructures—I: Bilinear Elastic Grains. *Int. Journal of Non-Linear Mechanic*, **26**, pp. 659–669.

Ostoja-Starzewski, M. (1991b) Transient Waves in a Class of Random Heterogeneous Media. *Appl. Mech. Rev.* **44**, no. 11, Part 2, pp. S199–S209.

Ostoja-Starzewski, M. (1993) On the Critical Amplitudes of Acceleration Wave to Shock Wave Transition in White-Noise Random Media. *J. Appl. Phys. (ZAMP)*, **4**, pp. 865–879.

Ostoja-Starzewski, M. (1994) Transition of Acceleration Waves into Shock Waves in Random Media. *Appl. Mech. Rev.* (Special Issue: *Nonlinear Waves in Solids*), **47** (Pt. 2).

Ostoja-Starzewski, M. (1995) Wavefront Propagation in a Class of Random Microstructures—II: Nonlinear Elastic Grains. *Int. Journal of Non-Linear Mechanics*, **30**, no. 6, pp. 771–781.

Parker, D.F., and Seymour, B.R. (1980) Finite Amplitude One-Dimensional Pulses in an Inhomogeneous Granular Material. *Archive for Rational Mechanics and Analysis,* **72**, pp. 265–284.

Pendry, J.B., Holden, A.J., Robbins, D.J., and Stewart, W.J. (1999) Magnetism From Conductors and Enhanced Nonlinear Phenomena. *IEEE Transactions on Microwave Theory and Techniques*, **47**, no. 11, pp. 2075–2084.

Radjai, F., Jean, M., Moreau, J.-J., Roux, S. (1996) Force Distribution in Dense Two-Dimensional Granular Systems. *Phys. Rev. Letters,* **77**, no. 2, pp. 274–277.

Reddy, T.Y., Reid, S.R., and Barr, R. (1991) Experimental Investigation of Inertia Effects in One-Dimensional Metal Ring Systems Subjected to End Impact—II. Free-Ended Systems. *Int. J. Impact Engng.* **11**, no. 4, pp. 463–480.

Remoissenet, M. (1999) *Waves Called Solitons (Concepts and Experiments).* 3rd revised and enlarged edition. Springer-Verlag, Berlin.

Rosenau, P. (1986a) A Quasi-Continuous Description of a Nonlinear Transmission Line. *Physica Scripta*, **34**, pp. 827– 829.

Rosenau, P. (1986b) Dynamics of Nonlinear Mass-Spring Chains Near the Continuum Limit. *Physica Letters A*, **118**, no. 5, pp. 222–227.

Rosenau, P. (1988) Dynamics of Dense Discrete Systems. *Progress of Theoretical Physics*, **79**, no. 5, pp. 1028–1042.

Rosenau, P., and Hyman, J.M. (1993) Compactons: Solitons with Finite Wavelength. *Physical Review Letters*, **70**, no. 5, pp. 564–567.

Rosenau, P. (1994) Nonlinear Dispersion and Compact Structures. *Physical Review Letters*, **73**, no. 13, pp. 1737–1741.

Rosenau, P. (1996) On Solitons, Compactons, and Lagrange Maps. *Physical Letters A*, **211**, pp. 265–275.

Rosenau, P. (1997) On Nonanalytic Solitary Waves Formed by a Nonlinear Dispersion. *Physics Letters*, **230**, nos. 5–6, pp. 305–318.

Rossmanith, H.P., and Shukla, A. (1982) Photoelastic Investigation of Dynamic Load Transfer in Granular Media. *Acta Mechanica*, **42**, pp. 211–225.

Russel, J.S. (1838) Report of the Committee on Waves. *Report of the 7th Meeting of the British Association for the Advancement of Science*, Liverpool, pp. 417–496.

Russel, J.S. (1845) On Waves. *Report of the 14th Meeting of the British Association for the Advancement of Science*, York, pp. 311–390.

Sadd, M.H., Tai, QiMing, and Shukla, A. (1993) Contact Law Effects on Wave Propagation in Particulate Materials Using Distinct Element Modeling. *Int. J. Non-Linear Mechanics*, **28**, no. 2, pp. 251–265.

Samsonov, A.M. (1987) Transonic and Subsonic Localized Waves in Nonlinear Elastic Waveguide. In: *Proc. of Intern. Conference on Plasma Physics,* Naukova Dumka, Kiev, **4**, pp. 88–90.

Sander J., and Hutter, K. (1991) On the Development of the Theory of the Solitary Wave. A Historical Essay. *Acta Mechanica,* **86**, pp. 111–152.

Sander, J., and Hutter K. (1992) Evolution of Weakly Non-Linear Shallow Water Waves Generated by a Moving Boundary. *Acta Mechanica*, **91**, pp. 19–155.

Scott, D.R., and Stevenson, D.J. (1984) Magma Solitons. *Geophysical Research Letters,* **11**, no. 11, pp. 1161–1164.

Sen, S., Manciu, M., and Wright, J.D. (1998) Solitonlike Pulses in Perturbed and Driven Hertzian Chains and Their Possible Applications in Detecting Buried Impurities. *Phys. Rev., E,* **57**, no. 2, pp. 2386–2397.

Sen, S. (1998) Soliton-like Objects in Granular Beds. *Presentation at UCSD Seminar in Mechanical and Materials Engineering,* December 16, 1998.

Sen, S., Manciu, M., and Manciu, F.S. (1999) Ejection of Ferrofluid Grains Using Nonlinear Acoustic Impulses—A Particle Dynamical Study. *Applied Physics Letters,* **75**, no. 10, pp. 1479–1481.

Sen, S., and Manciu, M. (1999) Discrete Hertzian Chains and Solitons. *Physica A,* **268**, pp. 644–649.

Shukla, A., Sadd, M.H., Xu, Y., and Tai, Q.M. (1993) Influence of Loading Pulse Duration on Dynamic Load Transfer in a Simulated Granular Medium. *J. Mech. Solids, J. Mech. Solids,* **41**, no. 11, pp. 1795–1808.

Singh, R., Shukla, A., and Zervas, H. (1996) Explosively Generated Pulse Propagation through Particles Containing Natural Cracks. *Mechanics of Materials,* **23**, pp. 255–270.

Sinkovits, R.S., and Sen, S. (1995) Nonlinear Dynamics in Granular Columns. *Phys. Rev. Letters,* **74**, no. 14, pp. 2686–2689.

Sinkovits, R.S., and Sen, S. (1996) Sound Propagation in Impure Granular Columns. *Phys. Rev., E,* **54**, no. 6, pp. 6857–6865.

Smith, D.R., Vier, D.C., Padilla, W., Nemat-Nasser, S.C., and Schultz, S. (1999) Loop-Wire Medium for Investigating Plasmons at Microwave Frequencies. *Applied Physics Letters,* **75**, no. 10, pp. 1425–1427.

Sobczyk, K. (1992) Korteweg–de Vries Solitons in a Randomly Varying Medium. *Int. Journal of Non-Linear Mechanics,* **27**, no. 1, pp. 1–8.

Strenzwik, D.E. (1979) Shock Profiles Caused by Different End Conditions in One-Dimensional Quiscent Lattices, *J. Appl. Physics,* **50**, no. 11, pp. 6767–6772.

Takahashi, D., and Satsuma, J. (1988) Explicit Solutions of Magma Equation. *J. Phys. Soc. Japan,* **57**, no. 2, pp. 417–421.

Takahashi, D., Sachs, J.R., and Satsuma, J. (1990) Soliton Phenomena in a Porous Medium. In: *Research Reports in Physics, Nonlinear Physics* (Edited by Gu Chaohao, Li Yishen, and Tu Guizhang). Springer-Verlag, Berlin, pp. 214–220.

Takahashi, D., and Sato, Y. (1949) On the Theory of Elastic Waves in a Granular Substance. *Bulletin of the Earthquake Research Institute,* Japan, **27**, pp. 11–16.

Takahashi, D., and Sato, Y. (1950) On the Theory of Elastic Waves in a Granular Substance. *Bulletin of the Earthquake Research Institute,* Japan, **28**, pp. 37–43.

Thornton, C., and Barnes, D.J. (1986) Computer Simulated Deformation of Compact Granular Assemblies. *Acta Mechanica,* **64**, pp. 45–61.

Timoshenko, S.P., and Goodier J.N. (1987) *Theory of Elasticity,* 3rd ed. McGraw-Hill, New York, p. 414.

Toda, M. (1981) *Theory of Nonlinear Lattices.* Springer Series in Solid State Science, Vol. 20, Springer-Verlag, Berlin.

Travers, T., Ammi, M., Bideau, D., Gervois, A., Messager, J.C., and Troadec, J.P. (1987) Uniaxial Compression of 2-D Packings of Cylinders. Effects of Weak Disorder. *Europhys. Letters,* 4, no. 3, pp. 329–332.

Travers, T., Ammi, M., Bideau, D., Gervois, A., Messager, J.C., and Troadec, J.P. (1988) Mechanical Size Effects in 2-D Granular Media. *J. Phys. France,* 49, pp. 939–948.

Triantafyllidis, N., and Bardenhagen, S. (1993) On High Order Gradient Continuum Theories in 1-D Nonlinear Elasticity. Derivation from and Comparison to the Corresponding Discrete Models. *Journal of Elasticity,* 33, pp. 259–293.

Tsai, D.H., and Beckett, C.W. (1966) Shock Wave Propagation in Cubic Lattices. *Journal of Geophysical Research,* 71, no. 10, pp. 2601–2608.

Tsemahovich, B.D. (1988) Concept of Support Selection for Explosive Welding. *Proc. International Symposium on Metal Explosive Working,* Purdubice, Chehoslovakia, Vol. 1, pp. 82–89 (in Russian).

Walton, K. (1987) The Effective Elastic Moduli of a Random Packing of Spheres. *J. Mech. Phys. Solids,* 35, no. 2, pp. 213–226.

Whitham, G.B. (1970) Two-Timing, Variational Principles and Waves. *J. Fluid Mechanics,*.44, pp. 373–395.

Whitham, G.B. (1974) *Linear and Nonlinear Waves.* Wiley, New York.

Wright, T.W. (1984) Weak Shocks and Steady Waves in a Nonlinear Elastic Rod or Granular Material. *Int. J. Solids Structures,* 20, nos. 9/10, pp. 911–919.

Wright, T.W. (1985) Nonlinear Waves in a Rod: Results for Incompressible Elastic Materials. *Studies in Applied Mathematics,* 72, pp. 149–160.

Yablonovitch, E. (1993) Photonic Band-Gap Crystals. *J. Phys.: Condens. Matter,* 5, pp. 2443–2460.

Zabusky, N.J., and Galvin, C.J. (1971) Shallow-Water Waves, the Korteveg–de Vries Equation and Solitons. *J. Fluid Mech.,* 47, pp. 811–824.

Zaikin, A.D. (1990) Effective Elastic Moduli of Granular Media. *Prikl. Mekh. Tekh. Fiz.,* 31, no. 1, pp. 91–96 (in Russian). English translation: *Journal of Applied Mechanics and Technical Physics(JAM),* July, 1990, pp. 85–89.

Zakharov, V.E. (1972) Collapse of Langmuir Waves. *Sov. Phys. JETP,* 35, no. 5, pp. 908–914.

Zakharov, V. E. (1974) On Stochastization of One-Dimensional Chains of Nonlinear Oscillators. *Sov. Phys. JETP,* 38, no. 1, pp. 108–110.

Zhu, Y., Shukla, A., and Sadd, M.H. (1996) The Effect of Microstructural Fabric on Dynamic Load Transfer in Two-Dimensional Assemblies of Elliptical Particles. *J. Mech. Phys. Solids,* 44, no. 8, pp. 1283–1303.

2
Mesomechanics of Porous Materials Under Intense Dynamic Loading

2.1 Introduction

Intense dynamic loading of porous, granular materials results in a completely new class of phenomena, in comparison with their behavior considered in the previous chapter. It includes a complex geometry of plastic flow, as well as fracture and melting. The mechanical deformation (shock loading, plastic flow) of heterogeneous porous materials has an essentially multiscale nature. For example, the hierarchy of scales includes:

Macroscales. Length of shock impulse, typical length of deformable part of the material.

Mesoscales. Shock front thickness, shear band thickness, sizes of the particles and pores, sizes of fragmented material, size of plastic flow localized on the interfaces of particles, plastic flow around pores, spacing between shear bands, heat conductivity, and mass diffusion lengths. Very often this scale is close to the initial scale of material heterogeneity that makes the continuum approach problematic.

Microscales. Sizes of lattice defects (dislocations, point defects, twins), possible additional scales parameters reflecting the nonlinear interaction, and collective dynamic behavior in molecular systems.

Among the hierarchy of scales presented above the mesoscale is the crucial scale for the initiation and detonation build-up of energetic materials or for bonding during the dynamic loading of powders. It is well known that detonation phenomena, even in the most uniform initial state of matter, gas, are governed by mesoscale phenomena (Lee [2000]).

In general, mesoscale phenomena can be expected to play a dominant role in the initiation and propagation of detonations in composite energetic materials, which are heterogeneous and locally anisotropic. The fundamental reason for this behavior is the strongly nonlinear coupling between heat release due to mechanical work, chemical heat release, and the mechanical response of solid matter. The material mesostructure and its mechanical response are very important for the interpretation of the initiation/detonation phenomena (Howe [2000a,b]).

The important part of the research in this area should include the establishment of the critical conditions for macroparameters which will result in a qualitative change of meso (micro) behavior, like critical conditions for transition from a quasistatic to dynamic mode of porous material deformation on the mesoscale. The reason for this is that very often the transition from one mode of mechanical behavior to another, resulting in finer scales of mechanical energy dissipation and in accumulation of microkinetic energy, is accompanied by bonding between particles or phase (chemical) changes in reactant materials.

To accomplish this goal, it is necessary to develop physically based material models appropriate for each length scale, and to introduce a computational technique for bridging the length and time scales. The latter should be validated in special well-characterized experiments.

2.2   Dynamics of Pore Collapse: The Carroll–Holt Model

2.2.1   Dynamics of Visco-Plastic Pore Collapse

The simplest description of the plastic deformation of the porous materials can be provided in the frame of a single cell model. In this type of model all properties of porous material are represented by the characteristics of a single cell related to global material properties. The boundary conditions on the cell outer surface correspond to the average pressure in the medium. Different processes and phenomena, on different time and space scales developing in the material, are reflected in different approaches inside the single pore models.

Carroll and Holt [1972b] proposed a geometry of the single pore model (Fig. 2.1) for the description of the pressure–density relation in the dynamic process of powder densification in the shock wave. The geometrical model parameters and material properties were selected according to the following assumptions:

(1) The initial pore radius in the model (a_0) is equal to the characteristic size

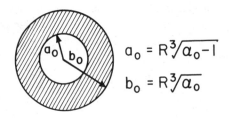

$$a_0 = R\sqrt[3]{\alpha_0 - 1}$$

$$b_0 = R\sqrt[3]{\alpha_0}$$

Figure 2.1. Geometry of the Carroll–Holt model.

of pores in a real porous material with initial density ρ_{p0}.

(2) The mass of hollow sphere in the model, divided by the total volume of this sphere $(4/3\pi b^3)$, is equal to the density of the real porous material ρ_p.

(3) The material of the model sphere is incompressible (ρ_s = const) and viscoplastic. The viscosity and yield strength are taken to be independent of the temperature and pressure.

(4) Spherical symmetry is maintained throughout the collapse process.

These assumptions reflect the main mechanism of viscoplastic deformation-pore closure, but neglect some features such as compressibility of the matrix, temperature dependence of material properties, and nonspherical modes of material flow. The geometrical characteristics of the model are selected to provide the correct time scale for the collapse process.

Carroll, Kim, and Nesterenko [1986] compared the results obtained within the frame of this model with experimental results on shock front thickness (Nesterenko [1975]). The following geometrical relations describe a radially symmetric motion of an incompressible hollow sphere:

$$r^3 - a^3 = r_0^3 - a_0^3, \qquad \Theta = \Theta_0, \qquad \varphi = \varphi_0, \tag{2.1}$$

where r, θ, and φ are spherical polar coordinates, r_0 corresponds to the initial radius of some point with current coordinate r, and a and a_0 are the current and initial inner radii of the model sphere (Fig. 2.1).

Based on the model assumptions, the initial (α_0) and current (α) porosities of the real material are connected with the model geometry by the relations

$$\alpha_0 = \frac{b_0^3}{b_0^3 - a_0^3} = \frac{\rho_s}{\rho_{p0}}, \tag{2.2a}$$

and

$$\alpha = \frac{b^3}{b^3 - a^3} = \frac{\rho_s}{\rho_p}, \tag{2.2b}$$

where b is the current outer radius and ρ_p is the current density of the porous material. The spherically symmetrical motion is a 1-D motion because the value of the inner radius a determines other parameters through the equations

$$r^3 = r_0^3 - a_0^3 + a^3, \tag{2.3a}$$

$$b^3 = a^3 + \frac{a_0^3}{\alpha_0 - 1}, \tag{2.3b}$$

$$\alpha = 1 + \frac{(\alpha_0 - 1)a^3}{a_0^3}. \tag{2.3c}$$

The equation of motion for spherical geometry is

$$\rho_s \ddot{r} = \frac{\partial \sigma_{rr}}{\partial r} + \frac{2}{r}\left(\sigma_{rr} - \sigma_{\theta\theta}\right). \tag{2.4}$$

The constitutive equation for viscoplastic flow can be written as

$$\sigma_{rr} - \sigma_{\theta\theta} = Y + 2\eta\left(\frac{\partial v}{\partial r} - \frac{v}{r}\right) = Y - 6\eta\,\frac{\dot{r}}{r} = 2\sigma_s, \tag{2.5}$$

where σ_s is the shear stress.

The boundary conditions on the surfaces of the model sphere are

$$\sigma_{rr}\,(r=a) = 0, \qquad \sigma_{rr}\,(r=b) = P_m(t), \tag{2.6}$$

where $P_m(t)$ is the pressure on the outer surface representing the collective actions of the neighboring pores in real material. The initial conditions are

$$a(0) = a_0, \qquad \dot{a}(0) = 0. \tag{2.7}$$

Integrating Eq. (2.4) over the radius from a to b, using the boundary conditions (Eq. (2.6)), and relations described by Eqs. (2.3), the dependence of the inner pore radius a on pressure $P_m(t)$ can be expressed in the form

$$P_m(t) = 2\int_a^b \sigma_s\,\frac{dr}{r} + P_{kin}\,(\ddot{a}, \dot{a}, a). \tag{2.8}$$

The second term, P_{kin} in Eq. (2.8), represents the inertia-dependent part of the total pressure and determines the main difference between the quasistatic and total pressure and determines the main difference between the quasistatic and dynamic cases

$$P_{kin}\,(\ddot{a}, \dot{a}, a) = -\rho_s\left[\left(a\ddot{a} + 2\dot{a}^2\right)\left(1 - \frac{a}{b}\right) - \frac{1}{2}\,\dot{a}^2\left(1 - \frac{a^4}{b^4}\right)\right]. \tag{2.9}$$

It is easy to find the first and second derivatives of the inner radius a with respect to time as functions of porosity from Eqs. (2.3):

$$\dot{a}^2 = \frac{\dot{\alpha}^2 a_0^6}{9\left(\alpha_0 - 1\right)^2 a^4}, \tag{2.10a}$$

$$a\ddot{a} + 2\dot{a}^2 = \frac{\ddot{\alpha}\left(\alpha - 1\right)^{-1/3} a_0^2}{3\left(\alpha_0 - 1\right)^{3/2}}. \tag{2.10b}$$

The pore collapse equation (2.8) may be written in the following form using Eqs. (2.10) if temperature dependence of yield stress and viscosity is ignored

$$P_m = \frac{2}{3} Y_0 \ln \frac{\alpha}{\alpha - 1} - \frac{4\eta_0}{3\alpha(\alpha - 1)} \dot{\alpha} - \frac{\rho_s a_0^2}{3(\alpha_0 - 1)^{2/3}} \frac{d}{d\alpha} \left\{ \frac{1}{2} \left[(\alpha - 1)^{-1/3} - (\alpha)^{-1/3} \right] \dot{\alpha}^2 \right\}. \tag{2.11}$$

It follows from Eq. (2.11) that, under the static loading,

$$P_m = \frac{2}{3} Y_0 \ln \frac{\alpha}{\alpha - 1}. \tag{2.12}$$

This equation coincides with one obtained by Torre [1948] for the model sphere under static compression. The distinguishing feature of Eq. (2.12) is its independence of the initial porosity and the geometrical size of the pore. It is not appropriate for modeling dynamic behavior, as will be shown later.

2.2.2   Equation for Shock Wave Structure

In view of the simplified geometry of the Carroll–Holt model, it is doubtful that this approach can describe a shock wave profile in detail. Nevertheless, it is instructive to apply the principles of conservation of mass

$$\rho_1 D = \rho(D - u) \tag{2.13}$$

and momentum

$$P - P_1 = \rho_0 D u, \tag{2.14}$$

to obtaining equations for the stationary shock wave profile for a high amplitude impulse for which the elastic and elastic–plastic parts of the deformation can be neglected (Carroll and Holt [1973]). The dependence of pressure on porosity in the stationary shock (at pressures above P_1) follows from Eqs. (2.13) and (2.14):

$$P - P_1 = \frac{\rho_0 D^2}{\alpha_0} (\alpha_1 - \alpha), \qquad P_1 = \frac{2}{3} Y \ln \frac{\alpha_1}{\alpha_1 - 1}, \qquad \alpha_1 = \frac{2G\alpha_0}{2G + Y}, \tag{2.15}$$

where P_1, ρ_1, α_1 are the pressure, density, and porosity ahead of the front of the plastic shock wave, ρ_0, α_0 are the initial density and porosity, and G is the shear modulus of the solid material. The porosity α_1 corresponds to the elastic–plastic/plastic transition. During further densification the material around a pore is in the plastic state. For metals, the porosity α_1 is actually very close to the initial porosity, α_0, because Y/G is about 10^{-2}.

In the case of a stationary shock, introduction of the new variable $\xi = x - Dt$ allows one to replace the partial derivative with respect to time with derivatives with respect to ξ:

$$\frac{\partial \alpha}{\partial t} = -D\frac{\partial \alpha}{\partial \xi}. \tag{2.16}$$

The difference between P_m and P can be introduced using the relation $P = P_m/\alpha$ (Carroll and Holt [1972a]). This connection between the global pressure in the powder, P, and the average pressure P_m on the outside surface of the model cell (or the average pressure in the material of the granules) is the simplest one. It is more reasonable for "zero tension" granular materials than averaging over the volume of particles and dividing by α (Attetkov et al. [1984]). This conclusion is based on the fact that measurable parameters in such materials (e.g., particle velocity or pressure) are naturally averaged over the contact surface of particles and over the global cross section of the sample. Also, the former approach is more suitable for powders with high initial porosity at pressures $P \gg Y$ where the correct description of the shock densification can be obtained based on the hypothesis of incompressible material collapsing into an open pore space. Nigmatullin [1978] considers the problem of averaging in the mechanics of heterogeneous materials in detail.

Finally, from Eqs. (2.11), (2.15), and (2.16), we obtain the following equation for shock wave structure using the boundary conditions $P = P_1$, $\alpha = \alpha_1$, and $d\alpha/d\xi = 0$:

$$\alpha P = \alpha \left\{ P_1 + \frac{\rho_0 D^2}{\alpha_0}(\alpha_1 - \alpha) \right\} \tag{2.17}$$

$$= \frac{2}{3}Y_0 \ln\frac{\alpha}{\alpha-1} + \frac{4\eta_0 D \frac{d\alpha}{d\xi}}{3\alpha(\alpha-1)} - \frac{\rho_s a_0^2 D^2}{3(\alpha_0-1)^{2/3}}\frac{d}{d\alpha}\left\{ \frac{1}{2}\left[(\alpha-1)^{-1/3} - (\alpha)^{-1/3}\right]\left(\frac{d\alpha}{d\xi}\right)^2 \right\}.$$

An analysis of similar equations was performed by Dunin and Surkov [1979] and results based on this equation will be presented in Section 2.3. It should be mentioned that a strain-rate dependent viscous term is of principal importance for describing shock wave structure in the frame of a single pore model. Neglect of this term results in a very instructive Carroll–Holt paradox for a partially compacted final state (see Problem 14). Another problem arises with stronger shocks because plastic flow can dissipate only a certain amount of energy in the single pore geometry (Dunin and Surkov [1982], Tong and Ravichandran [1993]). This problem will be resolved in numerical calculations taking into account a complex, not spherically symmetrical geometry of plastic flow during pore closer (see Section 2.6).

2.2.3 Comparison with Experiment

Equation (2.8) allows one to obtain the time dependence of the inner radius a and the porosity α based on known time dependence of the outside pressure P. Carrol, Kim, and Nesterenko [1986] used this approach to calculate pore collapse time and characteristic shock front thickness. For this purpose, the pressure

$$P_m(t) = \begin{cases} P_1 + (P^* - P_1)t/\tau & (0 \le t \le \tau), \\ P^* & (t > \tau), \end{cases} \qquad (2.18)$$

dependence was used, where P^* corresponds to the final pressure of the shock wave and P_1 is the pressure corresponding to the precursor, where the whole pore is subjected to plastic flow. The stage of pressure increase from zero to P_1 is neglected here because, at a strong shock wave, the elastic densification is negligible in comparison with the densification due to plastic pore collapse. In this equation, τ is the shock front thickness and can be taken, for example, from experiments with Cu/Ni powders (Nesterenko [1992]).

At shock amplitudes, $P^* > 1$ GPa, the final density of the copper powder is close to the solid density (Bakanova, Dudoladov, and Sutulov [1974]). That is why the final pressure P^* on the outer surface of the model sphere can be considered to be equal to the shock pressure in the powder. During the collapse process, the average pressure on the outside surface of the sphere should be higher than the corresponding average pressure in the shock wave in the powder (P). For example, $P_m = \alpha P$ (Holt, Carroll, and Kusubov [1971]; Carroll and Holt [1972a]). For the purpose of calculations of the shock front thickness, introduction of the difference between P_m and P is not important because it will result mainly in a change of the shape of the shock front and not the shock front thickness. Because $P_m(t)$ is approximated by linear time dependence it is possible to say that this corresponds to the concave down average shock pressure (P) profile.

Comparison with experiment should be made using reasonable values of material parameters measuring strength and viscosity. The strain rate prevailing in the process of pore collapse in granular material is of the order of magnitude 10^6–10^7 s^{-1} and the strain is about 1. Material property data for this range of parameters are not available in the literature. Experimental data for the copper viscosity for strain rates in the range 10^4–10^5 s^{-1} are presented in Table 2.1. There is a wide data spread for the viscosity coefficient. This is partially due to the essentially different conditions of spatial scale (shock front thickness or macroscopic size zone adjacent to an explosive weld) and the range of strains involved in the dynamic experiments. Nevertheless, Stepanov [1978] mentioned that at high strain rate, viscosity does not depend on strain level and metal behavior corresponds to that of a Newtonian liquid. The dependence of viscosity on the pressure is neglected because shearing occurs primarily near the surface of the pores where the pressure is equal to zero.

A very similar case to the densification process in powder is the collapse of a thin cylindrical copper tube in experiments performed by Matyshkin and Trishin [1978]. The measured average viscosity value is $\eta = 10^4$ Pa s. That is why the numerical value for the viscosity of copper is taken as $\eta = 10^4$ Pa s. The yield strength for copper is selected as 400 MPa to take into account the strain hardening effect (Slater [1977]).

Numerical calculations for $\alpha(t)$ using Eqs. (2.11) and (2.18) were performed

Table 2.1. High strain, high strain rate viscosity for copper.

Viscosity, η 10^3, Pa s	Reference
1	Saharov, Zaidel, Mineev et al. [1964]
	Mineev and Savinov [1967]
	Mineev and Zaidel [1968]
20–27	Godunov, Deribas et al. [1971]
3–30	Kumar and Kamble [1969]
3	Kanel [1978]
6–16	Matyshkin and Trishin [1978]

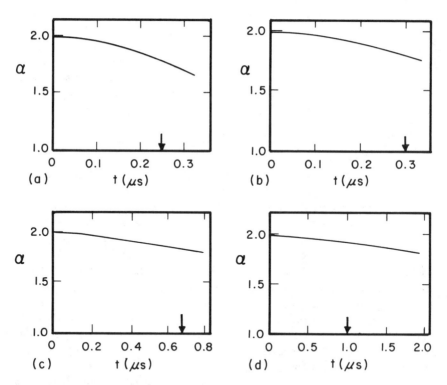

Figure 2.2. Time dependence of porosity for the Carroll–Holt model ($a_0 = 0.3$ mm) under different shock loading conditions: (a) $P^* = 14.3$ GPa, $\tau = 0.25$ μs; (b) $P^* = 10.0$ GPa, $\tau = 0.30$ μs; (c) $P^* = 2.40$ GPa, $\tau = 0.65$ μs; (d) $P^* = 1.0$ GPa, $\tau = 1.0$ μs. Arrows show the corresponding experimental shock front thickness τ.

by the fourth-order Runge–Kutta method with time step $\Delta t = 10^{-9}$ s, and are presented in Figure 2.2 for the corresponding values of the shock pressure P^* and the experimental shock front thickness τ for copper powder (Nesterenko [1992]).

It is clear that the model will be valid only if the characteristic time of pore collapse in the calculations is about time τ. The data presented in Figure 2.2 demonstrate the essential discrepancy between the kinetic of the pore collapse in the model and in the shock front—the collapse predicted by the model is much slower than that observed in the experiment. This means that the model used is not self-consistent ($\tau_{cal} \gg \tau$) and should be replaced by another approach.

2.3   Modified Carroll–Holt Model

2.3.1   Critical Analysis of the Carroll–Holt Model

The pore dynamics calculated using the Carroll–Holt model do not agree with experimental data on shock front thickness in powders with high initial porosity, as was demonstrated in the previous section. This raises a question of model validity. Before proceeding with the modification of this model, let us make some general remarks about the modeling of granular (porous) materials under dynamic loading (Nesterenko [1988a,b,c], [1992]).

First, it is clear that, in a medium with high porosity, where the volume occupied by pores is comparable with the volume of the solid material, it is impossible to apply first principles to the calculation of interactions among particles. This is the area where a priori approaches based on physical intuition should be applied and corrected by adequate experiments. The Carroll–Holt model is one of the examples of such an approach which was very successful in the description of static and dynamic pressure—density behavior of porous materials (Carroll and Holt [1972b]). Nevertheless, in many papers, this model is considered without justification as being identical to the porous material. This model (as any other model) can be considered as valid only if verified experimentally.

The experiments performed to check the model can be considered adequate if the results are sensitive to the crucial model parameters or model assumptions. For example, measurements of the shock Hugoniot in highly porous materials cannot be considered as adequate if they are applied to check micromechanical modeling. This is connected with negligible sensitivity of the Hugoniot on the details of particle deformation and on the kinetic processes in general. Parameters determining the time-dependent properties of material (like viscosity) are not reflected in Hugoniot behavior. In contrast, measurements of the shock rise-time could be considered to be adequate experiments for checking the kinetics models of particle deformation.

The main goal of the modeling of heterogeneous materials under intensive shock loading should be a correct identification of the main, qualitatively important, tendencies in behavior, in response to the changing of the loading conditions or material parameters, but not a detailed description of the deformation process. The goal of building a model intended to capture details of the mesoscopic deformation process is currently a completely unrealistic one. At the same time, prediction of the possible qualitative changes of material behavior, and the comparative behavior of different materials using a model approach, is very important from a practical point of view.

The reasons for the modification of the classical Carroll–Holt approach are:

(1) Due to high porosity, the shocked powders are heated up to 10^3 K even at relatively low shock pressures $P < 10Y$. This temperature increase can essentially change material strength and viscosity during the process of loading. It provides a strong coupling of the viscoplastic flow and heat release.

(2) It is difficult to identify the pore size to be used in the model with some characteristic pore size in granular material, because it is the size of the particles rather than that of the pores that is well determined.

(3) The plastic flow that occurs during the densification of granular materials is the essentially different from spherically symmetrical flow. Only a portion of the material of the particle is involved in the nonsymmetric flow process—features opposite to the Carroll–Holt model behavior (Figs. 2.3 and 2.4; see also Fig. 6.1).

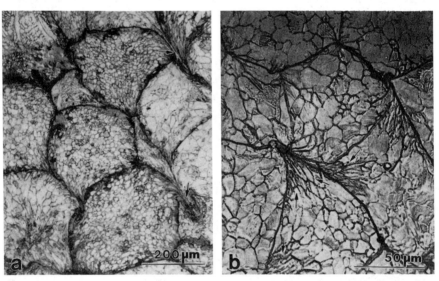

Figure 2.3. Concentration of plastic flow in the contact areas and triple points during the dynamic consolidation of copper powder with initial porosity $\alpha_0 = 1.88$. The distortion of originally equiaxed grains, inside initially spherical particles shown by arrows, mark the areas of intensive plastic flow: (a) particle size 297–420 μm, initial density $\rho_{p0} = 5.3$ g/cc, shock pressure 4.6 GPa; (b) particle size 106–125 μm, $\rho_{p0} = 5.6$ g/cc, pressure 4.6 GPa.

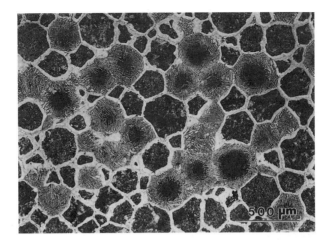

Figure 2.4. Microstructural picture resulting from deformation and preferable heating of material during shock consolidation localized at the outside layer of particles, which are seen as white regions. The material is a mixture of Ti–33Al with 30wt.% of titanium alloy BT25; its initial temperature was 1000 °C and α_0 = 1.88. The shock pressure is 2 GPa.

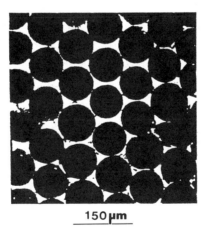

150 µm

Figure 2.5. Densification of glass (black) spheres uniformly covered by thin Al (white) film without extensive fracturing of glass particles. (Reprinted from *Shock Waves and High-Strain-Rate Phenomena in Metals. Concepts and Applications* (edited M.A. Meyers and L.E. Murr), Plenum Press, New York, paper by Staver, A.M., Metallurgical Effects under Shock Compression of Powder Material, pp. 865–880, 1981, with permission from Kluwer Academic/Plenum Publishers.)

Table 2.2. Shock front thickness for different materials.

Material of powder	Particle size (μm)	Pressure (GPa)	τ (μs)	Δ μm
Teflon	1500–2000	3.2 ± 0.5	0.80 ± 0.25	1350 ± 420
Glass	1450 (plates)	2.6 0.9	1.27 ± 0.3 2.20 ± 0.3	2000 ± 500 1900 ± 300
Tungsten	1000–2000	0.52 ± 0.05	4.0 ± 1.0	1000 ± 250
Copper– Constantan	40–70 30–90	1.3 3.5 6.0 9.4	0.067 ± 0.013 0.081 ± 0.013 0.045 ± 0.012 0.063 ± 0.031	40 ± 10 80 ± 13 60 ± 16 100 ± 50

Staver [1981] experimentally demonstrated that glass spheres covered by a thin layer of aluminium can be compacted to the solid density without extensive glass fracture (Fig. 2.5). It is clear from Figures 2.3 to 2.5 that densification can be achieved by the localized plastic flow of the periphery layer of the particles without significant involvement of the central part of the particle. This can be very important for the subsequent heat treatment and recrystallization of densified powders (see Figs. 6.11 and 6.14 (a),(b). The results of numerical calculations of pore collapse under plane shock loading demonstrate that the collapse process can proceed without the whole inner surface reaching the yield condition (Tang, Liu, and Horie [1999], [2000]).

(4) The width of the shock wave is comparable with particle size (approximately three particle diameters at $P \approx 2Y$) and approaches the particle size as the pressure increases to $P \approx 20Y$ for granular materials with initial porosity about 2 (see Table 2.2). This means that the plastic flow during pore collapse is nonuniform, not only as a result of the spherical symmetry of the pore but also as a result of the large pressure gradient along the shock propagation direction on a space scale comparable with particle size. The long wave approximation, that pressure which is uniformly distributed over the outside cell surface and slowly varies with time, is definitely disputable in this case. Additional mechanisms of compaction, for example densification due to the variation of particle velocity on the different parts of the pore surface, should be taken into account.

(5) The single pore model assumes that the dissipation occurs only during pore collapse. It is probable that a viscoplastic flow of material also exists after the porosity is eliminated. This can be confirmed by a jet-like material flow (Fig. 2.3(a), shown by arrow) which can persist for some time after pore closure.

(6) Spherically symmetrical geometry of the hollow sphere causes the nonphysical singularity when the inner radius a approaches zero (Attetkov,

Vlasova, Selivanov, and Solov'ev [1984]). For example, the density of dissipated energy (or the temperature T) produced due to the plastic work has a logarithmic singularity ($T \sim \ln a$) and a viscous dissipation tending to infinity ($T \sim a^{-5/2}$). Without dissipation, the density of kinetic energy develops a singularity as the inner radius of the pore approaches zero (Zababakhin [1970]). This behavior is a straightforward consequence of spherical geometry only and has nothing to do with real processes occurring during densification. These singularities actually raise a question of the correctness of the temperature calculations made in the frame of the hollow sphere model.

(7) The geometrical relationship between model parameters and real granular material is established by two assumptions (Eqs. (2.2), (2.3)): the densities of the powder and the model are equal, as are their characteristic sizes. These assumptions cannot, at the same time, provide the equality of specific surface areas where dissipation develops. In the hollow sphere model, the heat is concentrated near the center whereas, in the real material, it is concentrated along the particle boundaries (Figs. 2.3 and 2.4). Experiments demonstrated the creation of a new space scale during densification—the size of the zone of localized plastic flow on the particle interfaces (Figs. 2.3 and 2.4; see also Figs. 6.1(a),(b) and 6.14 (a),(b)) which, in general, is not scaled by the particle size and does not exist in the initial state of the granular material.

It is evident that the main drawbacks of the hollow sphere model are connected with its geometry. However, it is the spherical geometry that provides the tremendous simplification and permits analysis of the process of pore collapse, and allows great insight into the mechanics of this phenomenon.

The assumption of spherical symmetry not only simplifies the analysis but also has very important extreme properties. In particular, it provides the maximum dissipated energy during viscoplastic flow at a given initial porosity. As a result, the spherical symmetry provides a maximum time of collapse in comparison with other possible geometries subject to the same loading conditions. This is based on the fact that no other geometry will result in the uniform material involvement into viscoplastic flow. Based on these considerations, it is possible to conclude that the hollow sphere model provides a reasonable description of the experimental kinematics of powder deformation during shock loading. Nevertheless, it cannot describe either the distribution of internal energy throughout the volume of the individual particles or the characteristic temperature field in the densified powders.

2.3.2 Modification of the Carroll–Holt Model

It is clear from the previous discussion that it is impossible to develop a physically based and relatively simple model that can embrace all facets of granular material deformation during shock loading. Spherical geometry of the model is a very desirable feature for analysis, but should be used with some modifications. Such modifications were proposed by Carroll, Kim, and

Nesterenko [1986], and Nesterenko [1988a,b], 1992]. They included the following assumptions of a geometrical nature in addition to the equality of average densities of the real granular material and the model cell:

(1) For granular materials (see the simplified structure in Fig. 2.6(a)), the particle mass was used to define a size parameter to be used in place of the nonexistent pore size. This means that the characteristic sizes of the model are determined by the condition that masses of the model spherical cell and the mass of the particle in the granular material are equal.

(2) A nondeformable central core with radius c (Fig. 2.6(b)) is introduced in the center of the hollow sphere, which naturally restricts the stage of the spherically symmetrical collapse. This core prevents the convergence of the outer shell from developing to the center, as in the hollow sphere model (Fig. 2.1). It separates the dissipation process into two stages: the first during pore closure and the second after closure. It also allows introduction of the concept of microkinetic energy (Nesterenko [1988a,b]), which is the contribution to the microlevel energy dissipated by the viscoplastic flow of the material after the pore collapse. Also, the concentration of intense plastic deformation in the outer layer of the particle surface is a natural consequence of this initial geometry, in agreement with experiments (Figs. 2.3 to 2.5). Finally, this simple modification helps with the introduction of the important distinction between the quasistatic and the dynamic regime of shock deformation (see Section 2.4).

The third length appearing in the modified geometry is the core radius c (Fig. 2.6(b)). It is determined based on the assumption that the volume of the plastically deformed material (the volume of the outside shell) is equal to the volume of empty space if the initial porosity of the granular material, α_0, is smaller than the minimal porosity α^* achieved by dense packing of the particles.

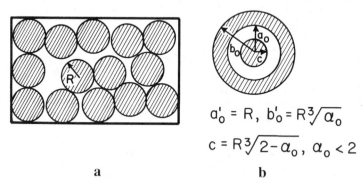

$$a_0' = R, \quad b_0' = R\sqrt[3]{\alpha_0}$$

$$c = R\sqrt[3]{2 - \alpha_0}, \quad \alpha_0 < 2$$

a   b

Figure 2.6. (a) A "real" granular material and (b) geometry of the modified model. (Reprinted from *Journal of the Mechanics and Physics of Solids*, **45**, no. 11/12, Benson, J.D., Nesterenko, V.F., Jonsdottir, F., and Meyers, M.A., Quasistatic and Dynamic Regimes of Granular Material Deformation under Impulse Loading, pp. 1955–1999, 1997, with permission from Elsevier Science.)

This is characteristic of predensified powder. For packing with a porosity larger than α^*, the volume of the outside shell is supposed to be equal to the volume of the empty space corresponding to the porosity α^*. This porosity is a function of particle morphology. For the random packing of spheres, $\alpha^* = 1.66$ (German [1996]).

These three assumptions result in the following relations between model parameters and properties of the powder if $\alpha^* \geq \alpha_0 < 2$:

$$b_0' = R \sqrt[3]{\alpha_0}, \tag{2.19a}$$

$$a_0' = R, \tag{2.19b}$$

$$c = R \sqrt[3]{2 - \alpha_0}, \tag{2.19c}$$

$$\alpha = \left(\frac{a'}{a_0'}\right)^3 + (\alpha_0 - 1), \tag{2.19d}$$

where R is the radius the particles forming the granular material.

For comparison, the radii for a hollow sphere model (Fig. 2.1), based on the equal masses of the hollow sphere and particle mass, are

$$b_0 = R \sqrt{\alpha_0}, \tag{2.20a}$$

$$a_0 = R \sqrt[3]{\alpha_0 - 1}, \tag{2.20b}$$

$$\alpha = \left(\frac{b}{b_0}\right)^3 \alpha_0 = \left(\frac{a}{a_0}\right)^3 (\alpha_0 - 1) + 1. \tag{2.20c}$$

For the initial porosity $\alpha_0 = 1.88$, that is, typical for tapped granular materials with different particle sizes, the internal radii a_0 and a_0' for corresponding models have a very small difference: $a_0 = 0.958 \, a_0'$. This means that the initial kinematics of pore collapse will be the same for both models at this specific initial porosity. It is important that the core radius c is not small in comparison with R, being approximately one-half of it ($c \approx 0.49R$). This estimate agrees with the structure of deformed particles represented in Figures 2.3 and 2.4.

Different radii, a_0' and c, should be introduced if the initial porosity, α_0, is larger than α^*. In this case, powder densification from α_0 to α^* can be achieved as a result of particle repacking without essential plastic flow. This is taken into account by the assumption that the initial volume of the outer shell is equal to the volume of empty space at the state when the porosity is equal α^*. Supposing that the mass of the model cell (core plus shell) is equal to the mass of the particle and that the average density of the model is equal to the initial density of the powder we obtain

$$b_0' = R \sqrt[3]{\alpha_0},$$

(2.21a)

$$a_0' = R \sqrt[3]{\alpha_0 - \alpha^* + 1},$$

(2.21b)

$$c = R \sqrt[3]{2 - \alpha^*}.$$

(2.21c)

Repacking without the essential plastic flow of particles in the porosity interval $\alpha_0 - \alpha^*$ can be modeled by taking the yield strength, Y, of the material to be equal to zero during this stage of compaction.

At the moment when the porosity is equal to α^* the outer and inner radii of the shell are equal

$$b^* = R \sqrt[3]{\alpha^*},$$

(2.22a)

$$a^* = R.$$

(2.22b)

Further densification resulting in porosity smaller than α^* must include viscoplastic dissipation.

As we saw in the previous section, the geometry and material properties of the Carroll–Holt model (Fig. 2.1) are too "hard" to achieve agreement with experiments on the shock front width (Fig. 2.2). Modification is needed for material properties, especially for the introduction of some softening mechanism. For this purpose, Carroll, Kim, and Nesterenko [1986] introduced a linear dependence of yield strength Y and an exponential dependence of viscosity η on temperature:

$$Y = \begin{cases} Y_1(1 - T/T_m), & T \le T_m, \\ 0, & T > T_m, \end{cases}$$

(2.23)

$$\eta = \eta_m \exp B\left(\frac{1}{T} - \frac{1}{T_m}\right),$$

(2.24)

where Y_1 is the yield strength at zero temperature, T_m is the melting temperature, η_m is a viscosity of the melt, and B is a constant. For copper, the relatively high constant yield strength $Y_1 = 400$ Mpa was selected to take into account the effect of strain hardening (which otherwise is not included in the description). The constant B was found from the conditions that η $(T = 293$ K$) = 10^4$ Pas (see Table 2.1) and the viscosity at the melting temperature is equal to $\eta_m = 2 \cdot 10^{-3}$ Pas, a typical value for molten metals. The value $B = 5765$ K^{-1} was used. Exponential-like dependence of the metal viscosity on temperature was used for the description of the high-strain-rate deformation of metals (Godunov et al. [1974]), condensed explosives (Khasainov at al. [1980]), and for crystallization processes in undercooled liquids (Spaepen and Turnbull [1975]). Eq. (2.24) for η

provided a very good agreement between experiments and numerical simulations of the shape of a copper shell during reverse cumulation (Jach [1989]).

These parameters provide a description essentially different from a constitutive equation for the copper obtained from Hopkinson bar experiments (Klopp, Clifton, and Shawki [1985]; Tong, Clifton, and Huang [1992]). It is worth mentioning that the experimental parameters should be selected from appropriate tests, like experiments on explosive welding, describing dissipation typical for high strains which are close to those encountered in pore collapse, or from data on cumulative phenomena. Using the more refined equations of state, with strain hardening and power law strain hardening terms, is possible but can hardly be considered as a more correct approach in conditions when the real state of strain is definitely far from spherically symmetrical. The use of the more exact models can be justified only when the geometry of deformation is close to a real one as in 2- or 3-D computer simulations. As will become clear later, the main reason for the qualitative change of material description in comparison with the classical single pore model is the need for a softening mechanism over the relevant range of temperature.

Tong and Ravichandran [1993], [1994], [1997] used the single hollow-sphere model with an elastic–viscoplastic material that included a more realistic power law constitutive equation with incorporated strain-rate sensitivity, strain hardening, and thermal softening. They obtained good agreement with the available data on the shock front thickness in copper powder measured by Nesterenko [1975], and with the data of Holman, Graham, and Anderson [1993] for $2Al + Fe_2O_3$ powder mixtures.

Equations (2.5), (2.8), and (2.9) can be transformed into the following form using Eqs. (2.20), (2.23), and (2.24)

$$P_m(t) = 2\int_{a'}^{b'} \left\{ Y_1\left(1 - \frac{T}{T_m}\right) - 6\eta_m \exp\left[B\left(\frac{1}{T} - \frac{1}{T_m}\right)\right]\left(\frac{\dot{r}}{r}\right) \right\} \frac{dr}{r}$$

$$- \rho_s\left[\left(a'\ddot{a}' + 2a'^2\right)\left(1 - \frac{a'}{b'}\right) - \frac{1}{2}a'^2\left(1 - \frac{a'^4}{b'^4}\right)\right],$$

(2.25)

$$\rho_s C_v T = -2Y_1\left(1 - \frac{T}{T_m}\right)\frac{\dot{r}}{r} + 12\eta_m \exp\left[B\left(\frac{1}{T} - \frac{1}{T_m}\right)\right]\left(\frac{\dot{r}}{r}\right)^2,$$

(2.26)

$$b' = \left[a'^3 + a_0'^3(\alpha_0 - 1)\right]^{1/3},$$

(2.27a)

$$r = \left[r_0^3 - a_0'^3 + a'^3\right]^{1/3},$$

(2.27b)

$$a' \geq c = R \sqrt[3]{2 - \alpha_0},$$
(2.27c)

$$\alpha_0 \leq \alpha^*.$$
(2.27d)

Initial conditions for the system of equations (2.25)–(2.27) are

$$a'(0) = a'_0, \quad \dot{a}'(0) = 0, \quad T(r_0, 0) = T_0.$$
(2.28)

Boundary conditions for the system of equations (2.25)–(2.27) are represented by Eqs. (2.6) and (2.13). The heat capacity for copper was constant and equal to 385 J/kg K.

It should be mentioned that the distinctive feature of stationary shock loading is not only the fast change of parameters but, mainly, the Raleigh relationship between pressure and density. The linear dependence of pressure on time closely matches this type of relation, as will be shown later.

2.3.3   Exact Solution for a Partial Case

The system of equations (2.25)–(2.28) has the following exact analytical solution in the case if $\eta = 0$ (Carroll, Kim, and Nesterenko [1986]):

$$T = T_m - (T_m - T_0)\left(\frac{r}{r_0}\right)^s, \qquad s = \frac{2Y_1}{\rho_s C_v T_m},$$
(2.29)

where r is the current radius of the point with initial radius r_0.

For the geometry of the modified model (Fig. 2.6), the temperature at the moment of pore closing is finite and equal to

$$T(a = c) = T_m - (T_m - T_0)(2 - \alpha_0)^{s/3} < T_m.$$
(2.30)

The physical meaning of this equation is that the melting temperature cannot be reached during pore closing as calculated within the frame of this model. This exact solution can be used for checking the procedure in computer calculations using single pore geometry.

2.3.4   Results of Computer Calculations and
Comparison with Experiments

The system of equations (2.25)–(2.28) was solved numerically using the fourth-order Runge–Kutta method with the time step $\Delta t = 10^{-9}$ s. The results are presented in Figure 2.7 for the corresponding values of shock pressure P^* and the experimental shock front thickness τ (Nesterenko [1992]). Pore collapse in the modified model with given values of α_0, and α^* from porosity α_0 to complete densification ($\alpha = 1$) is equivalent to the pore collapse of a hollow sphere starting with $\alpha_0 = 2.01$ and finishing at $\alpha = 1.01$. Despite this similarity, pore

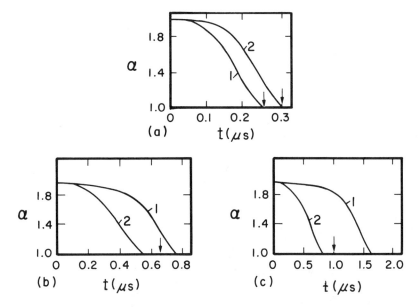

Figure 2.7. Time dependence of porosity for the modified Carrol–Holt model (a'_0 = 0.3 mm, $\alpha_0 = \alpha^* = 1.99$) under different shock loading conditions: (a) 1: $P^* = 14.3$ GPa, $\tau = 0.25$ μs, T_0= 293 K; 2: $P^* = 10.0$ GPa, $\tau = 0.3$ μs, T_0= 293 K; (b) 1: $P^* = 2.4$ GPa, $\tau = 0.65$ μs; T_0= 293 K; 2: $P^* = 2.4$ GPa, $\tau = 0.65$ μs, T_0= 900 K; (c) 1: $P^* = 1.0$ GPa, $\tau = 1.0$ μs, T_0= 293 K; 2: $P^* = 1.0$ GPa, $\tau = 1.0$ μs, T_0 = 900 K. Arrows show the corresponding shock front thickness τ.

collapse in the modified model is finished at an inner radius of the pore equal to c = 0.21R.

It is evident from Figure 2.7 that the modified model successfully describes the kinetics of pore collapse in the wide range of shock pressures 1–14 GPa. The results of calculations are self-consistent in that the collapse process is essentially completed during the shock front duration τ, especially taking into account the uncertainty in the parameters Y and η.

The increase of initial temperature from 293 K to 900 K results in a decrease of densification times at the same parameters τ. Thus, shock front thickness decreases with sample preheating. Self-consistence can be obtained by a slight adjustment of τ. The interesting difference between high (2.4 GPa, Fig. 2.7(b)) and low (1.0 GPa, Fig. 2.7(c)) shock pressures should be mentioned. Initial preheating to the same temperature, 900 K, results in a more significant change of densification kinetics for lower shock pressures.

Self-consistent calculations of shock front thickness as function of pressure were performed for two initial temperatures: T = 293 and 450 K (Fig. 2.8, Carroll et al. [1986]). In these calculations the pressure increase time τ was changed until it reaches τ_α, where τ_α is the time of complete densification.

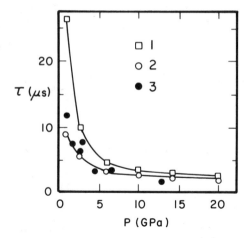

Figure 2.8. Dependence of self-consistent thickness of a shock front on shock pressure at different initial temperatures ($a'_0 = 0.3$ mm, $\alpha_0 = \alpha^* = 1.99$); 1: initial temperature 293 K; 2: initial temperature 450 K; 3: experimental data for initial temperature 293 K.

It is clear from Figure 2.8 that the influence of initial temperature on shock front thickness is significant only at relatively low shock pressures—for copper powder, less than 5 GPa. The explanation of this effect is connected with the dominance of temperature-independent inertial effects on shock front duration at high shock pressures. The theoretical curve for ambient temperature agrees with the experimental dependence of shock front thickness τ on shock pressure.

2.4   The Front Width of a Strong Shock

As the amplitude of a shock wave increases, the kinetics of densification in the shock front become less and less sensitive to the initial properties of the powder material, as was demonstrated in the previous section for changes in initial temperature. At high shock pressures, inertial effects dominate the kinetics of densification. In this case, only particle size, porosity, density, and pressure amplitude effect the shock width. We can say that the front width of a strong shock is a structure sensitive parameter. Experimental data for shock front thickness in granular materials will be considered first.

2.4.1   Experimental Results for Shock Front Thickness

The front thickness of high amplitude shocks should scale as particle size, the only material length scale present in this case. That is why, for reliable measurements of the front thickness, powders with relatively large particles (from 30 μm to 1500 μm) were selected. Shock front thickness was measured with the help of the thermoelectric method for a copper–nickel junction by Nesterenko [1975], [1988c], [1992] and for a copper–constantan junction by Schwarz, Kasiraj, and Vreeland [1986]. These data are presented numerically in Table 2.2 and graphically in Figure 2.8.

Particle velocity profiles also allowing measurements of shock front width were obtained with the electromagnetic method for several granular materials (see Fig. 2.9). The data are presented in Table 2.2. The gauges embedded in the powder were aluminum foil with thickness 80 μm and sizes in plane 8 × 8 mm area (Teflon powder) or 15 × 10 mm area (tungsten and glass powders). The gauge motion in the magnetic field generates an electrical signal proportional to the particle velocity. Dremin and Shvedov [1964] initially used the method for measurement of averaged particle velocity in detonation products. In the case of tungsten powder, the aluminum foil was insulated on both sides by Teflon ribbon with thickness 40 μm. Plane shock wave explosive generators were used to create the shocks (Roman, Nesterenko, and Pikus [1979]; Nesterenko [1992]).

The shock front width, Δ, was calculated using the equation

$$\Delta = (D - u_{av})\tau,$$ (2.31)

where D is the shock front speed, u_{av} is the average particle velocity in the front (which was taken to be equal half of the maximal particle velocity), and τ is the measured duration of the shock front. Eq. (2.31) takes into account that the methods used are Lagrangian ones that measure the particle velocity (or contact temperature) at the given material point.

It is important to mention that the experimental values of τ in (Eq. (2.31)) are affected by other factors in addition to the real shock front width. Nonplanar geometry of the shock front and the irregular shape of the contact between adjacent powder granules also contribute to the measured width.

For example, the thickness of the contact zone between two powders is in the order of magnitude of the particle size and thus affects the thermoelectric measurements. The inertial properties of the foil can also contribute to the measured values of τ and Δ. That is why the data presented in Table 2.2 should be considered as upper limits for the shock front thickness. It is evident that at high pressures for Teflon, tungsten, and glass powders, Δ is comparable to, or even less than, the average particle size, dav. At relatively low shock pressures ($P \approx 0.9$ GPa) for copper and nickel powders, the shock front thickness, Δ, is equal to 3dav and decreases with shock pressure, being less than particle size, for example at 13 GPa.

Sheffield, Gustavsen, and Anderson [1997] observed a strong dependence of shock front thickness on initial particle size in experiments with "weak" shock loading of fine and coarse HMX (the high explosive octotetramethylene tetranitramine) and sugar powders (particle sizes 10–15 μm and 120 μm correspondingly) at initial densities of 1.24 and 1.40 g/cm^3. In coarse HMX the estimated shock thickness was 3–5 and in fine powder it was 6–8 average

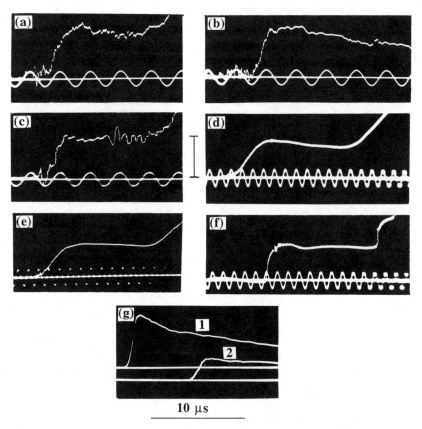

Figure 2.9. Particle velocity–time profiles for various granular materials: (a), (b), (c) Teflon powder, particle size 1.5–2 mm, shock amplitude 3.2 GPa; (d), (e) tungsten powder, particle size 1–2 mm, shock amplitude 0.52 GPa; f tungsten powder, particle size (5–10)·10–3 mm, shock amplitude 0.52 GPa; (g) glass powder, particle size 1.45 mm, shock amplitudes 2.6 GPa (curve 1) and 0.9 GPa (curve 2). Vertical scale is 960 m/s (a), (b), (c), 280 m/s (d), (e), (f), 750 m/s ((g), 1) and 1160 m/s ((g), 2). The frequency of the calibrating oscillations is 1 MHz, the horizontal scale 10 μs is only for part (g) of the figure.

particle diameters. The authors concluded that particle size is apparently setting a length scale for the width of the compaction wave in HMX powder at the conditions of loading investigated. This was generally valid for experiments with HMX and sugar, suggesting that reaction in HMX is not greatly affecting the shock rise time. They do not emphasize the observed difference in front thickness according to the number of particles in fine and coarse powder. One of the physical reasons for this different behavior may be the rate dependence of HMX mechanical properties caused by the order of magnitude difference in strain rates between fine and coarse particles in the compaction wave.

Davison [1971] considered the shock wave structure based on the rate-dependent collapse rule associated with a given departure from equilibrium. The corresponding function in the collapse rule can be selected to fit experimentally observed waveforms at the intermediate shock pressures.

2.4.2   Analytical Estimation of Shock Front Thickness

Mass and momentum conservation laws for stationary shock (Eqs. (2.13) and (2.14)), combined with the relationship between global pressure in the powder, P, and pressure on the outside surface of the model cell $P_m = \alpha P$, allow us to obtain the following equation for the shock wave structure in the frame of the Carroll–Holt model (Nesterenko [1988], [1992]):

$$\frac{\alpha}{\alpha_0^2}(\alpha_0 - \alpha) = \frac{-a_0^2}{3(\alpha_0 - 1)^{2/3}} \frac{d}{d\alpha}\left\{\frac{1}{2}\left[(\alpha - 1)^{-1/3} - (\alpha)^{-1/3}\right]\left(\frac{d\alpha}{d\xi}\right)^2\right\} + \psi(\alpha, \tfrac{d\alpha}{d\xi}, ...). \tag{2.32}$$

For the geometry of the modified model with parameters described by Eq. (2.21) and under the condition that $\alpha_0 \geq \alpha^*$, the equation for the shock front structure, is

$$\frac{\alpha}{\alpha_0^2}(\alpha_0 - \alpha) = \frac{-a_0^2}{3(\alpha_0 - 1)^{2/3}} \frac{d}{d\alpha}\left\{\frac{1}{2}\left[(\alpha + 1 - \alpha^*)^{-1/3} - (\alpha)^{-1/3}\right]\left(\frac{d\alpha}{d\xi}\right)^2\right\} + \psi_1(\alpha, \tfrac{d\alpha}{d\xi}, ...). \tag{2.33}$$

The functions ψ and ψ_1 in the right-hand members of Eqs. (2.32) and (2.33) represent the dissipative terms. In the case of strong shocks in highly porous materials the main input controlling the shock front thickness can be expected to come from the inertial term in Eqs. (2.32) and (2.33) because the characteristic time for pore collapse is determined by the inertial effects associated with the movement of the particle material into the pores. The dissipative term becomes important during the last stage of pore collapse but determines only the part of the shock front where the porosity, α, is close to 1. This is confirmed by the comparison between the shock front width in a solid (where there are no collapse processes involved) with the shock front in porous materials, for example, in solid and porous Teflon (compare Figs. 2.9(a),(b),(c) with Fig. 3.11(b)).

The front width is be determined as follows (Zel'dovich and Raiser [1963]):

$$\Delta = (\alpha_0 - \alpha_f) \Big/ \left(\frac{d\alpha}{d\xi}\right)_{\alpha = \hat{\alpha}}, \tag{2.34a}$$

$$\hat{\alpha} = \frac{1}{2}(\alpha_0 + \alpha_f). \tag{2.34b}$$

For high shock amplitude, material is densified close to the theoretical density, so $\alpha_f = 1$. According to Eq. (2.34), the shock front thickness is determined by the derivative of α with respect to ξ in the middle part of the front. If the dissipative effects are neglected then we obtain the following equations for Δ_{K-H} and Δ_M.

From the Carroll–Holt model,

$$\left(\frac{\Delta_{K-H}}{a_0}\right)^2 = \frac{4 \cdot 2^{1/3}(\alpha_0 - 1)\alpha_0^2}{(2\alpha_0^3 - 3\alpha_0^2 + 1)}\left[1 - \left(\frac{\alpha_0 - 1}{\alpha_0 + 1}\right)^{1/3}\right] \tag{2.35}$$

and from the modified model with $\alpha_0 > \alpha*$:

$$\left(\frac{\Delta_M}{a_0}\right)^2 = \frac{4 \cdot 2^{1/3}(\alpha_0 - 1)^2\alpha_0^2}{(2\alpha_0^3 - 3\alpha_0^2 + 1)(\alpha_0 - \alpha^* + 1)^{2/3}}\left[(\alpha_0 + 3 - 2\alpha^*)^{-1/3} - (\alpha_0 + 1)^{-1/3}\right] \tag{2.36}$$

and, finally, from the modified model with $\alpha_0 \le \alpha*$:

$$\left(\frac{\Delta_M}{a_0}\right)^2 = \frac{4 \cdot 2^{1/3}(\alpha_0 - 1)^2\alpha_0^2}{(2\alpha_0^3 - 3\alpha_0^2 + 1)}\left[(3 - 2\alpha^*)^{-1/3} - (\alpha_0 + 1)^{-1/3}\right]. \tag{2.37}$$

Calculations of the shock front width with the help of Eqs. (2.35) and (2.36) demonstrate that it is approximately one particle radius for porosity more than 2 for both the modified and the original Carroll–Holt models ($\alpha* = 1.82$ is a typical density of approximately spherical granules as obtained by the rotating electrode method).

An essential difference exists between these two models at porosity less than $\alpha*$, with the gap between these two values for shock width increasing with porosity decrease (see Problem 15). The decreased porosity corresponds to a predensified granular material. The modified model predicts a shock width of only one-fifth the particle radius, whereas the Carroll–Holt model gives a shock width of $1.6R$ at an initial porosity of 1.10. geometrical parameters for both models were selected based on the assumption that the mass of the model unit cell is equal to the particle mass. Introducing pore size in addition to the particle radius in the Carroll–Holt model can eliminate the difference. This pore size needs to be decreased with porosity decrease. At the same time, this additional scale is naturally introduced into the modified model at small porosity. Of course, special care should be exercised in dealing with results of the continuum approximation when the characteristic scale in the theory is comparable with the typical size of the structural element. Nevertheless, the unusual tendencies can prompt experimental investigation to support a final decision about the validity or failure of any model.

It is interesting to mention that, at a constant pore radius a_0, the shock front thickness for the Carroll–Holt model increases with a porosity decrease! This behavior is not physically appropriate and is connected with the increased mass and outside radius of the unit cell with a porosity decrease. It results in an artificial increase of mass which is involved in pore collapse, with a subsequent increase of the shock front thickness.

For a shock wave with relatively small amplitude, additional mechanisms can contribute to the shock front thickness. Strain-rate-dependent dissipative processes (Davison [1971]) are very important for the determination of the front thickness and will increase it in comparison with the thickness caused by the inertia effects.

Another important mechanism for weak shock front thickness is connected with a structural aspect of granular material—existence of one-dimensional force chains ("fabric") as was considered in Section 1.9.

2.5   Quasistatic and Dynamic Deformation of Powders Under Shock Loading

2.5.1   Quasistatic–Dynamic Transition Criterion and the Concept of Microkinetic Energy

The continuum description of granular material behavior ignores processes occurring on the mesolevel, e.g., on the scale of the particle contacts during void closure and the peculiarities of the particle deformation. This approach cannot be justified for explaining phenomena with scales comparable with the particle size. Nevertheless, these small-scale processes are extremely important for bonding between particles and the initiation of chemical reactions and phase transitions. Only at relatively large pressures (Pastine et al. [1970]; O'Keeffe [1971]) can the details of the process of compaction be ignored for prediction of the final state.

The dynamics of pore collapse in the shock front and the concentration of the microkinetic energy on the surface of the particles have a qualitative influence on the properties of the material after the dynamic compaction. Shock waves are classified by Nesterenko [1988a,b], [1992] into two different regimes. One is the quasistatic regime, in which geometry of the viscoplastic deformation of the particles is practically the same as obtained in a static press (despite the existing shock wave). The other is the dynamic regime, in which substantial morphological differences are produced in comparison to the static case.

In the dynamic regime, the particle contacts depart from the planar geometry found in statically compressed powders. For example, the triple contact areas (being point-like for the static deformation) have qualitatively different features (localized melting, jets, vortices, etc.) that are unique to dynamic loading. These features are consequences of the dissipation of the microkinetic energy during the last stages of the pore closure.

The quasistatic shock deformation does not result in bonding between the particles. The structurally sound compacts with high strength were obtained only in the fully developed dynamic regime of particle deformation, as is evident from experiments by Kasiraj et al. [1984a,b], Wang et al. [1988], Meyers and Wang

[1988], and Nesterenko [1992], [1995]. Horie [1993] connects the beginning of a chemical reaction in the shock loaded mixture of nickel and aluminium with an eddy-like flow of material within the localized regions exhibiting the dynamic type of shock deformation.

It is therefore important to establish physically based models of the dynamic regime of the particle deformation which result in mutual plastic flow and subsequent bonding of the particle interfaces (Bondar and Nesterenko [1991]).

The phenomenological approach to this problem was proposed by Nesterenko [1988a,b], [1992], [1995]. It is based on a comparison of the energy, E_d, which could be dissipated during dynamic pore collapse with the geometric conditions typical of static deformation, and the internal shock wave energy, E. The corresponding criteria for the transition from the static regime to the dynamic regime can be written as

$$E_d/E = K, \tag{2.38}$$

where K is a constant.

The physical meaning of Eq. (2.38) is that if, at the end of void collapse (into the quasistatic geometry), the amount of the dissipated energy is small enough in comparison to the total energy E, the remaining energy (represented by the microkinetic energy) will produce specific changes in the particle geometry and material structure. This may happen during and after the void closure. This provides the conditions for shear localization, jetting resulting in the pressure–shear deformation and local melting, which are favorable for the initiation of phase transformations, chemical reactions, and bonding between particles. In the quasistatic case, when the microkinetic energy is small during the collapse of the pore, there is no need for the material to undergo plastic deformation geometrically different from the static process of compaction.

It is easy to evaluate E_d with a spherically symmetrical model, e.g., the modified Carroll–Holt model, and for a cylindrically symmetrical model. The spherical symmetry ensures the maximum dissipation of energy for a given material at a specified porosity under quasistatic conditions, as was suggested by Nesterenko [1992]. If the material properties allow the viscous part of the dissipation to be neglected, and a constant magnitude of yield strength, Y, is introduced, then the equation for E_{ds} in the spherical case for the modified model can be written

$$E_{ds} = \frac{2}{3} \frac{Y}{\rho_s} \left[H(x_0) - H(x_1) \right], \tag{2.39a}$$

$$H(x) = x \ln x - (x-1) \ln (x-1), \tag{2.39b}$$

$$x_0 = \frac{\alpha_0}{\alpha_0 - 1}, \tag{2.39c}$$

$$x_1 = \frac{\alpha_1}{\alpha_0 - 1},$$
(2.39d)

$$\alpha = \frac{\rho_s}{\rho_p},$$
(2.39e)

where ρ_s and ρ_p are the density of the solid material and the density of the powder, and α_0 and α_1 are the density ratios (porosity) for the initial and densified states behind the shock. It is necessary to use the initial and final coefficients α_0 and α_1 for the original Carroll–Holt model (Fig. 2.1), instead of x_0 and x_1 for the modified Carroll–Holt model (Fig. 2.6b).

First we compare the energy of plastic deformation, E_{ds}, which is necessary to completely collapse the pore in the spherical geometry, with E_{dc}, the same quantity for the axisymmetric cylindrical model of porous material, to be sure that the two geometries do not produce qualitatively different results per unit mass.

The plastic work increment per unit volume is

$$dw = \sigma_{ef}\, d\varepsilon_{ef},$$
(2.40)

where σ_{ef} and ε_{ef} are the effective stress and effective strain

$$\sigma_{ef} = \frac{1}{\sqrt{2}}\left[(\sigma_1 - \sigma_2)^2 + (\sigma_2 - \sigma_3)^2 + (\sigma_3 - \sigma_1)^2\right]^{1/2},$$
(2.41a)

$$\varepsilon_{ef} = \frac{\sqrt{2}}{3}\left[(\varepsilon_1 - \varepsilon_2)^2 + (\varepsilon_2 - \varepsilon_3)^2 + (\varepsilon_3 - \varepsilon_1)^2\right]^{1/2}.$$
(2.41b)

For the axisymmetric (plain strain) model geometry

$$\varepsilon_1 = \varepsilon_{rr} = \ln\left(\frac{\rho_0}{\rho}\right),$$
(2.42a)

$$\varepsilon_2 = \varepsilon_{\varphi\varphi} = -\ln\left(\frac{\rho_0}{\rho}\right) = -\varepsilon_1,$$
(2.42b)

$$\varepsilon_3 = \varepsilon_{zz} = 0,$$
(2.42c)

$$\varepsilon_{ef} = \frac{\sqrt{2}}{3}\left(4\varepsilon_1^2 + \varepsilon_1^2 + \varepsilon_1^2\right)^{1/2} = \frac{2}{\sqrt{3}}\varepsilon_1.$$
(2.42d)

Using the von Mises criterion,

$$(\sigma_1 - \sigma_2)^2 + (\sigma_2 - \sigma_3)^2 + (\sigma_3 - \sigma_1)^2 = 2Y^2, \tag{2.43}$$

where Y is the yield stress in uniaxial tension, we obtain $\sigma_{ef} = Y$ for a cylindrical geometry. For the rigid–plastic material, the plastic deformation work, w_c, per unit volume at a point having a final radius of ρ is

$$w_c = \frac{2}{\sqrt{3}} Y\varepsilon_1 = \frac{2}{\sqrt{3}} Y \ln\left(\frac{(\rho^2 + R_0^2)^{1/2}}{\rho}\right), \tag{2.44}$$

where R_0 is the initial pore radius.

The work of the plastic deformation per unit mass, E_{dc}, is

$$E_{dc} = \frac{2Y}{\sqrt{3} R_1^2 \rho_s} \int_0^{R_1} \ln\left(1 + \frac{R_0^2}{\rho}\right) \rho d\rho, \tag{2.45}$$

where R_1 is the final outer radius.

Introducing the expression

$$\alpha_0 = 1 + \frac{R_0^2}{R_1^2}, \tag{2.46}$$

for the initial density ratio α_0, and using a table of definite integrals (Gradshteyn and Ryzhik [1980]), the final expression for E_{dc} is

$$E_{dc} = \frac{Y}{\sqrt{3} \rho_s} \left[\alpha_0 \ln \alpha_0 - (\alpha_0 - 1) \ln \alpha_0 - 1\right]. \tag{2.47}$$

It is important that E_{dc} and E_{ds} depend only on the initial density ratio α_0 (porosity). A comparison of Eqs. (2.39) and (2.47) reveals that the geometrically necessary dissipation energies have an identical dependence on the porosity for both geometries, and their ratio is

$$\frac{E_{dc}}{E_{ds}} = \frac{\sqrt{3}}{2} \approx 0.87. \tag{2.48}$$

This difference between the characteristic energies E_{dc} and E_{ds} is too small to qualitatively change the dynamic behavior of the powders.

The expression for the total internal energy per unit mass, E, supplied to the material during shock densification, can be easily obtained if the final density is close to the theoretical solid density ρ_s:

$$E = \frac{1}{2} u^2 = \frac{P}{2\rho_s} (\alpha_0 - 1), \tag{2.49}$$

where u is the particle velocity behind the shock front. From Eqs. (2.38), (2.39), and (2.49), a criterion for the transition from the quasistatic to the dynamic regime is

$$P \geq P_c = \frac{2\rho_s E_d}{K(\alpha_0 - 1)}, \tag{2.50a}$$

$$P \geq P_c = \frac{4}{3} \frac{Y}{K(\alpha_0 - 1)} \left[H(x_0) - H(x_1) \right], \tag{2.50b}$$

where E_d is the geometrically necessary energy for a complete pore collapse in quasistatic geometry, and P_c is the transition pressure. The potential energy is ignored in this consideration because, at initial porosity of approximately 2 and at technologically important pressures (<10 GPa), it is negligible. Equation (2.50a) does not depend on a specific model for the calculation of the geometrically necessary energy E_d. For example, it may be calculated from quasistatic densification data if the strain rate effects are not essential during shock loading. Equation (2.50b) is applied for specific cases of the Carroll–Holt model or the modified model with the corresponding functions $H(x)$ determined by Eq. (2.39b).

Nesterenko and Lazaridi [1990] took Y equal to its initial value Y_0. In this case, all materials can be compared according to their initial yield strength. This approximation is reasonable for some rapidly solidified granular materials if their microhardness does not change drastically during compaction and if the viscous dissipation can be neglected. The coefficient K was taken to be 0.5 for the beginning of the dynamic deformation regime at a critical pressure P_{c1}. Use of the lower value $K = 0.25$ was proposed to describe the transition to the developed dynamic deformation regime at a critical pressure P_{c2}, resulting in structurally sound compacts (with strength larger than half of the value for the particle material). This selection of the coefficient K is rather arbitrary, and depends on the determination of what regime is considered to be truly dynamic. In practice, the existence of jetting, local melting and/or formation of a mushroom-like shape of particles manifest this qualitative transition. The classical transition from laminar to turbulent flow in liquids is a reasonable analogue of the criterion presented. The selection of coefficient K was based on experiments with rapidly solidified steel powders.

Equation (2.50b) is more convenient for practical applications, with an additional replacement of initial yield strength by Vickers microhardness $Y = HV/3$ (Dieter [1986]), because the latter is a measurable parameter for particles of powder. Simple calculations for the initial porosities $\alpha_0 = 1.65$ and $\alpha_1 = 1$ give $P_{c1} \approx HV$ (in kg/mm^2) for the modified Carroll-Holt model. In this case, the critical shock pressure for ensuring a good bonding is $P_{c2} \approx 2HV$ (Nesterenko and Lazaridi [1990]; Nesterenko [1995]). The traditional Carroll–Holt model gives values approximately 50% larger at the same porosity. It is difficult at this stage of research to make a final selection between these models when some uncertainty for distinguishing the dynamic regime is taken into account.

2.5.2 Densification Pressure and Transition to the Dynamic Regime

The criterion for transition from the quasistatic to the dynamic regime (data presented in Table 2.3) should not be confused with the "engineering" correlation

Table 2.3. Microhardness and regimes of deformations.

Material of powder	Initial (final) microhardness (GPa)	Shock pressure (GPa)	Type of regime	Source
AISI 9310 Steel 44 μm < d < 74 μm ($\alpha_0 = 1.65$)	3.44 ± 0.34 (4.13 ± 0.26)	5.9	Dynamic	Kasiraj et al. [1984a]
304SS-CA 100 μm < d < 150 μm ($\alpha_0 = 1.54$)	2.4 ± 0.2 (3.95 ± 0.16)	3.7	Dynamic	Flinn et al. [1988]
304SS-VGA 30 μm < d < 150 μm ($\alpha_0 = 1.54$)	2.0 ± 0.3 (3.30 ± 0.25)	3.7	Dynamic	Flinn et al. [1988]
IN 718 74 μm < d < 88 μm ($\alpha_0 = 1.67$)	2.3 (4.70)	3	Dynamic	Wang et al. [1988]
EP-450 Steel (as is) 310 μm < d < 440 μm ($\alpha_0 = 1.63$)	6.26 ± 0.6 (6.30 ± 0.05)	6	Quasistatic	Nesterenko et al. [1989]
EP-450 Steel (as is) 310 μm < d < 440 μm ($\alpha_0 = 1.63$)	6.26 ± 0.6	14	Dynamic	Nesterenko et al. [1989]
EP-450 Steel (annealed) 310 μm < d < 440 μm ($\alpha_0 = 1.63$)	2.9 ± 0.4 (4.7 ± 0.15)	6	Dynamic	Nesterenko et al. [1989], [1990]; Lazaridi [1990]
40 μm < d < 70 μm ($\alpha_0 = 1.60$)	2.9 ± 0.40 (4.4 ± 0.20)	6	Dynamic	

between the initial powder hardness and the pressure required for shock densification $P_{con} \approx HV$ (Prummer [1986]; Meyers [1994]). The former criterion, given by Eq. (2.50), is a pressure scaling based on providing a certain amount of excess total energy over that geometrically necessary for complete densification, and is a relatively strong function of the initial porosity. Only for the modified Carroll–Holt model, and for a material without strain hardening, and having the initial porosity $\alpha_0 = 1.65$, the critical pressure $P_c = HV$ coincides with P_{con}. This coincidence should be considered as accidental and may be partially based on the uncertainty in the determination of pressure in an engineering correlation between HV and P_{con}.

The general sequence of shock loading at a relatively long duration of high pressures, allowing plastic flow to be accomplished, can be presented as follows. At $P = P_{con}$ the powder attains a density close to the solid density, ρ_s, during the action of shock. The regime of deformation at this shock pressure is mainly quasistatic. At the higher pressure, P_c, which is approximately two times higher (if no strain hardening occurs), the transition to the dynamic regime can be observed. A further increase of shock pressure (approximately two-fold) results in a well-developed dynamic regime facilitating a strong bonding between particles or a chemical reaction in energetic materials.

2.5.3   Comparison with Experiments

Experimental data, from papers where shock pressure in the first wave was reported (or can be calculated), and microhardness and microstructural features allowing one to identify regimes of loading, are shown in Table 2.3.

The dynamic regime was identified by the presence of local jetting, melt, and/or formation of a mushroom-like particle shape. The existence of the dynamic regime of deformation agrees with the proposed criterion $P > P_{c1} \approx HV$ for different materials investigated by different authors.

The importance of this approach, which was based on experimental results for one material and works reasonably well for others, is in clarifying the tendencies that can be expected if the material properties are changed. For example, P_{c1} in Eq. (2.50) does not depend on the particle size, being in agreement with data for EP-450 (annealed) steel powder (see Table 2.3). This is also in agreement with the independence of the strength of the compact on the initial particle sizes of the powders (Nesterenko [1992]).

The critical pressure does not depend on the solid density but is a strong function of the initial porosity. It is necessary to mention that the criterion described by Eqs. (2.50) is valid only if viscous dissipation, strain and strain-rate hardening, and thermal softening can be ignored. For metals with essential strain hardening, this approach can be used to estimate (at least) a lower bound on the transition pressure, as is clear from Table 2.3.

The critical pressure is expected to be dependent on the particle size and ρ_s, if the viscous effects dominate the dissipation stage of the quasistatic deformation, which is natural for strain rate sensitive materials in general (e.g., for polymers) (Nesterenko [1992]).

2.5.4   Restrictions of Models Based on the Single-Pore Approach

The micromechanical models based on the single-pore approach play a very important role in the analysis of the dynamic processes, and they provide guidelines in understanding the densification kinetics of powders and granular materials. The single-pore models are useful when their application results in analytical equations and qualitative predictions that may be compared to experiments. Unfortunately, the single-pore approach has very strong geometrical constraints. It does not take into account the real geometry of the particles and the pores, the resulting nonuniform plastic flow, or the pore collapse in porous mixtures of particles with different mechanical properties. It also neglects the mutual interactions of the collapsing pores through the nonuniform pressure field surrounding the pores. The essential disadvantage of the single-pore models is that, for a strong shock, the actual shock front thickness is close to the particle size, which makes very questionable any continuum level approach to understanding the processes on the shock front. One of the principal problems of this approach is the impossibility of separating the two different mechanisms of pore collapse: that due to the gradient of particle velocity and that due to the plastic flow resulting from pressure-assisted collapse.

Moreover, even a "perfectly" cylindrically symmetrical pore collapse is accompanied by shear localization and by a subsequent break of the cylindrical symmetry during the last stages of pore closure. This was experimentally shown by Nesterenko and Bondar [1994], Nesterenko, Meyers, and Wright [1998], Stokes, Nesterenko, Shlacter et al. [2001] for several materials. The existence of strong shear localization during pore collapse was shown by Tang, Liu, and Horie [1999], [2000] in the frame of a 2-D Discrete Meso-Element Dynamic Method (Tang, Horie, and Psakhie [1997]).

Approaches that neglect the aforementioned problems, while trying at the same time to saturate the single-pore model with "exact" material behavior models, heat conductivity effects, melting, etc., have rapidly diminishing returns.

It has been demonstrated that the material structure is responsible for the energy distribution among the components of powder mixtures, and that an increase in particle size can result in an increase of the particle contact non-equilibrium temperature, as was demonstrated by Nesterenko [1975], [1992]. The particle morphology determines the thermomechanical response of the components in porous mixtures, and the possibility of chemical reactions among the components (Krueger, Mutz, and Vreeland [1992]; Tadhani, Dunbar, and Graham [1994]).

Meyers, Benson, and Olevsky [1999] presented an analysis of energy dissipation during shock consolidation and emphasized that reduction of particle size can be useful for avoiding cracks opening at the stage of pressure release. Therefore, it is very important to develop models that take into account the real geometry of the powder and the resulting nonuniform plastic flow. This can be very important for the identification of the dynamic mechanism of particle deformation.

Although the boundary between the quasistatic and dynamic regimes is of a qualitative nature and is determined a priori, the vastly different behavior of the granular material in these two regimes, resulting in different properties of the compacts, justifies their separation. The situation resembles the transition from laminar to turbulent flow or the onset of plastic flow in solid materials. It is important to establish the dependence of the transition pressure on the material properties and the morphology of powder particles, or at least to clarify the main tendencies.

Computer modeling can be a powerful instrument to determine the qualitative regimes of powder deformation as a function of their mechanical properties and particle morphology, and to clarify the main tendencies in powder behavior.

2.6   Two-Dimensional Numerical Calculations of Shock Densification

2.6.1   Numerical Modeling of the Ordered Packing of Particles

A qualitatively different approach to this problem, microlevel numerical modeling, was first introduced by Williamson and Berry [1986]. This modeling takes into account a more realistic geometry of the particles and the void space. They considered a two-dimensional model of a closest-packed unit cell, which was represented by three layers of stainless steel (304 SS) cylinders with diameters of 75 μm, placed within rigid boundaries on the bottom and on the sides. The compaction was initiated by the impact of a stainless steel flyer having an initial velocity of 1 km/s. An elastic/perfectly plastic model without work hardening or rate dependence was used to model the powder particles. The conduction heat transfer effects were included, along with the melt transition and the dependence of the material strength on temperature. The model predicted the localization of plastic deformation on the particle boundaries, resulting in the concentration of heat and local melting on the surfaces of the particle. Williamson and Berry recognized that the introduction of friction on the particle contacts is not necessary for plastic flow localization on the particle contacts.

Williamson, Wright, Korth, and Rabin [1989], [1990] investigated the dynamic compaction of a porous mixture of two different materials (SiC fiber-reinforced aluminum matrix composite) with a two-dimensional computer model and experiments. For impact plate velocities greater than 0.5 km/s, melting of the aluminum matrix was predicted in the regions of the greatest strains, which is in accord with experimental observation. The pores with the greatest volume (without the SiC fibers) are the sites of largest thermal energy deposition during consolidation. It is worthwhile mentioning that the pore collapse in this model results from a plastic flow of approximately one-half of the material (from the side of the shock propagation) around the pore. This is in disagreement with the predictions made using the spherical single-pore model.

The authors also mentioned the occurrence of local high-pressure spikes at the moment of pore closure, which clearly demonstrates that plastic flow during the pore closure is unable to dissipate all of the shock energy. The remaining energy is represented by the microkinetic energy, which is dissipated into the fully compacted material. These features cannot be taken into consideration within the limited framework of the hollow sphere model.

Flinn, Williamson et al. [1988] and Williamson [1990] have undertaken an extensive study of the dynamic compaction of a granular material (304 SS) with various particle morphologies (monosized and bimodal particle size distribution, and a matrix of identical hollow cylinders) within the framework of a two-dimensional model. The influence of the gas trapped in the interstitial space in a monosized system was also considered. The material strength in the solid phase is approximated by an elastic, perfectly plastic constitutive relation without work hardening, but which includes the effects of pressure and temperature on the shear modulus and flow stress. An important result of this paper is the demonstration (according to determination by Nesterenko [1988a,b], [1992], [1995], of the transition from quasistatic (projectile impact velocity 0.5 km/s) to the dynamic (impact at 1 km/s and 2 km/s) regime of powder compaction. At impact velocities of 1 km/s and 2 km/s, heating of the particles near their surfaces results from the plastic flow of the material into the empty space and, to a comparable extent, from the local high pressures resulting from pore collapse. It demonstrates the qualitative importance of the collective character of the pore dynamics on the heterogeneous heat release.

The system with a bimodal particle distribution has a decrease in the heating in the vicinity of the large particle surfaces in comparison with a monosized system (note that the porosities of the systems are different). Also, the relative degree of deformation of small particles is much greater than for the large ones, as was seen in experiments by Nesterenko [1985], [1992]. The analysis of the porous material with the hollow particles demonstrated that the final temperature distribution is qualitatively different from that of the solid particles, resulting in a greater energy density at complete consolidation in the case of the solid particles. The collapse of the cylindrical holes created hot spots separated at a distance of approximately the radius of the initial hole. A gas trapped in the interstitial spaces greatly increased the local temperature on the particle surfaces, but only slightly changed the internal energy and the size of the melted zones.

The results of these papers on the microlevel numerical modeling of ordered particle packings provide important, qualitatively new information about the material behavior during dynamic consolidation. This information is, in many respects, outside the range of that obtainable from single-pore models. Nevertheless, these results are only applicable for a powder element with a few cylindrical particles packed with an initial porosity 1.1, which is quite different from the typical porosity 1.5–1.6 for approximately spherical particles. Within the framework of this approach, it is impossible to consider the complicated morphology of real granules, which may result in qualitative differences with experiments. The small unit cell considered in their investigations cannot be used to obtain averaged parameters of the material (e.g., pressure or wave speed) which could be compared with the values measured in experiments. The geometry of the model does not permit comparison with the state behind a stationary shock wave and leaves open the question as to whether these results are applicable for stationary shocks, or if they are typical only for an experimental geometry close to the idealized one of the computer calculations. The latter permits only symmetric material flow according to the wave propagation direction.

2.6.2   Geometry of the Model with Randomly Packed Particles

One of the distinguishing features of real granular materials is that they are usually composed of particles with different sizes and are randomly packed even if the particles are of identical sizes. This determines the surface area of particles and the local variation of porosity affecting the energy deposition along the particle boundaries. It is very important to incorporate a more realistic particle geometry and packing into the computer modeling. Benson [1994] was the first to introduce random packing into the 2-D computer modeling of dynamic processes, and the main content of this section is based on a paper by Benson, Nesterenko et al. [1997].

For initial random packing, in contrast to ordered packing, a representative volume element must contain a moderate-to-large number of particles to represent the powder. A Monte Carlo technique, which may be based on an experimentally measured particle size distribution, was combined with a pseudo-gravity particle packing method to generate an initial particle distribution with over 100 particles (Benson [1994]).

The different initial geometry of the packing of cylindrically symmetrical particles with various sizes and porosities is shown in Figure 2.10. The pseudo-gravity method for cylindrically symmetric particles typically packs them to a density of approximately 80% of the solid density (Fig. 2.10(a),(c)). Most granular materials have a density closer to 50% of the solid density. This density gap may be overcome by randomly deleting individual particles until the correct density is obtained from the initial pseudogravity packing. This approach creates large localized pores representing a global porosity (as can easily be noticed, starting with Figs. 2.10(a),(c)) and each particle faces a large hole from one side and relatively dense packing from the other. Calculations based on this method show a large amount of jetting, even at low shock pressures. In the experiments, the geometry of particle deformation is very similar to that produced under quasistatic compaction by a weak shock. The reason for the discrepancy is that

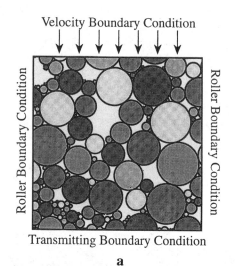

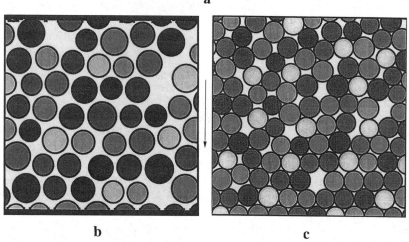

Figure 2.10. The initial geometry of particles in 2-D numerical calculations:
(a) particles with porosity $\alpha_0 = 1.23$ having a large spread in sizes (0.5–10 µm);
(b) particles with porosity $\alpha_0 = 1.67$ having a small spread in sizes (310–440 µm);
(c) particles with porosity $\alpha_0 = 1.23$ having a small spread in sizes (74–88 µm).
(Reprinted from *Journal of the Mechanics and Physics of Solids*, **45**, no. 11/12, Benson, J.D., Nesterenko, V.F., Jonsdottir, F., and Meyers, M.A., Quasistatic and Dynamic Regimes of Granular Material Deformation under Impulse Loading, pp. 1955–1999, 1997, with permission from Elsevier Science.)

this approach produces holes with sizes which are absent in real 3-D granular material, where "pores" are distributed more or less uniformly around a particle surface.

Benson [1994] and Benson et al. [1997] proposed another approach to create the initial packing of cylindrical particles with $\alpha_0 \sim 2$ and where the initial porosity is more uniformly distributed. The particles are initially packed using the pseudogravity method, and then the diameters of the packed particles are rescaled by $1/S$ while keeping their locations fixed. The factor S is chosen so that the final density after the rescaling is equal to the desirable powder density. The original size distribution may be taken from the experimental data for granular material. In this method, the cylindrical particles are not in contact each with other (Fig. 2.10(b)) despite having the same density as in 3-D granular material. This is due to the difference between the cylindrical and spherical geometry of particles. Nevertheless, this packing may be a reasonable approximation for strong compaction waves for which the final density behind the shock is comparable with the solid density and where the particles are heavily deformed. This conclusion is based on the fact that the total dissipated energy behind the shock front depends only on the initial and final densities, and not on the details of packing.

Of course, the initial packing represented in Figure 2.10(b) does not reflect some important qualitative features like the "force chains" present in 3-D granular materials. Therefore, it cannot be used for relatively weak shocks where the kinetics of the deformation is predetermined by the initial contacts between particles. Another important difference is that, in general, this geometry is not able to reproduce the specific surface of particles (surface per unit mass) despite having the same density as the real granular material. The main consequence of this discrepancy may be a numerical difference in the thickness of the heavily deformed surface layer or in the amount of localized shear strain on the particle surfaces. It may also affect the final distribution of contact temperatures in comparison with 3-D granular materials. Development of computer codes which will describe shock waves in 3-D random packing geometry of the spherical particles is very desirable and the first results were recently presented by Baer [1999], [2000] and Baer, Kipp, and van Swol [2000].

2.6.3 Boundary Conditions

A shock wave generated by a velocity boundary condition imposes the particle velocity **u** at one end of the powder (Fig. 2.10(a)). To minimize the numerical oscillations caused by an abrupt change in the velocity profile, the velocity of the piston generating the shock is ramped with the quadratic function

$$u(t) = u \min \left(1, (t/t_{\text{blend}})^2\right), \tag{2.51}$$

where u is the steady state particle velocity and t_{blend} is equal to the first output step time.

The left and right boundaries are symmetry planes, with normal and tangential unit vectors $\mathbf{n}$ and $\mathbf{t}$, respectively, which impose the following conditions

$$\mathbf{u} \cdot \mathbf{n} = 0, \quad \boldsymbol{\sigma} \cdot \mathbf{t} = 0 \tag{2.52}$$

on the particle velocity $\mathbf{u}$ and Cauchy stress $\boldsymbol{\sigma}$. A transmitting boundary condition, developed by McGlaun [1982], permits the wave to pass through the powder without reflection.

2.6.4   Material Modeling

The numerical solutions for the model problem were obtained with a multi-material Eulerian finite element program (Benson [1995]). The Eulerian mesh is fixed in space and the material is transported through it. A high-resolution interface reconstruction method developed by Youngs [1982] resolves the material boundaries within each element. The transport between the adjacent elements is computed with the second-order accurate monotonic MUSCL method developed by van Leer [1977]. A computational mesh of 100 elements on each side, for a total of 10,000 elements in two dimensions, is used in the calculations. Test calculations with 200 elements on each side have demonstrated that the higher mesh resolution is not needed for the present purposes.

The plasticity model of Steinberg and Guinan [1978], [1996] and Steinberg, Cochran, and Guinan [1980], and the Gruneisen equation of state, are used to describe the material behavior at high strains, strain rates, and pressures. The differential equations for the Steinberg–Guinan plasticity model follow the well-established framework of J_2 flow theory with the isotropic hardening defined by

$$\dot{\sigma}' = 2G\left(\dot{\varepsilon}' - \dot{\varepsilon}_p\right), \tag{2.53}$$

$$\varepsilon_{p,ef} = \int_0^t \sqrt{\tfrac{2}{3}\dot{\varepsilon}_p : \dot{\varepsilon}_p}\, d\tau, \tag{2.54}$$

$$\sigma_y = \sqrt{\tfrac{3}{2}\sigma' : \sigma'}, \tag{2.55}$$

where σ' is the deviatoric part of the Cauchy stress tensor, ε' is the deviatoric strain tensor, ε_p is the plastic strain, and $\varepsilon_{p,ef}$ is the equivalent plastic strain. A dots on the top of the symbols mean derivative of Cauchy stress tensor with respect to time, strain rate tensor and plastic strain rate correspondingly. These equations are integrated numerically using the radial return method, e.g., Krieg and Key [1976].

Both the shear modulus, G, and the yield strength, σ_y, are temperature and pressure dependent and have the same general functional form (Steinberg,

Cochran, and Guinan [1980]). The strain-rate dependence is not taken into account in these calculations, which can be reasonable in the case of high-strength materials (e.g., rapidly solidified granules and ceramics) and where large strains are involved, as is the case with powder densification. The temperature dependence is expressed in the constitutive model in terms of the internal energy, E, per initial volume, V_0. When the internal energy is below the "melting" energy, E_m, the shear modulus is

$$G(P, T) = G_0 \left\{ 1 + AP/\eta^{1/3} - B(T - T_{ref}) \right\} \exp\left(-fE/(E_m - E)\right), \tag{2.56}$$

where G_0, A, B, and f are material constants, T_{ref} is the initial temperature, and η is the compression V_0/V. The yield strength has the same functional form

$$Y = Y_0 f(\varepsilon_p) \left\{ 1 + A' P/\eta^{1/3} - B(T - T_{ref}) \right\} \exp\left(-f' E/(E_m - E)\right) \tag{2.57}$$

with the work hardening law

$$Y_0 f(\varepsilon_{p,ef}) = Y_0 \left[1 + \beta \left(\varepsilon_i + \varepsilon_{p,ef} \right) \right]^n \leq Y_{max}. \tag{2.58}$$

If $E > E_m$, then G and Y are set equal to zero.

The "melting" energy, E_m, the energy necessary to increase the temperature to the melting point, is defined as

$$E_m = E_c + c_p T_m. \tag{2.59}$$

This definition of melting energy does not include the latent heat of melting, so it does not account for the actual amount of the melted material, or yield the correct temperature when it is higher than the melting temperature. But, in our calculations, we are primarily interested in the internal energy dissipated during the plastic deformation, and suppose that the achievement of the melting temperature will result in material properties of the liquid phase even without an additional supply of latent heat. This is a reasonable approximation because the change of mechanical properties during transition, from solid at the melting temperature to liquid at the same temperature, is negligible.

In accord with this consideration, Eqs. (2.57) and (2.58) ensure the gradual approach to zero of Y and G as the internal energy, E, approaches E_m. Beyond the melting temperature, the internal energy is correct but the calculated temperature is inaccurate.

The melting temperature, T_m, calculated using a modified Lindemann law, following Steinberg, Cochran, and Guinan [1980], is approximated as

$$T_m = T_{m0} \exp\left\{ 2a(1 - 1/\eta) \right\} \eta^{2(\gamma_0 - a - 1/3)}, \tag{2.60}$$

where T_{m0} is the melting temperature at ambient pressure and a is the first-order volume correction factor for the Gruneisen parameter γ:

$$\gamma = \gamma_0 + a\left(\frac{1}{\eta} - 1\right) = \gamma_0 + a\frac{V - V_0}{V_0}. \tag{2.61}$$

The temperature T is defined by

$$T = \frac{E - E_c}{c_p}, \tag{2.62}$$

where E_c is the cold compression energy. The specific heat c_p is assumed to be

$$c_p = 3\frac{R\rho_0}{M}, \tag{2.63}$$

where R is the gas constant, M is the atomic mass, and ρ_0 is the reference (initial) density. For the calculation of E_c the following equation was used

$$E_c(\eta) = \int_1^\eta P(\eta)\frac{d\eta}{\eta^2} - 300c_p\exp\left[a\left(1 - \frac{1}{\eta}\right)\right]\eta^{\gamma_0 - a}, \tag{2.64}$$

where the first term is the integral on the zero Kelvin isotherm. The energy is supposed to be zero at $T = 300$ K in an uncompressed material where $E_c = -300c_p$.

The Gruneisen equation of state used for calculating the pressure in states of compression is

$$P = \frac{\rho_0 C_0^2 \mu\left[2 + (2 - \gamma_0)\mu - (\gamma_0 - a)\mu^2\right]}{2\left[1 - (S_1 - 1)\mu - S_2\mu^2/(\mu + 1) - S_3\mu^3/(\mu + 1)^2\right]^2} + \left[\gamma_0 + (\gamma_0 - a)\mu\right]E \tag{2.65}$$

and in tension ($\mu = \eta - 1 < 0$), the pressure is

$$P = \rho_0 C_0^2 \mu + \gamma_0 E, \tag{2.66}$$

where C_0 is the sound speed and S_1, S_2, and S_3 are the coefficients in the shock velocity (u_s)–particle velocity (u_p) relation:

$$u_s = C_0 + S_1 u_p + S_2\left(\frac{u_p}{u_s}\right)u_p + S_3\left(\frac{u_p}{u_s}\right)^2 u_p. \tag{2.67}$$

For metals, the linear approximation is very accurate in the pressure ranges considered in the present investigation. The parameters which were used in the calculations are presented in Table 2.4.

Details of the computational model can be found in Benson [1992]. Data for stainless steel were taken mostly from data for SS 304 (Steinberg, Cochran, and Guinan [1980]; Steinberg [1996]). The most important stainless steel parameters for this type of calculation were obtained with $Y = HV/3$. Measurements of initial microhardness were used to calculate Y_0. The maximum microhardness, corresponding to the heavily deformed peripheral layers of the particles of alloy EP-450 after densification without melting, was used to calculate Y_{max} (Nesterenko et al. [1989]). Data for the Ni-based alloy 1 are also taken from Steinberg [1996] data for Ni, except for the values for β, n, ρ_0, and Y_0 which were taken from the properties of the Ni-based alloy IN718 investigated by Wang, Meyers, and Szecket [1988].

Table 2.4. Materials parameters.

Parameters	Ni-based alloy 1	Stainless steel
G_0 (Mbar)	0.85	0.77
Y_0 (Mbar) ·	0.007136	0.0205
b	2.75	43
n	0.475	0.35
Y_{max} (Mbar)	0.012	0.024
A (Mbar^{-1})	1.63	2.6
A' (Mbar^{-1})	1	2.6
q	1	1
B (K^{-1})	0.000326	0.00045
f	0.001	0.001
M	58.7	55.35
T_{m0} (K)	2330	2380
γ_0	1.93	1.93
a	1.4	1.4
ρ_0 (g/cm^3)	8.26	7.77
C_0 (cm/ms)	0.465	0.457
S_1	1.445	1.49

Two-dimensional computer calculations were carried out for Ni-based alloy 1 at $\alpha_0 = 1.67$ and $\alpha_0 = 1.23$ for three different particle velocities behind the shock front: 0.22 km/s, 0.75 km/s, and 1.1 km/s. The particle sizes were chosen to have uniform distributions in the intervals 0.74–0.88 µm, 2.22–2.64 µm, and 74–88 µm.

According to the phenomenological criterion of Eq. (2.38), this set of particle velocities guarantees the transition from the quasistatic to the well-developed dynamic regime of particle deformation for the Ni alloys.

The shock densification of the stainless steel powders was investigated at α_0 = 1.64 for two particle velocities behind the shock front: 0.7 km/s and 1.29 km/s, and for particle sizes 310–440 µm.

Additional computer calculations were made for artificial materials with one of the parameters different from the set presented in Table 2.4:

(1) Y_0 was decreased by one order of magnitude from 0.0205 to 0.00205 Mbar for stainless steel.

(2) Y_0 was set to 0 retaining the other parameters for the IN-718 (fluid model).

(3) The Ni-based alloy 2 with density 8.9 g/cm³ and with the other parameters for the Ni-based alloy 1.

(4) The density was increased to 19.35 g/cm³ with the other parameters for the Ni-based alloy 1.

It is important to mention (this will be discussed in detail later) that no qualitative dependence of the powder behavior on the initial particle size was observed for a fixed particle velocity in the frame of the material model presented.

2.6.5   Results of the Numerical Calculations

To evaluate the results of the computer calculations, the mean values of the parameters were defined by

$$\Phi_m(y) = \frac{1}{\ell} \int_0^\ell \Phi(x, y) \, dx, \tag{2.68}$$

where $\Phi_m(x, y)$ is a solution variable, x is parallel to the shock front, and y is the distance from the wall at which the velocity boundary condition is imposed. These data are plotted as functions of y in Figures 2.11 to 2.18.

The parameters calculated in every element include the total kinetic energy

$$E_{\text{Tkin}} = \sum_{i=1}^{NMAT} \frac{1}{2} \rho_i u_i u_i V_{f,i}, \tag{2.69}$$

the internal energy

$$E_{int} = \sum_{i=1}^{NMAT} \left[E_{c,i} + \int_0^t \sigma_i' : \dot{\varepsilon}_{p,i} \, dt \right] V_{f,i},$$ (2.70)

the pressure

$$P = \sum_{i=1}^{NMAT} P_i(\rho_i, E_i) V_{f,i},$$ (2.71a)

and the density

$$\rho = \sum_{i=1}^{NMAT} \rho_i V_{f,i}.$$ (2.71b)

Here $NMAT$ is the number of materials in the element at the current time, ρ_i and $V_{f,i}$ and $V_{c,i}$ are, correspondingly, the current density, the volume fraction, and the current volume of the ith material. The two latter parameters are connected as follows:

$$V_{f,i} = \frac{V_{c,i}}{\sum_{j=1}^{NMAT} V_{c,j}},$$ (2.72a)

$$\sum_{i=1}^{NMAT} V_{f,i} = 1.$$ (2.72b)

Note that $V_{c,i}$ is not the specific volume of the ith material. Additionally, two mesoscopic parameters that were calculated are the microkinetic energy

$$E_{mkin} = E_{Tkin} - E_{Mkin},$$ (2.73a)

$$E_{Mkin} = \frac{1}{2}\rho u_m u_m,$$ (2.73b)

$$u_m(y) = \frac{1}{l} \int_0^l u_y(x, y) \, dy,$$ (2.73c)

and the spin

$$\omega = \frac{1}{2}(\frac{\partial u_2}{\partial x_1} - \frac{\partial u_1}{\partial x_2}),$$ (2.74)

where E_{Mkin} is the macrokinetic energy.

Note that for the calculation of the microkinetic energy, the mean value of the velocity u_m in its definition is used. These parameters are not measurable in the experiments but they are very important to the overall material behavior, e.g., bonding between particles or the initiation of chemical reactions under shock loading.

2.6.5.1 Observed Stationary Shock Profile Versus Results of Numerical Calculations

The traditional approach to calculation of the material parameters behind the shock waves is based on the stationary condition for the shock front. The compression impulse (Figs. 2.11(a),(b)) in the computer calculations, as described by the total kinetic energy and the internal energy profiles, has the typical features of a shock profile. But, with the relatively small number of particles (in comparison with the experiments), the interaction of the shock front with the boundary, which separates the rigid piston from the compacted powder, can result in significant differences in comparison to stationary shocks.

Therefore, it is desirable to compare the averaged values of the maximum pressures in the computer calculations, P_{cal}, with the calculated values of the pressures, P_{st}, based on data for the particle velocity, the final density from the computer model, and the conservation laws for mass and momentum in a stationary shock (Benson, Nesterenko, and Jonsdottir [1996]; Benson, Nesterenko Jonsdottir, and Meyers [1997]):

$$P_{st} = \frac{\rho_f u^2}{\left(\dfrac{\rho_f}{\rho_0} - 1\right)}. \tag{2.75}$$

Additional data for different particle sizes, strengths and porosities can be found in a paper by Benson, Nesterenko et al. [1997].

For the pressure, the average values can be significantly different from the maximum and minimum values (Fig. 2.12(b),(d)). Arrows show the maximal (P_m) and minimal (P_{min}) pressures used to calculate the P_{cal}. In this case, P_{cal} was defined as $(P_{max} + P_{min})/2$, where the corresponding maximum and minimum pressures (shown by arrows in Fig. 2.12(b),(d) were taken at the moment after complete compaction ($\alpha = 1$).

Table 2.5 presents the values of the pressures, P_{cal}, obtained in computer calculations and the values of the pressures P_{st}, for the stationary shock for a material having small particles and porosity $\alpha_0 = 1.67$.

From a comparison of P_{cal} and P_{st} in Table 2.5, we conclude that there is qualitative correlation between them and a satisfactory quantitative agreement for the range of pressures investigated (see also data from Benson, Nesterenko et al. [1997]. The relatively large pressure spread characteristic of the computer results reflects the actual large spread of local pressures in and immediately following the shock front that was found in the experimental results.

In the conditions investigated, the pressure spread from the calculated average values was in the interval of 7–34%, except in the case of large particles, where at the particle velocity 0.75 km/s, this value was 52%. In contrast to the pressures, the spread in the densities and the average particle velocities in the direction of shock propagation was much less (5–10%). This is due to the strong nonlinear dependence of the cold (and overall) pressures on the density, which

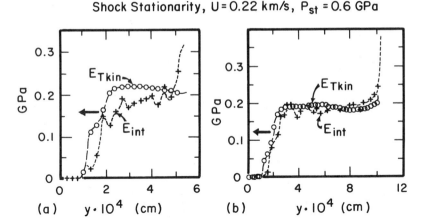

Figure 2.11. Stationarity of the shock in computer calculations for Ni-alloy 1. Shock profiles for the total kinetic energy E_{Tkin} and for internal energy E_{int} at different distances from shock entry: (a) profiles for "short" system; (b) profiles for the "long" system. The porosity $\alpha_0 = 1.67$, and the particle size fell in the range 0.74–0.88 μm. (Reprinted from *Journal of the Mechanics and Physics of Solids*, **45**, no. 11/12, Benson, J.D., Nesterenko, V.F., Jonsdottir, F., and Meyers, M.A., Quasistatic and Dynamic Regimes of Granular Material Deformation under Impulse Loading, pp. 1955–1999, 1997, with permission from Elsevier Science.)

Table 2.5. Data for shock in an Ni-alloy 1 powder with small particles with diameters in the range $d = 0.74$–0.88 μm and an initial porosity of 1.67.

u (km/s)	ρ_1 (g/cc)	P_{cal} (GPa)	P_{st} (GPa)
0.22	8.3	0.9	0.6
0.22(long)	8.3	0.6	0.6
0.75	8.3	4.5	6.9
1.1	8.9	12.2	13.5

magnifies the oscillations of the density behind the shock wave. In the cases of the total kinetic energy, internal energy, and particle velocity of the porous material, the computer calculations provide a relatively small data spread (5–10%), as shown in Figure 2.11(a),(b).

Another important feature of the stationary shock conditions is that the internal energy per unit mass must be equal to the corresponding macrokinetic

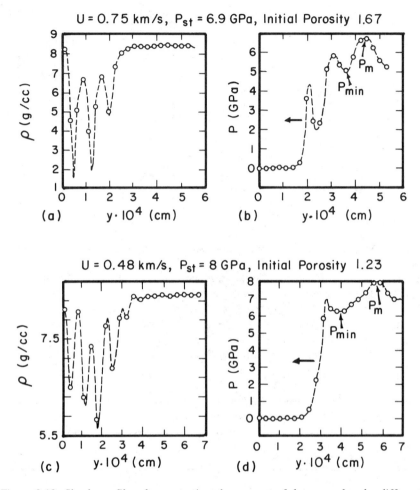

Figure 2.12. Shock profiles demonstrating the amount of data spread under different conditions: (a) density profile, and (b) pressure profile for Ni-alloy 1, for which the solid density 8.26 g/cc and $\alpha_0 = 1.67$. (c) density profile, and (d) pressure profile for Ni-alloy 2, for which the solid density is 8.9 g/cc and $\alpha_0 = 1.23$. The particle size was in the range 0.74–0.88 μm in both cases. (Reprinted from *Journal of the Mechanics and Physics of Solids, 45*, no. 11/12, Benson, J.D., Nesterenko, V.F., Jonsdottir, F., and Meyers, M.A., Quasistatic and Dynamic Regimes of Granular Material Deformation under Impulse Loading, pp. 1955–1999, 1997, with permission from Elsevier Science.)

energy ($E_{\text{Mkin}} = E_{\text{Tkin}} - E_{\text{mkin}}$). The total kinetic energy, internal energy, and microkinetic energy were calculated from Eqs. (2.69), (2.70), and (2.73) and the typical behavior is represented in Figure 2.11(a),(b).

A comparison of "short" and "long" systems, consisting of 5 and 11 particles in the shock direction, respectively (Fig. 2.11(a),(b)), shows that the density, particle velocity, and total and microkinetic and, correspondingly, the macrokinetic energy, are close for these two cases. This behavior shows that the shock wave is nearly stationary after covering the distance of a few particle diameters. Nevertheless, we notice the elevated values of the dissipated energy at the piston boundary, a result of the nonstationary wave propagation (Nesterenko [1992]).

The numerical values of the pressures do not reveal a significant dependence on the particle size, which agrees with the independence of the experimentally measured shock Hugoniots of granular materials on the initial particle size (Benson, Nesterenko et al. [1997]).

The main conclusion of the discussion in this section is that the two-dimensional computer calculations produce averaged data for shocked powders that are in close agreement with the stationary shock conditions for a large variety of materials, particle geometries, porosities, and pressures. This result allows us to interpret the data from the computer calculations as representative for stationary shock loading and not just as reflecting the loading of a finite cell of powder with a few particles inside.

2.6.5.2   Microkinetic Energy and Transition from the Quasistatic to the Dynamic Regime

Nesterenko [1988a,b,c], [1992] introduced the term "microkinetic energy" to describe the qualitative transition from the quasistatic type of particle deformation to the dynamic type of deformation during the shock wave loading of powders. One of the most interesting qualitative results of the two-dimensional computer calculations is the potential to explain this transition. The energy explanation for this transition is the adjustment of the geometry of the plastic flow of the particles to balance the amount of energy supplied by the shock loading. The manifestation of this transition can be followed by introducing a new parameter, the microkinetic energy.

This parameter E_{mkin} was calculated using Eq. (2.73). The maximum values of E_{mkin} and internal energy E_{int}, at the moment of complete pore collapse for different conditions, are shown in Table 2.6. The data for the energies were converted into kJ/kg by dividing the numerical results by the density of the completely consolidated material.

The increase of shock pressure results in a qualitative change of the mode of particle deformation. This transition is very important for the creation of bonds between the particles and the initiation of chemical reactions (Nesterenko [1992]). Data illustrating this transition are presented in Table 2.6 and in Figure 2.13. Arrows in Figure 2.13(b),(d),(f) mark the points of complete densification. Additional data can be found in the paper by Benson, Nesterenko et al. [1997].

One of the important results of these calculations is that the internal energy follows the dependence of the macrokinetic energy on the shock pressures. This

Table 2.6. Microkinetic and total energies for a shock in a Ni-alloy 1 powder formed from small particles with diameters in the range $d = 0.74$–0.88 μm and $\alpha_0 = 1.67$.

P_{st} (GPa)	u (km/s)	E_{mkin} (kJ/kg)	E_{int} (kJ/kg)	Type of regime
0.6	0.22	6.0	19.3	Quasistatic
0.6	0.22 (long)	4.8	21.1	Quasistatic
6.9	0.75	66.3	217	Dynamic
13.5	1.1	157.3	449	Dynamic

means that the strain rate independent material model in the two-dimensional geometry provides an adequate mechanism to dissipate exactly the amount of energy required by the stationary conditions.

The kinematics of the energy dissipation in the two-dimensional computer calculations is connected with the localized material flow on the interfaces. It allows an increase in the energy dissipation with the increasing amplitude of the shock pressure above the limit corresponding to complete densification. This behavior is in quantitative and qualitative disagreement with the predictions made using the spherically symmetrical single-pore models without rate dependence. In the latter case, the energy dissipated during the complete pore collapse does not depend on the pressure because the geometry of the deformation is initially fixed and does not depend on pressure. In contrast, in the two-dimensional computer calculations, the geometry of deformation is determined not only by initial-pore geometry but also, to a great extent, by the pressure amplitude.

Quasistatic particle deformation (Fig. 2.13(a),(b)) is characterized by the microkinetic energy which, for low pressures, is approximately one-third to one-forth (Table 2.6) of the energy dissipated up to the moment of complete densification. This dissipated energy, at a pressure that is just sufficient to produce complete densification, can be considered to be close to the geometrically necessary energy for quasistatic pore collapse. Therefore, we can conclude that quasistatic deformation of the particles correlates with the microkinetic energy, being small in comparison to the geometrically necessary energy of the plastic deformation, in this case, about 20 kJ/kg.

An increase in particle velocity (Table 2.6 and Fig. 2.13(c) to (f)) results in an increase in the microkinetic energy and a transition to the dynamic regime of particle deformation. It is characterized by deviations of the particle contact geometry from straight lines, and with the experimentally observed peculiarities (melts, jets) in the joint triple points of neighboring particles. At large pressures (Figs. 2.13(c) to (f)), the material cannot dissipate all the internal energy during the densification stage and some additional movement occurs in the compacted state. The geometry of particles is noticeably distorted from the quasistatic

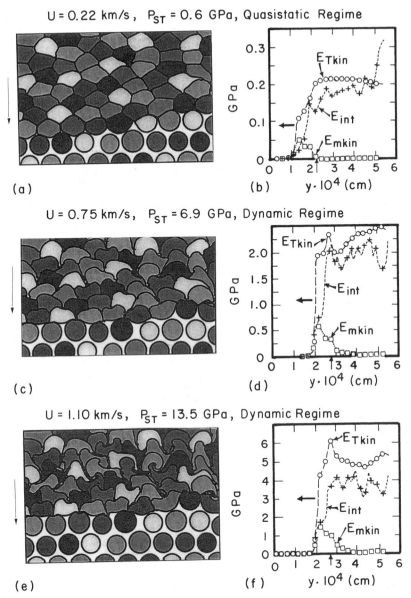

Figure 2.13. The pressure-induced transition from quasistatic to dynamic deformation. The particle sizes are in the range 0.74–0.88 μm and $\alpha_0 = 1.67$. (Reprinted from *Journal of the Mechanics and Physics of Solids*, **45**, no. 11/12, Benson, J.D., Nesterenko, V.F., Jonsdottir, F., and Meyers, M.A., Quasistatic and Dynamic Regimes of Granular Material Deformation under Impulse Loading, pp. 1955–1999, 1997, with permission from Elsevier Science.)

geometry when the microkinetic energy is three times larger than the energy geometrically necessary for complete densification. This additional movement is very important for the bonding process, and for the initiation of chemical reactions in the compressed state, because it provides favorable conditions for the combined pressure and shear deformation that is paramount for the initiation of chemical transformations.

From Figs. 2.13(b),(d), and (f), it is evident that the microkinetic energy can be considered to be a "virtual" parameter. We use the word "virtual" because, in general, for the final state after the passage of a steady shock wave, E_{mkin} cannot be simply added as a separate part to the internal energy. This is evident because the dissipation processes will finally transform the microkinetic energy into thermal energy.

The microkinetic energy exists in some portion on the shock front, and just behind the densification front, but is not present in the final state, which is determined by the Hugoniot relation. As usual, the profile of the microkinetic energy is relatively symmetrical for the quasistatic regime, and asymmetric for the dynamic regime. But, in both cases, the build-up time and the decay time have the same order of magnitude. This means that the additional motion in the completely densified powder continues on a scale comparable with the thickness of the densification front.

Unlike the internal energy, the microkinetic energy is not important for shock relations. But it is important as an indication of the existence of a qualitatively important channel for energy transformation from the initial macrokinetic energy ahead of the shock into heat behind it. This transformation is accomplished through viscoplastic deformation. This channel of energy dissipation is responsible for the new space scale in the compacted powders, namely, the width of the shear localization zones on particle interfaces, the jet thickness, and finally, the compact quality. A relatively small value of this parameter identifies the shock wave consolidation process as the quasistatic one (Fig. 2.13(a),(b)).

The existence of E_{mkin} after the full pore collapse has a fundamental importance because it demonstrates a mechanism of energy dissipation that the framework of the hollow sphere model does not represent, even qualitatively. In the case of a viscoplastic material, the hollow sphere model can dissipate any amount of energy since the strain rate approaches infinity as the pore radius approaches zero.

This result creates a problem for modeling the micromechanical motion after the pore collapse within the framework of the phenomenological relations. The process of the pore collapse only prepares the material for this final stage, as was emphasized by Nesterenko [1992]. The modified Carroll–Holt model, where the process of symmetrical collapse is stopped by the central undeformable core, allows us to find microkinetic energy that should be dissipated in nonporous material. Single-pore models cannot model the mesoscale plastic flow that occurs in the compressed state after complete densification.

2.6.5.2.1   Particle Size and Qquasistatic–Dynamic Transition

Strain Rate Insensitive Materials. The particle size is an important structural parameter that can effect the temperature at particle interfaces and determines the width of the shock transition (Nesterenko [1975], [1992]). That is why it is very interesting to investigate how this parameter affects the transition from the quasistatic to the dynamic deformation regime. It should be emphasized that particle size does not affect the macroscopic shock response of powder represented by the Hugoniot curve at the same initial density (Boade [1970]; Butcher and Karnes [1969]).

The data for powders with particles that are larger, in comparison to those associated with the data of Table 2.6 and Figure 2.13, are presented in Table 2.7. The shapes of large particles in the corresponding moments are identical to those shown in Figure 2.13 (see, e.g., Benson, Nesterenko et al. [1997], Figs. 5 and 6). The particles are the same except for their sizes, and have the same relative size distribution.

From Table 2.7 we can see that the two-dimensional computer model predicts, for IN-718 ($\alpha_0 = 1.67$), the transition from quasistatic to the dynamic deformation regime in the shock pressure interval $P_{st} = 0.6$–6.9 GPa.

A well-developed dynamic regime exists at $P_{st} = 14.1$ GPa, independent of the particle size. A change of the particle size does not result in a change of the deformation regime or the particle shape after densification.

These data are in qualitative agreement with the compacted material structures produced in experiments on a nickel-based superalloy IN 718 powder having an initial porosity 1.67 that were conducted by Meyers and Wang [1988]. In these experiments the shock pressures were 3 GPa (mainly in the quasistatic regime) and 15 GPa (in the dynamic regime).

The independence of the deformation regime on the particle size in numerical modeling is also in agreement with the observed independence of the strength of the consolidated material on particle size. It is true for a wide range of rapidly solidified stainless steel particle sizes, ranking from less than 40 μm to 310-440 μm, as demonstrated by Nesterenko, Lazaridi et al. [1989], [1990], [1992].

Table 2.7. Microkinetic and total energies for a shock in powder formed of large Ni-alloy 1 particles with diameters $d = 2.22$–2.64 μm and initial porosity 1.67.

P_{st} (GPa)	u (km/s)	E_{mikin} (kJ/kg)	E_{int} (kJ/kg)	Type of regime
0.6	0.22	3.6	19.3	Quasistatic
6.9	0.75	48.2	150.6	Dynamic
14.1	1.1	93.0	464.9	Dynamic

It is worthwhile mentioning that, in some models, as well as in experiments, the local temperatures are very sensitive to the particle size. Nevertheless, criteria based on the critical interface temperature have failed to explain the observed independence of the compact strength on the initial particle size, as demonstrated by Nesterenko [1992], [1995] for rapidly solidified stainless steel particles. Strain rate hardening of this material can be considered negligible due to its high initial strength.

Perfect scaling of the geometry of the residual particle deformation, as well as the localized areas on their interfaces, was observed in the calculations with the strain rate insensitive model. This result is natural since no scale in addition to the particle size exists in this model. The reason why the particle size does not affect the deformation regime is the independence of the plastic work on the pore size for a model without strain rate sensitivity. Thus, this insensitivity of the numerical results to the initial particle size demonstrates the consistency of the computer modeling. This is an important observation for validation of the computer results because no exact solutions of the problem exist.

Strain Rate Sensitive Materials. The strain rate sensitivity of the powder material will definitely change its behavior associated with a change of particle size, including a transition from the quasistatic to the dynamic regime of particle deformation (Nesterenko [1992]). This can be explained by the fact that, for relatively strong shocks, the compaction front thickness is about the same as the particle size. If the porosity for powders with different particle sizes is the same, then the Hugonoit and the shock speed should also be the same at the identical loading conditions. The effective strain rates during the densification process within the shock front will vary, being larger for smaller particles. Strains will be about the same for identical porosities and the compaction time will be proportional to particle size. If the strain rate hardening is included into the modeling, it will increase the amount of geometrically necessary energy for complete densification, and will make the regime of particle deformation closer to the quasistatic one for smaller particles.

The experimental results for copper powders with different grain sizes under identical conditions of loading (the shock amplitude in the first wave is 4.6 GPa) support this point: large particles are deformed in the dynamic regime and small particles in the quasistatic regime (Fig. 2.14). This means that, for strain rate sensitive materials, variation of the particle size may be a very important parameter to control local jetting and shear localization on the interfaces. For example, a granular explosive with smaller particles may be less sensitive that one with larger particles. This is consistent with the data on the sensitivity of porous explosives (Howe [2000a,b]).

For viscous granular materials the transition from the quasistatic to the dynamic regime of particle deformation cannot be determined solely by the single parameter of shock pressure. This critical pressure also depends on the particle size and the density of material.

Benson and Conley [1999] presented very interesting results of the numerical

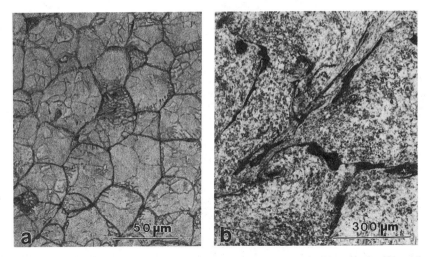

Figure 2.14. The increase in copper particle size facilitates transition to the dynamic regime at the same shock pressure, 4.6 GPa: (a) spherical particles having a particle size of 38–53 μm and $\alpha_0 = 1.6$; (b) cubical particles with a particle size 600–850 μm, $\alpha_0 = 1.6$.

modeling of viscous and inviscid granular materials with properties of HMX explosive under the same conditions of loading. They found that the introduction of viscous dissipation changed the dynamic regime of deformation into the quasistatic regime, with a corresponding change of deformed particle morphology and the profile of microkinetic energy.

2.6.5.2.2 Material Strength and Quasistatic–Dynamic Transition

The mechanisms of energy transformation can be investigated via numerical calculations by changing the initial material strength of the particles, and even setting the strength of the material equal to zero (fluid model). Parameters of the shock profile for these cases are presented in Table 2.8 and Figure 2.15 for stainless steel particles (solid density 7.77 g/cc, initial porosity 1.67, particle size 310–440 μm) with different initial strengths. Arrows mark the point of complete densification and the direction of shock wave propagation. Some very interesting features are evident from the table and figure.

The values of pressures P_{st} and P_{cal} (Benson, Nesterenko et al. [1997]) do not depend on the material strength. Also, the amount of the dissipated energy does not change significantly when the initial strength is decreased by one order of magnitude. This occurs despite the fact that, in each case, the porosity changes by the same amount during shock densification.

The microkinetic energy increases with a decrease of the initial material strength. For $u = 1.29$ km/s the microkinetic energy is asymmetric for the material with low initial strength: it has a fast growth with a subsequent slow decay behind the shock.

Table 2.8. Microkinetic and total energies for a shock propagating in a powder formed from large stainless steel particles with diameters $d = 310\text{--}440$ μm having an initial porosity 1.67, and various initial strengths Y_0.

Y_0 (Mbar)	P_{st} (GPa)	u (km/s)	E_{mkin} (kJ/kg)	E_{int} (kJ/kg)	Type of regime
$2.05 \cdot 10^{-2}$	5.7	0.7	48.4	184.3	Quasistatic
$2.05 \cdot 10^{-2}$	17.9	1.29	164.5	598	Dynamic
$2.05 \cdot 10^{-3}$	5.7	0.7	78.7	189.3	Dynamic
$2.05 \cdot 10^{-3}$	17.7	1.29	221.9	486	Dynamic

It is natural that the lower initial strength results in a much more turbulent plastic flow of the particles at the same level of dissipated energy. The reason that the dissipated energy does not depend significantly on the initial strength is connected with the physics of the stationary shock. As was discussed in the previous section, the shock parameters in the calculations are close to those expected for stationary shock conditions. In this case, the amount of the internal energy is equal to the macrokinetic energy and does not depend on the initial material strength if the particle velocity is kept the same. It is interesting that this property of the stationary shock is also valid for an assemblage of a relatively small number of particles and for shock propagation distances of only 5–7 particle diameters.

Experiments by Nesterenko et al. [1989], [1990] with rapidly solidified stainless steel powders, having different initial microhardnesses (EP-450 as is and annealed, see Table 2.3), clearly demonstrate that a decrease of the initial microhardness can result in the transition from a quasistatic to a dynamic regime at the same loading conditions. The same conclusion can be made based on the residual structure of stainless steel powders CA 304 ($HV = 240$) and VGA 304 ($HV = 200$) shock loaded at the same pressure, 3.7 GPa presented in the paper by Flinn et al. [1988, Figs. 3(a) and 4(a) correspondingly]). Despite the evident jetting and particle mushrooming present in both cases, the amount of quasistatic-type contact deformation (mechanical abutments without bonding) is more pronounced for powder with higher microhardness.

The extreme case is the material with zero initial strength (a fluid), for which data are presented in Table 2.9 and in Figure 2.16.

For the same loading conditions, the density in the zero strength model is the same as in the model where dissipation due to the plastic flow was included (Fig. 2.13(a)). The residual shapes of particles in these two cases are dramatically different, even at the same P_{st} (compare Figs. 2.13(a) and 2.16(a)). Despite the absence of plastic dissipation during the fluid flow at $Y_0 = 0$, a significant increase in the internal energy during shock compression is observed. The amount of internal energy cannot be attributed to the cold energy at a calculated

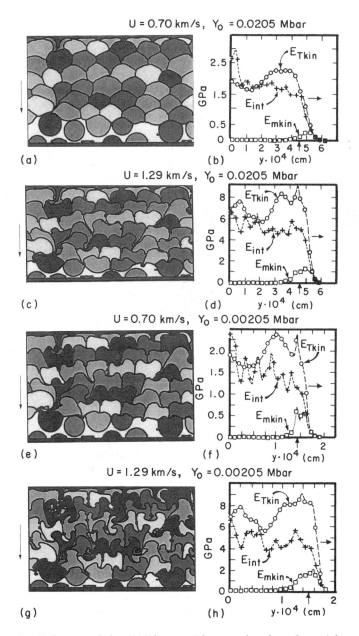

Figure 2.15. Influence of the initial material strength of steel particles on the transition from the quasistatic to the dynamic regime. (Reprinted from *Journal of the Mechanics and Physics of Solids, 45*, no. 11/12, Benson, J.D., Nesterenko, V.F., etal. Quasistatic and Dynamic Regimes of Granular Material Deformation under Impulse Loading, pp. 1955–1999, 1997, with permission from Elsevier Science.)

Table 2.9. Microkinetic and total energies for shock in a powder, particles with diameters $d = 2.22$–2.64 µm, $\rho_s = 8.26$ g/cc, $\alpha_0 = 1.67$, and zero initial strength Y_0.

u (km/s)	P_{st} (GPa)	E_{mkin} (kJ/kg)	E_{int} (kJ/kg)	Type of regime
0.22	0.6	4.3	15.0	Dynamic

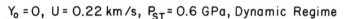

$Y_0 = 0$, $U = 0.22$ km/s, $P_{ST} = 0.6$ GPa, Dynamic Regime

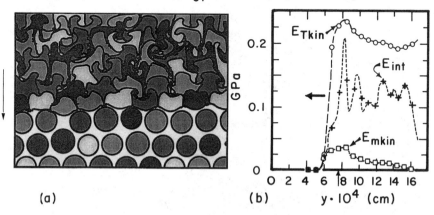

(a)   (b) $y \cdot 10^4$ (cm)

Figure 2.16. Shock parameters in the fluid model: (a) geometry of particle deformation; (b) profiles of total kinetic energy, internal energy, and microkinetic energy. The material had a solid density 8.26 g/cc, $\alpha_0 = 1.67$, particle size 2.22–2.64 µm. (Reprinted from *Journal of the Mechanics and Physics of Solids,* **45**, no. 11/12, Benson, J.D., Nesterenko, V.F., Jonsdottir, F., and Meyers, M.A., Quasistatic and Dynamic Regimes of Granular Material Deformation under Impulse Loading, pp. 1955–1999, 1997, with permission from Elsevier Science.)

density of 8.28 g/cc. The only mechanism for dissipation in the fluid is the shock reverberation (cycles of shock loading and isentropic rarefaction) inside the particles during the pore collapse. In this case, the shock viscosity, which is introduced in the calculations, provides the energy dissipation. This behavior is analogous to the dissipation process in the Thouvenin model (Thouvenin [1966]), in which large plastic deformation of the particles (layers) is absent. In this case, the necessary amount of dissipated energy is provided by the many reverberations at a relatively low strain in each step, as was shown by Hofmann, Andrews, and Maxwell [1968]. In the one-dimensional geometry, this process takes a long time (as was demonstrated in experiments and computer calculations) in comparison with the experiments on powders with a spherical or cubic geometry (Nesterenko [1992], see Section 3.3).

A very unusual result is that, in the two-dimensional geometry (even at low pressures), this process can result in very rapid dissipation of the macrokinetic energy on the scale of the particle size. This dissipation does not produce equality of the macrokinetic and internal energies within the propagation distances from the shock entrances that were investigated, as would be the case for a stationary shock, but it is comparable with the macrokinetic energy. The mechanism of the shock wave dissipation can, therefore, provide a significant proportion of the total energy dissipation in the dynamic regime of particle deformation.

It is worthwhile emphasizing that, in the one-dimensional phenomenological single-pore models, the material behavior with the proposed equations for plastic flow (Eqs. (2.53)–(2.58)) will not result in the required amount of dissipated energy for high shock pressures (see Eqs. (2.39)–(2.47)). That is why, for single-pore models, it is critical to include the strain-rate sensitivity to provide the dissipation required by the conservation laws.

Including the strain-rate sensitivity (viscous dissipation) in the two-dimensional computer code will only change the energy distribution inside the localized plastic flow along the particle interfaces, but not the amount of total energy dissipated. At the same time, including the strain-rate effect can result in a dependence on the particle size of the transition pressures from the quasistatic to the dynamic regime for materials where this is the dominant mode of dissipation.

The fluid demonstrates the existence of the jet formation ahead of the shock compaction front (see the middle of the compaction front in Figure 2.16(a)). This jetting also occurs in materials with normal strength (see Figure 2.15(c) for stainless steel). It can increase the local pressure and provide sites for the initiation of chemical transformations in porous explosives.

Do [1999], in two-dimensional numerical calculations, found that chemical transformations can increase the microkinetic energy in comparison with inert behavior.

2.6.5.2.3 *Porosity and the Quasistatic–Dynamic Transition*

The quasistatic–dynamic transition, according to its qualitative definition, should occur for all porosities as the shock pressure increases. Indeed, for any initial porosity, the increase of the energy supply to some threshold value will result in a qualitative change in the powder behavior, but the character of this change depends on the initial porosity.

For example, for very small porosities, we cannot expect large changes in the particle geometry in comparison to the quasistatic geometry just because of the small volume of empty space. Local melting will be the main hallmark of the dynamic regime in this case.

The evaluation of the capacity of powders with different initial porosities to accumulate microkinetic energy is important, because a small initial porosity will provide better conditions to reach high shock pressures. A decrease of the

Table 2.10. Microkinetic and total energies for a shock in a Ni-alloy 1 powder with small particles having diameters $d = 0.74$–0.88 μm and $\alpha_0 = 1.22$ (compare with Table 2.5).

P_{st} (GPa)	u (km/s)	E_{mkin} (kJ/kg)	E_{int} (kJ/kg)	Type of regime
7.8	0.48	18.4	54.7	Quasistatic dynamic locally
14	0.67	32.1	172.0	Dynamic

initial porosity will decrease the internal energy, and probably the microkinetic, energy as well, at the same type of loading.

To clarify this point, computer calculations for an initial porosity $\alpha_0 = 1.22$ were made at the same pressures as for the $\alpha_0 = 1.67$ calculations. Figure 2.17 presents the particle deformations, and the total kinetic, microkinetic, and internal energy profiles, for two particle velocities. The corresponding data are presented in Table 2.10.

These data should be compared with the data for similar pressures (6.9 and 13.5 GPa) and for different initial porosities presented in Table 2.6 and Figure 2.13 (c) to (f).

From a comparison of the data at similar pressures and different initial porosities (Figs. 2.13(c),(d) and Figs. 2.17(a),(b); Figs. 2.13(e),(f) and Figs. 2.17 (c),(d)), it is evident that the larger initial porosities definitely enhance the dynamic behavior of the powder at the same pressures.

It is more difficult to judge the difference in the critical pressures for these two cases because the local hallmarks of the dynamic regime (i.e., localized melting) can occur at lower pressures for the lower initial porosities. But we can conclude that the intensive plastic flow of particles in the consolidated state is more pronounced for the higher initial porosities. The comparison of the microkinetic energies in the cases of these two initial porosities (1.67) and (1.22), see Tables 2.10 and 2.6 for the corresponding pressures) demonstrates that the microkinetic energy increase has an approximately quadratic dependence on the initial porosity.

This increase in the microkinetic energy is qualitatively different from the linear dependence of the internal energy on the initial porosity at the same pressure (Eq. (2.49)). It is worthwhile mentioning that this strong dependence of the microkinetic energy on the initial porosity cannot be predicted from the existing phenomenological models.

Comparison of the shapes of the deformed particles in the corresponding figures (Figs. 2.13 and 2.17), and the differences in the microkinetic energies, show that higher initial porosities are preferable for the initiation of chemical reactions in energetic materials and particle bonding in inert materials at given pressures.

U = 0.48 km/s, P_{ST} = 7.8 GPa, Quasistatic Regime

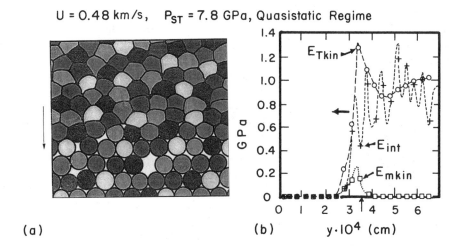

(a)   (b) $y \cdot 10^4$ (cm)

U = 0.67 km/s, P_{ST} = 14.0 GPa, Dynamic Regime

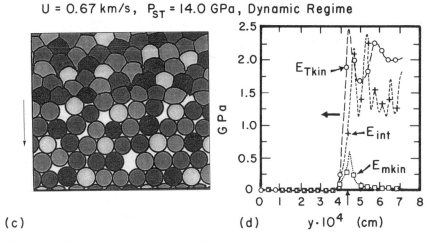

(c)   (d) $y \cdot 10^4$ (cm)

Figure 2.17. The dependence of particle deformation on shock pressures for Ni-alloy 1 at $\alpha_0 = 1.22$. Compare with Figure 2.13 (c),(d),(e),(f) for $\alpha_0 = 1.67$ and similar shock pressures). The solid density is 8.26 g/cc and the particle size is in the range 0.74–0.88 μm. Arrows mark the points of complete densification and the direction of the shock wave propagation. (Reprinted from *Journal of the Mechanics and Physics of Solids,* **45**, no. 11/12, Benson, J.D., Nesterenko, V.F. et al. Quasistatic and Dynamic Regimes of Granular Material Deformation under Impulse Loading, pp. 1955–1999, 1997, with permission from Elsevier Science.)

2.6.5.2.4   *Dependence of the Transition on Particle Density*

The solid density of the powder particles is a very important parameter that determines the pressure for given initial porosities and internal energy (Eqs. (2.49) and (2.75)). Therefore, it is important to understand the effect of this parameter on the quasistatic–dynamic transition in the two-dimensional numerical calculations.

To clarify its role, the calculations were performed for an artificial material with the same properties as IN-718, except for the solid density of the particles. The results are presented in Table 2.11 and Figure 2.18. The data in Figure 2.18 correspond to the solid density of the particles equal to 9.35 g/cc, with the particle size in the range 0.74–0.88 μm and $\alpha_0 = 1.67$. Stationary pressure for parts (a) and (b) of the figure is $P_{st} = 0.6$ GPa. For parts (c) and (d), $P_{st} = 6.9$ GPa, and for parts (e) and (f), $P_{st} = 14.9$ GPa. Arrows mark the moments of complete densification. Figure 2.18 should be compared with the results obtained for particles with the same properties (except for a lower density) for similar values of P_{st} as given in Figure 2.13 and Table 2.6.

We can see that the regimes of particle deformations are practically identical at similar pressures, including the details of their flow, in the quasistatic and in the dynamic regimes (compare the same particles in Figs. 2.13 and 2.18 for similar pressures). This means that the initial density of the particles is not an important parameter for the quasistatic–dynamic transition, despite the dependence of the shock parameters on the solid density.

This is in agreement with the transition criteria proposed by Nesterenko [1988], [1992], [1995] in the frame of single-pore model based on the ratio of the shock internal energy and the geometrically necessary energy for the complete densification in the static regime. We emphasize, however, that if viscous deformation is the main mechanism of the energy dissipation process, then the solid density is an important parameter for the transition pressure (Nesterenko [1992]).

Table 2.11. Microkinetic and total energies for a shock in a powder with heavy particles of diameters $d = 0.74$–0.88 μm, solid density 19.35 g/cc, and initial porosity 1.67.

P_{st} (GPa)	u (km/s)	E_{mkin} (kJ/kg)	E_{int} (kJ/kg)	Type of regime
0.57	0.14	2.8	7.8	Quasistatic
6.9	0.49	32.9	94.3	Dynamic
14.9	0.72	77.4	170.7	Dynamic

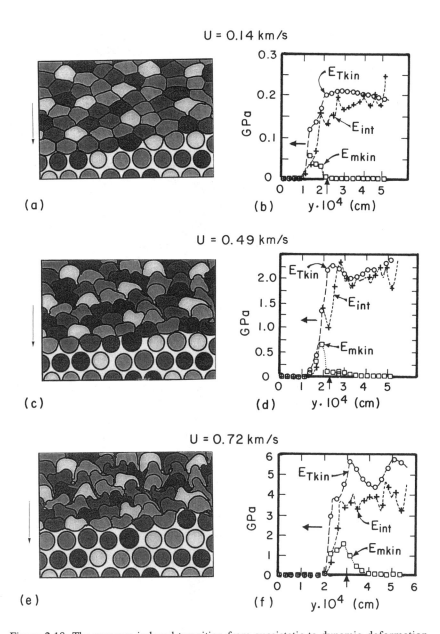

Figure 2.18. The pressure-induced transition from quasistatic to dynamic deformation. (Reprinted from *Journal of the Mechanics and Physics of Solids,* **45,** no. 11/12, Benson, J.D., Nesterenko, V.F. et al. Quasistatic and Dynamic Regimes of Granular Material Deformation under Impulse Loading, pp. 1955–1999, 1997, with permission from Elsevier Science).

2.6.5.2.5   Heterogeneity of the Contact Deformation in the Dynamic Regime

A qualitatively important experimental feature of the shock wave deformation of granular materials is the heterogeneity of the contact deformation in the dynamic regime (Fig. 2.19). This figure corresponds to the pressure at the first shock (responsible for densification) equal 28 GPa, with powder having initial porosity 1.67 and a solid density of 7.77 g/cc. Figure 2.19 (a),(b),(c) shows the distribution of deformation between neighboring particles (for part (a), $d =$ 310–440 µm, for parts (b) and (c), $d = 90$–145 µm). Note the molten "pockets," shown by the arrows. Figure 2.19(d) demonstrates the localized shear (shown by the arrows at the side of the picture) on the boundary between adjacent particles ($d = 310$–440 µm), and Figure 2.19(e) represents the "uniform" plastic deformation of the contact zone ($d = 310$–440 µm).

Even for adjacent particles, the deformation does not necessarily have the same character. In some locations, the intensive localized deformation on the contact surfaces is initiated without involving the adjacent particle contacts (Nesterenko [1985], [1986]). This feature was absent in the quasistatic regime, where all contacts were deformed in the same manner. This resulted in the concentration of intensive shear on the individual, separate interfaces and the absence of this behavior on neighboring particles (Figs. 2.19(a),(b),(c)). It is especially visible for large particles.

The intensive shear deformation (Fig. 2.19(d)), partially under high shock pressure, results in good bonding without significant melting. Only local melting is observed in these conditions of loading (shown by arrows in Figs. 2.19(a),(b). Pressure by itself is not enough to produce good bonding without shear localization on the particle interfaces (Fig. 2.19(e)).

The inherently highly heterogeneous contact deformation in granular material (Figs. 2.19 and 2.20) makes impossible the goal to achieve high strength and ductility of compacts after shock compaction even in the developed dynamic regime of deformation (Nesterenko [1986], [1992]).

Flinn, Williamson et al. [1988] obtained highly nonuniform temperature distribution under the shock densification of 304 stainless-steel powder in both experiments and in 2-D numerical calculations with ordered particles.

This behavior apparently cannot be explained by the single-pore models, but it is adequately described by the two-dimensional computer model of stainless steel particles (Fig. 2.20(a), particle size 310–440 µm) and Ni-alloy 2 (Fig. 2.20(b), particle size 0.74–0.88 µm). The arrows mark the direction of the shock wave propagation. As in the experiments, the adjacent particle interfaces in the computer calculations can have qualitatively different localized plastic deformations. This qualitatively important feature of the two-dimensional model provides the natural introduction of the new deformation scale, namely, the size of the localized deformation zone on the particle interfaces. The central part of the particle undergoes much less plastic deformation, and the size of this region agrees with the modified model (Nesterenko [1992]), decreasing with an increase of the initial porosity.

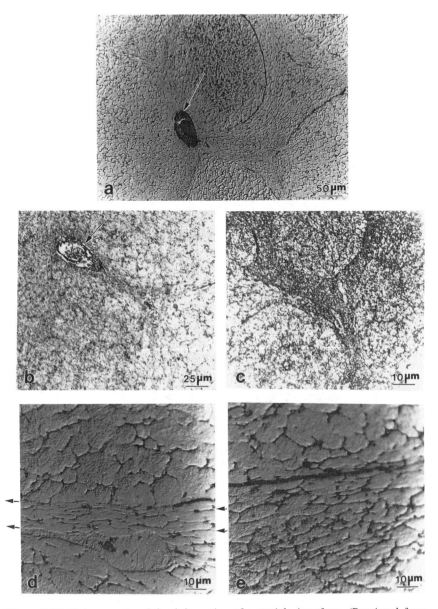

Figure 2.19. Heterogeneity of the deformation of a particle interface. (Reprinted from *Journal of the Mechanics and Physics of Solids,* **45**, Benson, J.D., Nesterenko et al. Quasistatic and Dynamic Regimes of Granular Material Deformation under Impulse Loading, pp. 1955–1999, 1997, with permission from Elsevier Science.)

U = 1.29 KM/S, P_{ST} = 17.9 GPa, STAINLESS STEEL,
INITIAL POROSITY 1.67

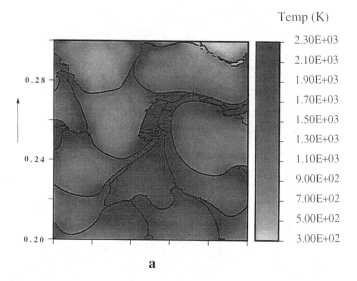

a

U = 0.67 KM/S, P_{ST} = 13.7 GPa, Ni ALLOY 2,
INITIAL POROSITY 1.23

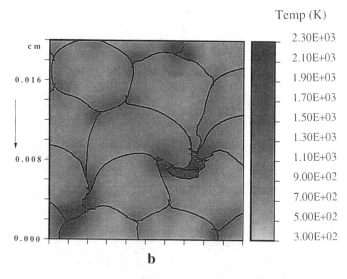

b

Figure 2.20. Distribution of the deformation for adjacent particles in 2-D computer calculations in the dynamic regime for various porosities. (Reprinted from *Journal of the Mechanics and Physics of Solids,* **45**, Benson, J.D., Nesterenko, V.F. et al. Quasistatic and Dynamic Regimes of Granular Material Deformation under Impulse Loading, pp. 1955–1999, 1997, with permission from Elsevier Science.)

2.6.5.2.6 *Spin in the Shocked Powders*

Two-dimensional computer modeling permits the calculation of new parameters in the shocked powders. For example, microkinetic energy is very important for the global material behavior, such as bonding between particles or the initiation of the chemical reactions. Unfortunately, its measurement is beyond the capability of any straightforward experimental method.

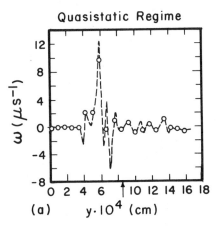

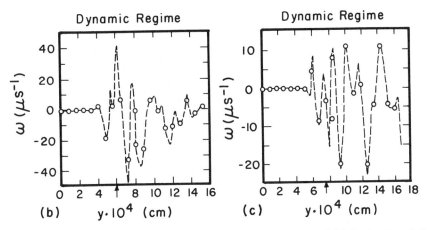

Figure 2.21. Spin in the quasistatic and dynamic regimes: (a) $u = 0.22$ km/s, $P_{st} = 0.6$ GPa, Ni-alloy 1; (b) $u = 1.10$ km/s, $P_{st} = 13.5$ GPa, Ni-alloy 1; (c) $u = 0.22$ km/s, $P_{st} = 0.6$ GPa, fluid model. Solid density 8.26 g/cc, initial porosity 0.4, particle size 2.22–2.64 µm. Arrows mark the point of complete densification. (Reprinted from *Journal of the Mechanics and Physics of Solids*, **45**, no. 11/12, Benson, J.D., Nesterenko, V.F. et al. Quasistatic and Dynamic Regimes of Granular Material Deformation under Impulse Loading, pp. 1955–1999, 1997, with permission from Elsevier Science.)

One of the other parameters is the spin, which is represented by Eq. (2.74) and describes the rotation in the material of the particles that is important for material mixing. The spin profiles are presented for the quasistatic and dynamic regimes in Figure 2.21.

It is easy to see the correlation between the microkinetic energy and the spin for these two qualitatively different regimes. Vanishing of the microkinetic energy is in good agreement with vanishing of the spin immediately after the complete densification for the quasistatic regime (Fig. 2.21(a)), in which no additional motion in the compacted material is observed.

A completely different behavior of the spin and microkinetic energy is evident in Figures 2.21(b),(c) for the dynamic regime. In the latter case, the magnitude of the spin does not change immediately after full compaction (compare Figs. 2.21(a),(b), and (c)).

The spin in the fully compacted material ensures the mutual plastic flow of the particle contacts, which is favorable for good bonding and the initiation of chemical reactions. This is especially important for reactive mixtures where mixing of the different components is promoted by the local rotations.

2.6.5.3   Shock Front Thickness

Shock front thickness is an especially interesting parameter to consider when studying micromechanical behavior because, for relatively strong shocks, the experimentally measured front width is close to the particle size (Nesterenko [1975], [1992]). The single-pore models provide this result for strong shocks (Carroll, Kim, and Nesterenko [1986]), but the validity of this approach on a space scale close to that of the particle diameter is questionable.

The shock front thicknesses (measured from 10 to 90% of the final particle velocity) that are obtained from the numerical calculations for different particle diameters, porosities, and pressures (Benson, Nesterenko et al. [1997]) are presented in Table 2.12. The particle diameters of small granules for porosities 1.23 and 1.67 were the same, and the diameters of large granules were three times larger.

The main features of the shock front dependence on the porosity, and the initial particle sizes in numerical calculations, can be summarized as follows:

(1) The shock front width, Δ, depends on the initial particle size.

(2) The shock front width is approximately equal to the particle diameter and decreases with increases in pressure.

(3) In the range of pressures and porosities investigated, Δ does not reveal an essential dependence on the initial porosity. This is in agreement with data by Sheffield, Gustavsen, and Anderson [1997] obtained in experiments on the "weak" shock loading of course HMX powder (particle size 120 μm) at density 1.24 and 1.40 g/cm^3.

A very important result is that, in the experiments (Table 2.2), the relationships between the shock front thickness and the initial particle size are

Table 2.12. Shock front thickness (Δ) for different pressures (P_{st}), initial porosities (α_0), and particle sizes (d).

$\alpha_0 = 1.23$ Small particles		$\alpha_0 = 1.67$ Small particles		$\alpha_0 = 1.67$ Large particles	
P_{st} (GPa)	Δ/d	P_{st} (GPa)	Δ/d	P_{st} (GPa)	Δ/d
2.4	1.5	0.6	2.0	0.6	1.5
		0.6 (long)	2.5		
8.0	1.5	6.9	1.5	6.9	1
13.9	0.8	14.1	1.3	14.1	1

close to the values in Table 2.12. This demonstrates that the kinetics of densification, as described by the two-dimensional model, is quantitatively correct.

This cannot be achieved using any single-pore model on the size scale of the particle diameter. The final stage for the stationary shock wave is insensitive to the type of dissipation process, but the shock front thickness can be dependent on the details of the material behavior and particle morphology. As was observed, the dynamic regime of the particle deformation correlates with the shock front thickness being close to the particle size (Nesterenko and Lazaridi [1990]; Nesterenko [1992]).

This numerical analysis does not take into account the possibility of elastic wave propagation due to elastic interaction at the contact points. These disturbances may propagate with a speed of about 1000 m/s in metal powders and form an elastic precursor in the "force chains" present in the material ahead of the compaction front, as was considered in Chapter 1 (Section 1.8).

The spatial shock width for a mixture of particles with different particle sizes is comparable to the largest particle size, and this is in good agreement with the results of experiments (Nesterenko [1975], [1992]).

Menikoff and Kober [1999] performed similar calculations with material properties similar to HMX high explosive and emphasized the precursor phenomena for shock fronts corresponding to weak compaction waves. The particle velocity profiles for granular Teflon with large particles (see Fig. 2.9(a),(b),(c)) have a precursor probably connected with elastic waves in the granular skeleton.

Conley [1999] and Conley and Benson [1999] presented a very detailed analysis of shock wave thickness for porous HMX. They obtained a reasonable estimate of the viscosity based on the shock front thickness.

2.6.5.4   Tensile Stresses in Compression Shock

The steep increase of pressure in a shock front is comparable with particle size, and creates conditions for tensile stresses due to the interaction of the shock wave free surfaces of the particles. This may result in microspall inside particles as illustrated for the dynamic deformation of copper powder with an embedded nickel foil of similar thickness as a particle diameter in Figure 2.22 (Nesterenko [1992]).

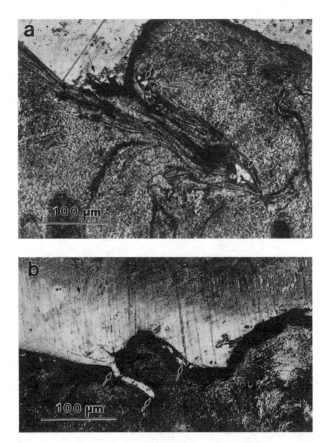

Figure 2.22. Microspall (locations and detached fragments are shown by arrows) in a nickel particle (top white) embedded into copper powder under shock loading within the dynamic regime. The initial porosity was 1.98, the pressure was 4.6 GPa, the particle size was 100–200 μm for the copper, and the thickness of the nickel foil was 140 μm.

The angle between the plane of the foil and the shock wave was 66° and the first shock amplitude was 4.6 GPa. In the same experiment, a plane spall was also detected (Fig. 2.22(b)). Spalled layers of a similar scale were observed under laser-induced shock loading by Gilath [1995].

This result clearly demonstrates the existence of microtensile stresses and subsequent particle fracture in the compression wave in granular materials subject to complete densification behind the shock. The movement of spalled fragments creates additional shear flow and may be significant for the initiation of a chemical reaction in granular energetic materials.

It is difficult to resolve this microtension in numerical simulations because the spall thickness is about 10 μm, which is less than one-tenth of the particle diameter. Nevertheless, in numerical calculations, Menikoff and Kober [1999] (using only 14 cells to resolve a grain) were able to show that portions of some grains go into tension for low shock pressures when a material is partly compacted behind the shock. The same approach did not allow them to resolve local tensions for large shock amplitudes producing compaction shock front widths of approximately one particle diameter. A more refined mesh is necessary to do this.

Williamson et al. [1989], [1990] investigated the dynamic compaction of a porous mixture of two different materials (SiC fiber-reinforced aluminum matrix composite) with a two-dimensional computer model and experiments. Fracture was taken into account by using a simple maximum tension criterion and introducing a localized crack resulting in the relief of the tension. During the initial stress wave propagation through the particles, tensile stresses in excess of the estimated dynamic fracture stress of SiC were obtained in the numerical calculations, thus predicting the fracture within the SiC fibers, also detected in the experiments.

The existence of large tensile stresses, as is experimentally revealed by the fracture of SiC particles in shock densified initially porous composite materials (Ti–SiC), was emphasized by Tong and Ravichandran [1994b] and Tong, Ravichandran, Christman, and Vreeland [1995].

2.7 Metallization of Dielectric Powders Under Shock Loading

2.7.1 Metallographic Observations

The previous sections addressed two qualitatively different mechanisms of particle deformation in shock waves—quasistatic and dynamic. The latter is characterized by the intensive localized flow of material in surface layers of the particles, instead of the relatively uniform contact deformation typical for compression in static conditions. This difference is apparently very important for the behavior of the particles.

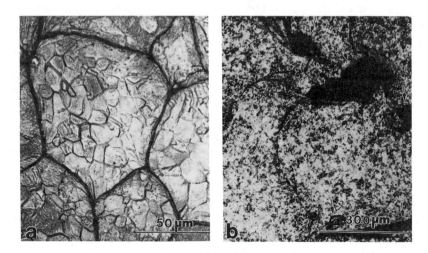

Figure 2.23. Behavior of oxidized layers on the surfaces of copper particles in (a) quasistatic and (b) dynamic regimes.

For example, the micrographs of Figure 2.23 show samples of copper granules coated with oxide layers compacted in the quasistatic (Fig. 2.23(a)) and the dynamic (Fig. 2.23(b)) regimes (Kusubov, Nesterenko et al. [1989]). The initial powder was oxidized in air at a temperature of 220 °C for 6 hours. The initial densities were approximately $\rho_{00} = 5.6$ g/cc, and the particle sizes were 106–125 μm (Fig 2.23(a)) and 600–850 μm (Fig. 2.23(b)). Despite shock loading and complete densification to theoretical density the oxide layers were not disrupted in the quasistatic regime of shock deformation. In the dynamic regime (Fig. 2.23(b)) the localized plastic flow on the surfaces of particles and the jetting at particle contacts result in disruption of the oxide layer, providing purified contacts of the particles. Oxides are concentrated in separated "pockets" shown by the arrows in Figure 2.23(b).

It is worthy of mention that, according to computer calculations by Wilkins (Kusubov, Nesterenko et al. [1989]), shock pressures in both cases were practically equal (4.6 GPa). Increasing the initial particle size from 106–125 μm (spherical particles, Fig. 2.23(a)) to 600–850 μm (cubic particles, Fig. 2.23(b)) resulted in the transition from the quasistatic to the dynamic regime of particle deformation. This is a result of a particle size effect on the transition pressure (from the quasistatic to the dynamic regime) due to the strain rate sensitivity of copper. Benson [1994] found from computer calculations that, in the 2-D packing of rectangular particles, there is much less evidence of turbulent flow exhibited in the final configurations, in comparison with the cylindrical particles for particles of strain-rate independent material.

2.7.2 A Noncontact Method for Detection of Metallization

It is important to determine the actual moment of oxide layer break-up. A noncontact electromagnetic method for measuring the velocity of conductive material was proposed by Nesterenko [1979], [1980], [1992]. In the case of powder initially in a dielectric state, the method allowed measurement of the particle velocity at which a macroscopically conductive state was established in the shocked powder. The experimental set-up is shown in Figure 2.24.

Two types of material were tested. One was copper powder with particle size 100–500 μm that was nonconductive after oxidation in air at 200 °C for 1 hour and the other was copper powder with fine particles (1–20 μm) without any special oxidation step. The density of both powders was 2.2 g/cc, the diameters of the powder samples were 20 mm, and they were placed into a copper ring with outside diameter 50 mm. The top and bottom parts of the powder samples were covered by plastic film of 200 μm thickness. The powders were shock loaded by the contact explosion of a TNT/RDX cast charge with a plane wave generator. The shock pressures in both samples were 16 GPa, which corresponds to the regime of well-developed dynamic powder deformation.

To make sure that the measured signal is connected with the motion of the conductive central part of the set-up, the experiments were performed with replacement of the copper powder by solid paraffin. The latter has the same density and remains dielectric under these shock wave conditions. The amplitude of the electric signal was very small with the paraffin sample. It points toward the conclusion that movement of the peripheral part of the copper ring and air ionization do not contribute to the measured signal.

If the porous media was initially nonconductive and became metallized at some point on the shock front (as in the case of oxidized copper powder, Fig. 2.25), then the electric signal will be determined by the motion of material in

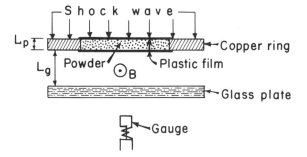

Figure 2.24. An experimental set-up for noncontact measurements of the dielectric–metal transition using a static magnetic field.

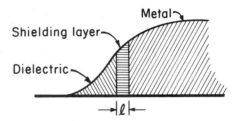

Figure 2.25. Schematic representation of metallization in a shock wave propagating into a dielectric powder.

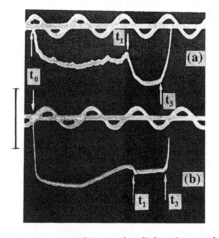

Figure 2.26. Oscillograms corresponding to the dielectric–metal transitions in coarse ((a), 100–500 μm) and fine ((b), 1-20 μm) copper powders under identical loading conditions.

the vicinity of this transition point. The amplitude of this signal is proportional to the particle velocity in the vicinity of the transition point.

It was determined in separate experiments with different metal specimens that changing of the conductivity from copper to constantan did not result in the change of signal amplitude at the same particle velocity. The signal depends on the geometry of the experiment.

The motion of the back part of the metallized powder is shielded by a layer with thickness l, that is determined by the coefficient of magnetic diffusion κ and shock speed D: $l \sim \kappa/D$. For copper $\kappa \sim 10^2$ cm²/s and for $D = 0.3$ cm/s the relation gives $l \sim 10$ μm. This value is essentially less than the size of large particles in the specially oxidized powder. Therefore, it is also less than the shock front thickness, and results in the shielding of motion of the back part of the sample by some intermediate layer inside the shock front in this case. On the contrary, for powder with small particles, l is of the same order of magnitude as the shock front thickness (particle size ~ 20 μm). This means that the shock front will not be able to shield the motion of the back part of the sample for powder with small particles.

Figure 2.26 shows two typical oscillograms corresponding to different powders (Fig. 2.26(a) corresponds to large oxidized particles, and Fig. 2.26(b) to small particles without the special oxidation step) in the same experimental geometry (Fig. 2.24). In Figure 2.26, the time t_0 corresponds to the shock entrance into the powder, and t_1 corresponds to the arrival of the shock wave on the free surface of the powder covered by the plastic film (checked in different experiments by an increasing thickness of the powder sample), and t_3 corresponds to the time of powder impact onto the glass plate. The electric signals were completely reproducible in different experiments with the same conditions.

The comparison of Figure 2.26(a) and (b) shows an identity of signals according to the characteristic time intervals t_1–t_0, t_3–t_1, and equal signal amplitudes at the time t_3. It reflects the macroscopically identical behavior of these powders (the same shock and particle velocity of the free surface of the powder) because of the equal initial density and identical loading conditions.

However, the signal amplitudes are different at time t_0. Because the signal amplitudes are proportional to the particle velocities at the shock front where the powder becomes an electric conductor, the difference in signal amplitudes at time t_0 can be interpreted only as a result of metallization in oxidized copper powder at the point in the shock front with particle velocity equal one-half of its value behind the front (Fig. 2.25).

This is also in complete agreement with the signal behavior at the subsequent time t_1 (Fig. 2.26(a)): after shock arrival at the free surface, all powder is accelerated by the back part of the sample up to the particle velocity behind the shock. The increase of particle velocity due to the rarefaction wave from the free surface is relatively small in both powders. This coincides with a small amplitude increase at time t_1 for powder with small particles (Fig. 2.26(b)) where the back motion is not shielded by the shock front, and also with a large amplitude increase at time t_1 for powder with a large oxidized particles where the shielding effect was substantial during the time interval t_1–t_0.

These results demonstrate the significant feature of powder shock compression that, in the well-developed dynamic regime of deformation, oxide layers are destroyed in the leading part of the shock front. The dynamic densification at the last stage develops with partially purified contacts at which mutual plastic flow can result in bonding of the particles.

2.7.3   A Noncontact Method of Hugoniot Measurements

This method allows recording of the times of shock entrance (t_0) and exit (t_1) from the sample, and also a moment of impact (t_3) of the free surface with the dielectric obstacle (glass plate in Fig. 2.24). Knowledge of the first two times allows calculation of the shock velocity in the sample

$$D = \frac{L_p}{t_1 - t_0}, \tag{2.76}$$

where L_p is the thickness of the porous sample.

For powders with high porosity, the particle velocity increase, due to the rarefaction wave from a free surface, can be considered small in comparison with the particle velocity behind incident shock in the given conditions of loading. Then, the velocity of the free surface, u_{fs}, may be taken approximately equal to the particle velocity behind the shock

$$u_{fs} = \frac{L_g}{t_2 - t_1} \approx u, \tag{2.77}$$

where L_g is the gap between the powder free surface and the glass plate.

If the electrical signal is experimentally calibrated to measure the particle velocity, then the velocity increase due to the rarefaction wave can be obtained. In the experiments with small particles of copper powder, the velocity increase at t_1 (Fig. 2.26(b)) was a few times less than that expected for a rarefaction wave from a shock pressure of 16 GPa in solid copper. This means that the state of the powder behind the shock cannot be described as solid copper.

The shock Hugoniot of a powder can be found from Eqs. (2.76) and (2.77). For example, it is evident from Figure 2.26 that the shock parameters of powders made of particles with different sizes, but the same initial density, are the same. The method also provides information on shock attenuation in powders with small particles where the signal is determined by the particle velocity behind the shock (Fig. 2.26(a)). Shock attenuation in powders made from large particles is less noticeable because the electrical signal is created by the leading part of the front, which does not change significantly due to the shock attenuation.

Pai, Yakovlev, and Kuz'min [1996] developed a technique based on a similar approach to measure the shock Hugoniots of powders.

2.7.4   Noncontact Measurements of the Sound Speed Behind Shock

A modification of this method, in which the copper ring was replaced by dielectric material (e.g., solid alumina or Teflon) to reduce the effects of the side release wave, was used to measure the sound speed in shock compressed metal samples (Fig. 2.27; Nesterenko [1979], [1980]). Several solid metallic samples (Cu, Al, SS, and constantan) with diameter 20 mm were investigated in an experimental set-up similar to that illustrated in Figure 2.24 (except for

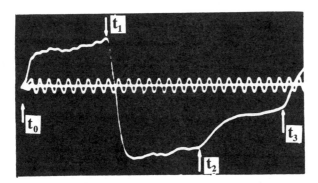

Figure 2.27. An oscillogram yielding sound speed measurements in shock compressed copper by the noncontact method.

replacement of the copper ring with Teflon).

Application of this method allowed measurement of the time, t_0, of the shock entrance into the sample, the time of shock arrival, t_1, at the free surface, and the time of impact, t_3, of the metal free surface onto the dielectric plate, and the time of arrival of the rarefaction wave from the free surface to the back free surface of the sample (seen as a positive kink at instant t_2 between t_1 and t_3, Fig. 2.27). The measured time intervals $(t_1-t_0,\ t_2-t_1)$ make it possible to find, in one experiment, the value of the shock speed and the particle velocity of the free surface.

If the shock Hugoniot is found or known, then the calculated thickness of the compressed sample and time interval t_3-t_1 allows calculation of the sound speed in the shocked state. The sound speed for copper at 40 GPa, corresponding to Figure 2.27, was close to the value predicted on the base of the equation of state (5.4 km/s, Kinslow [1970]).

The unique feature of this method is that the sound speed can be calculated from the time of the single passage of the rarefaction wave through the shock compressed sample. A second wave passage through the unloaded sample is not required. This provides measurements based only on shock compressed state and avoids effects of the release wave, which complicate other noncontact methods based on the free surface velocity measurements.

2.7.5 Compression of a Magnetic Field by Shock-Metallized Powder

The metallization in a shock wave was proposed for magnetic field compression by Bichenkov, Gilev, and Trubachev [1980], and by Nagayama [1981]. Based on the experimental results described, only powders with small particle sizes (less than 10 μm) can be used for this purpose in an efficient way. In powders with larger particle sizes, the magnetic flux compression will be affected by some intermediate stage on the shock front, which will have a smaller particle velocity (Fig. 2.25).

The experimental set-up proposed by Gilev and Trubachev [1996] has been most successful, resulting in the generation of a magnetic field of 4 MGs. Thus, this method may be used for the generation of a superhigh magnetic field. For this purpose any material with shock induced dielectric –metal transition can be used instead of oxidized metal powder.

2.8   Boundary Layers

2.8.1   The "Cold" Boundary Layer at a Powder–Solid Interface

2.8.1.1   Experimental Results

A very interesting effect of the particle size scale in shock-loaded powders in the vicinity of a powder–solid interface was experimentally observed by Kusubov, Nesterenko et al. [1989]. The experiments were conducted in the cylindrical geometry shown in Figure 2.28. Nitromethan-based explosives with detonation velocities of 2.4 km/s and 3.6 km/s were used. Five copper powders with the relatively narrow particle size distributions, 38–53 μm, 106–125 μm, 297–420 μm, 600–850 μm, and 1160–1700 μm, were investigated at initial densities in the range ρ_{p0} = 5.3–5.6 g/cc. In addition, a powder with a large difference in particle size, 1–400 μm, and density, ρ_{p0} = 6.48 g/cc, was investigated. In all the experiments, the radius of the central copper was 0.318 cm, the radii of the container tubes were 1.11 and 1.27 cm, respectively, and the explosive radius was 3.81 cm.

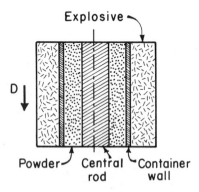

Figure 2.28. Scheme of the experimental set-up for detection of a "cold" boundary layer on the interface between a central rod and powder. The arrow shows the direction of detonation propagation along the cylindrical axis.

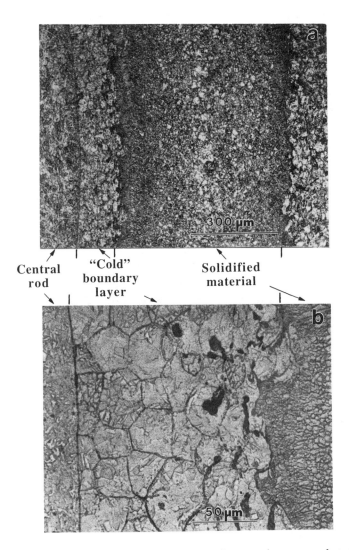

Figure 2.29. The "cold" boundary layer adjacent to the central copper rod at different magnifications for a powder having a particle size in the range 38–53 µm, and a detonation speed $D = 3.6$ km/s. Micrographs are taken in the plane perpendicular to the axis of the experimental set-up shown in Figure 2.28.

In this geometry, the shocks approach the central rod at angles to the axis of symmetry of $\phi = 38°$ ($D = 2.4$ km/s) and $\phi = 24°$ ($D = 3.6$ km/s) according to the results of the 2-D computer calculations presented in these papers. The estimate of the first shock pressure in each case was the same, 4.6 GPa.

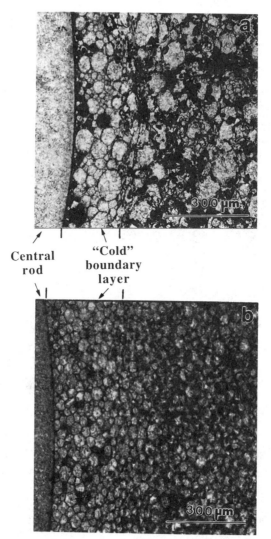

Figure 2.30. The "cold" boundary layer adjacent to the central copper rod (a) for powder with the wide size distribution 1–400 μm at detonation speed $D = 3.6$ km/s and (b) for the lower detonation speed $D = 2.4$ km/s and particle size 38–53 μm. Micrographs are taken in the plane perpendicular to the axis of the experimental set-up shown in Figure 2.28.

At this shock pressure, the critical angle, $\phi*$, of transition from regular to irregular wave reflection from the copper rod (with Mach step creation) is 16° (Kostyukov [1991]). This means that, in the experiments described by Kusubov, Nesterenko et al. [1989], Mach reflection from the rod is taking place.

In Figures 2.29 and 2.30 the "cold" boundary layer is shown adjacent to the central rod. It was structurally a very well-identified phenomena for the powder with the narrow range of particle size 38–53 μm (Fig. 2.29(a),(b)), and for the powder with the wide size distribution 1–400 μm (Fig. 2.30(a)) when the detonation speed was $D = 3.6$ km/s. It has also been observed in the powder having particles in the size range 38–53 μm when the detonation speed was $D = 2.4$ km/s (Fig. 2.30(b)).

The thickness of this layer was in the interval 150–300 μm, increasing with the decrease of the detonation wave speed (the same shock pressure in the incident wave and larger φ). This phenomenon was not clearly identified in the pictures for powders with larger particle sizes and narrow distribution: 297–420 μm, 600–850 μm, and 1160–1700 μm. It was shown that the irregular Mach reflection is the probable origin of melting in the area adjacent to the "cold" boundary layer at detonation speeds 3.6 km/s and higher (Nesterenko [1992]).

The term "cold" boundary layer is used to emphasize the relatively lower temperature of this area where particles are deformed in the quasistatic regime in comparison with the more remote areas from the axis where the dynamic regime of particle deformation prevails (Fig. 2.30(b)), or even where melting at higher detonation velocity (Fig. 2.29(a),(b)), is observed.

Another important feature of this phenomenon is the formation of an anomalous contact discontinuity in temperature and velocity, which is apart from the physical powder–solid interface comprising the outer boundary of the "cold" boundary layer (Nesterenko [1992]; Nesterenko, Luk'yanov, and Bondar [1994]). Kostyukov [1991] experimentally observed the same phenomenon in experiments with different powders (tin bronze) and a different material of the central rod. The thickness of the "cold" layer exceeded the particle size by an order of magnitude in some cases, and depended on the material of the rod.

2.8.1.2   Proposed Mechanisms

2.8.1.2.1   *Nonstationarity Induced by a Finite Shock Front*

Currently, the mechanism of cold boundary layer formation is not known, except for the general understanding that it is caused by a local nonstationary regime of powder flow in the vicinity of the rod. Originally, it was proposed that it was connected with the nonstationary shock interaction with the interface on the length scale comparable to the shock front thickness (Kusubov, Nesterenko et al. [1989]). The nonstationarity was attributed to the finite thickness of the shock front in the powder. It is just as if the final state of the medium adjacent to the interface of the two different materials is different from that calculated, based on the Hugoniot states for each of them even for one-dimensional shock waves on this length scale. The reason for this behavior is connected with the fact that these parts of materials, adjacent to the contact, are not loaded along the Rayleigh lines corresponding to the stationary shocks.

The effects caused by local shock nonstationarity may result in significant differences in the local temperature with respect to the temperature in the bulk caused by stationary shock loading (lower or higher depending on the specific shock configuration). For example, in computer calculations where an artificial viscosity was incorporated into the flow description, the temperature in the nonstationary regime in the vicinity of a piston ("wall") moving into the gas was two times higher than the temperature behind a stationary shock (Samarski and Popov [1972]). A similar "wall" effect for the shock loading of powder was reported in the numerical calculations by Benson, Tong, and Ravichandran [1995].

The temperature effects in shock-loaded porous materials are much more pronounced than in solids due to essential differences in the densities before and behind the shock front. That is why the shifting of the loading of porous material from single shock loading in the bulk (Hugoniot state) to the loading in the nonstationary regime in the vicinity of the interface may result in a dramatic difference in temperatures.

The estimate of the effect caused by the finite thickness of the shock front in powder can be clearly illustrated using the laminar (Thouvenin [1966]) model of powder densification. The incident shock wave approaches the interface from the powder side. In this model the first particle (layer) adjacent to the contact with the solid material is loaded by the single shock, in sharp contrast with the loading of layers in the remote areas by the sequence of shocks (see Fig. 5.9). As a result the temperatures in the vicinity of the interface due to this mechanism can be one-seventh or one-third the temperature in the remote areas of powder for incident shock pressures 1 or 10 GPa, respectively (Problem 11, Nesterenko [1992]). The area with the temperature anomaly in powder is apparently comparable to the shock front thickness in powder in the frame of the laminar model (about 10 layers, see Fig. 5.9).

Shock front thickness in powders is comparable with particle size (Sections 2.4 and 2.6.5.3). Two-dimensional computer modeling of shock densification shows that the compaction front has a precursor propagating on the distance of 4–5 particle diameters and the compaction wave lags slightly behind the pressure wave (Benson [1995]; Benson, Tong, and Ravichandran [1995]). Interaction of this leading disturbance in the shock wave in powder with the wall can provide some precompaction before the arrival of the main shock front reducing temperature in the area close to the interface in comparison with the powder bulk.

We conclude that there can be essential differences in the temperature in the stationary and nonstationary regimes of shock interaction with powder–solid interfaces. The real extent of this type of nonstationary behavior is still unclear for the shock loading of granular materials when the shock wave approaches the solid from the powder side, and when heat release results from the complex geometry of the plastic flow of the particles.

It is difficult to explain, in the frame of this mechanism, why the thickness of the "cold" layer increases with an increase of angle ϕ at the same shock pressure in the incident wave (see Figs. 2.29(a),(b) and 2.30(b)), and why the material of the rod essentially affects it.

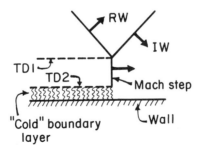

Figure 2.31. The schematic representation of irregular shock reflection of an incident wave (IW) in powder from a rigid wall. Two tangential discontinuities, TD1 and TD2, form.

Figure 2.31 presents a highly schematic picture of irregular shock reflections in the vicinity of an interface (Nesterenko [1992]), illustrating the incident (IW) and reflected (RW) shock waves, the Mach step, the "cold" boundary layer, and the two tangential discontinuities. The tangential discontinuity TD2 is very unusual feature which is not observed for shock wave reflection in a gas. The deformation of the surface of the wall is considered negligible in this mechanism (see Figs. 2.29(b) and 2.32), and it is not shown in Figure 2.31. No mechanism for the contact discontinuity departure from the physical powder–solid interface was proposed. It is possible to conclude only that the "cold" boundary layer is created before the hydrodynamic flow with the Mach reflection is formed. The dynamically formed "cold" boundary layer sets the conditions for Mach reflection.

The partially melted zone outside the "cold" layer (Fig. 2.29(a),(b)) is connected with the Mach step, which is consistent with conditions for copper melting (Nesterenko [1988a,b,c], [1992]).

2.8.1.2.2 Precompaction Due to the Precursor Wave in the Rod

Another origin of the nonstationary loading of a powder in the vicinity of its boundary with a solid was proposed by Kostyukov [1991]. It is based on the fact that a "cold" boundary layer was observed at the detonation velocities of 2.4 and 3.6 km/s, which are lower than the sound speed in the central copper rod. This was the reason for the hypothesis that a small "bump" propagates in the front of the main shock wave inside the copper rod (before the Mach step in Fig. 2.31) and predensifies the powder before the arrival of the main shock. Because the final energy of the powder is very sensitive to the initial density, this densification by a "double shock" will result in a smaller temperature increase than would be produced by a single shock. The condition for "bump" propagation resulting in the "cold' layer (that the speed of the incident shock along the rod surface is smaller than the speed of the plastic shock wave in the rod) was formulated by Kostyukov [1991]. This hypothesis, at least qualitatively, can explain why the thickness of the layer depends on the rod sound speed and strength.

This origin of nonstationary behavior resulting in the "cold" layer is different from that considered previously in the sense that it is principally a 2-D phenomenon and demands the plastic deformation of the rod.

It should be emphasized that calculations based on the Hugoniot states (characteristic for steady shock loading) for the material points adjacent to the interface of the powder and rod on the length scale comparable to the shock front thickness, where the effects of nonstationarity can be very strong, cannot be justified. No noticeable precursor was found in the rod or the powder in 2-D computer calculations (Kusubov, Nesterenko, Wilkins, Reaugh and Cline [1989]). Numerical calculations with the mesh smaller or comparable with particle size in the powder are necessary to make a final judgement about this mechanism.

Nesterenko, Luky'anov, and Bondar [1994] tried to identify plastic deformation due to possible "bump" propagation in a thin surface layer using twins in the copper rod. The geometry and conditions of the experiment were similar to the experiments by Kusubov, Nesterenko et al. [1989]: detonation speed 3.6 km/s, rod diameter 8 mm, inner and outside diameters of the steel

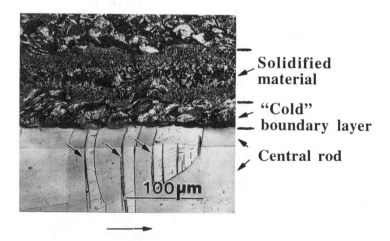

Figure 2.32. The "cold" boundary layer adjacent to the central copper rod for powder with particle sizes 40–90 μm at detonation speed D = 3.6 km/s. The micrograph is taken in a plane parallel to the axis and across the center of the experimental set-up shown in Figure 2.28. Twins and the oxidized layer inside the inner oxidized Cu–Al alloy rod serve as microscale marks for the plastic deformation of the rod in the vicinity of its interface with the powder.

container are 14 and 16 mm, respectively, and an explosive diameter of 66 mm, explosive density of 1 g/cc, copper powder with particle sizes 40–90 µm, and initial density 5.5 g/cc. The difference is that, in the former experiments, the central rod was made from Cu–Al (0.2 wt.% Oxygen) alloy with subsequent inner oxidation. This procedure resulted in the creation of an inner oxidized layer adjacent to the surface of the rod with thickness about 20 µm which is well distinguished from the bulk of material (see Fig. 2.32). The line separating this inner oxidized layer from the rest of the rod served as a marker of plastic flow together with twins, which usually have parallel straight boundaries.

This method allows the detection of details of the plastic flow on the scale of 10 µm in conditions of impact during explosive welding with severe tangential discontinuity of the flow (Bondar and Ogolikhin [1985]). No macroscopic change of the geometry of the inner oxidized layer which could accompany the "cold" boundary layer was observed. Only slight bending of twins in one place of the rod was detected, and it can be attributed to the friction between the copper powder and rod. This distortion of twins is the largest observed in experiments. The conclusion is that the "cold" boundary layer is not accompanied by the essential plastic flow of the rod in the vicinity of the interface (Fig. 2.32).

It should be mentioned that no specific speed for the plastic bulk and surface waves exists, creating uncertainty about how to apply this criterion for the "cold" boundary layer. For example, under the conditions of explosive welding, no plastic flow in the copper, before the contact point moving with velocity 900 m/s, was observed (Nesterenko, Luky'anov, and Bondar [1994]).

These experimental results do not support the hypothesis of the "precursor" bump as a mechanism for an explanation of the "cold" boundary layer. The 2-D computer calculations, with a refined space mesh near the interface, which would allow for the detection of small strain deformations, are necessary to prove that this type of mechanism can explain the observed phenomenon.

2.8.1.2.3 Inhomogeneity of the Initial Powder Density

Nesterenko et al. [1994] proposed an additional mechanism involving an initial density heterogeneity in the wall vicinity for producing this phenomenon. The "cold" boundary layer was observed for powders formed from granules having a regular, approximately spherical, shape. In the case of spherical particles, the flat surface of the central rod can provide some ordering of particles during filling of the container by powder. This results in a higher powder density near the interface.

For example, the packing fraction of steel balls is observed to have a maximum (0.7 of theoretical density ρ_t) at a distance of three particle diameters from a wall, and to approach the bulk density (0.65 of ρ_t) at a depth of six diameters under certain conditions of deposition (Gray [1968]). The flat wall may encourage the particles in the first layer to be aligned as shown in Figure 2.33. Then this layer may provide the closest packing of the following layer. This process can hardly continue for a distance of more than a few particle diameters.

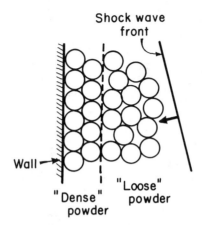

Figure 2.33. Schematic representation of a density gradient in the powder layer adjacent to the rod, which may effect shock reflection and contribute to the mechanism of a "cold" boundary layer.

The difference between the most dense packing of equal spheres (0.74 of solid density, or 6.6 g/cc for copper) and the average density of powders in the experiments described (5.3–5.6 g/cc) is large enough to cause the observed phenomenon of "cold" boundary layer formation (Nesterenko, Luky'anov, and Bondar [1994]). The existence of this intermediate layer with higher density may help to understand the formation of two tangential discontinuities, TD1 and TD2, on its boundaries with the powder and the rod, respectively (Fig. 2.31).

Again, interaction of the shock wave with such a layer, having the length scale comparable with the shock front thickness, cannot be considered based on Hugoniot states.

This influence of the wall on the packing efficiency of the spherical particles is an intrinsic property of the powder and certainly should be taken into account for the explanation of the solid–powder interface phenomena. It is necessary to notice that the local "porosity" in the layer directly adjacent to the wall is higher than the average in the densely packed layer.

The same mechanism can also be responsible for a well-known anomaly that occurs at a contact interface between two batches of powders after successive tamping by flat-ended rams: even after explosive loading, the compact will disintegrate into pieces corresponding to the original powder batches. The well-densified surface layers stay "cold" and do not provide bonding between particles due to the lack of internal energy.

Finally, the phenomenon of the "cold" boundary layer formation has solid experimental proof, by different authors in independent experiments with different materials and conditions. But its mechanism(s) are not yet established and deserve further investigation. It can have an important application in the validation of two- and three-dimensional computer algorithms intended to explain the dynamic behavior of granular materials at the scale of particle size.

2.8.2 Boundary Layer on a "Soft–Hard" Powder Interface

The Hugoniot of powders under strong shock loading does not depend on the initial strength of the powder material or on the morphology of the particles. This means that the final state of densified material behind the steady shock will be very much the same for powder materials with different initial strength or particle sizes if other characteristics (initial density, porosity, and heat capacity) are the same if the amplitude of shock is high enough. The material behavior in the nonstationary stage of shock propagation does not have the same constraints resulting from the conservation laws, and may be more sensitive to the strength of the particles or the particle size distribution. Therefore, it can be expected that powder behavior on the scale of particle size in the vicinity of an interface between two powders with different internal characteristics like material strength, particle size, or Hugoniot can be different than in the bulk.

In Figure 2.34 the contact layer is formed when a shock wave is propagated from a bronze powder (microhardness $197HV$) into a titanium alloy powder (microhardness $490HV$). It is clear that the upper parts of titanium alloy particles contacting the softer bronze particles are practically undeformed in comparison with the same particles in the bulk. Because the main mechanism of heating is plastic flow, these parts of particles will have a lower temperature than that which was evaluated from the Hugoniot conditions valid for a stationary shock in the bulk. Because the bronze particles are deformed in a well-developed dynamic regime of deformation (Section 2.5), their temperature may be expected to be close to the bulk temperature of the same powder.

Figure 2.34. Different deformation of hard particles (Ti alloy, bottom part of the figure) near to, and away from, the interface with softer particles (bronze, top part of the figure).

Finally, the contact temperature between the two powders may be expected to be close to the temperature between Ti-based solid material at the initial temperature and bronze powder at the equilibrium shock temperature.

This example demonstrates that local behavior in the vicinity of the powder–powder interface may be essentially different from that predicted on the basis of behavior in the bulk. Computer modeling on the scale of particle sizes is the most appropriate instrument for the investigation of these special class of boundary layer phenomena in shock compression physics and mechanics.

2.9   Separation of Components in Powder Mixtures

Shock loading of powder mixtures with components having different physical properties is used in technological applications such as the synthesis of new materials and explosive cladding. Nevertheless, a description of the mechanics of shock compaction of the mixtures presents a fundamental problem connected with the unknown interaction forces between components. If the particle interaction is weak, a relatively long propagation distance is required to establish velocity equilibrium between components, and the lamination process (concentration of a particle of the same type in some area of the sample) can take place on a macroscopic scale comparable with the sample size.

Deribas, Nesterenko, and Staver [1976] observed the phenomenon of component lamination in porous mixtures BN + Cu (weight fractions of 1:37, 3:7) with the initial density equal to half of the solid density and with particle sizes 1–10 μm (BN) and 1–20 μm (Cu). The experimental set-up had a cylindrical geometry similar to that shown in Figure 2.28. Experiments without the central rod were also conducted. In experiments with the mixture BN + Cu (1:37 weight fraction) without a central rod, Mach reflection took place in the center. Three zones were clearly distinguished—central zone 1, peripheral zone 3, and intermediate zone 2, with an increased content of Cu in comparison with zones 1 and 3 (Nesterenko [1992]). In the case of shock loading of the mixture BN + Ni–Cr alloy with the central rod, lamination was observed in the vicinity of the rod–powder interface. Lamination in this mixture was observed in wide range of weight ratios including the case when none of the components can be considered as a continuous matrix. No phase lamination was observed in the mixture Fe + Cu (with the weight ratio 1:1).

This phenomenon was explained by the large radial gradient of velocity connected with a tangential discontinuity of particle velocity in the case of Mach reflection. Some estimates qualitatively confirming this hypothesis were made based on the known mechanism of particle drag in a Poiseuille flow (Eichhorn and Small [1964]). The same mechanism can be responsible for the symmetric concentration of solid particles in the middle of partially melted areas associated with the Mach step in the copper powder where a "cold" boundary layer was observed (Fig. 2.29(a)).

The lamination effect can also be observed without Mach reflection resulting in the concentration of light or heavy components near the axis of the cylindrical specimen (Kostyukov [1990]). The proposed inertial mechanism of lamination does not need gradient flow of one of the components around the other, considered as inclusion. The main reason for lamination is that heavier particles are slowed less than light particles when they both pass through the shock fronts (Kostyukov [1990]).

The effect depends on the mass ratio of particles in the mixture, particle morphology, strength, and the duration of the shock pulse. Lamination was absent in mixtures with high-strength components, or when the strengths of the components were similar.

It is interesting that the effect of lamination disappeared when the ampoule diameter decreased, for example, for a mixture of magnesium–tungsten (from 15 to 3.7 mm) and of graphite–titanium (from 15 to 6 mm) with the same type of explosive (Kostyukov [1990]). The natural explanation of this scale effect can be that the time of shock-induced flow was not enough to develop the lamination process in samples with small diameters.

The phenomenon of lamination can be considered as the sole opportunity to get an insight on the interaction forces between components in heterogeneous porous mixtures under shock loading, and to adjust the parameters of computer models based on well-characterized experiments on different scales and type of loading.

2.10   Three-Dimensional Simulations

The current state of the art in numerical modeling allows researchers to begin performing three-dimensional simulations of complex processes during the shock loading of granular materials. Baer [2000], and Baer, Kipp, and van Swol [2000] were the first to perform the numerical modeling of an impact loading of a collection of inert and reactive polycrystalline grains with polyhedral and spherical particles having frictionless material interfaces. The initial density of the granular porous materials was 65–68% of theoretical density. The calculations were made using CTH, an Eulerian shock physics code (McGlaun, Thompson, and Elrick [1990]).

One of the advantages of three-dimensional modeling is the potential to work with closely packed particles beds of relatively large porosity that cannot be studied with two-dimensional models. The main results of these calculations are similar to those of the 2-D modeling considered in previous sections. Of particular note, it was shown that the material behind the shock wave is in a highly heterogeneous state with highly fluctuating stresses and temperatures far different from those prescribed by single jump conditions, as was emphasized earlier based on experimental results (Nesterenko [1992]). It is interesting that, in a mixture of large and small particles, rapid reaction occurs first in the large particles despite the fact that plastic flow in such mixtures is concentrated mainly in small particles (see Chapter 5, Fig. 5.20). These types of calculations represent the real processes in porous heterogeneous materials in the most complete manner.

Nevertheless, the three-dimensional modeling cannot be considered as a complete replacement for other approaches. For example, the calculations of a macroscopical assembly of modest size: a square copper plate with thickness 0.2 mm and side 1.2 mm in contact with a porous crystal array of 190,000 HMX particles packed to 68% of the theoretical density with thickness 0.9 mm, requires the entire Sandia TFLOP parallel computer with 4500 processors each operating at 104 Mb/s (Baer [1999], [2000])!

2.11   Interaction of Plane Shock Waves with a Cavity on the Free Surface

One subject related to the dynamics of porous materials is the interaction of a shock wave with defects like spherical or cylindrical cavities on free surfaces. These interactions can result in new phenomena, which may also take place during shock interaction with pores inside the porous materials. The advantage of experiments where a shock interacts with surface defects is that the details of material behavior are more accessible to observation than in the case with pores in the material bulk. At the same time, these phenomena have a practical importance for controlled fracture and jet formation, depending on the relative values of shock wave amplitude and materials strength. We shall consider these phenomena separately.

2.11.1   "Elastic" Material: Fracture Initiation

There are many practical problems that can be addressed by explosively producing a fracture using relatively weak shock waves, for which the material behavior before the fracture is dominated by elastic properties. One extremely important example is the application of explosive methods to cutting radiation-contaminated metal structures from nuclear power stations. This is important for recovering from disastrous accidents like that at the Chernobyl Atomic Power Station. In addition, many existing nuclear power stations will soon exhaust their limits, and the problem of dismantling them will be a serious one.

Explosive methods may provide a reasonable strategy for avoiding the prolonged presence of technical personnel in the contaminated zone. Currently, thick-walled metal structures are often cut using cumulative jets. But this is a very expensive method demanding a precise explosive charge of large mass.

That is why the shock wave interaction with surface artificial defects like cavities with sizes much smaller than the thickness of the metal structure, which can result in crack formation, is of great practical importance.

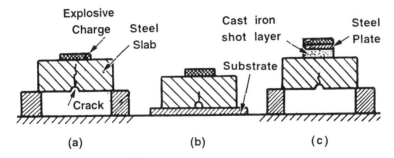

Figure 2.35. Interaction of a "weak" shock wave generated by a contact explosive charge on a simply supported steel slab with a surface groove: (a) general set-up; (b) fracture initiation by the shock without macrobending of the steel slab; and (c) dynamic loading by a wave dispersed using an intermediate layer of iron shot.

Ipatiev, Kosenkov, Nesterenko et al. [1986], performed experiments on the shock loading of artificial cavities on the free surface of thick-walled steel (Steel 3) slabs. The experimental set-up is presented in Figure 2.35. The thickness of the slab was 150 mm, its width 500 mm (in the plane of the figure), and its length was 190 mm. The shock wave was generated by contact detonation of ammonite, a commercial powder explosive having a density of 1 g/cc and a detonation speed of 4 km/s. The charges had a rectangular shape with various heights (70 to 8 mm) and widths (50 to 200 mm in the plane of the figure) and a fixed length of 190 mm (in the plane perpendicular to the figure). In each case, the charge was placed symmetrically with a 10 mm radius milled cylindrical groove, and was initiated by a detonator placed at one of its ends.

In some experiments the slab was placed on supporting metal plates on the ends of the side opposite the detonator, allowing bending of the slab and crack propagation after shock interaction with the groove (Fig. 2.35(a)). The optimized geometry of the charge resulted in an explosive consumption of only 22 g/m to fracture a 150 mm thick steel slab. Table 2.13 shows the experimental results, and the specific consumption of the explosive is calculated as the ratio of the explosive mass to the length of the charge.

In a few experiments the total surface of the slab was supported by a steel plate, practically eliminating the bending of the slab and leaving a "fracture signature" of the shock wave intact (Fig. 2.35(b), Table 2.14). This experimental set-up allows us to determine a qualitative mechanism of the shock-induced fracture because the post-shock bending of the slab was arrested. Nevertheless, after this test a macrocrack with a length of 26 mm was detected. This means that macrocrack initiation is connected to shock wave interaction with the surface defect, and is not affected by macrobending. The macrocrack length is measured from the groove bottom using liquid–penetrant testing.

Table 2.13. Experimental data on the fracture of simply supported slabs using a contact explosive charge.

Charge width (mm)	Charge height (mm)	Charge mass (kg)	Charge mass per unit length (g/cm)	Macrocrack length (mm)
70	50	0.805	35	90
70	50	0.805	35	110
70	50	0.805	35	100
50	70	0.805	35	65
100	35	0.805	35	140
100	17	0.395	17	110
100	17	0.395	17	110
100	22	0.505	22	135
100	24	0.550	24	130
100	25	0.590	25	125
200	13	0.600	26	140
200	11	0.506	22	140
200	11	0.506	22	140
200	8	0.380	16	No macrocrack
150	11	0.380	16	77

Nevertheless, bending of the slab is crucial for post-shock propagation of the crack. This was evident when a cracked slab, recovered after an experiment with a rigid substrate (Fig. 2.35(b)), was loaded by the same explosive charge, with a layer of iron shot between the explosive and the plate placed on the slab (Fig. 2.35(c)). This layer reduced the peak pressure in the shock wave without changing the total momentum supplied to the slab. Following the experiment, a macrocrack with length 140 mm was detected in the slab. Even in a case where no macrocrack was detected (charge height 8 mm, Table 2.13), microcracking apparently took place in the vicinity of the groove. That is because, after secondary loading of this slab by an explosive charge with a damping layer of iron shot, macrocracking had developed (Table 2.14).

No macrocracking was detected when a simply supported slab, without any preliminary fracture, was loaded by an explosive charge with a damping layer of iron shot, resulting in a dispersive shock wave front (Table 2.14). This experiment suggests that an amplitude of local transient tensile stresses depend on the incident shock front width.

The experimental results and numerical modeling of the process (the shock impulse was approximated by a triangular shape with amplitude 10 GPa and duration 5 μs) suggested the following mechanism of the fracture (Ipatiev, Kosenkov, Nesterenko et al. [1986]). First, shock wave interaction with the groove results in local tensile stress at the top of the groove (its amplitude was calculated to be 1.6 GPa in the direction perpendicular to the direction of shock

Table 2.14. Experimental data on slab fracture under various conditions of explosive loading, including sequential loading.

Charge width (mm)	Charge height (mm)	Charge mass (kg)	Charge mass per unit length (g/cm)	Macrocrack length (mm)	Comments
200	11	0.506	22	26	Slab on rigid substrate, contact explosion
200	11	0.506	22	140	Slab with free span, explosion damped by layer of iron shot, slab was previously loaded in test with rigid substrate
200	11	0.506	22	140	Slab with free span, explosion damped by layer of iron shot, slab was cracked in previous test (Table 2.13, charge height 8 mm)
200	11	0.506	22	0	Slab with free span, explosion damped by layer of iron shot, slab was not preloaded

wave propagation). This phenomenon resembles the tensile stresses at a cavity tip under static compression loading (Timoshenko and Goodier [1970]) but, in the investigated case, the dynamic local tensile stress is induced by a shock wave interacting with the cavity. The propagation of this macrocrack is controlled by the post-shock macrobending of the slab. No cracking in the direction parallel to the free surface (spall) was detected.

2.11.1.1   Influence of the Shape of the Surface Groove

Experiments on the shock loading of the artificial cavities of different shape (hemicylindrical, rectangular, and triangular) on the surface of thick-walled steel (Steel 20) slabs with a thickness of 50 mm, a width of 150 mm, and a length of 75 mm were performed by Ipatiev [1990].

It was determined that the shape of defects at the same depth (about 3 mm) did not change the explosive mass necessary for fracture. For example, for a slab with such dimensions, the explosive mass necessary for fracture initiation varied from 120 to 135 g, depending on the geometry of the groove. The smallest mass corresponds to a triangular groove and the largest to a rectangular groove. The mass of the contact explosive charge, in the same geometry of loading necessary to fracture the same slab without a surface defect, was 750 g.

The mass of explosive charge necessary for fracture does depend on the depth of the groove. For example, for semicylindrical grooves with radii 3.3 mm and 1.65 mm, the required explosive mass was 120 g and 180 g, respectively.

For a groove radius of 0.4 mm, the critical mass of explosive was about the same as the required mass without a groove. Interestingly, loading by a 500 g explosive charge using this small radius of defect did not result in a macrocrack, nor did sequential loading by two charges with mass 150 g each. However, sequential loading by two explosive charges with masses 150 g and 320 g resulted in complete fracture of the plate. These experiments demonstrate that even small explosive charges, which did not create macrofracture, created microcracks in the vicinity of the groove. These microcracks can result in macrofracture in a subsequent explosive event.

2.11.1.2   Pore in the Vicinity of a Free Surface

Macrocracking was observed in a geometry of explosive loading similar to that of Figure 2.35, but with a pore placed in the material bulk close to the free surface instead of the surface cavity (Ipatiev [1990]). In this case, slab fracture depends on the pore distance from the free surface. For example, in the case of a hole radius 3.3 mm and a mass of charge 120 mm, where the distances of the hole from the surface were 5.3 mm and 8.3 mm, macrocracks running from both sides of the hole in the direction of shock propagation were detected. At a distance 13.3 mm, macrocracking was not observed. The same explosive loading (explosive mass 120 g) resulted in complete fracture of the slab when a surface cavity having a radius of 3.3 mm was introduced.

The existence of local tensile stresses at the tip of the cavity under compressive loading is in accord with stress distributions in plates under static compression (Timoshenko and Goodier [1970]). The difference with the phenomena of cracking under wave loading is connected to the dynamic character of the tensile stress, which depends on the duration of the shock front. Also, it should be noted that the existence of local transient tensile stresses is not a sufficient condition for macrocrack propagation. Nemat-Nasser and Hori [1987] analytically showed the development of high tensile stresses on unloading at the tips of a dynamically collapsed void under plane shock wave loading. These stresses result in crack initiation within a region of heavy plastic deformation, which propagated in the direction perpendicular to shock wave propagation. Nemat-Nasser and Chang [1990] confirmed this mode of dynamic cracking in experiments using the Hopkinson bar technique. This phenomenon is apparently different from that which was observed in the experimental set-up corresponding to Figure 2.35.

2.11.1.3   Method of Fracture

Ipatiev, Afanasenko, and Nesterenko [1993] patented the method of the explosive fracture of large metal slabs where, at the first stage, surface grooves were formed by small explosive charges on one side of the slab. Charges consisted of thin-walled conical or wedge-shaped steel elements surrounded by an explosive. Upon detonation, these charges generated irregular jets which created surface grooves. For example, charges of this sort produced continuous hemispherical, or an interrupted sequence of hemispherical, defects with diameter 5–10 mm sufficient for subsequent fracture of a steel slab 80 mm thick. These quasicumulutive charges need not be manufactured with high precision, and use inexpensive industrial explosive (ammonite powder).

In the second step, a contact explosion on the opposite side of the slab initiated fracture along the line connecting the surface defects. This line of fracture is determined by the alignment of the initial defects and need not be a straight line. The distance between the grooves should be 40–60 mm. The sequence of explosive events can be performed in one shot. The total mass of industrial explosive (ammonite powder) used for the fracturing of slabs 75 mm and 80 mm thick was 23 g/cm. This mass is much lower than that required for cutting by cumulative jet and the method is less expensive.

2.11.2   Viscoplastic Material: Jetting

Strong shock interaction with a semicylindrical, hemispherical or apex cavity on a free surface results in high velocity jetting (Fig. 2.36). In porous material, this phenomenon can induce hot spot formation (Bourne and Field [1992]). Cooper, Benson, and Nesterenko [2000] studied the effect of initial cavity shape on the jetting parameters.

Computer calculation is very powerful, and in many cases, is the only method for investigating processes like pore collapse or the densification of granular materials. Large deformations and gradients of physical parameters are typical for these phenomena. Physical measurements in these cases are impossible because of the small scale of the processes and the larger space and time scales involved. Unfortunately, analytical solutions are available only for a very restrictive symmetrical geometry of collapse. The computer algorithms also have some restrictions. Therefore, it is very desirable to validate them, if not

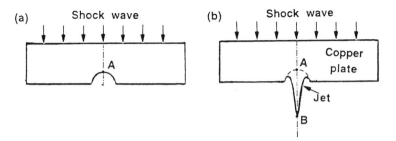

Figure 2.36. Interaction of a strong shock wave with (a) a surface groove, general set-up, and (b) jetting induced by the shock interaction.

using exactly the same type of process, then using a similar process where experimental data or analytical solutions are available.

The closest process to pore collapse for which experimental data exist is the cumulative phenomenon arising from the interaction of a shock wave with a macroscopical (hemispherical, semicylindrical, or triangular) groove on a free surface (Mali [1973], Kozin and Simonov [1973]). This phenomenon is geometrically identical to pore collapse, and involves a high strain and high strain rates typical of the collapse process.

The major results of the experiments by Mali [1973] for copper, which were compared with numerical calculation, are summarized as follows:

(i) For constant loading conditions, the jet velocity U (velocity of the leading part of the jet at point B, Fig. 2.35(c)) did not depend on the radius of the groove if it is larger than some critical size. For a hemispherical groove of 4 mm radius, the measured velocity was $U = 0.25$ cm/μs, and for a radius of 1.5 cm, $U = 0.27$ cm/μs.

(ii) The jet diameter δ_c (taken at the middle of the jet's length) was proportional to the initial radius of the groove, R, according to the equation

$$\delta_c = k_1 R, \tag{2.78}$$

where k_1 is a material constant ranging between 0.4 and 0.6.

(iii) The jet velocity U is constant after the jet flow is stabilized.

The comparison between the computer calculations by Cooper, Benson, and Nesterenko [2000] and the experimental results included quantitative values of the jet velocities: It was found that U was independent of R and time, and δ_c was dependent on R. Comparison between experimental and computed results demonstrated their qualitative and quantitative agreement, which can be considered as validation of the numerical procedures used and the modeling of material behavior.

2.12   Interaction of a Long Rod Penetrator with a Porous Target

In many practical cases, pores are loaded by a plane shock produced by a projectile impact. For example, the ballistic performance of Ti–6Al–4V-based homogeneous and high-gradient porous composite materials, produced by hot isostatic pressing (HIPing) against impact by a long rod, was investigated by Nesterenko, Indrakanti et al. [2000a,b] (Fig. 2.37(a)). Penetrators made of Tungsten (93%W) alloy with L/D = 10 (mass 16.8 g, diameter 4.93 mm) were launched using the 20 mm smooth bore powder gun at the UDRI. The inclination (pitch and yaw) of the projectile was measured prior to impact with a flash X-ray system and a fiducial setup in each experiment.

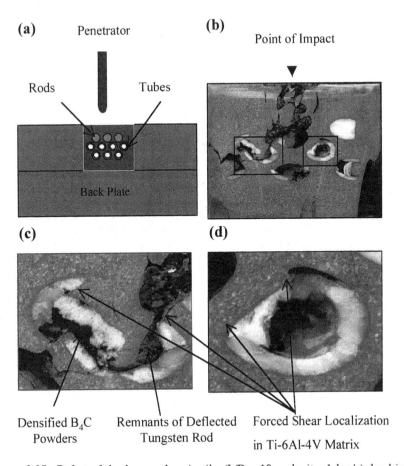

(a) Penetrator

Rods   Tubes

Back Plate

(b) Point of Impact

(c)

(d)

Densified B$_4$C   Remnants of Deflected   Forced Shear Localization
Powders   Tungsten Rod   in Ti-6Al-4V Matrix

Figure 2.37. Defeat of the long rod projectile (L/D = 10, velocity 1 km/s) by high gradient composite armor: (a) general set-up; (b) projectile deflection and hole sealing; and (c),(d) forced shear localization caused by nonsymmetrical collapse of alumina tubes filled with B$_4$C powder.

In general, the ballistic performance of Ti-based HIPed homogeneous material is better than the baseline material, Mll-T-9047G, at the impact conditions investigated (Nesterenko, Indrakanti et al. [2000a,b]; [2001]. The distinguished feature of HIPed material was the absence of adjacent cracks running ahead of the crater as was a case for the baseline material. In the former case the bottom of the crater was flat probably due to the destabilized penetration and fracture of long rod. That is why HIPing is a useful technique for the manufacture of "high-gradient" composite materials from powders characterized by a large gradient of density on a scale comparable with the penetrator diameter, a property that is crucial for projectile deflection and defeat.

Powder filled voids and rods were intended to induce a volume distributed, highly heterogeneous pattern of damage of both the penetrator and target. The damage in this case is initiated by cavities and their interactions, and replaces a few dominant shear bands often causing plugging in homogeneous targets. The result of a test with composite material is shown in Figure 2.37(b),(c),(d). It is clear that the projectile was deflected by the rows of rods and pores. The entrance hole was similar to a case with a baseline homogeneous target made of MIl-T-9047G. Heavily deformed remnants of the projectile (Fig. 2.37(b)) sealed the crater.

In addition to shear bands, spontaneously formed as a result of the instability of the matrix material, "forced" shear bands initiated by the fracture of alumina tubes on the early stage of penetration process were also present (Fig. 2.37(c),(d)). These shear bands were clearly transgranular and grain boundaries did not influence their propagation. The interface layer on the boundary, between the alumina tubes and the matrix during penetration, had multiple cracks which were arrested by the matrix material. Chen, Meyers, and Nesterenko [1999] investigated the relation between spontaneous and forced shear localization.

Another important features of the "high gradient" composite material are heavy plastic deformation of long rod penetrator and sealing of the crater by its remnants (Fig. 2.37(b)). Such materials are able to deflect and to stop long rod projectile within the target reducing the absorbed linear momentum.

2.13   Validation of Numerical Modeling Against Experiments

Perhaps the most important practical application of experimental research in the shock dynamics of granular materials is the validation of computer codes and models by comparison with the carefully conducted experiments. This is absolutely necessary for reaching conclusions about the reliability of the models developed and the global approach as a whole, in conditions where no analytical solutions are available, and where the system is multiscale with a wide range of parameter changes and high gradients. It is of paramount importance for the dynamics of heterogeneous materials, where homogenization procedures and a large amount of hypotheses about components interaction is inevitably introduced.

It is useful to divide the validation procedure into two steps—comparison of some experiments with a single event where the initial geometry is simple (like single pore collapse) and a second step involving comparison of essentially collective phenomena (like densification of granular materials) with experiments.

A comparison of experimental data by Mali [1973] on jetting from a cavity in free surface under plane shock loading to finite element predictions was used to validate the accuracy of the computational modeling by Cooper, Benson, and Nesterenko [2000]. The study by Mali [1973] on the formation of jets in semicircular grooves on the surface of copper specimens was chosen because it is one of the few truly two-dimensional experimental studies in the literature. It therefore permits a direct assessment of computational accuracy without introducing the issue of comparing three-dimensional experiments to two-dimensional calculations.

Among other possible approaches is the use of experiments involving the collapse of thick-walled cylinders for the validation of computer results. This provides the simplest two-dimensional geometry during the period before the onset of localization due to material instability (see Chapter 4).

Measurements of shock wave thickness in materials with relatively large particles, component separation in mixture of different materials, microspall in the particles in powder, and interface phenomena like the "cold" boundary layer are examples where the collective behavior of collapsing pores may result in new effects. Comparison of experiments of this type with computer predictions may serve for validation of the numerical modeling of mesomechanics of the dynamic behavior of heterogeneous materials.

2.14 Conclusion

Peculiar phenomena on the scale of particle size are experimentally well identified under the intense shock loading of granular materials. They should be explained by mechanisms different from the continuum approach, or from an approach based on single pore (particle) models. These include the thickness of strong shock fronts, formation of layers adjacent to solid–granular material or powder/powder interfaces, behavior of particle contacts including shear localization and oxide layers fracture, plastic flow of particles with different strengths, and separation of particles in mixtures. In some cases, the mechanisms of these phenomena are far from being understood, even from a qualitative point of view—the "cold" boundary layer is one of the examples. Simple qualitative models should be developed together with numerical codes for an explanation of the mechanisms of these phenomena.

The applications of two-dimensional computer codes and the first results of three-dimensional numerical simulations for the mesomechanics of granular flow under intense loading created a very promising example of a new approach. It resulted in the correct prediction of qualitatively different particle geometry under different conditions of loading. The highly heterogeneous state behind shocks in granular materials is now a well-established fact. This means that the difference between averaged parameters like pressure and temperature and their local values is larger than the averaged value itself. The inherently highly heterogeneous contact deformation in granular material makes impossible the achievement of high strength and ductility of compacts after shock compaction even in the developed dynamic regime of deformation. The complexity of granular media demands a truly synergetic approach where experimental work, analysis, and computer modeling are strongly coupled.

Problems

1. Derive Eq. (2.9) from Eq. (2.8) using Eq. (2.3).

2. What properties of porous material important for shock loading response are not included into the single pore model?

3. Derive Eq. (2.11).

4. Obtain an exact solution for the temperature on the inner surface of the pore for the power-law temperature dependence of a yield strength similar to Eq. (2.30).

5. Explain why the single pore model results in Eq. (2.12), which does not depend on particle size in the static case.

6. Why does the modified Carroll–Holt model result in a different dependence of shock front thickness in comparison with the original Carroll–Holt model?

7. Evaluate the maximal local pressures in porous material at the beginning of the dynamic regime for a rigid plastic material.

8. What transition in porous material behavior can you expect if the initial temperature is increased at the same type of shock loading.

9. What changes in the material structure can you expect if porous material is loaded in double shock up to the same pressure as in the single shock?

10. If material is very sensitive to the strain rate, what type of change can you expect due to an increase of initial particle size for the same shock loading?

11. Find the difference in powder temperature in the vicinity of the powder/solid copper interface and in the bulk of the powder, using a laminar model of powder (powder is represented by a periodical sequence of plates and gaps) when a shock is approaching the interface from the powder side. For calculation use the copper powder with the initial density equal to half of the density of the solid, incident shock pressures in powder are 1 or 10 GPa.

12. Evaluate the difference in temperature of a "cold" layer at a wall vicinity and the rest of the powder due to a difference in the initial density resulting from "wall enforced" packing. Suppose that its initial density is close to 0.74 of the solid density and the density of the remote material is equal to 0.5 of the solid density. Use properties of the copper and Hugoniot states of shocked powder. The shock pressure in the incident one-dimensional shock wave is 5 GPa and the wall is solid copper.

13. Find shock pressures in stainless steel powder being in contact with a copper plate and with an initial density equal to $0.65 \, \rho_s$. Pressures of the incident shock in the copper plate are 12, 15, and 21 GPa. For the P–u relation for powder use Eq. (2.75) and suppose that $\rho_f = \rho_s = 8$ g/cc.

14. Explain the Carroll–Holt paradox (Carroll and Holt [1973]): porosity α^* in the final state behind a shock wave with pressure P^*

$$P^* = -P_1 + 4Y \frac{\left[H(\alpha_1) - H(\alpha^*)\right]}{3\left(\alpha_1 - \alpha^*\right)},$$

$$H(\alpha) = \alpha \ln\alpha - (\alpha - 1)\ln(\alpha - 1).$$

in the case of partial compaction is smaller than in quasistatic compression (Eq. (2.12)), a result opposite to that expected for a dissipative shock process. The corresponding relation, derived by Carroll and Holt [1973] is Eq. (2.17) without the viscous term (and without the multiplier α on the left-hand side) can be used to derive this equation. The final state corresponds to a zero spatial derivative of porosity α. Why does addition of viscosity change this situation?

15. Calculate the shock width for a strong wave in a predensified granular material using modified and original Carroll–Holt models at the initial porosity 1.1 (Eqs. (2.35)–(2.37)). Geometrical parameters for both models should be selected based on the assumption that the mass of the model unit cell is equal to the particle mass. Explain the appearance of a new porosity-dependent scale parameter in the modified model.

References

Attetkov, A.V., Vlasova, L.N., Selivanov, V.V., and Solov'ev, V.S. (1984) Effect of Nonequilibrium Heating on *the* Behavior of a Porous Material in Shock Compression. *Zhurnal Prikladnoi Mekhaniki i Tehknicheskoi Fiziki*, **25**, pp. 120–127 (in Russian).
English translation: *Journal od Appled Mechanics and Technical Physics*, 1985, May, pp. 914–921.

Baer, M.R. (1996) Continuum Mixture Modeling of Reactive Media. In: *High-Pressure Shock Compression of Solids IV: Response of Highly Porous Solids to Shock Loadning* (Edited by L. Davison, Y. Horie, and M. Shahinpoor). Springer-Verlag, New York, Chapter 3.

Baer, M.R. (2000) Computational Modeling of Heterogeneous Reactive Materials at the Mesoscale. In: *Shock Compression of Condensed Matter—1999* (Edited by M.D. Furnish, L.C. Chhabildas, and R.S. Hixson). AIP, New York, pp. 27–33.

Baer, M.R., Kipp, M.E., and van Swol, F. (2001) Micromechanical Modeling of Heterogeneous Energetic Materials. In: *Proceedings of the Eleventh International Detonation Symposium* (Edited by J.M. Short and J.E. Kennedy). Thomson-Shore, Dexter, pp. 788–797.

Bakanova, A.A., Dudoladov, I.P., and Sutulov, Yu.N. (1974) The Shock Compressibility of Porous Tungsten, Molibdenum, Copper and Aluminum in the Range of Low Shock Pressures. *Zhurnal Prikladnoi Mekhaniki i Tehknicheskoi Fiziki*, **15**, no. 2, pp. 117–122 (in Russian).
English translation: *Journal of Applied Mechanics and Technical Physics*, 1975, October, pp. 241–245.

Benson, D.J. (1992) Computational Methods in Lagrangian and Eulerian Hydrocodes. *Computer Methods in Applied Mechanics and Engineering*, **99**, pp. 235–394.

Benson, D.J. (1994) An Analysis by Direct Numerical Simulation of the Effects of Particle Morphology on the Shock Compaction of Copper Powder. *Modelling and Simulation in Materials Science and Engineering*, **2**, pp. 535–550.

Benson, D.J., and Nellis, W.J. (1994) Numerical Simulation of the Shock Compaction of Copper Powder. In: *Proceedings of the Joint Meeting of the International Association for the Advancement of High Pressure Science and Technology and the American Physical Society Topical Group on Shock Compression of Condensed Matter* (Edited by S.C. Schmidt J.W. Shaner, G.A. Samara, and M. Ross), AIP, New York, pp. 1243–1245.

Benson, D.J., and Nellis, W.J. (1994) Dynamic Compaction of Copper Powder: Computation and Experiment. *Appl. Phys. Lett.*, **65**, 418–420.

Benson, D.J. (1995) A Multi-Material Eulerian Formulation for the Efficient Solution of Impact and Penetration Problems. *Computational Mechanics,* **15**, pp. 558–571.

Benson, D.J., Nesterenko, V.F., and Jonsdottir, F. (1995) Numerical Simulations of Dynamic Compaction. In: *Net Shape Processing of Powder Materials* (Edited by S. Krishnaswami, R.M. McMeeking, and J.R.L. Trasorras). ASME, New York, AMD **216**, pp. 1–8.

Benson, D.J. (1995) The Calculation of the Shock Velocity—Particle Velocity Relationship for a Copper Powder by Direct Numerical Simulation. *Wave Motion*, **21**, pp. 85–99.

Benson, D.J., and Conley, P. (1999) Eulerian Finite-Element Simulations of Experimentally Acquired HMX Microstructures. *Modelling and Simulation in Materials Science and Engineering*, 7, pp. 333–354.

Benson, D.J., Tong, W., and Ravichandran, G. (1995) Particle-Level Modeling of Dynamic Consolidation of Ti–SiC Powders. *Modelling and Simulation in Materials Science and Engineering*, 3, no. 6, pp.771–796.

Benson, D.J., Nesterenko, V.F., and Jonsdottir, F. (1996) Micromechanics of Shock Deformation of Granular Materials. In: *Shock Compression of Condensed Matter—1995*: Proceedings of the Conference of the American Physical Society Topical Group on Shock Compression of Condensed Matter (Edited by S.C. Schmidt and W.C. Tao). AIP Press, New York, pp. 603–606.

Benson, D.J., Nesterenko, V.F., Jonsdottir, F., and Meyers, M.A. (1997) Quasistatic and Dynamic Regimes of Granular Material Deformation under Impulse Loading. *Journal of the Mechanics and Physics of Solids,* **45**, no. 11/12, pp. 1955–1999.

Bichenkov, E.I., Gilev, S.D., and Trubachev, A.M. (1980) Magnetic Course Generators Using the Transition of a Semiconductor Material into a Conductrive State. *Zhurnal Prikladnoi Mekhaniki i Tehknicheskoi Fiziki*, **21**, no. 5, pp. 125–129 (in Russian). English translation: *Journal of Applied Mechanics and Technical Physics*, 1981, March, pp. 678–682.

Bichenkov, E.I., Gilev, S.D., Ryabchun, A.M., and Trubachev, A.M. (1987) Shock-Wave Method of Generating Megagauss Magnetic Fields. *Zhurnal Prikladnoi Mekhaniki i Tehknicheskoi Fiziki*, **28**, no. 3, pp. 15–24 (in Russian). English translation: *Journal of Applied Mechanics and Technical Physics*, 1987, November, pp. 331–339.

Boade, R.R. (1970) Principal Hugoniot, Second-Shock Hugoniot, and Release Behavior of Pressed Copper Powder. *Journal of Applied Physics*, **41**, no. 11, pp. 4542–4551.

Bondar, M.P., and Ogolikhin, V.M. (1985) Plastic Deformation in the Joint Zone with Cladding by Explosion Clading. *Fizika Goreniya I Vzryva*, **21**, no. 2 pp. 147–151 (in Russian). English translation: *Physics of Explosion, Combustion and Shock Waves*, 1985, July, pp. 266–270.

Bondar, M.P., and Nesterenko,V. F. (1991) Contact Deformation and Bonding Criteria under Impulsive Loading. *Fizika Goreniya I Vzryva*, **27**, no. 3, pp. 103–117 (in Russian). English translation: *Physics of Explosion, Combustion and Shock Waves*, 1991, November, pp. 364–376.

Bourne, N., and Field, J. (1992) Shock Induced Collapse of Single Cavity in Liquids, *J. Fluid Mechanics*, **244**, pp. 225–240.

Butcher, B. M., and Karnes, C.H. (1969) Dynamic Compaction of Porous Iron. *Journal of Applied Physics*, **40**, no. 7, pp. 2967–2976.

Butcher, B.M., Carroll, M.M., and Holt, A.C. (1974) Shock-Wave Compaction of Porous Aluminium. *J. Appl. Phys.* **45**, no. 9, pp. 3864–3875.

Carroll, M.M., and Holt, A.C. (1972a) Suggested Modification of the *P*-α Model for Porous Materials. *J. Appl. Phys.* **43**, no. 2, pp. 759–761.

Carroll, M.M., and Holt, A. C. (1972b) Static and Dynamic Pore-Collapse Relations for Ductile Porous Materials. *J. Appl. Phys.* **43**, no. 4, pp. 1626–1635.

Carroll, M.M., and Holt, A.C. (1973) Steady Waves in Ductile Porous Solids. *J. Appl. Phys.* **44**, no. 10, pp. 4388–4392.

Carroll, M.M., Kim, K.T., and Nesterenko, V.F. (1986) The Effect of Temperature on Viscoplastic Pore Collapse. *J. Appl. Phys.* **59**, no. 6, pp. 1962–1967.

Chen, Y.-J., Meyers M.A., and Nesterenko, V.F. (1999) Spontaneous and Forced Shear Localization in High-Strain-Rate Deformation of Tantalum. *Material Science and Engineering A*, **268**, pp. 70–82.

Conley, P.A. (1999) Eulerian Hydrocode Analysis of Reactive Micromechanics in the Shock Initiation of Heterogeneous Energetic Materials. PhD Dissertation, University of California, San Diego, p. 263.

Conley, P.A. (1999) An Estimate of the Linear Strain Rate Dependence of Octahydro-1,3,5,7-Tetranitro-1,3,5,7-Tetrazocine. *Journal of Applied Physics*, **86**, no. 12, pp. 6717–6728.

Cooper, S.R., Benson, D.J., and Nesterenko, V.F. (2000) The Role of Void Geometry on the Mechanics of Void Collapse and Hot Spot Formation in Ductile Materials. *Int. Journal of Plasticity*, **16**, no. 5, pp. 525–540.

Davison, L., (1971) Shock-Wave Structure in Porous Solids, *J. Applied Physics*, **42**, no. 13, pp. 5503–5512.

Dieter, G.E. (1986) *Mechanical Metallurgy*, McGraw-Hill, New York.

Deribas, A.A., Nesterenko, V.F., and Staver, A.M. (1976) The Separation of Components in Explosive Compaction of Multicomponent Materials. In: *Proc. III International Symposium on Metal Explosive Working*, Marianske Lazni, Chehoslovakia, Semtin, Pardubice, Chehoslovakia, Vol. **2**, pp. 367–372 (in Russian).

Do, Ian Phuc Hoang (1999) Shock Induced Chemical Reactions of Multi-Material Powder Mixtures—An Eulerian Finite Element Computational Analysis. PhD Dissertation, University of California, San Diego, p. 158.

Dremin, A.N., and Shvedov, K.K. (1964) Determination of the Chapmen–Juge Pressure and Reaction Time at the Detonation Wave of High Explosives. *Zhurnal Prikladnoi Mekhaniki i Tehknicheskoi Fiziki*, **5**, no. 2, pp. 154–159 (in Russian).

Dunin, S.Z., and Surkov, V.V. (1979) Structure of a Shock Wave Front in Porous Solid. *Zhurnal Prikladnoi Mekhaniki i Tehknicheskoi Fiziki*, **20**, no. 5, pp. 106–114 (in Russian).
English translation: *J. Appl. Mech. and Tech. Phys.*, 1980, March, pp. 612–618.

Dunin, S.Z., and Surkov, V.V. (1982) Effects of Energy Dissipation and Melting on Shock Compression of Porous Bodies. *Zhurnal Prikladnoi Mekhaniki i Tehknicheskoi Fiziki*, **23**, pp. 131–142 (in Russian).
English translation: *J. Appl. Mech. and Tech. Phys.*, 1982, July, pp. 123–134.

Eichhorn, R., and Small, S. (1964) Experiments on the Lift and Drag of Spheres Suspended in a Poiseuille Flow. *J. Fluid Mech.*, **20**, no. 3, pp. 513–521.

Flinn, J.E., Williamson, R.L., Berry, R.A., Wright, R.N., Gupta, Y.M., and Williams, M. (1988) Dynamic Consolidation of Type 304 Stainless-Steel Powders in Gas Gun Experiments. *Journal of Applied Physics*, **64**, no.3, pp. 1446–1456.

German, R.M. (1996) *Sintering Theory and Practice.* Wiley, New York.

Gilath, I. (1995) Laser-Induced Spallation and Dynamic Fracture at Ultra High Strain Rate. In: *High-Pressure Shock Compression of Solids II, Dynamic Fracture and Fragmentation* (Edited by L. Davison, D.E. Grady, and M. Shahinpoor). Springer-Verlag, New York, pp. 90–120.

Gilev, S.D., and Trubachev, A.M. (1996) Shock-Induced Conduction Waves in Solids and Their Applications in High Power Systems. In: *Shock Compression of Condensed Matter—1995* (Edited by S.C. Schmidt, W.C. Tao). AIP Conference Proceedings 370. AIP Press, New York, Part 2, pp. 933–936.

Godunov, S.K. Demchuk, A.F., Kozin, N.S., and Mali, V.I. (1974) Interpolation Formulae for the Dependence of Maxwellian Viscosity of Some Metals on the Tangential Stress Intensity and on Temperature. *Zhurnal Prikladnoi Mekhaniki i Tehknicheskoi Fiziki*, **15**, pp. 114–118 (in Russian).

Gourdin, W.H. (1986) Dynamic Consolidation of Metal Powders. *Progress in Materials Science*, **30**, pp. 39–80.

Gradshteyn, I.S., and Ryzhik, I.M. (1980) *Table of Integrals, Series, and Products* (Corrected and Enlarged Edition). Academic Press, Inc., San Diego, 205 p.

Gray, W.A. (1968) *The Packing of Solid Particles.* Chapman and Hall, London.

Hofmann, R., Andrews, D.J., and Maxwell, D.E. (1968) Computed Shock Response of Porous Aluminium. *J. Appl. Phys.* **39**, pp. 4555–4562.

Holman, G.T., Graham, R.A., and Anderson, M.U. (1993) Shock Response of Porous $2Al + Fe_2O_3$ Powder Mixtures. In: *Proceedings of the Joint Meeting of the International Association for the Advancement of High Pressure Science and Technology and the American Physical Society Topical Group on Shock Compression of Condensed Matter* (Edited by S.C. Schmidt, J.W. Shaner, G.A. Samara, and M. Ross), pp. 1119–1122. AIP Press, New York.

Holt, A.C., Carroll, M.M., and Kusubov A. (1971) A Simple Constitutive Relation for Porous Materials. Livermore. Preprint/Lawrence Radiation Laboratory, UCRL-73057.

Holt, A.C., Carroll, M.M., and Butcher, B.M. (1974) Application of a New Theory for the Pressure-Induced Collapse for Pores in Ductile Materials. In: *Pore Structure and Properties of Materials* (Edited by S. Modry), Vol. **5**, pp. D63–D68. Academia, Prague.

Horie, Y. (1993) Shock-Induced Chemical Reactions in Inorganic Powder Mixtures. In: *Shock Waves in Materials Science* (Edited by A.B. Sawaoka). Springer-Verlag, Tokyo, Chapter 4, pp. 67–100.

Howe, P.M. (2000a) Explosive Behaviors and the Effects of Microstructure. In: *Solid Propellant Chemistry, Combustion, and Motor Interior Ballistics,* Chapter 1.6, (Edited by V. Yang, T.B. Brill, and W.Z. Ren). Progress in Astronautics and Aeronautics. Vol. **185**, American Institute of Aeronautics and Astronautics, Reston, VA.

Howe, P.M. (2000b) Trends in the Shock Initiation of Heterogeneous Explosives. In: *Proceedings of the Eleventh International Detonation Symposium* (Edited by J.M. Short and J.E. Kennedy). Thomson-Shore, Dexter, pp. 670–678.

Ipatiev, A.S., Afanasenko, S.I., and Nesterenko, V.F. (1993) Method of Fracture. Russian patent #1434594, 18 August, 1993.

Ipatiev, A.S., Kosenkov, A.P., Nesterenko, V.F., Meshcheriakov, Y.P. and Afanasenko, S.I. (1986) Experimental Investigation of Fracture Characteristics of Metal Plates with Hemicylindrical Grooves at Explosive Loading with Contact Surface Charges. In: *Proc. IX Intern. Conf. on High Energy RateFabrication* (Edited by V.F. Nesterenko and I.V. Yakovlev). Lavrentyev Institute of Hydrodynamics and Special Design Ofice of High-Rate Hydrodynamics, Novosibirsk, pp. 64–70 (in Russian).

Ipatiev, A.S., Kosenkov, A.P., Nesterenko, V.F., Meshcheriakov, Y.P., and Afanasenko, S.I. (1986) Experimental and Numerical Investigation of Influence of Hemicylindrical Groove on the Fracture Characteristics of Metal Plates at Contact Explosive Loading. In: *Proc. IX All-Union Conf. on Numerical Methods of Solving Problems in Elasticity and Plasticity* (Edited by V.M Fomin). Institute of Theoretical and Applied Mechanics, Novosibirsk, pp. 64–70 (in Russian).

Ipatiev, A.S. (1990) Experimental Investigation of Deformation and Fracture of Metal Plates with Macrodefects at Contact Explosive Loading, *Impulse Treatment of Materials* (Edited by A.F. Demchuk, V.F. Nesterenko, V.M. Ogolikhin, A.A. Shtertzer, Y.V. Kolotov, S.A. Pershin, and V.I. Danilevskaya). Special Design Ofice of High-Rate Hydrodynamics and Institute of Theoretical and Applied Mechanics, Novosibirsk, 1990, pp. 145–155 (in Russian).

Ishutkin, S.N., Kuz'min, G.E., and Pai, V.V. (1986) Thermocouple Measurements of Temperature in the Shock Compression of Metals. *Fizika Goreniya i Vzryva*, **22**, no. 5, pp. 96–104 (in Russian).

Jach, K. (1989) Numerical Modeling of Two-Dimensional Elastic-Visco-Plastic Deformation of Materials at Dynamic Loads. In: Proceedings of XI AIRAPT International Conference on High Pressure Science and Technology (Edited by N.V. Novikov). Naukova dumka, Kiev, pp. 198–200.

Kasiraj, P., Vreeland, T. Jr., Schwarz, R.B., and Ahrens, T.J. (1984a) Shock Consolidation of a Rapidly Solidified Steel Powder, *Acta Metall.*, **32**, no. 8, pp. 1235–1241.

Kasiraj, P., Vreeland, T. Jr., Schwarz, R.B., and Ahrens, T.J. (1984b) Mechanical Properties of a Shock Consolidated Steel Powder. In: *Shock Waves in Condensed Matter: Proc. Amer. Phys. Soc. Topical Conference* (Edited by J.R Asay, R.A. Graham, and G.K. Straub). Elsevier Science, Amsterdam, pp. 435–438.

Khasainov, B.A., Borisov, A.A., Ermolaev, B.S., and Korotkov, A.I. (1980) Closed Model of Shock Initiation of Detonation in High Density Explosives. In: *Chemical Physics of Combustion and Explosion: Detonation*. Institute of Chemical Physics, Chernogolovka, pp. 52–55 (in Russian).

Kinslow, R. (Editor) (1970) *High-Velocity Impact Phenomena*. Academic Press, New York, p. 532.

Klopp, R.W., Clinton, R.J., and Shawki, T.G. (1985) Pressure-Shear Impact and the Dynamic Viscoplastic Response of Metals. *Mechanics of Materials*, **4**, no. 3/4, pp. 375–385.

Kostyukov, N.A. (1990) Mechanism of Lamination of Particulate Composites in Shock Loading. *Zhurnal Prikladnoi Mekhaniki i Tehknicheskoi Fiziki*, **31**, no. 1, pp. 84–91 (in Russian).
English translation: *Journal of Applied Mechanics and Technical Physics*, 1990, July, pp. 79–85.

Kostyukov, N.A. (1991) Physical Causes and Mechanisms of the Formation of Boundary Regions in the Two-Dimensional Explosive Compaction of Powdered Materials. *Zhurnal Prikladnoi Mekhaniki i Tehknicheskoi Fiziki*, **32**, no. 6, pp. 154–161 (in Russian).
English translation: *Journal of Applied Mechanics and Technical Physics*, 1990, July, pp. 967–973.

Kozin, N., and Simonov, V.A. (1973) Shock Interaction with a Wedge Cavity. *Fizika Goreniya i Vzryva*, **9**, no. 4, pp. 551–558 (in Russian).
English translation: *Combustion, Explosion, and Shock Waves*, 1975, May, pp. 477–482.

Kreig, R.D., and Key, S.W. (1976) Implementation of a Time Dependent Plasticity Theory into Structural Programs. In: *Constitutive Equations in Viscoplasticity: Computational and Engineering Aspects*, Vol. **20**, ASME, New York, pp. 125–137.

Krueger, B.A., Mutz, A.H., and Vreeland, T. Jr. (1992) Shock-Induced and Self-Propagating High-Temperature Synthesis Reactions in Two Powder Mixtures: 5:3 Atomic Ratio Ti/Si and 1:1 Atomic Ratio Ni/Si. *Metallurgical Transactions A*, **23**, pp. 55–58.

Kusubov, A.S., Nesterenko, V.F., Wilkins, M.L., Reaugh, J.E., and Cline, C.F. (1989) Dynamic Deformation of Powdered Materials as a Function of Particle Size. In: *Proc. International Seminar on High-Energy Working of Rapidly Solidified Materials and High-T_C Ceramics* (Edited by V.F. Nesterenko and A.A. Shtertzer). Special Design Office of High-Rate Hydrodynamics, Siberian Branch USSR Academy of Sciences, Novosibirsk, pp. 139–156.

Lazaridi, A.N. (1990) The Influence of Initial Characteristics of Steel Granules and Regimes of Loading on the Strength of Explosively Densified Powders. In: *Impulse Treatment of Materials* (Edited by A.F. Demchuk, V.F. Nesterenko, V.M. Ogolikhin, A.A. Shtertzer, Y.V. Kolotov, S.A. Pershin, and V.I. Danilevskaya). Special Design Ofice of High-Rate Hydrodynamics and Institute of Theoretical and Applied Mechanics, Novosibirsk, 1990, pp. 70–86 (in Russian).

Lee, J.H.S. (2000) Initiation and Propagation Mechanisms of Detonation Waves. In: *Proceedings of Third International High Energy Materials Conference and Exhibition* (in press).

Mali, V.I. (1973) Flow of Metals with a Hemispherical Indentation Under the Action of Shock Waves. *Fizika Goreniya I Vzryva*, **9**, no. 2, pp. 282–286 (in Russian). English translation: *Physics of Explosion, Combustion and Shock Waves*, 1975, January, pp. 241–245.

Matyushkin, N.I. and Trishin, Yu.A. (1978) Some Effects Which Arise in Connection With the Explosive Squeezing of a Viscous Cylindrical Shell. *Zhurnal Prikladnoi Mekhaniki i Tehknicheskoi Fiziki*, **19**, no. 3, pp. 99–112 (in Russian). English translation: *Journal of Applied Mechanics and Technical Physics*, 1978, 19, no. 3, pp. 362–371.

McGlaun, J.M. (1982) Improvements in CSQII: a Transmitting Boundary Condition. Technical Report SAND82-1248, Sandia National Laboratories, Albuquerque, NM.

McGlaun, J.M., Thompson, S.L., and Elrick, M.G. (1990) CTH: A Three-Dimensional Shock Wave Physics Code. *Int. J. Impact Eng.*, **10**, pp. 351–360.

Menikoff, R., and Kober, E. (1999) Compaction Waves in Granular HMX. Report, Los Alamos National Laboratory, January 1999, LA-13546-MS, pp. 1–68.

Meyers, M.A., and Wang, S.L. (1988) An Improved Method for Shock Consolidation of Powders. *Acta Metall.*, **36**, pp. 925–936.

Meyers, M.A. (1994) *Dynamic Behavior of Materials*. Wiley, New York.

Meyers, M.A., Benson, D.J., and Shang, S.S. (1994) Energy Expenditure and Limitations in Shock Consolidation. In: *Proceedings of the Joint Meeting of the International Association for the Advancement of High Pressure Science and Technology and the American Physical Society Topical Group on Shock Compression of Condensed Matter* (Edited by S.C. Schmidt, J.W. Shaner, G.A. Samara, and M. Ross). AIP Press, New York, pp. 1239–1242.

Meyers, M.A., Benson, D.J., and Olevsky, E.A. (1999) Shock Consolidation: Microstructurally-Based Analysis and Computational Modeling. *Acta Mater.*, **47**, no. 7, pp. 2089–2108.

Nagayama, K. (1981) New Methods of Magnetic Flux Compression by the Propogation of Shock-Induced Metallic Transition in Semiconductors. *Appl. Phys. Letters*, **38**, no. 2, pp. 109–110.

Nemat-Nasser, S., and Hori, M. (1987) Void Collapse and Void growth in Crystalline Solids. *Journal of Applied Physics*, **62**, no. 7, pp. 2746–2757.

Nemat-Nasser, S., and Chang, S.N. (1990) Compresion-Induced High Strain Rate Void Collapse, Tensile Cracking, and Recrystallization in Ductile Single and Polycrystals, *Mechanics of Materials*, **10**, pp. 1–17.

Nesterenko, V.F. (1975) Electrical Effects under Shock Loading of Metals Contact. *Fizika Goreniya i Vzryva* **11**, 444–456 (in Russian). English translation: *Physics of Explosion, Combustion and Shock Waves*, 1976, July, 11, pp. 376–385.

Nesterenko, V.F. (1979) Noncontact Method of Measurement of Parameters of Shocked Metals. In: *Abstracts of III All Union Symp. on Impulse Pressures* (Edited by S.S. Batsanov), Institute of Standards, Moscow, pp. 14–15.

Nesterenko, V.F. (1980) Method of Measurement of Parameters of Shock Compression of Conductive Plate. Patent USSR (A.C. 717656 SSSR, MKI 01P3/52), Claimed: 28 March 1978; Published: *Discoveries, Inventions*, 1980, no. 7, p. 96 (in Russian).

Nesterenko, V.F. (1985) Potential of Shock-Wave Methods for Preparing and Compacting Rapidly Quenched Materials. *Fizika Goreniya i Vzryva*, **21**, no. 6, 85–98 (in Russian).
English translation: *Physics of Explosion, Combustion and Shock Waves*, 1986, May, pp. 730–740.

Nesterenko, V.F. (1986) Heterogeneous Heating of Porous Materials at Shock Wave Loading and Criteria of Strong Compacts. In: *Proc. IX International Conf. on High Energy Rate Fabrication* (Edited by V.F. Nesterenko and I.V. Yakovlev). Lavrentyev Institute of Hydrodynamics and Special Design Office of High-Rate Hydrodynamics, Novosibirsk, pp. 157–163 (in Russian).

Nesterenko, V. F. (1988a) Micromechanics of Powders under Strong Impulse Loading. In: *Computer Methods in Theory of Elasticity and Plasticity: Proceedings of X All-Union Conference* (Edited by F.M. Fomin). Institute of Theoretical and Applied Mechanics, Novosibirsk, pp. 212–220.

Nesterenko, V.F. (1988b) Influence of the Parameters of Powder Internal Structure on the Process of Explosive Compaction. In: *Proc. International Symposium on Metal Explosive Working*. Semtin, Purdubice, Chehoslovakia, Vol. **3**, pp. 410–417 (in Russian).

Nesterenko, V.F. (1988c) Nonlinear Phenomena under Impulse Loading of Heterogeneous Condensed Media. Doctor in Physics and Mathematics Thesis, Academy of Sciences, Russia. Lavrentyev Institute of Hydrodynamics, Novosibirsk, Siberian Branch.

Nesterenko, V.F., Lazaridi, A.N., Pershin, S.A., Miller, V.Y., Feschiev, N.H., Krystev, M.R., Minev, R.M. and Panteleeva, D.B. (1989) Properties of Compacts from Rapidly Solidified Steel Granules of Different Sizes After Shock-Wave Consolidation. In: *Proc. International Seminar on High-Energy Working of Rapidly Solidified Materials and High-T_C Ceramics* (Edited by V. F. Nesterenko and A. A. Shtertzer). Special Design Office of High-Rate Hydrodynamics, Siberian Branch USSR Academy of Sciences, Novosibirsk, pp. 118–126 (in Russian).

Nesterenko, V.F., and Lazaridi, A.N. (1990) Regimes of Shock-Wave Compaction of Granular Materials. In: *High Pressure Science and Technology: Proceedings of XII AIRAPT and XXVII EMPRG International Conference* (Edited by W.B. Holzapfel and P.G. Johansen). Gordon and Breach, New York, pp. 835–837.

Nesterenko, V.F. (1992) *High-Rate Deformation of Heterogeneous Materials*. Nauka, Novosibirsk. (in Russian).

Nesterenko, V.F., and Bondar, M.P. (1994) Investigation of Deformation Localization by the "Thick-Walled Cylinder" Method. *DYMAT Journal*, **1**, 245–251.

Nesterenko, V.F. (1995) Dynamic Loading of Porous Materials: Potential and Restrictions for Novel Materials Applications. In: *Metallurgical and Materials Applications of Shock-Wave and High-Strain-Rate Phenomena: Proceedings of the 1995 International Conference EXPLOMET-95* (Edited by L.E. Murr, K.P. Staudhammer, and M.A. Meyers). Elsevier, Amsterdam, pp. 3–13.

Nesterenko, V.F., Luk'yanov, Y.L., and Bondar, M.P. (1994) Deformation of the Contact Zone in the Formation of a "Cold" Boundary Layer. *Fizika Goreniya i Vzryva*, **30**, no. 5, pp. 126–129 (in Russian).
English translation: *Physics of Explosion, Combustion and Shock Waves*, 1994, **30**, no. 5, pp. 693–695.

Nesterenko, V.F., Meyers, M.A., and Wright T.W. (1998) Self-Organization in the Initiation of Adiabatic Shear Bands. *Acta Materialia,* **46**, no. 1, pp. 327–340.

Nesterenko, V.F., Indrakanti, S.S., Brar, S. and Gu, Y. (2000a) Long Rod Penetration Test of Hot Isostatically Pressed Ti-Based Targets. In: *Shock Compression of Condensed Matter—1999, AIP Conference Proceedings* (Edited by M.D. Furnish, L.C. Chhabildas, and R.S. Hixson). AIP, New York, Vol. **505**, pp. 419–422.

Nesterenko, V.F., Indrakanti, S.S., Brar, S. and Gu, Y. (2000b) Ballistic Performance of Hot Isostatically Pressed (Hiped) Ti-Based Targets. In: *Key Engineering Materials,* Vols. **177–180**. Trans Tech Publications, Switzerland, pp. 243–248.

Nesterenko, V.F., Indrakanti. S.S., Goldsmith, W., and Gu, Y. (2001) Plug Formation and Fracture of Hot Isostatically Pressed (HIPed) Ti-6Al-4V Targets. In: *Fundamental Issues and Applications of Shock-Wave and High-Strain-Rate Phenomena.* (Edited by K.P. Staudhammer, L.E. Murr, and M.A. Meyers). Elsevier Science, Amsterdam, pp. 593–600.

Nigmatullin, R.I. (1978) *The Fundamentals of Mechanics of Heterogeneous Media.* Nauka, Moscow (in Russian).

O'Keeffe, D.J. (1971) Theoretical Determination of the Shock States of Porous Copper. *J. Appl. Phys.,* **42**, pp. 888–889.

Pai, V.V., Yakovlev, I.V., and Kuz'min, G.E. (1996) Investigation of Shock Compression of Composite Porous Media by a Nondisturbing Electromagnetic Technique. *Fizika Goreniya i Vzryva,* **32**, no. 2, pp. 124–129 (in Russian). English translation: *Combustion, Explosion, and Shock Waves,* **32**, no. 2, pp. 225–229.

Pastine, D.J., Lombardi, M., Chatterjee, A., and Tchen, W. (1970) Theoretical Shock Properties of Porous Aluminum. *J. Appl. Phys.,* **41**, pp. 3144–3147.

Prummer, R. (1986) Explosive Compaction of Powders. State of Art. In: *Proc. IX International Conf. on High Energy Rate Fabrication* (Edited by V.F. Nesterenko and I.V. Yakovlev). Lavrentyev Institute of Hydrodynamics and Special Design Office of High-Rate Hydrodynamics, Novosibirsk, pp. 169–178.

Roman, O.V., Nesterenko, V.F., and Pikus, I.M. (1979) Influence of the Powder Particle Size on the Explosive Pressing. *Fizika Goreniya i Vzryva,* **15**, no. 5, pp. 102–107 (in Russian). English translation: *Combustion, Explosion, and Shock Waves,* March 1980, pp. 644–649.

Samarskii, A.A., and Popov, Y.P. (1972) *Discrete Schemes of Gas Dynamics.* Nauka, Moscow, p. 351.

Schwarz, R.B., Kasiraj, P., Vreeland, T. Jr., (1986) Temperature Kinetics During Shock-Wave Consolidation of Metallic Powders. In: *Metallurgical Applications of Shock Waves and High-Strain-Rate Phenomena* (Edited by L.E. Murr, K.S. Staudhammer, and M.A. Meyers). Marcel Dekker, New York, pp. 313–327.

Shang, S.S., Benson, D.J. and Meyers, M.A. (1994) Microstructurally-Based Analysis and Computational Modeling of Shock Consolidation. *Journal de Physique,* **4**, C8-521–C8-526.

Sheffield, S.A., Gustavsen, R.L., and Anderson, M.U. (1997) Shock Loading of Porous High Explosives. In: *High-Pressure Shock Compression of Solids IV/Response of Highly Porous Solids to Shock Loading* (Edited by L. Davison, Y. Horie, and M. Shahinpoor). Springer-Verlag, New York, pp. 23–61.

Slater, R.A.C. (1977) *Engineering Plasticity.* McMillan, London.

Spaepen, F., and Turnbull, D. (1975) Formation of Metallic Glasses. In: *Rapidly Quenched Metals*, MIT, Cambridge, pp. 205–229.

Staver, A.M. (1981) Metallurgical Effects under Shock Compression of Powder Materials. In: *Shock Waves and High-Strain-Rate Phenomena in Metals. Concepts and Applications* (Edited by M.A. Meyers and L.E. Murr). Plenum Press, New York, pp. 865–880.

Steinberg, D.J., and Guinan, M.W. (1978) A High-Strain-Rate Constitutive Model for Metals. University of California, Lawrence Livermore National Laboratory. Rept. UCRL-80465.

Steinberg, D.J., Cochran, S.G., and Guinan, M.W. (1980) A Constitutive Model for Metals Applicable at High-Strain Rate. J. *Appl. Phys.*, **51**, pp. 1498–1504.

Steinberg, D.J. (1996) Equation of State and Strength Properties of Selected Materials. University of California. Lawrence Livermore National Laboratory. Rept. UCRL-MA-106439.

Stepanov, G.V. (1978) Coefficient of Viscosity of Metallic Materials at High-Strain Deformation in Elastic-Plastic Waves. In: *Detonation. Critical Phenomena. Physical-Chemical Transformations in Shock Waves*. Institute of Chemical Physics, Chernogolovka, pp. 106–111 (in Russian).

Stokes, J.L., Nesterenko, V.F., Shlachter, J.S., Fulton, R.D., Indrakanti, S.S., and Gu, Y. (2001) Comparative Behavior of Ti and 304 Stainless Steel in Magnetically-Driven Implosion at the Pegasus-II Facility. In *Fundamental Issues and Applications of Shock-Wave and High-Strain-Rate Phenomena* (Edited by K.P. Staudhammer, L.E. Murr, and M.A. Meyers). Elsevier Science, Amsterdam, pp. 585–592.

Tadhani, N.N., Dunbar, E., and Graham, R.A. (1994) Characteristics of Shock-Compressed Configuration of Ti and Si Powder Mixtures. In: *Proceedings of the Joint Meeting of the International Association for the Advancement of High Pressure Science and Technology and the American Physical Society Topical Group on Shock Compression of Condensed Matter* (Edited by S.C. Schmidt, J.W. Shaner, G.A. Samara, and M. Ross). AIP Press, New York, pp. 1307–1310.

Tang, Z.P., Horie, Y., and Psakhie, S.G. (1997) Discrete Mesoelement Dynamic Simulation of Shock Response of Reactive, Porous Solids. In: *High-Pressure Shock Compression of Solids IV: Response of Highly Porous Solids to Shock Loading* (Edited by L. Davison, Y. Horie, and M. Shahinpoor). Springer-Verlag, New York, pp. 143–176.

Tang, Z.P., Liu, W., and Horie, Y. (1999) *Bulletin of the American Physical Society*, **44**, no. 2, p. 21.

Tang, Z.P., Liu, W., and Horie, Y. (2000) Numerical Investigation of Pore Collapse under Dynamic Compression. In: *Shock Compression of Condensed Matter–1999* (Edited by M.D. Furnish, L.C. Chabildas, and R.S. Hixson). AIP, New York, pp. 309–312.

Thouvenin, J. (1966) Action d'une onde de choc sur un solide poreux. J. *Physics, 27*, pp. 183–189.

Timoshenko, S.P., and Goodier, J.N. (1970) *Theory of Elasticity*, 3rd ed., McGraw-Hill., New York, pp. 35–97.

Tong, W., Clifton, R.J., and Huang, S. (1992) Pressure–Shear Impact Investigation of Strain Rate History Effects in Oxygen-Free High-Conductivity Copper. J. *Mech. Phys. Solids*, **40**, pp. 1251–1294.

Tong, W., and Ravichandran, G. (1993) Dynamic Pore Collapse in Viscoplastic Materials. *J. Appl. Phys.* **74**, pp. 2425–2435.

Tong, W., and Ravichandran, G. (1994a) Rise Time in Shock Consolidation of Materials. *Appl. Phys. Lett.* **65**, pp. 2783–2785.

Tong, W., and Ravichandran, G. (1994b) Effective Elastic Moduli and Characterization of a Particulate Metal-Matrix Composite with Damaged Particles. *Composites Science and Technology,* **52**, no. 2, pp. 247–252.

Tong, W., Ravichandran, G., Christman, T., and Vreeland, T., Jr. (1995) Processing SiC-Particulate Reinforced Titanium-based Metal Matrix Composites by Shock Wave Consolidation. *Acta Metallurgica et Materialia,* **43**, no. 1, pp. 235–250.

Tong, W., and Ravichandran, G. (1997) Recent Developments in Modeling Shock Compression of Porous Materials. In: *High-Pressure Shock Compression of Solids IV. Response of Highly Porous Solids to Shock Loading* (Edited by L. Davison, Y. Horie, and M. Shahinpoor). Springer-Verlag, New York, pp. 177–203.

Torre, C. (1948) Theorie und Zusammengeprebteer Pulver. *Berg; Huttenman. Monatsh. Montan. Hochschule Leoben,* **93**, pp. 62–67.

van Leer, B. (1977) Towards the Ultimate Conservative Difference Scheme. IV. A New Approach to Numerical Convection. *Journal of Computational Physics* **23**, pp. 276–299.

Wang, S.L., Meyers, M.A., and Szecket, A. (1988) Warm Shock Consolidation of IN718 Powder. *J. Mater. Science,* **23**, pp. 1786–1804.

Williamson, R.L., and Berry, R.A. (1986) Microlevel Numerical Modeling of the Shock Wave Induced Consolidation of Metal Powders. In: *Proceedings of the Fourth American Physical Society Topical Conference on Shock Waves in Condensed Matter* (Edited by Y.M. Gupta). Plenum Press, New York, pp. 341–346.

Williamson, R.L., Wright, R.N., Korth, G.T., and Rabin, B.H. (1989) Numerical Simulation of Dynamic Consolidation of SiC Fiber-Reinforced Aluminum Composite. *J. Appl. Phys.,* **66**, pp. 1826–1831.

Williamson, R.L., and Wright, R.N. (1990) A Particle-Level Numerical Simulation of the Dynamic Consolidation of a Metal Matrix Composite Material. In: *Shock Waves of Condensed Matter—1989: Proceedings of the American Physical Society Topical Conference* (Edited by S.C. Schmidt, J.N. Johnson, and L.W. Davison). Elsevier Science, Amsterdam, pp. 487–490.

Williamson, R.L. (1990) Parametric Studies of Dynamic Powder Consolidation Using a Particle-Level Numerical Model. *J. Appl. Phys.,* **68**, pp. 1287–1296.

Youngs, D.L. (1982) Time Dependent Multi-Material Flow with Large Fluid Distortion. In: *Numerical Methods for Fluid Dynamics* (Edited by K.W. Morton and M.J. Baines). Pergamon Press, Oxford, pp. 273–285.

Zababakhin, E.I. (1970) Phenomena of Nonrestircted Cumulation. In: *Mechanics in USSR for 50 Years.* Nauka, Moscow, Vol. **2**, pp. 313–342 (in Russian).

Zel'dovich, Ya.B. and Raiser, Yu.P. (1963) Physics of Shock Waves and High-Temperature Hydrodynamic Phenomena. Fizmatgiz, Moscow, p. 632 (in Russian).

3
Transformation of Shocks in Laminated and Porous Materials

3.1 Introduction

Impulse transformation in laminates and porous materials has important practical applications connected with impact and blast mitigation. The investigation of strong shock-wave dynamics in laminar media is focused mainly on two important aspects. One is wave transformation, for example, attenuation or amplification of the shock amplitude as a function of the laminar system structure. Zababakhin [1965], [1970] predicted that as the layer thickness continuously decreased with distance for a one-dimensional system of alternating materials with different densities, the phenomenon of unlimited cumulation of the shock wave could be obtained. This was a first demonstration of unrestricted cumulation which is not related to the centripetal motion of matter. Later, this tendency was numerically and experimentally confirmed (Kozyrev, Kostyleva, and Ryazanov [1969]; Ogarkov, Purygin, and Samylov [1969]; Fowles [1979]). The successful application of this idea was demonstrated by the launching of the thin titanium plate to velocities of 16 km/s using a multiply graded-density impactor with increasing shock impedance from the impact surface in sequence TPX-plastic, magnesium, aluminum, titanium, copper, and tantalum (Chhabildas, Kmetyk, Reinhard, and Hall [1996]). This system tailors the time-dependent stress pulse to launch the flier plate intact by using a precisely controlled thickness of each layer.

Formulas for the pressure and particle velocity based on the consideration of the head–wave interaction with interfaces in a linear acoustic approximation were obtained by Laptev and Trishin [1974] for laminar material (LM). It was shown that an increase or decrease in the shock pressure and particle velocity is determined by the ratio of the acoustic impedance of the layers. This fact was later confirmed for the nonlinear behavior for media consisting of two or three different layers (Kroshko and Chubarova [1980]).

However, these results are based on the consideration of the leading shock-wave interaction with interfaces. It was demonstrated by Nesterenko, Fomin, and Cheskidov [1983a,b] that the growth in amplitude of the head shock is a consequence not only of the wave splitting on the boundary of materials with different acoustic impedances. The other reason for it is the origination of compression waves behind the leading front that later overtake the head wave. This is essentially a nonlinear phenomenon and does not exist in linear acoustic approximation.

A strong nonlinear shock with pressure varying periodically in the front can propagate in a periodic laminar system with constant absolute thicknesses of light and heavy layers, if the piston generating the motion moves with constant velocity (Zababakhin [1965], [1970]). The cell dimension here determines only the scale of the phenomenon and the build-up time of the stationary wave pattern. It does not influence the pressure amplitude.

For finite impulse duration, it was shown experimentally by Nesterenko [1977], that the head wave amplitude of a strong shock at the identical depth is independent of the number of cells in the periodic laminar material. This contradicts the approach following only the interaction of the head wave with interfaces. This approach yields strong damping of the head wave amplitude as the number of crossings of the interfaces increases as result of the shock decomposition on each of them (Akhmadeev and Bolotnova [1985], [1994]).

On the basis of an analysis of the nonlinear wave equation derived in continuum approximation, it is shown that an increase due to geometrical dispersion occurs in the amplitude of a weak nonlinear shock as it propagates in a periodic laminar material (Lundergan and Drumheller [1971]; Peck [1971]).

Another area of research is connected with the application a laminar system as a model for transient processes in heterogeneous porous and solid media (Thouvenin [1966]; Hofmann, Andrews, and Maxwell [1968]; Nigmatullin, Vainshtein et al. [1976]). This modeling is necessary because micromechanisms, of the material transition from the initial into the final state in the heterogeneous materials under shock wave, are not sufficiently understood. The laminar model provides the natural way to take into account the multiple wave reflections phenomena driving material into the final state. Yet the validity of such modeling should be established.

There are some other important aspects in the dynamic behavior of the composite laminates, such as damage caused by mesoscopic tensile stresses under impact, including spall phenomena. The corresponding results may be found in papers by Jih and Sun [1993], and by Gupta, Pronin, and Anand [1996]. The use of the laminar materials as the meteor bumpers is also described by Riney [1970].

There are a large quantity of papers on the investigation of shocks in laminar materials (see, for example, the reviews by Bedford, Drumheller, and Sutherland [1976], and by Gupta, Pronin, and Anand [1996]). Nevertheless, the propagation of finite-duration strong shock waves, in periodic LM during the unsteady motion phase with nonlinear effects taken into account, is not extensively studied. At the same time, nonlinear effects can result in qualitatively new phenomena outside of description by linear models, and they need to be studied more thoroughly. This can provide some additional insight into the optimization of strong shock damping by laminar materials. This is important for lightweight armor design, as well as for other objects with a potential danger of shock loading (composite turbine blades, wings).

Shock transformation by porous materials is recognized as a method for blast mitigation or for the damping of shock loading on structures. But in some cases, porous layers may amplify the load. In general, there is a lack of information for the optimization of porous materials properties and structures for this purpose.

3.2   Experiments on Shock Transformation in Solid Laminar Materials (LMs)

To investigate the complex mechanism of the shock-wave transformation, in LMs as a function of their mesoscopic structural characteristics, the method allowing for detection of the complete wave profile is needed. The electromagnetic technique (Dremin and Shvedov [1964]) allows a relatively long detection time and is free from the complications connected with the gauge records from the multiple wave loading inherent for laminates. This technique is most suitable for materials that are not only dielectric in the initial state, but also preserve this property under shock-wave loading. That is why the LM Teflon–paraffin which remains a dielectric under the high pressure being investigated (Davison and Graham [1979]) was selected.

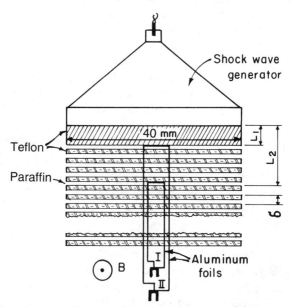

Figure 3.1. An experimental set-up for the measurement of the particle velocity profile in laminates using an electromagnetic method.

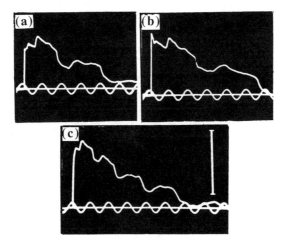

Figure 3.2. Particle velocity in solid LM Teflon–paraffin: (a) $L = 10$ mm, $\delta = 5$ mm; (b) $L = 25$ mm, $\delta = 5$ mm; (c) $L = 25$ mm, $\delta = 2.5$ mm. The frequency of the timing mark is 1 MHz. The vertical scale corresponds to 1 mm/μs and refers to (b) and (c), while for (a) it is twice as large.

The laminar solid Teflon-paraffin material was assembled from individual Teflon (thickness h_1) and paraffin (h_2) plates which were glued together with ratios of corresponding thicknesses equal to $h_1/h_2 = 3/7$. The specimen was a cylinder of diameter 40 mm. An experimental set-up for particle velocity measurements is presented in Figure 3.1.

Values of the velocities u were found from the formula

$$u = \varphi\, \frac{10^8}{Bd},\tag{3.1}$$

where $B = 396$ Oe is the magnitude of the permanent magnetic field intensity used in the experiments, d is the sensor crossbar width equal to 7.6 mm, and φ is the voltage (in Volts) on the sensor generated as it moves perpendicular to the magnetic field B.

Shock loading was performed along the cylinder axis so that the plane of the shock was parallel to the plane of the plates comprising the LM. Shocks in Teflon–paraffin laminate were generated by using a plane shock generator assembled with the plate of a cast mixture TNT + RDX, of thickness 15 mm and diameter 40 mm, and a composite explosive lens (see Fig. 3.1). This corresponds to the effective charge thickness 16.8 mm in one-dimensional geometry and is designated as a shock-wave generator 1 (Deribas, Nesterenko et al. [1979]).

Typical oscillograms for the particle velocity for the LM with $\delta = h_1 + h_2 = 5$ mm between the second and third cells, and between the fifth and sixth cells (starting from contact with the explosive) are presented in Figures 3.2(a) and (b).

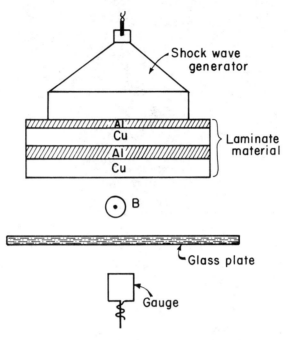

Figure 3.3. Measurements of a free surface velocity in composite material Al–Cu by the contactless electromagnetic method.

Figure 3.2(c) represents particle velocity at the point between the tenth and eleventh cells for LM with $\delta = 2.5$ mm.

Wave transformation in metal-based solid laminar material Al–Cu was investigated by Nesterenko, Fomin, and Cheskidov [1983a,b]. It was also shock loaded with the help of generator 1. The diameter of the specimens comprised of aluminum and copper plates was 50 mm. Measurement of the LM Al–Cu free surface velocity was made by a contactless method shown in Figure 3.3 (Deribas, Nesterenko et al. [1979]). This method is based on the measurements of the electromagnetic perturbations generated due to the motion of the free metal surface in the magnetic field B. The axis of the cylindrical specimen in this case is placed perpendicularly to the vector B. The shock propagated along the specimen axis, in this case, a contour of several turns (1–10) of copper wire of diameter 0.1 mm wound on an organic glass holder of dimensions 1.25×6 mm. The magnitude of the signal depends on the specimen geometry and the loading conditions in this measurement scheme. A linear dependence of the emf being measured in the contour was found by special tests with copper specimens in an analogous geometry from the flight velocity of the central part (diameter 20 mm) of the free surface with coefficient 60 ± 4 mV (mm/μs). In this geometry, a constant emf for ~5 μs corresponds to a constant flight velocity. The scheme assured good time resolution and is not sensitive to electrical noise under explosion. The free surface velocity on a 3–4.5 mm base was checked in all tests using the moments at the beginning of the free surface motion and of its impact with a glass plate placed in front of the measuring element (Fig. 3.3). This check aids in additional verification of the results obtained on the basis of a continuous velocity measurement.

3.3   Numerical Calculations of Shock Attenuation, the Mechanism of the Anomalous Effect of Cell Size

3.3.1   Basic Equations

The main attention in this section will be focused on the influence of the mesoscopic composite laminar structure (e.g., size of the cell) on the behavior of the strong shock waves with finite length (Nesterenko [1977]; Nesterenko, Fomin, and Cheskidov [1983a,b]). If this parameter can essentially change the wave characteristics, then it can be used to design the composite laminates, for example, according to the shock damping effect.

The laminar materials consisting of n identical cells will be considered. Each of these cells contains a layer of the first substance thickness h_1 and of the second substance of thickness h_2. The layer thickness is considered to be much smaller than the diameter of the LM. This permits solving the problem in a one-dimensional approximation. The loading of this system is performed by an explosive charge being in contact with one end of the system, analogously to the experimental scheme described in the previous section.

An elastic-plastic model is used to describe the behavior of solid material. The equations are written in the Euler coordinates (t, x) and have the following form for one-dimensional flow with plane symmetry

$$\frac{d\rho}{dt} + \rho \frac{\partial u}{\partial x} = 0, \tag{3.2a}$$

$$\rho \frac{du}{dt} + \frac{\partial \sigma}{\partial x} = 0, \tag{3.2b}$$

$$\frac{de}{dt} + \sigma \frac{\partial v}{\partial t} = 0, \tag{3.2c}$$

$$P = P(\rho, e), \tag{3.2d}$$

$$\sigma = P + \frac{4}{3}\tau, \tag{3.2e}$$

$$\frac{\partial \tau}{\partial t} = \begin{cases} -\mu \frac{\partial u}{\partial x}, & |\tau| < Y_0, \\ 0, & |\tau| \geq Y_0, \end{cases} \tag{3.2f}$$

$$\frac{d}{dt} = \frac{\partial}{\partial t} + u\frac{\partial}{\partial x},$$

(3.2g)

where ρ is the materials density of the substance, P is the pressure, e is the specific internal energy, u is the particle velocity, v is the specific volume defined by the relationship $v = 1/\rho$, μ is the shear modulus, and Y_0 is the yield strength. In the case $\tau = 0$ ($\mu = 0$, $Y_0 = 0$) we arrive at the hydrodynamic model which is used to describe the detonation products (DP).

At time $t = 0$ a detonation wave emerges on the contact surface between the explosive charge and the laminar material. The geometry of the explosive was approximated by a plate with an effective thickness accounting for an additional amount of energetic material in the explosive lens. The current coordinate of the DP free surface is denoted by A_0, the free LM surface by A_{2n+1}, and its contact surfaces by A_i, $i = 1, ..., 2n$. Then this problem is formulated mathematically as follows: Find the functions u, P, ρ, e, τ in the domain $z = \{A_0 (t) < x < A_{2n+1} (t), 0 < t < \infty\}$, that satisfy the system of differential equations (3.2) with the following initial and boundary conditions in the domains $A_i (t) < x < A_{i+1} (t)$, $i = 0, ..., 2n$.

Initial Conditions.   The distribution of the detonation product parameters behind the detonation wave front at the time $t = 0$ is found from a self-similar solution describing the Chapman–Jouget detonation wave (Chelyshev et al. [1970]; Baum, Orlenko, Stanykovich et al. [1975]) and in the laminar material the following conditions are assumed at $t = 0$:

$$P = 0, \quad \tau = 0, \quad u = 0, \quad \rho = \rho_i \quad \text{for } A_i < x < A_{i+1}, \quad i = 1, ..., 2n.$$

(3.3)

Boundary Conditions.   For $t \geq 0$ the stresses are equal to zero on the free surfaces $x = A_0(t)$ and $x = A_{2n+1}(t)$, and compliance with the continuity conditions for the normal stresses and velocities is required for the contact surfaces $x = A_i(t)$ ($i = 1, ..., 2n$). These boundary conditions do not include the possible shear damage of the interface as well as the debonding of layers. We can expect that these two effects will not qualitatively influence the final result of the transmitted compression wave profile calculations.

The formulated problem was solved numerically by using a Wilkins-type finite-difference scheme (Wilkins [1964], [1999]) in which artificial viscosity was used for stable computation of the compression waves. Linear viscosity

$$q = \begin{cases} -q_0 h c_0 \rho_0 \dfrac{\partial u}{\partial x} & \text{if} \quad \dfrac{\partial u}{\partial x} < 0, \\ 0 & \text{if} \quad \dfrac{\partial u}{\partial x} \geq 0, \end{cases}$$

(3.4)

was chosen for the computation of the dissipative term q, where h is the mesh spacing, and q_0 is a constant.

Methodological questions associated with the numerical solution of the problem formulated by this finite-difference method are considered by Deribas, Nesterenko et al. [1979] and by Sapojnikov and Fomin [1980].

In numerical calculations, the equation of state for the detonation products is selected in the form

$$P = (\gamma - 1)\rho e, \tag{3.5}$$

and for the materials comprising the laminar material:

$$P = a_0^2(\rho - \rho_0) + (\gamma - 1)\rho e. \tag{3.6}$$

3.3.2   Interaction of Leading Shock with Interfaces

Before starting to analyze a complex picture of the wave interaction obtained in numerical calculations, let us qualitatively consider the case when the deviator of the stress tensor is small in comparison with the shock amplitude. Then a hydrodynamic model can be a good approximation for material behavior.

Shock waves in the absence of phase transitions have a one-wave structure in the homogeneous material. Upon loading of the LM by detonation products, a leading shock wave is propagating in composite. Behind it there is a complex wave pattern that varies with the time. It is formed because of the compression and rarefaction waves interaction with each other and with materials interfaces. The amplitude of the leading wave also varies with time.

The first factor in this variation is shock decay on the contact interfaces of the different materials, whereupon the formation of the shock and rarefaction waves moving in the forward and reverse directions occurs.

The second, less evident, but also important factor affecting the amplitude of the leading wave is its interaction with the overtaking waves moving at a higher phase speed over the previously loaded material.

The influence of the first factor is conveniently demonstrated on the P–u diagram (Fig. 3.4). Let curve 1 be the Hugoniot of the first material, and curve 2 the Hugoniot of the second one which compose the periodic LM. The point O corresponds to the state before the leading wave.

There is also a natural reason for shock decay due to the elastoplastic behavior of material, considered in great detail by Davison [1998] for small strain, uniaxial longitudinal pulses. For strong shocks, this mechanism can be small in comparison with the previous two.

Let the initial shock move over the layer of material 1, and let point B correspond to the state behind it. After a certain time, the shock will reach the interface between the first and second materials. In the case under consideration, a transmitted shock changes the state of material 2 from O to C. A reflected shock wave is formed in material 1 bringing it from B to C. After interaction between the leading wave in material 2 and the next contact boundary (now between the

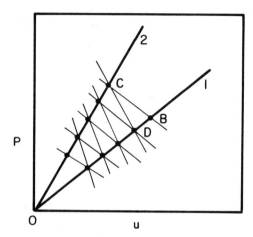

Figure 3.4. The decay of the leading shock wave due to the interaction with interfaces with different acoustic impedance (similar to Al–Cu LM).

second material and the first one), the state marked by point D, which lies in our case below point B, is formed. This second interaction results in the transmitted shock wave into material 1 (change of state from O to D) and the reflected rarefaction wave back into material 2 (change of state from C to D). At this moment the situation is qualitatively analogous to the starting point B, and the process of wave interaction with the next interfaces can be repeated. The result is the decay of the amplitude of the leading wave. A further pattern of the change in the leading wave amplitude is seen in Figure 3.4.

From Figure 3.4, it is evident that this damping effect occurs because of the discrepancy in the values of the acoustic impedances of the materials of which the layers consist. This damping effect does not involve any dissipation additional to the dissipation processes in the shock fronts and is also effective for acoustic waves without any dissipation at all. The more different the acoustic impedances of the layers are, then the lower point D lies on the shock Hugoniot relative to B. This means that the leading wave decay will be stronger as well. The different laminar acoustic dampers and antimeteorite shields are based on this property (Riney [1970]).

The influence of the second factor is impossible to analyze by such a simple method as a P–u diagram. In this case, the amplitude of the leading wave is additionally damped by the interaction with the rarefaction waves that occur, due to flow of the detonation products into a vacuum and as a result of rarefaction waves originating on the interfaces. At the same time, the amplitude of the leading shock wave can be increased by the reflected shock waves overtaking it.

3.3.3   Experimental and Numerical Results for Solid Teflon/Paraffin Laminate

The influence of both factors on the leading wave amplitude was estimated in numerical calculations for the laminar material Teflon-paraffin in accord with the experimental results described in Section 3.1. The ratio of the acoustic impedances is 1.46 in this case. Experiments, as well as computer calculations, were performed with LMs having the same macrocharacteristics and geometrical sizes but with different mesoscopic structural characteristics (cell sizes and amount of interfaces). The explosive thickness was 16.8 mm in the computations, which took into account the influence of the plane wave lens in the experimental set-up. The following parameters of the cast explosive charge TNT–RDX were used: density 1.68 g/cm^3, detonation speed 7.6 mm/μs, γ = 3 (Baum, Orlenko, Stanykovich et al. [1975]). The data for the material parameters in Table 3.1 are obtained using the results published by Pavlovskii [1976] and by Chelyshev, Shehter, and Shushko [1970].

Figure 3.5 presents experimental and calculated profiles of particle velocities for LM Teflon–paraffin with different sizes of the cells (δ = 5 mm and 2.5 mm) in the points with an initial distance from the explosive charge equal to L = 10 mm. This corresponds to a different number of interfaces, four and eight, which the shock wave meets before arriving at a given point.

Comparison of the particle velocity curves, obtained experimentally and numerically for the cell size δ = 5 mm, shows their good quantitative and qualitative agreement in complex details of the wave process (Fig. 3.5). A certain deviation of the computed data from the experimental data for times greater than 4.5 μs can be related to the influence of side unloading, i.e., in this case the physical process in the experiment becomes substantially different from the one-dimensional model used in calculations. It is seen in Figure 3.5 that the calculated amplitudes of the particle velocity of the leading wave in LM, with δ = 5 mm and 2.5 mm, agree within the limits of accuracy of the computations. This is despite that, in the second case, the head wave encountered twice as many interfaces in its path. This is in strong disagreement with the mechanism of attenuation shown in Figure 3.4.

Table 3.1. Materials properties used in computer modeling.

Material	ρ (g/cc)	γ	μ (GPa)	a_0 (km/s)	Y_0 (GPa)
Teflon-4	2.19	6.9	—	1.8	—
Paraffin	0.9	4.1	—	3.3	—
Aluminum	2.79	4.2	25	5.39	0.25
Copper	8.9	4.8	46	3.99	0.3

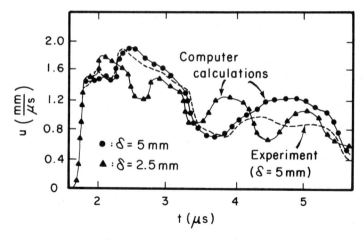

Figure 3.5. Experimental ($\delta = 5$ mm) and calculated ($\delta = 5$ mm and 2.5 mm) particle velocity–time profiles in LM Teflon–paraffin, $h_1/h_2 = 3/7$, the first layer facing the explosive charge is Teflon.

The dependence of the amplitude of the particle velocity in the leading wave on the cell size was not observed in experiments (Nesterenko [1977]). The same result was obtained in numerical calculations for the identical materials for the depth 25 mm ($\delta = 25$ mm, 5 mm and 2.5 mm) and in the case when $L = 40$ mm ($\delta = 5$ mm and 2.5 mm) (Nesterenko, Fomin, and Cheskidov [1983a,b]). This allows us to conclude that under strong shock-wave loading, the increase of the interfaces crossed by a leading shock wave by one order of magnitude (from 2 to 20) did not result in the amplitude damping.

This phenomenon is in drastic contradiction to the approach based solely on consideration of the interaction between the leading wave and the interfaces which yield strong damping of the wave amplitude as the number of crossings of the interfaces increase (see Fig. 3.4).

To eliminate the specific loading conditions created by contact explosive loading the impact by the plate was also considered. In computer calculations, the LM Teflon–paraffin with a cell $\delta = 25$ mm and 2.5 mm ($L = 25$ mm) was loaded by a flat impactor that produces initial shock waves with parameters close to the loading by a contact explosive charge. The thickness of the impactor was selected to be adequate to eliminate the influence of the rarefaction wave from its free surface on the motion of the contact boundary A_l, meaning the wave pattern in the LM as well. In this case, the amplitude of the leading shock wave did not depend on the number of interfaces crossed by the shock wave. Therefore, the observed independence of the leading wave amplitude from the number of interfaces at a given depth is not related to the influence of the rarefaction wave, but to the nonlinear effects due to the overtaking of the leading wave by secondary shock.

3.3.4   Shock Decay in Aluminum–Copper Laminate

3.3.4.1   "Strong" and "Long" Shock Wave

How general is the result obtained with Teflon and paraffin laminate? To answer this question, the combination Cu–Al was selected to represent the essentially higher differences in acoustic impedances (the relation of acoustic impedances are correspondingly 1.46 and 2.38) and with lower compressibility.

The investigations were performed on the LM Al–Cu with the cell sizes δ = 4 mm and 2 mm. Each cell contained an aluminum plate and a copper plate of identical thickness. Here we follow the same approach to compare the free surface velocity profiles for different cell sizes (δ = 4 mm and 2 mm) at the same macrosizes (depth L) of the composites under similar loading conditions.

The aluminum layer was the closest to the explosive charge, which was identical to the charge used for the Teflon–paraffin composite (generator 1). The total LM length was 12 mm in the experiments, and 12 mm and 16 mm in the numerical computations. The free surface velocities were measured and numerically calculated.

The results of the experiments and calculations based on the hydrodynamic model are sufficiently close to each other (Fig. 3.6(a)). The velocity profiles of the free surface have a three-wave configuration for both δ = 4 mm and 2 mm. The amplitude of the second and third waves is somewhat lower than in experiments, which could be explained by the fact that the plates in the experiment recoiled from each other and then subsequently snapped, which was not taken into account in computer simulation. The calculations based on the elastic–plastic model resulted in wave profiles (Fig. 3.6(b)) similar to the hydrodynamic model for these strong shock waves (Nesterenko, Fomin, and Cheskidov [1983b]).

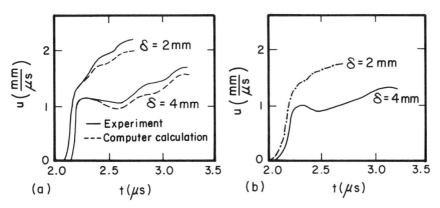

Figure 3.6. The calculated and experimental profiles of the free surface velocities in the composite material Al–Cu with δ = 4 mm and 2 mm: (a) hydrodynamic model and experiment; (b) an elastic–plastic model. The thickness of the composite is equal to 12 mm, the aluminum plate is facing the explosive.

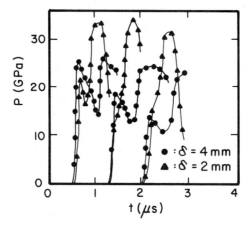

Figure 3.7. Calculated shock pressure profiles in solid LMs Al–Cu for different values of L (4 mm, 8 mm and 12 mm) and $\delta = 2$ mm and 4 mm.

In the case of the Al–Cu composite, an even more striking result than for Teflon–paraffin is demonstrated in Figure 3.6. The amplitude of the first jump in the free surface velocity for the composite with a smaller size of cell is essentially higher than one for the composite with a larger cell size! This means that the amplitude of the leading wave in the LM with the finer cell is higher than in the LM with the large cell at the same thickness of the composite. These similar results were obtained for LMs Ti–Al and Cu–Al for different conditions of shock loading (Benson and Nesterenko [2001]).

Analysis of the P–u diagram (Fig. 3.4) without considering the influence of the overtaking secondary waves on the leading shock wave, shows just the opposite result. This demonstrates that the leading particle velocity amplitude in the LM Al–Cu with $\delta = 2$ mm should be almost half that in the LM with $\delta = 4$ mm. It is especially interesting that this phenomenon is obtained for loading by a triangular pressure pulse. This result indicates the governing role of the effects of the secondary overtaking waves associated with the nonlinearity in material behavior in the formation of the leading wave.

To confirm this conclusion, the dynamic of the wave profile was investigated in computer calculations for a different geometry of the Al–Cu laminar system. Figure 3.7 displays the calculated time dependencies of the pressures for material points at various depths L in LM with different δ ($\delta = 4$ mm and 2 mm). In Figure 3.7 we can clearly see the dynamics of wave formation and the influence of the cell size and the traveling distance L on the change of the leading shock-wave amplitude.

Thus, we see a significant decrease in the amplitude of the leading wave as the number of crossings of the interfaces increases at the depths $L = 4$ mm and 8 mm. At these traveling distances, the influence of the overtaking effects from the secondary shock waves is still not observed. In this range of the distances for a shock of given amplitude and duration, the pressures and particle velocities in the leading wave can be determined with the help of the $P–u$ diagrams if the attenuation effects resulting from the rarefaction wave from detonation products are not essential. Consequently, a computation of the cumulating effects in inhomogeneous LMs, based on the examination of $P–u$ diagrams for a leading wave for a few contact boundaries, may be possible for a small number of layers, as was done by Laptev and Trishin [1974].

At the same time, the extension of this analysis to a larger number of layers is doubtful. Indeed, in our case, even at 8 mm depth in an LM with $\delta = 2$ mm it is clearly seen that a second loading wave is very close and ready to overcome the leading wave. At $L = 12$ mm the leading wave amplitude in an LM with $\delta = 2$ mm is already considerably greater than in a LM with $\delta = 4$ mm. In addition, a comparison of the maximal values of the pressure shows that it is 40% higher in the LM Al–Cu with $\delta = 2$ mm than in the LM with $\delta = 4$ mm. This is the result of the secondary wave reflections as is evident from Figure 3.7. Thus the distances L between 10 mm and 12 mm are critical for the behavior of the leading wave amplitude, depending on the cell size. Comparison of the calculated (Fig. 3.7) and experimental results (Fig. 3.6) demonstrates that the nonlinear effect of the secondary wave interactions with the leading wave is the reason for the anomalous decay of shock waves in LMs with a smaller cell size.

3.3.4.2 "Weak" Shock Decay

From previous analysis it is evident that secondary waves can qualitatively change the wave transformation properties of LMs and result in behavior different than that expected from linear acoustic analysis. This is an essentially nonlinear phenomenon and that is why its significance should be sensitive to the shock amplitude. For example, in the case of weak shock at the same distances, the influence of the secondary shocks should be lowered because of the diminution in the nonlinearity of the material behavior.

Figure 3.8 presents the time dependencies of the pressure in an LM Al–Cu under a loading by a weaker triangular pulse (compare with Fig. 3.7). We notice the strong damping of the leading wave amplitude with the increase of the number of crossed interfaces. The anomalous high values of the pressure in the LM with a finer cell are absent.

In this specific case the simple approach based on a $P–u$ diagram can give a reasonable evaluation of the leading shock decay due to interaction with the interfaces.

3.3.4.3 "Short" Shock Decay

As the loading wave duration diminishes (even though its amplitude is high) it can be expected that the considered effects of anomalous decay appear weaker. In particular, if the spatial size of the compression pulse is less than the size of one plate in the LM, then the overtaking effects cannot appear at all. Then the

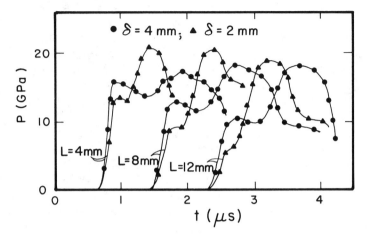

Figure 3.8. The attenuation of a "weak" shock in solid LM Al–Cu with a different cell size δ.

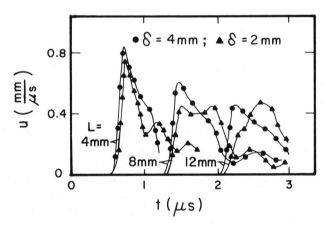

Figure 3.9. The attenuation of a "short" shock in solid LM Al–Cu with a different cell size δ.

"geometrical" damping (dispersion) of even a strong shock can be estimated on the basis of the interaction just between the leading wave and the contact surfaces, and due to the influence of the rarefaction wave.

The computer calculations were performed (Fig. 3.9) for a shock produced in an LM Al–Cu by the detonation of a layer of the same explosive charge but with a smaller (4.2 mm) thickness (Nesterenko, Fomin, and Cheskidov [1983a]). This essentially reduced the shock duration. From a comparison of the results presented in Figures 3.7 and 3.9, it is seen that the effect of the anomalous decay, which was observed for a long wave, disappeared. In particular, for $L = 12$ mm the leading wave in both LM with $\delta = 2$ mm, and with $\delta = 4$ mm, is clearly isolated in Figure 3.9. The amplitude in the LM with the finer cell is considerably lower than in the LM with the coarse cell.

There are no high-pressure spikes behind the shock in the laminar material with a smaller cell. Benson and Nesterenko [2001] obtained the similar results for Ti–Al and Cu–Al laminates under plate impact loading. Bonding between layers does not change attenuation of the shock.

3.3.4.4   Attenuation in Laminate Versus Homogeneous Solid

The presented results raise a question of the optimal material design that will ensure the most rapid attenuation of the shock wave. The decay of the shock wave in LMs with different cell sizes and the results for homogeneous material can be compared (Cheskidov [1984]). In Figure 3.10 the data for the leading shock wave amplitude decay and maximum pressures (in general, the latter can be attained in the secondary shock) for a given coordinate are presented for LM Al–Cu with different cells. Results for the homogeneous copper (solid line) at the same distances from the explosive charge are shown for comparison.

The values of the maximal pressure and pressure in the leading wave in the composite with $\delta = 4$ mm coincide. The particle velocity behind the leading shock in the last copper plate in this composite for $L = 12$ mm was 0.55 mm/μs and for solid copper 0.66 mm/μs. It is important that the mass of the LM Al-Cu is essentially less than the mass of the solid copper. Also the less pronounced attenuation is observed in solid copper.

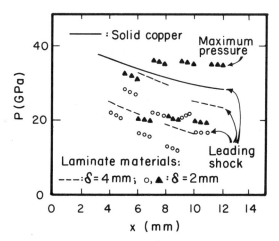

Figure 3.10. The dependence of the leading shock wave pressure amplitude ($\delta = 4$ mm and 2 mm) and maximum pressure ($\delta = 2$ mm) in solid LM Al–Cu and in solid copper at the same distance from the explosive charge.

Quite different behavior is observed for LM with $\delta = 2$ mm. The leading wave is attenuating faster up to the travel distance 9–10 mm in this composite. This is apparently connected with a larger number of interfaces crossed by leading shock. At the same time, the secondary shock waves behind the leading wave build the maximum pressures a few times larger than the pressure in the leading wave especially at distances more than 6 mm (Fig. 3.10). In the last copper plate in laminar material, the maximal pressure is attained in the leading wave and its magnitude is 20% *larger* than the corresponding shock amplitude in homogeneous copper. The particle velocity amplitude is also larger, and is equal to 0.81 mm/μs. Thus, the intention of using periodic LM with a smaller cell size, to increase the number of interfaces and to facilitate the damping, led to the opposite result. The amplitude of the leading shock wave, in comparison with a composite with a larger cell size, as well as in comparison with solid copper, was increased! This example demonstrates that laminar materials can be used for attenuation purposes only if they are optimized according to their structure and taking into account the initial impulse parameters.

3.4   Experimental Comparison of Shocks in Laminar Material and in a Random Heterogeneous Medium

3.4.1   Laminar Materials as a Model of a Heterogeneous Mixture

The laminar model was used to explain some key phenomena in the shock loading of porous materials. Thouvenin [1966] was the first who used this model to predict the Hugoniot curve for powder. One of the interesting applications of laminar models is their use in one-dimensional numerical calculations of the complex wave processes in mixtures (solid and porous). It was originally demonstrated by Hofmann, Andrews, and Maxwell [1968] for porous aluminum that this approach allows us to achieve the final state behind a shock front in agreement with the conservation laws based on the hypothesis of stationary shock. Results of the shock interaction with the powder–solid interface were explained, based on this model, by Voskoboinikov, Gogulya, Voskoboinikova, and Gel'fand [1977]. This model allows us to overcome the fundamental problem (Bogachev and Nikolaevskii [1976]) existing in the shock dynamics of mixtures in the continuum approach: how to distribute mechanically the thermal energy between components. In the laminar model, this problem is solved naturally—the circulation of the shock and rarefaction waves creates the path which each point in the mixture follows traveling from the initial to the final state. Nonequilibrium shock thermodynamics of powders and mixtures were explored by Nigmatullin, Vainshtein et al. [1976] in the frame of a laminar model.

To some extent, this approach was opposed to phenomenological considerations like the principle of additivity (Dremin and Karpuhin [1960]). In the latter each component is compressed according to its own single Hugoniot up to the final equilibrium pressure in a mixture. This approach is very successful to obtain the Hugoniot curves for mixtures based on their individual Hugoniot. Apparently it is far from the correct representation of the real state of the components in the mixture. The reason for its success is a relatively weak dependence of the specific volumes of components on temperature variations.

The main problem with the laminar model is that it is not clear what parameters of the LM are suitable to represent the real mixtures. Also there is a shortage of experiments with a direct comparison between chaotic mixtures (solid and porous) and their periodic laminar models. For this comparison the mixture Teflon + paraffin was selected (Nesterenko [1977]). The reason for this selection was partially connected with a relatively simple preparation, both of a mixture (adding Teflon into paraffin melt) and a laminar model. Teflon plates are commercially available and paraffin plates can be carefully rubbed by a glass edge up to the desirable thickness. Also, both materials are dielectric under relatively high shock pressures, which allows us to use the electromagnetic method for particle velocity detection (Figure 3.1). Another important property is connected with their large compressibility, which can promote the manifestation of differences in the thermal energy distribution between the mixture and laminar model into noticeable changes of the particle velocity profile. The selected materials have a ratio of the acoustic impedances close to mixtures of BN + water and Cu + W which are important from an application point of view.

3.4.2   Experimental Set-Up and Results

3.4.2.1   Teflon–Paraffin Solid Mixture and Laminate

The electromagnetic method with a geometry similar to Figure 3.1 was used to obtain the particle velocity profile (Fig. 3.11) for a solid mixture Teflon + paraffin (mass ratio of components 1.03 : 1) with average density 1.3 g/cc. The Teflon powder with particle size 20 µm was used to prepare a solid mixture. The particle velocity profiles for the corresponding periodic LM with cell sizes δ = 5.0 mm and 2.5 mm, and with the same density and under the same loading conditions, are shown in Figure 3.2. Each cell is composed from one Teflon plate and one paraffin plate having a ratio of thicknesses h_1/h_2 = 3/7. Two distances L from the explosive charge (generator 1) were selected: L = 10 and 25 mm. Experimental data for the parameters of the leading waves in the Teflon–paraffin (solid mixture and LMs) on different depths L from the same explosive charge are presented in Table 3.2.

3.4.2.2   Teflon Powder and Laminate

The behavior of granular Teflon in comparison with its laminar counterpart model was experimentally investigated by Nesterenko [1977]. Powders with average density 1.17 g/cm^3 and two different particle sizes, large cubic particles

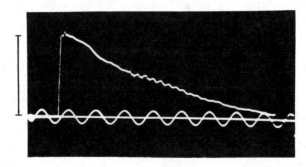

Figure 3.11. The experimental profiles of particle velocity in a solid mixture Teflon + paraffin, $L = 25$ mm. Oscillation frequency is 1 MHz, the vertical scale is 1 mm/μs. The loading conditions are the same as for Figure 3.2.

Table 3.2. Experimental data for shock parameters in the mixture Teflon–paraffin, Teflon powder, and in their laminar counterparts at the same loading by shock generator 1.

u (mm/μs)	P (kbar)	L (mm)	Material description
1.75 ± 0.08	119 ± 7	10	Solid mixture
1.05 ± 0.05	57 ± 5	25	Teflon–paraffin
1.36 ± 0.11	124 ± 16	10	LM Teflon–paraffin,
1.05 ± 0.06	76 ± 8	25	$\delta = 5$ mm
0.90 ± 0.05	67 ± 5	25	LM Teflon–paraffin,
			$\delta = 2.5$ mm
1.15 ± 0.15	32 ± 5	18.8	Teflon powder,
			particle size 1.5–2 mm
1.22 ± 0.24	35 ± 8	18.8	Teflon powder,
			particle size 20 μm
1.20 ± 0.10	102 ± 12	18.8	LM Teflon–air,
			$\delta = 2.9$ mm

with sizes 1.5–2 mm and small particles of the irregular shape with the sizes 20 μm, were used. The LM was represented by a periodic system of the Teflon plates with a thickness of 1.5 mm separated by air gaps (1.4 mm) with an average density the same as one of the powders. The diameter of the samples was 40 mm, a shock wave was created with the help of an explosive plane wave generator 1. Experimental data for the parameters of the leading waves in the Teflon (granular and laminar systems) on different depths from the explosive charge are presented in Table 3.2. The experimental profiles for Teflon powder

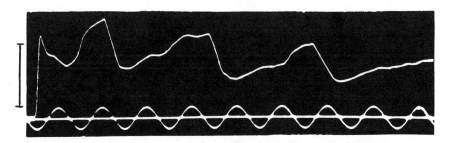

Figure 3.12. Particle velocity profile in Teflon–air LM with $\delta = 2.9$ mm. Oscillation frequency is 1 MHz, the vertical scale is 0.93 mm/μs. The shock loading conditions are the same as for Figures 3.2 and 3.11.

are presented in Figure 2.9(a),(b),(c), and for its laminar model are shown in Figure 3.12.

The data for pressures were calculated from the measured values of the particle velocity and the Hugoniot for mixtures obtained with the additivity principle (Dremin and Karpuhin [1960]; Alekseev, Al'tshuler, and Krupnikova [1971]).

3.4.2.3   Glass Powder and Laminate

Other experiments were performed to compare the behavior of brittle glass powder with glass/air laminates having the same density 1.24 g/cc under similar loading conditions (Table 3.3). The powder was prepared by fracturing the glass plates with a thickness of 1.45 mm. That created particles with the same thickness and with sizes in the plane 2–3 mm. Its laminar models with cell sizes 2.9 mm and 5.8 mm (glass plates separated by an air gap having the same thickness) and average densities similar to one of the powders were also investigated. Shock loading of these systems was performed with the help of a plane explosive charge of RDX (initial density 1 g/cc, diameter 100 mm, and height 40 mm) initiated by a plane wave generator. Between the explosive charge and the investigated systems, the buffer plate from two layers of Teflon (25 mm) and glass (36 mm) was placed.

One of the reasons for selection of the glass powder was connected with the intention to look at the wave profile in granular materials with a qualitatively different type of particle deformation in comparison with the Teflon powder. It is possible to speculate that the mechanism of glass particle deformation will be the brittle fracture of the particles with subsequent repacking in comparison with some type of plastic flow in the case of the Teflon powder.

The particle velocity profiles for glass systems are presented in Figure 3.13 and data in Table 3.3. In the experiments with granular glass gauges were placed on the contact with buffer plate and on the depth 14.5 from it (Fig. 3.13(a),(b)).

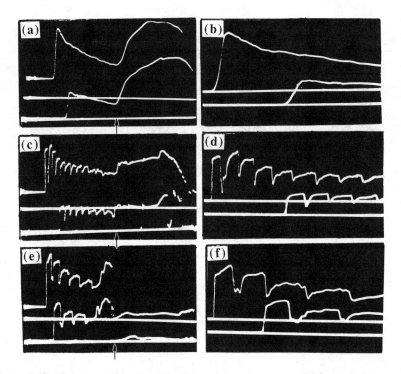

Figure 3.13. The experimental profiles of particle velocity in glass powder and in the different laminar models at the same loading conditions: (a), (b) powder, $L^1 = 0$ mm, $L^2 = 15$ mm; (c), (d) LM, $\delta = 2.9$ mm, $L^1 = 1.45$ mm, $L^2 = 16$ mm; (e), (f) LM, $\delta = 5.8$ mm, $L^1 = 2.9$ mm, $L^2 = 14.5$ mm. Each pair of pictures represents the same experiment. For (a), (c), (e) the whole horizontal line equals 100 μs; for (b), (d), (f) 30 μs. Velocities corresponding to the first maximums are presented in Table 3.3; amplitude jumps at the moments shown by the arrows correspond to the arrival of shock on the horizontal shoulders of the gauges.

In laminar materials particle velocity gauges 1 were placed on the free surface of the first glass plate (L^1), and gauges 2 were placed on the free surface of the sixth (Fig. 3.13(c),(d)) and third (Fig. 3.13(e),(f)) plates (L^2). Electrical signals from gauges 1 are recorded on the upper beam of the oscilloscope, from gauges 2 on the lower beam. The loading geometry that was used provided a relatively long shock pulse sufficient to ensure the equilibrium state behind the leading shock wave (Fig. 3.13, compare with Fig. 3.12 for a relatively shorter loading pulse).

The experimental data for the amplitudes of the particle velocities and the corresponding pressures are presented in Table 3.3. The pressure data in brackets correspond to the impact of the glass plate moving at velocity u along the

Table 3.3. Shock parameters in the glass powders and in their laminar counterparts. Zero pressure corresponds to the free surface with particle velocity from the left column and the data in parentheses corresponds to the pressure at the gap closure.

u_{lm} (mm/μs)	P (kbar)	L (mm)	Material description
1.00	26	0.0	Glass powder, $\rho_{00} = 1.24$ g/cm^3
0.62	9	15.0	
0.87	0(40)	1.45	LM, $\delta = 1.45 + 1.45$ mm
0.54	0(24)	16.0	
0.83	0(37)	2.9	LM, $\delta = 2.9 + 2.9$ mm
0.64	0(28.5)	14.5	

nearest front plate. The experimental spread of the data was below 10%, data are averaged over two to three experiments.

3.4.3   Comparison of Shock Loading of Mixtures and Laminates

3.4.3.1   Solid Mixture and Laminate

The solid system Teflon–paraffin can be considered as an example for understanding the general relations between behavior of the mixture and its laminar counterpart. The data for both systems are presented in Table 3.2. It can be concluded that the laminar system Teflon–paraffin has the peak values of the particle velocity and pressure in the leading wave, close to the values for the shock wave in the mixture at the same loading conditions and similar overall geometry and density of the samples. Some decrease in the pressure in the mixture in comparison with the LM can be mentioned. This behavior is observed for the cell sizes of the laminar material (5 mm and 2.5 mm) being two order of magnitudes larger than the particle sizes (~20 μm) in the solid mixture. The pressures in the leading wave for LM with different cell sizes at the same distances from the shock entrance are close. This behavior is connected with essentially nonlinear wave dynamics, as was discussed in the previous section. Here we can only mention that it doesn't agree with the mechanism of wave attenuation proposed by Barker [1971], where the characteristic time of decay is proportional to cell size.

The differences in the thermodynamical state of the components behind the shock wave, in a mixture and in its laminar model during the loading by the impulse of finite duration, can be expected. As was already mentioned, the amplitude values of the pressures in the leading shocks are relatively close for both cases. But compression of the components in the mixture is different than the single shock loading characteristic for a leading wave in LM, which can result in the qualitative difference of their thermodynamical behavior, depending on the structure of the mixture.

Let us consider two cases, which demonstrate the possible relationship between a thermodynamic mixture and the results of the laminar model approach.

I. Imagine that components in the mixture are presented as thin plates (characteristic thickness l) oriented by their planes mainly perpendicular to the direction of the shock wave propagation. If their characteristic sizes are much larger than the shock front thickness (Δ) in the corresponding solid material, then the components are loaded in accord with the laminar model. For the particles with sizes of 20 μm, the whole process of establishing the final equilibrium state as a result of wave reverberations (about 10 cycles, see e.g., Fig. 3.13(c)–(f)) takes place during ~0.1 μs. During this time the macroscopically measured pressure in the solid mixture behind the shock front is practically constant (Fig. 3.11). That is why the temperature increase for each component in the considered mixture at the given pressure P (which is close to the pressure in LM) can be a few times larger than the temperature increase in LM with δ = 2.5–5 mm in the single shock with the same peak pressure P.

The layers in the given LM are additionally shock loaded on the stage of the pressure decay (see Figs. 3.2 and 3.5). This increases the temperature of the layers additionally to the temperature rise in the leading shock. But the influence of this secondary shock loading is moderated by the overall pressure decay because the duration of shock t in the considered case (Fig. 3.2) is apparently less than the time of establishing the final state in a laminar material T for a given cell thickness. It is also clear that the final temperature increase in LM components with δ = 5 mm is less than the temperature rise in the LM with δ = 2.5 mm because of the same amplitude of the leading shocks and less numbers of the secondary shocks on the rarefaction stage (Fig. 3.2). The higher residual temperature (the difference is 16% under considered loading) in fine laminar material, in comparison with a coarse one (Cu–Al) at the same condition of loading was found numerically by Benson and Nesterenko [2001].

Thus, for a presumably laminar geometry of the components in a mixture with relatively small particles in comparison with a shock length, but large enough in comparison with the shock front, the correct description of the thermodynamic mixture through modeling by a laminar system can be ensured only at the condition $t > T$. Practically, this means that the cell size should be small enough. The real relation of the cell size and the size of the components in the mixture under this condition is irrelevant if the cell size is larger than the shock front thickness.

II. Let us consider the opposite case when the shock front thickness in the mixture is essentially wider than R/c, where R is the characteristic particle size in the mixture and c is the bulk sound speed in the particle's material. In this case no stationary shocks arise in the particles and the heat release in them is close to isotropic compression. At the same time, in the laminar model the temperature increase of the separate layer, if its thickness allows propagation of stationary shock with the pressure P, is higher than that corresponding to the isotropic compression of this material with the same pressure. The difference between the temperatures in these two cases is emphasized during the unloading process, because the layers in the laminar model are additionally loaded by secondary waves (Fig. 3.2). Finally, application of the laminar model results in an essentially higher temperature than the temperature of the component in the mixture. To be able to describe correctly the thermodynamics of the mixture, the cell size in the laminate should be smaller than the shock width. This eliminates the multiple shock loading of the layers in the laminar model.

Our analysis does not take into account the additional heat release connected with the plastic flow of the particles and the friction on their interfaces. It can be suggested that in solid mixtures, at relatively high shock pressures, the role of this effect is negligible.

Thus to describe the thermodynamics of the solid heterogeneous mixtures with the help of a laminar model, the real structure of the mixture should be taken into account. The correct scaling between the main parameters (particle sizes in the mixture, cell size in the laminar model, duration of the shock, and typical time of establishing the final equilibrium state in LM, as well as the thickness of the shock front) must be ensured.

3.4.3.2   Granular Materials and Laminate

3.4.3.2.1   Teflon Systems

Particle Velocity on Shock Fronts.   The values of the maximal particle velocity in powder and laminar materials are close and they do not depend on the particle size in the powder (two orders of magnitude difference, Table 3.2). This result reflects the independence of the linear momentum on the details of the inner structure under the same conditions of loading. Laminar models may be more relevant for the description of local high-pressure spots with 2-D geometry resulting from plate impact (see Fig. 2.22(b)). A good qualitative coincidence between the wave structure in laminar material composed from Teflon layers and computer calculations was obtained by Cheskidov [1984].

Pressure on Shock Fronts.   It is apparent from the data of Table 3.2 that the average pressure in the shock front in granular materials has nothing to do with the pressure in the leading wave for the corresponding laminar model. This latter pressure corresponds to the impact of the leading moving layer on the neighboring one. The difference is more than three times. This comparison demonstrates that maximal pressures in the laminar model cannot be used to predict the average peak pressure in the powder.

Width of Shock Fronts.   Another very important difference between granular material and the laminar model is connected with an essentially different time of the establishment of the final equilibrium state behind the shock front. The shock front duration in the powder is about 1 µs and the time of particle velocity equilibration in LM is 10 μs. The width of the shock front in powder is about Δ = 1.35 mm, which is close to the initial particle size (Table 2.2, Chapter 2).

These differences allow us to conclude that the laminar model does not even qualitatively describe the path which the powder follows during shock loading. This conclusion does not contradict the results of the computer calculations (Hofmann, Andrews, and Maxwell [1968]) which demonstrated that the final state of material in the laminar system and powder was the same. The coincidence of the final states in the powder and laminar system is just the simple consequence of impulse and energy conservation. This is similar to the kinetic energy loss for the inelastic collision of two balls. It does not depend on the details of real interaction processes that could be completely different for the same masses and velocities of the balls. Thus, shock transition kinetics in granular materials implemented by the laminar model is not correct.

3.4.3.2.2   Glass Systems

Particle Velocity on Shock Fronts.   Experimental results for the glass powder (Table 3.3) demonstrate that the particle velocity on the shock front is in agreement with the maximal particle velocity in the laminar model (velocity of the free surface layer) at the same loading conditions. It was valid also for Teflon powder and its laminate. The decay of the maximum velocity with a shock traveling distance for the laminar material and powder is similar, in addition to the equilibrium values of the particle velocity behind the shock. There is a slight tendency to less attenuation of the leading particle velocity with the increase of the cell size in the laminar materials at about the same depths of 16.0 mm and 14.5 mm, correspondingly (see Table 3.3).

Pressure on Shock Fronts.   The fact that particle velocities on the shock fronts are close does not mean that the corresponding shock pressures are also equal. This is apparent because the particle velocity of the layer in the laminate generates higher pressures due to impact with other solid layers than shock pressure in the granular material. As a result, there is up to three times difference between the magnitudes of average pressures in the leading waves in powder and the corresponding laminate model, as was the case for Teflon systems.

Experimental results show that the laminar model can be used to predict the particle velocity on the shock front in powders for stationary shocks and for nonstationary shock regimes. This is true even if the space size of the material involved in motion behind the shock is only one order of magnitude larger than particle size. This is in agreement with experimental results on the pressure transmission by the porous layers explained by Voskoboinikov, Gogulya et al. [1977] in the frame of the laminar model.

Another possibility for laminar model applications can be the estimation of the local maximal pressures existing on the shock front in powder particles, supposing that P_{lm} can represent it. These pressure spikes can result in microspall phenomena inside the particles (Fig. 2.22). In this case, if $u_{lm} \approx u_{lf}$

then the local pressure inside the shock front corresponds to the Hugoniot for the layer material with the particle velocity $u_{lm}/2 = u_f/2$. This results in a few time differences between final pressure in the powder P_f and local pressure spikes inside the shock front. For many condensed materials at the particle velocities of about 1 mm/μs and with tapped density close to one-half of the solid density, $P_{lm}/P_f \sim c/u_{lm}$ (see Problem 7 of this chapter).

Width of Shock Fronts. The shock front duration in powder corresponds to the ramping time of the particle velocity profile, resulting in establishing a final state. The corresponding thickness of a stationary shock in the laminate is the duration of the damped oscillatory regime approaching the final equilibrium state. A great difference between the duration of these fronts in powder and in the laminar model (2–3 and 30 μs, correspondingly, compare Fig. 3.12(a),(b) with others) was observed for brittle glass. It was also a characteristic feature for the "plastic" material Teflon. The value $\Delta = 1.7$ mm for the shock front thickness in granular glass for a pressure of 0.9 GPa is close to the initial particle size (for the initial density equal to one-half of the solid density and supposing that behind the shock we have completely compacted glass, $\Delta \sim u\tau$). So, the shock front thickness for a strong shock is close to the initial particle size for distended powders which don't depend on the mechanism of the densification in the shock front (the "plastic" flow for Teflon or brittle fracture and repacking for glass). This may suggest that inertia is the main factor in establishing the shock front thickness for strong shocks.

The observed large qualitative difference between transition processes in granular materials and their laminar counterparts (more than one order of magnitude difference in the transition time from initial to final state) cannot allow use of this model for thermodynamical calculations such as the rate of heat release, thermal energy distribution between components, etc.

Relations Between Maximal Particle Velocities in Stationary Shocks. It is interesting to compare relations between particle velocities in the granular material and in the laminate model, based on the hypothesis that the same loading results in stationary shocks with the same shock speeds $D_l = D_p$ (Thouvenin [1966]; Hofmann, Andrews, and Maxwell [1968]). This hypothesis is in agreement with experiments for granular material (Fig. 3.13(a),(b)) and for a laminate with the thickness of layers equal to the thickness of particles in powder (Fig. 3.13(c),(d)), as follows from a comparison of intervals between the shocks arrival on gauges sitting at different depths. At the same time, the average shock speed in laminate with an increased layer thickness ($\delta = 5.8$ mm) is apparently larger than in granular materials (compare the corresponding time intervals in Fig. 3.13(a),(b) and Fig. 3.13(c),(f)).

If the initial density of the powder $\rho_{00} = 1/2\rho_s$ (ρ_s is the density of the solid glass) and the density behind the shock is close to ρ_s, then from the mass conservation law for the stationary shock, with front speed D_p and particle velocity u_p, behind it follows

$$D_p = 2u_p. \tag{3.7}$$

The speed of the stationary shock in the laminar material D_1 with the layer thickness equal to the thickness of the gap is

$$D_1 = \frac{2u_{lm}}{1 + u_{lm}/c}, \tag{3.8}$$

where c is the shock speed in the layer corresponding to the impact of the previous one moving at the velocity u_{lm} to the neighbor ahead. Particle velocity behind this shock inside the layer is apparently equal to $u_{lm}/2$.

If the amplitude of the shock is relatively small ($u_{lm} \ll c$), then it follows from Eq. (3.8) that $D_1 = 2u_{lm}$. In this case, from the comparison of Eqs. (3.7) and (3.8), one can conclude that if $D_1 = D_p$, then $u_p = u_{lm}$. The estimations of c from the known Hugoniot for the glass (Dremin and Adadurov [1964]), corresponding to the particle velocity 0.5 mm/μs, show that this equality is valid to within 10% accuracy. For the powders and laminates with different ratio ρ_s/ρ_{00} from 2 we can also expect that $u_p = u_{lm}$ based on the coincidence of their shock velocities and under condition $c \gg u_{lm}$. The experimental data in Table 3.3 support this conclusion to within 10% accuracy of the measurements.

For strong shocks ($u_{lm} \sim c$) the amplitude of the particle velocity in laminar material, ensuring the same shock speed as in powder, will be essentially higher than the corresponding particle velocity on the shock front in powder at the same loading, as follows from comparison of Eqs. (3.7) and (3.8).

3.4.3.2.3   Maximal and Final Particle Velocities (Pressures) in Periodic Laminar Materials

The laminar materials demonstrate the fundamental difference between the maximal parameters attained in shock loading in the front and final parameters determined by Hugoniot relations.

There are two independent ways to obtain the shock speed for laminar material following the hypothesis of the existence of the stationary shock. The first one is based on the consideration of the leading first impact and on the geometry of the layers (Eq. (3.8)). Another one uses the mass conservation equation. It connects the state before shock with density ρ_{00} and behind it where the equilibrium state is established and the material has density ρ_f. The shock front thickness for the laminar material is the transition zone (between the initial and final steady states) with oscillations of pressure and particle velocity (about 10 cycles, see Fig. 3.13(c)–(f)).

The comparison between these different methods allows us to get a general relationship between the maximal velocity in the head of the wave in laminar material u_{lm}, and the final particle velocity in the equilibrium state u_{lf}. This is the exact relation for stationary shock in a laminate where c is the shock wave speed inside the layer corresponding to particle velocity $u_{lm}/2$:

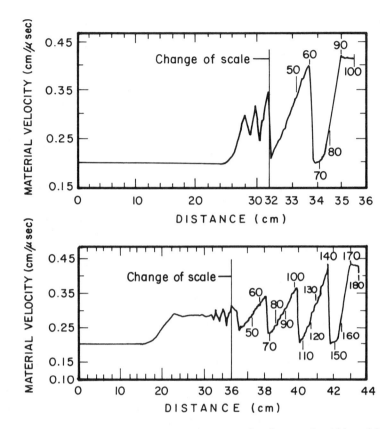

Figure 3.14. The dependence of particle velocity on the distance for Al-based laminar material (1-D model for porous aluminum) for two different times: (a) 9.833 μs and (b) 24.257 μs corresponding to the moments of closure of the fifth and thirteenth gaps. (Reprinted from *J. Appl. Phys.*, **39**, no. 10, Hofmann, R., Andrews, D.J., and Maxwell, D.E., Computed Shock Response of Porous Aluminum, pp. 4555–4562, 1968, with permission from the author.)

$$u_{\mathrm{lf}} = u_{\mathrm{lm}} \left(2c_0 + S_1 u_{\mathrm{lm}}\right) \frac{\rho_s}{\rho_f} \frac{\left(\rho_f - \rho_{00}\right)}{\left\{\left(2c_0 + S_1 u_{\mathrm{lm}}\right)\left(\rho_s - \rho_{00}\right) + u_{\mathrm{lm}} \rho_{00}\right\}}. \tag{3.9}$$

Parameters c_0 and S_1 represent the Hugoniot relation for material of the layers ($c = c_0 + S_1 u_{\mathrm{lm}}/2$).

If the shock wave is weak and the particle velocity is relatively low ($u_{\mathrm{lm}} \ll c$) and if $\rho_f = \rho_s = 2\rho_{00}$, then $u_{\mathrm{lm}} \approx u_{\mathrm{lf}}$ follows from Eq. (3.9). This case is close to the observed behavior of a glass-based laminar system for relatively weak shocks corresponding to $L = 14.5$ mm and 16 mm (Fig. 3.13(c)–(f), lower beams) when the duration of the shock is long enough. Of course, the maximal pressure at the shock front is much higher than the final equilibrium pressure (see Problem 6).

Equation (3.9) is also in good agreement with numerical calculations for the laminar system composed from aluminum layers in the case of a strong shock if u_{lm} is comparable with c (Hofmann, Andrews, and Maxwell [1968], Fig. 3.14). In this case the numerical result for the final equilibrium density $\rho_f = 2.78$ g/cc is close to the initial solid density $\rho_s = 2.7$ g/cc ($\rho_{00} = 1.35$ g/cc). The amplitude of the particle velocity at the moment of the gap closure is equal to $u_{lm} = 0.42$ cm/μs (at the end of the closure of the thirteenth gap). From the Hugoniot of the solid aluminum ($c_0 = 0.5386$ cm/μs, $S_1 = 1.339$, Steinberg [1996]) the shock speed c at the particle velocity $u_{lm}/2$ is equal to 0.82 cm/μs. With these data, one can calculate, from Eq. (3.9), $u_{lf} = 0.286$ cm/μs. The numerical result for the equilibrium particle velocity gives 0.285 ± 0.006 cm/μs. A comparison between these data for particle velocities demonstrates excellent agreement with the calculations based on Eq. (3.9).

The relations for the maximum pressure P_{lm} at the head of the stationary shock in laminar material and the final equilibrium pressure P_{pf} can be obtained using Eq. (3.9) (see Problems 6 and 7 of this chapter). The difference between P_{pf} and P_{lm} can be very large, depending on the shock amplitude and the initial powder density. It should be clearly stated that the hypothesis that was used of the leading stationary shock in laminar material has no regular proof. The computer calculations demonstrate that the regime close to the stationary one is established on the first five cells (Hofmann, Andrews, and Maxwell [1968]).

It is interesting that the same phenomenon, particle velocity (pressure) spike on the shock front, was also typical for another one-dimensional periodic system represented by the chain of elastic particles (see Fig. 1.36(a)) and was practically absent for the same system composed from mainly plastic particles (Fig. 1.36(b)). Compare these pictures with the shock profiles for the laminar system presented in Figure 3.13(c)–(f). Nevertheless, the underlying mechanisms resulting in similar shock profiles are quite different.

3.5   Damping of Shock Waves by Porous Materials

3.5.1   Porous Shock Attenuators

The application of the porous materials as buffers, between the wall to be protected and the shock wave in the air (e.g., resulting from explosion), can lead to an unexpected increase of the shock pressure on the obstacle (Gel'fand, et al. [1975]; Bajenova et al. [1986]; Nesterenko et al. [1988]) and also in a decrease (Nesterenko et al. [1988], [1992]). Ben-Dor et al. [1997] performed a detailed experimental study of a long weak shock interaction with granular layers of different structure. They observed the amplification effect for all investigated materials under their conditions of loading, and demonstrated that shock-wave behavior depends on the material's permeability and initial effective density. It is very important to clarify the difference between these types of behavior.

3.5.1.1 Porous Layer on Rigid Wall, Amplification versus Attenuation

Experiments on the shock protection of the rigid and deformable walls were conducted by Nesterenko et al. [1988] using the experimental set-up shown in Figure 3.15. Experimental results are presented in Table 3.4.

Table 3.4 shows that under explosion in air, the layer of Porolon with a thickness of 30 mm, instead of serving as a buffer, increases the shock amplitude on the wall two times, in comparison with the shock amplitude without it (Nesterenko et al. [1988]).

Such behavior is associated with the differences of the acoustic impedances of the air, Porolon, and the wall, and with the rapid increase of the acoustic impedance of Porolon under shock compression (Bajenova, Gvozdeva et al. [1986]). The reflection of the air shock wave from Porolon can be considered as a reflection from the rigid wall, as was experimentally shown by Anikiev [1979]. That is why the introduction of the Porolon layer should double the amplitude of the reflected weak shock with a subsequent doubling after reflection from the interface Porolon–rigid wall. At the same time porous layers are effective for the elimination of spall phenomena (Akhmadeev et al. [1985]; Nesterenko et al. [1986]). Kostykov [1980] considered the conditions for an amplitude increase for tangential loading.

It is important to find out what is the optimal structure of the porous media for shock attenuation and for decreasing the dynamic load on the wall. The general suggestion is that for the best results, it is necessary to select a porous material with a small sound speed in the initial state as well as with a slight increase under its dynamic compression.

Experiments with different porous materials (sawdust with a different density, Porolon, and laminar materials) were conducted to check the influence of the material structure on the amplitude of the transmitted impulse on the wall behind

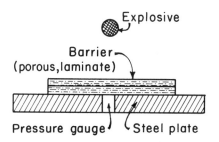

Figure 3.15. The experimental set-up for experiments on shock transformation by porous, laminar layers adjacent to the rigid wall.

Table 3.4. The maximal pressures on a rigid wall for explosion in the air (explosive mass 150 g, RDX, distance from wall 0.5 m) for different materials.

Material of the layer	Thickness H of the layer (mm)	Maximal pressure P_{max} on wall (MPa)
Porolon (40 kg/m^3)	0	5.8
	30	13
	50	9.3
	60	8.5
	70	7.6
	100	1.92
	150	1.12
Sawdust 260 (kg/m^3)	60	0.9
183 (kg/m^3)	60	1.6
170 (kg/m^3)	60	1.14
Steel + Porolon two cells ($l_1 = 2.5$ mm, $l_2 = 30$ mm)	65	0.58
Steel + Porolon one cell ($l_1 = 4$ mm, $l_2 = 60$ mm)	64	0.28
Steel + Porolon three cells ($l_1 = 2.5$ mm, $l_2 = 30$ mm)	97.5	0.65

the barrier. In these experiments, the incident impulse was the same and comparison of experiments with the same thickness of porous layers is very instructive.

For example, application of the layer of sawdust with a thickness of 60 mm results in the amplitude of the reflected shock wave from the contact porous layer/rigid wall being up to six times less than the amplitude of the reflected pulse in the air (5.8 MPa). This amplitude decrease is in striking contrast to an increase in the impulse amplitude for the Porolon layer with the same thickness, 60 mm (8.5 MPa, Table 3.4). These results confirm the qualitative importance of the porous materials characteristics to ensure a decay of the impulse amplitude on the rigid wall. Evidently, simple criteria, based on the relation of the length of the incident impulse and the porous layer thickness proposed by Gel'fand et al. [1983], and Bajenova et al. [1986], are not sufficient to ensure the decrease of the load by the porous material.

The influence of a porous (foam-type) material structure (Porolon and Perlite (similar to vermiculite)) on shock decay is illustrated by the experimental data in Table 3.5 (Afanasenko et al. [1990]).

The geometry of experiment is presented in Figure 3.15. These foams have a qualitatively different structure of the condensed skeleton. Porolon has an open porosity, being actually a cellular material. Perlite is composed from separate

Table 3.5. Dependence of the pressure duration and its amplitude on a rigid wall coated by layers of different porous materials with different structure. Explosion in the air, explosive mass 150 g, RDX, distance from wall 0.5 m.

Thickness H (mm)	Impulse parameters Porolon, (40 kg/m³)		Impulse parameters Perlite (60 kg/m³)	
	P_{max} (MPa)	τ (µs)	P_{max} (MPa)	τ (µs)
30	13	128	7.0	115
50	9.3	100	2.87	210
60	8.5	240	—	—
70	7.6	630	—	—
80	—	—	2.2	400
100	1.92	1060	1.35	420
120	—	—	1.24	390
150	1.12	1260	0.66	660

particles containing micropores. The parameters of the air shock reflected from the wall generated by the explosion of the spherical charge RDX with mass 150 g in all experiments were the same (P_{max} = 5.8 MPa, duration 200 µs). This does not depend on the material of the wall (steel or Porolon) because both materials have shock impedance much larger than the air, and shock reflection can be considered as reflection from the rigid wall in both cases. This corresponds to the distance between explosive and the wall equal to 0.5 m. It is seen from Table 3.5 that a relatively small thickness of foam results in the amplitude increase of the reflected shock wave.

There are interesting differences between these foams. Porolon has a much more profound effect, according to the amplitude increase which is detected up to 70 mm thickness of the layer. Perlite exhibits an essentially smaller amplitude increase, and effectively mitigates the shock wave at a thickness of 50 mm.

Another difference is connected with quite different impulse durations in Perlite, in comparison with Porolon (see, e.g., data for H = 100 mm). At the layer thicknesses when shock decay was detected, the duration of the reflected main shock in Perlite was two times smaller than in Porolon.

The interesting peculiarity of the experiments with sawdust is a two-wave structure with the pressure drop between them practically to zero (Fig. 3.16). In experiments with a higher initial density (260 kg/m³), and at the same loading conditions and thickness of the layer, the amplitude of the second pulse can be lower than the amplitude of the first pulse (difference of about 25%, Nesterenko [1988]). In all cases, the duration of the second pulse was three times larger than the duration of the first one.

A two-wave structure of reflected impulse was also observed in Perlite at a layer thickness of 150 mm. In sawdust and in Perlite, two-wave configuration was typical for a relatively weak amplitude of impulse about 1 MPa (1.2 MPa in sawdust and 0.66 MPa in Perlite). Such behavior might be explained by the

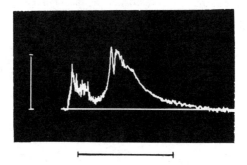

Figure 3.16. The double-wave type pressure dependence on time on the contact sawdust–rigid wall. Thickness of the porous layer 60 mm, initial density 183 kg/m^3. Vertical scale 1.5 MPa, horizontal scale 1 ms.

shock propagation through gas inside a solid skeleton and its reflection from the wall (first impulse) and subsequent movement of the sawdust material with impact on the wall (second impulse).

Another reason for this two-wave structure could be a "threshold" dependence of the sawdust densification on pressure. This can dynamically generate a layered material. The relatively dense compacted layer of sawdust will be near the shock entrance (where the amplitude of the shock was large enough) contacting the less compacted material in the depth. Then, the first shock wave reflected from the wall can again be reflected from this interface of compacted and less dense sawdust.

The two-wave structure of the impulse transmitted to the wall was also observed for relatively thick layers of the Porolon and Perlite (150 mm). The two-wave structure could also be accounted for by the anomalous compressibility of this material under dynamic loading, resulting in a disperse nonstationary wave structure. Further research is needed to understand the relative importance of these mechanisms.

3.5.1.2   Laminar Solid/Foam Buffers on a Rigid Wall

Another interesting possibility, according to shock attenuation, is the use of layers composed from materials having very high differences of acoustic properties like Porolon and steel (high-gradient buffers) with various numbers of cells (Fig. 3.15). The basic idea prompting this approach was explained at the beginning of this chapter and is illustrated in Figure 3.4.

It is seen from the data presented in Table 3.4, that the single-cell material (steel plate (14 mm) + layer of Porolon (60 mm)) is more effective in comparison with the two-cell material and even with the thicker (total $H = 97.5$ mm) three-cell structure. The reason for this behavior can be connected with the nonlinear effects and secondary wave reflections, as in the case of the laminar materials discussed in Section 3.3.

The laminar periodic system composed from heavy metal plates and light layers of foam can also be a good example of the strongly nonlinear discrete system described in Chapter 1, with a useful property to split the initial impulse into the sequence of solitary waves on very short distances from the entrance. To ensure this behavior the foam should have essentially nonlinear compressibility and dissipation should not be strong.

It is worth mentioning that a shock wave in air after reflection transmits the same linear momentum I to all the investigated laminar materials, irrespective of their inner structure. But the energy transmitted into material depends on the mass of the material of the first buffer plate. In fact, the kinetic energy E_k, which is obtained by the first steel buffer plate with the mass M is

$$E_k = I^2/2M. \tag{3.10}$$

Here the total momentum is supposed to be absorbed initially by the first plate only. Changing the plate mass does not change the transmitted momentum I, because the displacement of the plate during wave interaction can be neglected, and reflection can be considered as from a rigid wall. Thus the plate with less thickness (mass) will get more kinetic energy and velocity after shock reflection, resulting in a higher wave amplitude in the material. From this point of view, the most effective system should concentrate the most massive elements on the surface facing the incident shock impulse from the air (Nesterenko et al. [1988]). The case with a one-cell system in comparison with other systems clearly demonstrates this behavior.

If the main parameter is the mass per unit surface but not thickness, then the effectiveness of the barriers can be ordered in this way: Porolon, sawdust, steel–Porolon (one cell), steel–Porolon (two cells), steel–Porolon (three cells).

If the main parameter is the barrier thickness, then the order of effectiveness is like this (for a thickness of 60 mm): steel–Porolon (one cell), steel–Porolon (two cells), sawdust. Porolon as a single layer in this case is excluded from consideration at this thickness, as it produces an amplitude increase (Table 3.4).

The impulse transformation by porous barriers can be important for the modification of the structures behavior. But if the structure reacts only on the momentum, then its behavior does not change, even if the amplitude of the impulse increases under conditions of constant momentum. This can happen for the impulse loading of the deformable obstacles. Their plastic deformation ensures a long reaction time of the structure and sensitivity only to the momentum, and not to details of the impulse structure.

Polymer foam is not the only material for blast mitigation. For example, metallic cellular materials demonstrate a high-energy absorption capability. This property is based on their large compressive strains achievable at nominally constant stress before material compaction. This property suggests their implementation in ultralight blast amelioration structures (Evans, Hutchinson, and Ashby [1999]; Ashby, Evans, Fleck et al. [2000]). The foam thickness D needed to absorb the energy transmitted to the material as a result of the reflected shock can be calculated in a quasistatic limit, assuming that the foam is uniformly compressed:

$$D \geq I^2/2\Omega_\beta d_b U, \tag{3.11}$$

where $2\Omega_b$ is the density of the buffer plate, d_b is its thickness, and U is the energy absorbed by foam per unit volume (kJ/m^3).

The energy absorption per unit volume is equal to

$$U = \sigma_p \varepsilon_d, \tag{3.12}$$

where σ_p is the plateau compression strength of the foam (for aluminum foam it is about 2 MPa) and ε_d is about 0.6.

The minimum combined weight of the foam plus the buffer plate W_{min} can be calculated based on Eq. (3.11) for the minimum foam thickness D_{min}:

$$W_{min}/W_b = 1 + (I/d_b)^2/2\Omega_b U_\rho, \tag{3.13}$$

where W_b is the weight of the buffer and U_ρ is the energy absorbed per unit mass (for aluminum foams it is about 6 J/g)).

The map in the coordinates explosive mass–distance from the blast is presented for the desirable foam thickness needed to absorb the energy of the blast (Evans, Hutchinson, and Ashby [1999]). According to this map, the foam with a plateau flow stress of 1 MPa should be 0.5 m thick to absorb the blast energy from 50 kg explosive, located 3 m from a structure that can sustain a pressure of 1 MPa.

It should be mentioned that this approach is valid if the loading conditions and mass of the buffer plate ensure a quasistatic type of deformation. If the shock wave will propagate in the foam, then the situation can be quite different from the quasistatic approach.

3.5.1.3   Porous, Laminar Layer on Deformable Wall

Large plastic deformation of the wall can change the reflected shock in porous materials if a characteristic time of shock action is comparable or larger than the deformation time. To investigate the possible influence of porous barriers on the reaction of the deformable wall, experiments with the thin quadrant duralumine flat plates with fixed (clamped) perimeter (200 * 200 mm) and thickness 1.5 mm (Fig. 3.17) were performed by Nesterenko et al. [1988]. The final plastic deformation of the plate and maximal deflection in the center serves as an indicator of the porous barrier influence. The plates were deformed into a pyramid shape, indicating localized shear deformation as the basic mechanism. The central displacement of the plate did not depend on the application of the Porolon layer with a thickness 50 mm under the same explosion conditions.

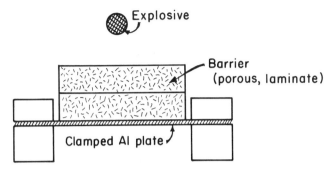

Figure 3.17. Experimental set-up for the investigation of the role of porous barriers on the plastically deformed aluminum plates with a fixed perimeter.

The laminar materials essentially decreased the central displacement, but no difference between steel–Porolon (one cell) and steel–Porolon (two cells) was observed, despite the difference in impulse amplitude.

The phenomenology and mechanism of the plastic shear deformation of thin steel plates under blast loading was studied experimentally and numerically by Guerke [1995]. The loading conditions were close to the conditions in our experiments, with a larger pressure amplitude (10 MPa instead of 5.8 MPa in our experiments with the same duration of 0.2 ms of the positive phase).

For practical applications, it is important that the pyramid shape of the initially flat plate can be considered as a signature of the blast event and cannot be reproduced by impact or other methods. The final deflection is proportional to the impulse of the shock wave. The impulse is the function of the charge mass, distance, and type of explosive. If two plates after the blast event can be identified with the known positions according to the explosive charge, then the explosive charge and its mass can be calculated in its TNT equivalent. This procedure was experimentally tested in the early 1980s by the author's research team, being involved in real scale experiments with out-of-date aircraft in the framework of the USSR antiterrorist program. The influence of the porous layer with a thickness of up to 50 mm can be neglected in the first approximation.

3.5.2 Numerical Simulations

3.5.2.1 Model Description and Validation Experiments

The development of analytical and computer modeling of the protective porous barriers is of practical interest. Such models tested on small-scale experiments can be used to optimize large-scale structures. Gel'fand et al. [1983] conducted computer calculations of the air shock interaction with the porous barrier.

Gvozdeva and Faresov [1985], used an approximate method for the calculation of the stationary shocks in porous media with compressibility determined by gas in the pores. Nesterenko et al. [1990] applied a two-phase model of the equilibrium mixture of an ideal gas and incompressible particles for the description of shock waves in Porolon.

The system of equations in Lagrange coordinates for one-dimensional flow with plane symmetry is

$$\frac{1}{v}\frac{\partial v}{\partial t} = \frac{\partial u}{\partial r}, \tag{3.14a}$$

$$\frac{1}{v}\frac{\partial u}{\partial r} = -\frac{\partial P}{\partial r}, \tag{3.14b}$$

$$\frac{\partial e}{\partial t} = -P\frac{\partial v}{\partial r}, \tag{3.14c}$$

$$\frac{dr}{dt} = u, \tag{3.14d}$$

$$T_1 = T_2 = T, \quad P_1 = P_2 = P, \quad u_1 = u_2 = u, \tag{3.14e}$$

$$P_1 = \rho_{11}RT, \tag{3.14f}$$

$$e_i = c_i T, \tag{3.14g}$$

$$v = \frac{1}{\rho}, \quad \rho_i = \alpha_i \rho, \quad \alpha_1 + \alpha_2 = 1, \quad \rho_{22} = \text{const}, \tag{3.14h}$$

$$\rho_i = m_i \rho_{ii}, \quad m_1 + m_2 = 1, \quad e = \alpha_1 e_1 + \alpha_2 e_2, \tag{3.14i}$$

where v, e, and ρ are the specific volume, energy, and density of the mixture; u, particle velocity; P, pressure; T, temperature; R, gas constant; c_i, specific heat of each phase; ρ_{ii}, ρ_i, real and average density of the phases; and m_i, α_i, volume and mass phase concentrations, $i = 1, 2$. The initial density of the Porolon $\rho_0 = 40$ kg/m³, the real initial density of the condensed phase $\rho_{22,0} = 1.2 \cdot 10^3$ kg/m³, the initial density of the gas phase $\rho_{11,0} = 1.29$ kg/m³.

This system of equations was numerically solved using Wilkins's method (Wilkins [1964], [1999]). The comparison of numerical result with experimental data on the shock decay in Porolon was used as a validation procedure for the model. The shock in numerical calculations was generated by a triangle pressure impulse on the free surface of the Porolon, modeling the loading from the explosive charge. The pressure–time history of load on the substrate and the dependence of amplitude of the reflected shock in the interval 13–2 MPa on the Porolon thickness (30–150 mm) were in a good agreement with experimental results (experimental set-up is shown in Fig. 3.15) for the initial shock amplitude 5.8 MPa and duration 0.2 ms (Nesterenko et al. [1990]). This was considered as validation of the proposed model. It was subsequently used to optimize the low-density and high-gradient porous barriers in the pressure range 1–15 MPa.

3.5.2.2   Critical Thickness of Foam for Shock Mitigation

The dependence of the threshold thickness $H*$ of the porous layer responsible for the transition from the amplitude increase to its decrease on the loading conditions (amplitude and duration of the shock impulse) is of great interest. The corresponding data from the numerical analysis are presented in Table 3.6. The initial impulse on the free surface of Porolon had a triangular shape with amplitude P_m and duration t. The critical thickness corresponds to the layer when the amplitude of the reflected shock from the rigid wall is equal to P_m. At larger thickness of the foam layer, the amplitude of the reflected shock is less than P_m decaying approximately linearly with distance. The important result is that the critical thickness $H*$ depends both on the pressure amplitude and duration. An increase of the impulse duration with the same initial amplitude in one order of magnitude resulted in an $H*$ increase approximately of one order of magnitude. At the same time, an increase of the initial maximal pressure of order of magnitude with the same duration also resulted in a six times increase of the $H*$.

The equation for the critical thickness of Porolon ensuring shock mitigation for high shock pressures (Table 3.6, bottom three lines) may be written as

$$H^* = 216 P_m^{0.6} t, \tag{3.15}$$

where $H*$ is in mm, the pressure is in MPa, and the time is in milliseconds.

The critical thickness for low pressure (1 MPa) does not obey this law, which predicts an essentially higher value of $H* = 28$ mm instead of 18 mm obtained in the numerical calculations.

The increase of the layer of Porolon thickness beyond $H*$ results in the fast decay of the loading on the substrate. Based on the results of numerical calculations (Nesterenko, Afanasenko, Cheskidov, and Grigoriev [1990]), the

Table 3.6. Dependence of the critical Porolon thickness $H*$ on the initial impulse parameters resulting in the decreasing shock amplitude transmitted on the wall.

Impulse parameters		Critical thickness
Pressure P_m (MPa)	Duration t (ms)	$H*$ (mm)
1	0.13	18
6.6	0.13	87
10	0.13	112
6.6	1.30	850

increase of the Porolon layer from H^* to $2H^*$ results in the decrease of maximum pressure on the substrate from P_m to $P_m/5$.

3.5.2.3   Modeling of High-Gradient Solid/Foam Barriers

The high-gradient laminar steel–Porolon system implying the incompressibility condition for metal was also modeled in a similar way, using the system of equations (3.14) for Porolon. The loading conditions correspond to the explosion of 0.15 kg of RDX at a distance 0.5 m from the barrier (amplitude of incident shock 5.8 MPa, and duration 0.2 ms).

It was shown in numerical calculations that a one-cell system is more effective than two- and three-cell systems in complete accord with experiments (see Table 3.4, Nesterenko, Afanasenko, Cheskidov et al. [1990]). It is important that the steel plate on the top of the barrier must face the shock in air, as was discussed earlier.

The results of computer modeling and this experiment are presented in Figure 3.18 for a one-cell system. It is evident from Figure 3.18 that the relatively simple approach represented by Eqs. (3.14) can simulate the peak response, and reasonably well describe the complicated wave structure of the impulse, including the rarefaction stage in the laminar solid/porous materials.

It was also demonstrated in calculations that the maximal damping effect is reached when the ratio of the steel plate thickness to the Porolon thickness is 0.8 (at a given overall barrier thickness) and this optimal ratio did not depend on the amplitude and duration of the initial shock impulse. The large increase of the Porolon density from 40 kg/m^3 to 120 kg/m^3 in a single porous layer and in a single cell laminar material did not also result in the essential change of the damping effect.

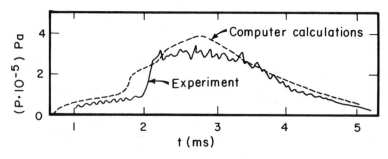

Figure 3.18. The experimental and calculated time dependence of pressure on the rigid wall with most efficient for shock damping one cell laminar barrier (steel (4 mm) + Porolon (60 mm)).

3.6   Blast Mitigation by Porous Materials, Criteria for Damping

3.6.1   Reduction of Blast Noise in Air by Porous Media

Several different porous materials were investigated for noise reduction produced by the detonation of high explosives in the atmosphere (Raspet, Butler, and Jahani [1987]). The explosive charge was centered in a cube of blast-reducing material—steel wool, fiberglass, straw, and plastic bubble pack. Experiments were done for relatively large distances (38 m and 76 m) from the explosive charge (explosive C-4 with mass 0.57 kg or 0.275 kg) which correspond to the relative radius (876–2203) with respect to the explosion radius in the far field area.

For all investigated materials, the noise reduction and particularly the peak of the sound level at constant distances from the charge are scaled fairy well with the dimensionless material depth X^*, a measure of the material mass per charge mass

$$X^* = l \left(\frac{\rho_b}{C_w} \right)^{1/3}, \tag{3.16}$$

where l is the geometrically averaged material depth

$$l = 0.5(abc)^{1/3}, \tag{3.17}$$

and a, b, and c are the cubic enclosure dimensions, ρ_b is the material bulk density, and C_w is the explosive mass (kg-TNT equivalent). The explosive C-4 has an effective energy yield (5.86 MJ kg^{-1}), approximately 1.36 times that of TNT.

These results are consistent with the data by Raspet and Griffiths [1983] on the reduction of blast noise with aqueous foam. In both papers, the lightweight polyethylene plastic sheeting was used to hold the blast-reducing material. The role of this type of container should be negligible for noise reduction.

The role of a relatively rigid container (a cylindrical metal culvert open at one end to the atmosphere) confining the blast-reducing material (aqueous foam) on the attenuation effectiveness was investigated by Raspet, Butler, and Jahani [1987] for the same range of charge weights (0.57–0.28 kg). They concluded that the rigid confinement contributes to the attenuation process, probably due to the additional dissipation connected with wave reflections off the internal wall of a cylindrical metal culvert. The system of culvert plus ground (the culvert sections were sunk into the ground) has a larger reduction on the energy than on the peak levels of noise. They have shown that noise reduction can be scaled with a modified foam depth, which includes the surface area of foam and the depth of foam.

Air shock damping with the help of an advanced explosively pulverized water "curtain" was investigated by Kravets et al. [1962], and by Bagrovskii, Fruchek, and Korzun [1976]. The detailed analysis of the efficiency of this air–water drop curtain, with consideration of the possible physical mechanisms that ensure the protective effect and corresponding references to original results, can be found in a paper by Buzukov [2000].

3.6.2   Reduction of Shock Effects on a Rigid Confinement (Explosive Chambers)

3.6.2.1   Stationary Explosive Chambers

The closed metallic vessels with semiautomatic control systems are widely used for explosive experiments in the laboratory and in industry. The dynamic behavior of these vessels under internal explosive loading is determined by the value of the linear momentum delivered to the wall by the reflected shock wave if condition $t \ll T$ is fulfilled (t is duration of pressure in reflected wave, and T is characteristic time of vessel vibration). Baker [1960] was the first who considered the behavior of a spherical shell (elastic and plastic) caused by internal blast loading.

Demchuk [1968], [1971] introduced a simple and useful engineering criterion for designers for the reliability of the explosive chambers which depends on the explosive mass, geometry of the vessel, and their elastic properties. Restrictions on the maximal explosive charge that can be safely confined in the chamber were proposed for the simple geometries of vessels such as spherical or cylindrical. He also courageously took responsibility for the design of explosive chambers for different industrial and research purposes, their manufacturing, and subsequent performance. It is perhaps not surprising, then, that Demchuk was ridiculed sometimes by colleagues for some inevitable imperfections, a common fate of pioneers in any field.

It should be mentioned that the maximal strain in the chamber can be achieved not in the first cycle of vibrations, as expected, based on the exciting of a single main mode. This may happen essentially later, after 10–20 cycles (Buzukov [1976]). This effect depends on the geometry of the chamber (Adishchev and Kornev [1979]; Kornev et al. [1979]; Whenhui et al. [1997]). Buzukov [1976] attributed this effect not to the repeated shock loading of the wall, but to the "complex vibrational process of the whole construction" connected with superposition of the radial and longitudinal modes of vibrations. The more correct explanation appears to be connected with the presence of similar frequences in the spectrum of the free vibrations of the shell (Adishchev and Kornev [1979]; Kornev et al. [1979]; Whenhui et al. [1997]). This effect may cause an increase by a few times a maximal strains in the shell in comparison with calculations, based on the consideration of shell as a system with one degree of freedom.

If the duration of shock pressure on the wall is larger than the main period of the chamber vibrations ($t > T$), then the vessel reaction does not depend on the momentum any more. This is mainly determined by the characteristic pressure in the reflected wave. By tuning the impulse parameters, with the help of different porous barriers, it is possible to increase by a few times the maximal mass of the explosive charge which can be safely confined in a given closed vessel (Nesterenko et al. [1979], [1984]; Demchuk, Kuznetsov, and Nesterenko [1982]).

Kudinov, Palamarchuk et al. [1976] suggested a filling of the explosive chamber with water foam. Neal [1980] used vermiculite with a density of 100 kg/m^3 and obtained a 50% decrease of the maximum strain in the explosive chamber. The reason for shock damping was considered in these two papers as being connected with the kinetic energy losses in porous media during its shock loading and subsequent impact with the wall.

On the other hand, Gel'fand, Medvedev et al. [1987] emphasized that the covering of the inner surface of the explosive chamber by porous material increased the dynamic strains in the chamber wall.

The attempts in this direction are mainly empirical, and the mechanism of shock damping is not well established. It does not allow an optimization of the porous medium parameters for damping purposes. Nesterenko et al. [1979], [1984], and Demchuk, Kuznetsov, and Nesterenko [1982] explained the strain reduction effect by impulse transformation by powder with a subsequent changing of the dynamic mode of chamber loading to the quasistatic one. The optimal density for blast damping was suggested by Demchuk, Kuznetsov, and Nesterenko [1982].

Let us discuss the importance of the kinetic energy losses in powders affecting shock damping in explosive chambers. The situation is not clear, as would appear at first glance. As was already mentioned, the main parameter for the chamber reaction under condition $t \ll T$ is not connected with the energy of the air shock wave or the shock wave in powder. Instead, it is fully determined by the momentum transmitted to the wall in reflected shock. The energy losses from the heat exchange between detonation products and porous media (sand in paper by Kuznetsov and Shatsukevich [1977]), and kinetic energy transformation into heat in the shock front and behind it do not automatically result in a momentum decrease (Nesterenko [1992]). To make any conclusion about the momentum change, we need to consider also the change of the mass involved in the interaction with the wall. The mass of the porous media surrounding the explosive charge can increase the specific linear momentum per unit surface of the chamber, even taking into account energy losses.

For example, in the experiments with sand the kinetic energy of the sand surrounding the explosive charge dropped down to 2% from the initial energy of the detonation product. At the same time, the total momentum of the moving sand is up to 20 times more than the initial momentum of the detonation products $I_0 = (2QM)^{0.5}$ where Q is the energy of explosive released during detonation and M is the mass of the explosive charge (Zel'manov, Kolkov, Tihomirov, and Shatsukevich [1968]). So, instead of a damping effect, we get a

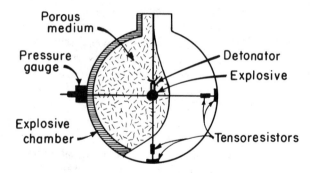

Figure 3.19. The geometry of the tensogauges on the chamber wall.

dramatic increase in momentum per unit surface due to the increase of the mass involved in motion despite the energy dissipation. The increase in the mass simply dominates the kinetic energy loss for condensed porous materials. And if the vessel responds to the momentum the maximum strain will increase!

It is necessary to emphasize that attenuation of the shock wave in porous materials such as foam (Kudinov, Palamarchuk et al. [1976]) does not necessarily lead to damping of the shock load on the chamber wall, because the latter is determined by the parameters of the reflected wave.

So the conditions of shock-wave damping on the wall of the explosive chamber with porous filling inside can be different from that for blast mitigation in the open atmosphere.

Nesterenko et al. [1979], [1984] and Demchuk, Kuznetsov, and Nesterenko [1982] conducted experiments with simultaneous measurements of the strains on the outer side of the chamber wall and the pressure on the inside. The experiments were performed in air and by filling the chamber with sawdust (density 80–150 kg/m^3, snow with a density of 300 kg/m^3, and copper powder with a density of 1940 kg/m^3). The small scale models of cylindrical explosive chambers (model VR, diameter 460 mm, wall thickness 7 mm) and spherical vessels (model KVG-10, diameter 455 mm, wall thickness 7 mm) for explosive masses (M_{ex}) up to 50 g, designed by Yu. Kuznetsov, were used in this research. For strain measurements the gauges 2PKB-20-200-HB were used, which were glued, in pairs, in mutually perpendicular directions (Fig. 3.19). A permanent voltage (9 V) was provided according to a potentiometric scheme. The strains were calculated from the electrical signal resulting from the change in the gauge length. The pressure on the chamber walls was measured with the standard piezoelectric gauges (Zagorel'skii, Stolovich, and Fomin [1982]).

The ratio of the radius of the porous shell around the charge to its effective radius (η) was changed from 2 to 16.7, and the corresponding ratio of the radius of the explosive chamber to the charge radius (k) was inside the range 11–20. The RDX porous explosive with tapped density (0.9–1 g/cm^3) having the shape of the cylinder, with height equal to diameter, was used in experiments. The

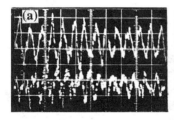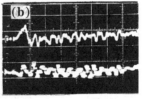

Figure 3.20. Strains in the spherical chamber for explosions in air and in snow: (a) explosion in air, $m_{ex} = 25$ g, upper beam, gauge II, low beam, gauge I; (b) explosion in snow, $m_{ex} = 50$ g, upper beam, gauge IV, low beam, gauge III; vertical scale 25 MPa/div, horizontal scale 0.5 ms/div.

explosive was placed in the geometrical center of the chambers (Fig. 3.19). The examples of the experimental records for the spherical chamber are presented in Figure 3.20. The gauges I–IV were placed near the equatorial plane of the sphere, gauges with even numbers being parallel to the sphere vertical axis and with odd numbers perpendicular to it. The comparison between Figures 3.20(a) and (b) demonstrates that snow can drastically decrease the maximal strains in the explosive chamber wall. In the case of an explosive mass equal to 50 g in snow (Fig. 3.20(b)) the maximum strains are equal to the strains for explosion in air with $M_{ex} = 10$ g. They are two times less for an explosion in air for $M_{ex} = 25$ g (Fig. 3.20(a)). The maximum strains for $M_{ex} = 30$ g in snow were two times less than for an explosion in air with $M_{ex} = 10$ g.

Thus filling the explosive chamber with snow can increase the maximal explosive charge in the investigated chamber geometry up to five times. This damping property of snow was successfully and routinely used by the author's team for experimental research in a simple explosive chamber—a pit in the ground. It was closed by an iron plate with a mass of 5 tons without any clamping for $M_{ex} = 1$ kg in the vicinity of ordinary buildings. Of course, to use this advantage you need to be in Siberia, where snow is available six months a year.

The experimental data for strains in a cylindrical explosive chamber are shown in Figure 3.21. The gauges I–IV were glued to the outer surface of the chamber in a parallel direction to the vertical axis (II, IV) and in a perpendicular direction (I, III). The qualitative behavior of the strains in the chamber lid demonstrates the same damping behavior of the strains as in the cylindrical part of the chamber after the chamber was filled with sawdust.

The large amount of experimental data obtained for explosions in sawdust by the author's research team (Nesterenko et al. [1979], [1984], [1988], [1992]) demonstrated a relatively large spread of experimental data. For example, data for different gauges placed in equivalent positions in the same experiment can differ by one and a half to two times. At the same time, there are qualitatively different

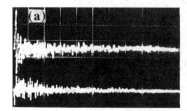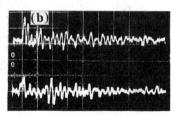

Figure 3.21. Strains in the cylindrical chamber for explosion in air and in sawdust: (a) explosion in air, $m_{ex} = 30$ g, upper beam, gauge I, low beam, gauge II; (b) explosion in sawdust, $m_{ex} = 60$ g, upper beam, gauge III, low beam, gauge IV; vertical scale 50 MPa/div, horizontal scale (a) 2 ms/div and (b) 0.5 ms/div.

and repeatable features in experiments with the explosion in a sawdust in comparison with the explosion in air:

(1) in the case of sawdust even a double mass of explosive results in a one and a half to two times in decrease the strains (averaging over five or six experiments) in the chamber wall;

(2) the second difference is the asymmetrical oscillation of the strains (during the first periods), for example, registered by the gauge placed in the direction perpendicular to the axis of the cylindrical chamber (gauge III, Fig. 3.21(b)).

These features are connected with the qualitative change of the chamber loading conditions for the explosion in sawdust, in comparison with the explosion in air that will be discussed later.

The partial filling of the explosive chamber with sawdust by the gap between the chamber wall and sawdust (an explosive charge of 60 g in close contact with the sawdust with a diameter and height of 240 mm, and a diameter of the cylindrical chamber of 460 mm) was investigated to clarify the damping conditions. No difference was found in this case, in comparison with a complete filling of the chamber with sawdust. The same result was obtained for partial filling of the explosive chamber with the air gap between the explosive charge and the sawdust shell equal to the charge diameter and in the case where the sawdust just covered the chamber wall.

To understand the mechanism of the damping, simultaneous measurements of the strains and pressures were performed. Figure 3.22 presents data of the synchronized records (two oscilloscopes were simultaneously started by the explosive event) of the strains in the cylindrical chamber wall and the pressure inside the chamber completely filled with sawdust. From a comparison between the strain and pressure history, it is clear that the characteristic time of the

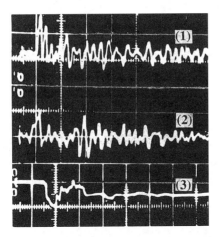

Figure 3.22. Synchronized records of the strain in the cylindrical chamber wall and the pressure inside the chamber completely filled with sawdust, m_{ex} = 60 g, RDX; (1) strain history recorded by gauge I, vertical scale, 50 MPa/div, horizontal scale, 1 ms/div; (2) strain history recorded by gauge II, vertical scale, 50 MPa /div, horizontal scale, 1 ms/div; (3) pressure history recorded by a piezogauge, vertical scale, 4 MPa/div, horizontal scale, 0.5 ms/div.

pressure increase and duration of the first part of the impulse acting on the inner wall are comparable with the characteristic period of the chamber vibrations T. The pressure impulse has a tail that is essentially longer than T.

This time dependence of the pressure impulse is the reason for the already-mentioned asymmetrical oscillations of the strains. If $t \sim T$, then as is well known from the theory, the vibrations are characterized by the shifting of their center from the equilibrium position (Landau and Lifshitz [1976]).

Another important feature of the chamber oscillations under this condition is that in comparison with the situation when $t \ll T$ (typical for an explosion in air, Demchuk [1971]), the maximum strain amplitude is determined not by the linear momentum but by the pressure of the reflected wave. The latter feature is of principal importance for the damping process in the explosive chamber filled by porous materials (Nesterenko et al. [1979], [1984], [1988], [1992]; Demchuk, Kuznetsov, Nesterenko [1982]). The shock pressures on the interface of the porous materials and chamber wall (explosive mass 30–60 g, chamber radius 225 mm, sawdust with density 80–200 kg/m^3) can have a typical ramp time of ~ 1 ms and a duration of about a few ms. This is one order of magnitude larger than typical for an explosion in air (Afanasenko [1990]). It certainly shifts the behavior of the chamber (wall thickness 7 mm) into a practically static mode.

Some additional relevant experimental results are to be mentioned:

(1) the use of the porous layer with a thickness equal to the effective radius of the explosive charge did not result in the damping of the chamber strains (η = 2,

$k = 11$). This was characteristic for the chamber filling with sawdust as well as as with copper powder in comparison with the air explosion;

(2) at $\eta = 4$–12.5, and for different values of coefficient k, the reduction of the wall strain, by two to five times without a noticeable change in the attenuation time of the chamber vibrations, is observed;

(3) the copper powder and sawdust with the same values of η and k provide the same damping effect, according to the reduction of the wall strains. This is despite the difference in density of these materials by more than one order of magnitude. This important observation demonstrates that the reduction of stresses in the wall may not be scaled by some parameter similar to dimensionless depth (Eq. (3.13)).

The heat conductivity of the porous media does not influence the reduction of the chamber's strains. Compare the poor heat conductivity of the sawdust particle with the good heat conductivity of the metallic copper particle. This demonstrates that heat losses due to the contact of cold powder and detonation products are of secondary importance.

The main conclusion may be drawn from the discussed experimental results: the damping effect is mainly connected with the qualitative change in the chamber loading regime caused by porous media. It is not essentially affected by the energy dissipation due to the heat conduction between hot detonation products and cold porous media and on the shock front as a result of plastic flow and friction.

The porous materials ensured a quasistatic type of chamber loading ($t \geq T$) which make unimportant the momentum increase due to the increase of the mass involved in the motion. For this type of loading, momentum is no longer the main parameter (as for air shock loading). Instead, the pressure in the reflected wave completely determines the chamber behavior (Nesterenko et al. [1979], [1984], [1988], [1992]; Demchuk, Kuznetsov, Nesterenko [1982]). Glenn [1986] arrived at the same conclusion.

Another important effect of porous material inside the chamber is connected with the suppression of "swinging" strains (Buzukov [1976]) caused by the superposition of similar frequencies after a few cycles. As a rule, in all experiments with filled explosive chambers of given geometry, the maximal strains were achieved in the first cycle of vibrations. This is due to the increased damping of oscillations due to the wall interaction with the adjacent porous media.

The simple criterion of damping (Problem 10) was proposed by Nesterenko [1988] based on the hypothesis that porous material ensures the transition of the spherical chamber into a quasistatic mode of loading (Figure 3.23):

$$P_f < \pi t_a/2TP_a, \tag{3.18}$$

where P_f is a pressure acting on the chamber wall being filled with powder, t_a and P_a are the duration of the pressure peak and its maximal value for explosion

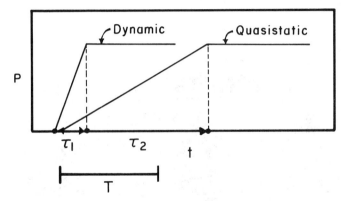

Figure 3.23. Pressure versus time dependence on the chamber wall for (1) dynamic and (2) quasistatic type of loading. Scales for pressure profiles are different.

of the same charge in air, and T is the characteristic time of chamber vibrations. It is supposed that the pressure profile can be approximated by a triangle. From this criterion it follows that if porous media result in a damping effect on one scale, then the damping effect also can be expected for a larger scale of the explosive chamber.

The effect of the porosity on the damping effect was investigated experimentally and analytically in the framework of the Kompaneetz model (Kompaneetz [1956]) by Nesterenko, Afanasenko et al. [1988] and by Afanasenko [1990]. There is a wide minimum for the loading pressure on the chamber wall approximately between the initial porosities 0.3 and 0.8. This result helps us to recognize that there is no sharp transition from the damping property of the porous media to the amplification behavior according to density variation. It can ensure the practical use of porous materials inside a large range of densities. In experiments with density variations of three times (sawdust, 60–240 kg/m³; Perlite foam, 60 kg/m³) no essential change of pressure on the wall was observed (Afanasenko [1990]).

Gerasimov [1997] demonstrated that porous shields can effectively reduce the peak stresses in the walls of explosion chambers. They are more effective than layered structures of the same thickness. This can help to avoid a spalling phenomenon and reduce the velocity of the chamber wall.

3.6.2.2 Disposable Structures

Thin-walled disposable metal containers filled with a porous medium can also be an effective method of blast localization. They can be designed to work in the essentially plastic region, drastically reducing the weight of the container. Baker [1960] considered an elastic–plastic response of the spherical shells filled with air under internal blast loading. Nesterenko [1988], [1992] evaluated the effect of light granular material (sawdust, density 100 kg/m³) by carrying out two separate

Figure 3.24. The comparison of the results of the explosion of the 0.5 kg RDX in a cubic metal container (wall thickness 3 mm, side 0.7 m) filled with (a) air and with (b) sawdust, density 100 kg/m^3.

explosive events using an explosive charge of 0.5 kg within steel cubic containers. They had wall thicknesses of 3 mm, side 0.7 m, and were reinforced at the edges and in the middle of each side by steel angles (side 50 mm, thickness 5 mm). The total weight of the container filled with sawdust M_{cf} is about 153 kg (without sawdust $M_c = 129$ kg). It is worth mentioning that the damping effect on the strain in the elastic mode of the chamber deformation cannot be automatically extended to the behavior of the plastically deformed chamber.

The results of the experiments are presented in Figure 3.24. The air-filled container was totally destroyed, while the container with filler (sawdust) was only plastically deformed, completely mitigating the blast. This demonstrates the effectiveness of a simple light granular material in damping the effect of an explosion. The typical weight ratio of the filled container to the charge weight M_{cf}/m_e in this experiment was 300 or about three times less than that used for the explosive chambers working in the elastic regime of deformation (see Problem 13). A similar approach may result in the optimized disposable systems even with an essentially lower M_{cf}/m_e ratio equal to 100–200 for complete blast mitigation.

From Figure 3.24, it is easy to conclude that the damping properties of the porous materials can be used for a relatively large scale of explosive mass within small distances from the charge (0.5 kg is sufficient for some applications). Lee, Wang, Sung et al. [1986] obtained the analogous result for small scale experiments, explosive mass 45 g.

The damping properties of the investigated porous media can be ensured starting from the thickness of the porous shell equal to or greater than six to eight characteristic diameters of the explosive charge (Nesterenko et al. [1984], [1988], [1992]). The opposite effect of the amplification of the load is possible if the thickness of the porous shell, used as a cover on the inner surface of the chamber, is small enough (Gel'fand, Medvedev, Polenov et al. [1987]).

3.6.3   Blast Mitigation by Perforated Structures

Perforated structures represent another type of "porous material" which may alter the transmitted shock characteristics. In comparison with hermetic explosive chambers, they are more easily constructed, essentially cheaper, simpler to repair and may drastically reduce the shock amplitude in air. The main idea of employing them is to force gas flow through repeated compression and rarefaction cycles with a change of flow directions, causing an enhanced dissipation of shock energy. Results of this type of blast mitigation barriers were reported by Grigoriev, Klapovskii [1987]; Grigoriev, Klapovskii, Koren'kov et al. [1987] and by Grigoriev, Klapovskii, and Vershinin [1990].

In the first paper, the authors presented equations for the dependence of the amplitude of the first shock transmitted through the perforated wall on the relative distance x:

$$\Delta P_{max} = 0.09214x^3 - 0.4429x^2 + 1.323x - 0.02232, \tag{3.19}$$

where the shock pressure is measured in MPa, and $x = R/(m)^{1/3}$ is the normalized distance from the chamber wall with respect to the explosive mass m (kg). Eq. (3.19) was obtained from experimental data with a perforated cylindrical chamber of height and diameter equal to 6 m. Its wall was manufactured from the I-beams (width 220 mm arranged in a vertical line with a distance of 160 mm centers) combined with the vertical rails between them. Both structural elements are installed in the reinforced concrete floor. The coefficient of perforation φ (ratio of openings area to the total area of wall) for this wall was equal to 0.08. The explosive mass was equal to 1, 3.75, 6.25, 20, 40, and 60 kg, corresponding to the air shock pressure incident on the wall in the interval 0.1–2.9 MPa. The radius of the danger-free zone with a shock amplitude less than 0.01 MPa for a 60 kg explosive charge was about 30 m. The air shock pressure at the same distance, 30 m, without the chamber would be twice as high (attenuation coefficient $K = 2$). The coefficient K depends on the coefficient of the wall perforation and on the shock amplitude.

The efficiency of the perforated barriers of different geometry with a size in plane of 1.1 m × 1.1 m and with various coefficients φ to reduce the amplitude of the leading air shock was investigated by Grigoriev, Klapovskii, Koren'kov et al. [1987]. The barriers were assembled from steel angles, channel bars, perforated plates, and I-beams with and without rails between them. This allows us to change the coefficient φ in the interval 0.08–0.25. The explosive charge had a mass in the range of 0.135–0.96 kg and was placed at a distance 1 m behind the barrier opposite to its center. This corresponds to the amplitude of incident shock 0.247–2.18 MPa. The investigated barriers provided up to an eightfold reduction of the shock pressure, depending on their geometry and wave amplitude. In most cases K was about three to three and a half in an area closer to the wall and decreases with distance from the barriers approaching value of two to three depending on the shock amplitude. There is a range of shock amplitudes where the attenuation is maximal for every barrier. Two-dimensional numerical calculations of flow through half of the unit cell of barrier composed from an I-beam and rail presented a detailed picture of the dissipation process due to the flow inside the barrier. The data enabled the evaluation of the radius of the danger zone ($\Delta P > 0.01$ MPa) for different cases. For example, a perforated barrier with $\varphi = 0.11$ reduced it to 18–20 m for an explosive charge of 100 kg (Grigoriev, Vershinin, and Klapovskii [1990]). Future research in this area should include the scaling of explosive events shielded by different perforated structures and the further optimization of their design.

Analysis of the behavior of structures under blast loading is presented in the book authored by Wierzbicki and Abramowicz [1994].

3.7   Conclusion

Material with a complex structure with large gradients of initial properties (laminar, granular, porous) may serve for a very efficient damping of shock waves in a wide range of amplitudes and duration. Some structures that have a complex inner geometry may also perform the same function and serve as blast mitigation barriers.

The mechanism of the damping effect may be very different. In some cases this is a result of pulse tailoring and modification of the structure behavior from dynamic (responding to linear momentum) to quasistatic (responding to pressure amplitude). It is also quite possible to obtain an opposite effect of a shock amplitude increase when powder, foam, or laminate materials are applied to protect a structure from the blast effects. The optimization process is very important to ensure the best performance of explosive chambers, disposable structures, or perforated barriers as blast mitigation devices.

Problems

1. Calculate graphically the decay of the leading shock wave amplitude in the periodic system Teflon–paraffin after traveling through five cells. Use the following equations for Hugoniots:

Teflon, $D = 1.98 + 1.71u$ (mm/μs); initial density 2.19 g/cm^3.

Paraffin, $D = 3.3 + 1.31u$ (mm/μs), initial density 0.9 g/cm^3.

2. Derive the following relations between incident, transmitted, and reflected impulses for the linear acoustic approximation for stresses (σ_i, σ_t, and σ_r) and particle velocities (u_i, u_t, and u_r) using the conditions of the continuity of the stresses and displacements (velocities) on the left and right sides of the interface between materials 1 (initial density ρ_1, sound speed c_1) and 2 (initial density ρ_2, sound speed c_2). The impulse propagates from left to right ($1 \rightarrow 2$):

$$\sigma_r = \sigma_i \frac{\rho_2 c_2 - \rho_1 c_1}{\rho_2 c_2 + \rho_1 c_1}, \qquad \sigma_t = \sigma_i \frac{2\rho_2 c_2}{\rho_2 c_2 + \rho_1 c_1},$$

$$u_r = u_i \frac{\rho_1 c_1 - \rho_2 c_2}{\rho_2 c_2 + \rho_1 c_1}, \qquad u_t = u_i \frac{2\rho_1 c_1}{\rho_2 c_2 + \rho_1 c_1}.$$

3. Derive the equation for the cumulation process in the laminar material by arithmetically decreasing (increasing) the acoustic impedances of the layers using linear acoustic approximation. Can you use this result for strong shock waves?

4. What is the main drawback of the laminar Thouvenin model for the shock compression of powders? What parameters for shock-loaded powders can be correctly calculated based on this model?

5. Prove that $u_p = u_{lm}$ if the D–u relations for powder and the corresponding laminar model coincide and the condition $c(\rho_s/\rho_{oo} - 1) \gg u_{lm}$ is valid and the shock waves provide complete densification.

6. Find the following relationship between local pressure spikes inside the weak shock P_{pm} in powder ($c_0(\rho_s/\rho_{oo} - 1) \gg u_{pf}$) and the final equilibrium pressure P_f using a laminar model consideration for a given particle velocity in the powder u_{pf} supposing that it is equal to u_{lf} in the corresponding laminar model:

$$\frac{P_{pm}}{P_{pf}} = \frac{c_0}{2 u_{pf}} \left(\frac{\rho_s}{\rho_{oo}} - \frac{\rho_s}{\rho_f} \right),$$

where c_0 is the sound speed in the layer material.

7. Find the following relationship between local pressure spikes inside the strong shock P_{pm} in powder and the final equilibrium pressure P_f using a laminar model consideration for a given particle velocity in the powder u_{pf}, supposing that it is equal to u_{lf} in the corresponding laminar model:

$$\frac{P_{pm}}{P_{pf}} = \frac{c_0 + S_1 \dfrac{u_{lm}}{2}}{2 u_{lm}} \rho_s \left(V_{00} - V_f \right),$$

$$u_{lm} = \frac{b - \sqrt{b^2 + 2 S_1 \left(V_{00} - V_s \right) c_0 u_{pf}}}{S_1},$$

$$b = -c_0 \left(V_{00} - V_f \right) + u_{pf} \left[S_1 \frac{V_{00}}{2} + V_s \left(1 - \frac{S_1}{2} \right) \right],$$

where V_f is the final equilibrium specific volume behind the shock, V_{00} is the initial specific volume of the powder, V_s is the initial specific volume of the layer material, and ρ_s is its solid density. Compare the numerical result of the calculations with computer calculations by Hofmann, Andrews, and Maxwell

[1968], for Al-based laminar material supposing that $\rho_f = 2.78$ g/cm^3, $\rho_s = 2.7$ g/cm^3, $\rho_{00} = 1.35$ g/cm^3, $u_{1f} = 0.285$ cm/μs and $c_0 = 0.5386$ cm/μs, $S_1 = 1.339$.

The calculated values of the maximal pressure in the laminar model is $P_{1m} = 0.444$ Mbar and the equilibrium pressure is $P_{1f} = 0.218$ Mbar.

8. Derive Eq. (3.9).

9. The main mode of vibrations of the spherical chamber in the elastic range of deformation are described by the following differential equation

$$\frac{d^2\sigma}{dt^2} + \frac{4\pi^2}{T^2}\sigma = \frac{2\pi^2 R}{T^2 \delta}P(t),$$

$$T = 2\pi R_0\left[\frac{(1-v)\rho}{2E}\right]^{1/2},$$

which is essentially the equation for the classical linear oscillator, where σ is a tensile stress in the chamber wall, δ and R_0 are the chamber wall thickness and initial radius, T is the main period of chamber vibrations, and ρ, E, and v are, correspondingly, the density, Young' modulus and the Poisson coefficient of the wall material. $P(t)$ is the pressure on the chamber wall shown in Figure 3.24.

Show that if the duration time of the pressure impulse with an abrupt front is less than $0.25T$, then the maximal stress σ_m in the wall will be determined solely by the momentum

$$I = \int_0^\tau P(t)\,dt,$$

transmitted to the wall in the reflected shock wave

$$\sigma_m = \left[\frac{2E}{\rho(1-v)}\right]^{1/2}\frac{I}{\delta}.$$

If the duration of the impulse is more than T $(t > T)$, then the maximal strains σ_m will be determined by the maximal pressure in the impulse P_m:

$$\sigma_m = \frac{P_m R_0}{\delta}.$$

10. Derive the criterion described by Eq. (3.15) using the equations from the previous problem. Suppose that the linear momentum delivered to the chamber wall in the reflected wave in air can be described as $I = P_a t_a/2$, where P_a is a maximal pressure in the reflected wave and t_a is a duration of the triangular pressure pulse in air. Suppose that the thickness of the porous layer is enough to ensure a quasistatic type of chamber loading.

11. Describe the influence of a buffer structure on shock attenuation, explain why a small thickness of the porous barrier may result in shock amplification.

12. What parameters of a porous layer can result in an unexpected shock pressure increase on the wall covered by this layer with the goal of mitigating the shock wave?

13. Find the following ratio of mass M_c of a spherical chamber (radius R and wall thickness δ ($\delta \ll R$)) filled with air at normal conditions which is able to safely localize an explosive charge of TNT of mass m_{ex}

$$\frac{M_c}{m_{ex}} \approx 2\sqrt{\frac{EQ\rho}{\sigma_{al}^2 (1-\nu)}},$$

where ρ, E, and ν are, correspondingly, the density, Young's modulus, and the Poisson coefficient of the wall material, Q is the effective energy yield of explosive (MJ/kg), and σ_{al} is the allowable elastic stresses in the wall. Use a simplified expression for momentum I based on the known Q, explosive mass m_{ex}, and ignoring the mass of the air in the chamber.

Show that this ratio is about 10^3 for structural steel with a yield strength of 250 MPa if the allowable stresses are one-half of the yield strength. Use $E = 200$ GPa, $\rho = 7860$ kg/m^3, and $\nu = 0.25$.

14. Evaluate the thickness of the Porolon layer with a free surface which will decrease by five times the amplitude of the shock pressure on the adjacent rigid wall caused by an explosion of 1 kg of TNT on the distance 1 m compared with an amplitude of reflected shock from the free surface of Porolon. For a front shock pressure in air P_+ use the following equation (valid at $1 < R/W^{1/3} < 10$, see data in Persson, Holmberg, and Lee [1994]):

$$P_+ = 10(W^{1/3}/R)^2,$$

where W is the mass of explosive in TNT equivalent and R is the distance from the charge. For an amplitude P_m of reflected shock wave from the free surface of Porolon apply an equation $P_m = 6P_+$, consider the duration of the reflected impulse equal to 1 ms.

15. Find the thickness of aluminium foam, with the typical properties presented in this chapter, which will protect a structure against an explosive damage caused by 100 kg of explosive located 5 m from the structure. Then thickness of the steel buffer plate is 25 mm. For momentum per unit area I, transmitted by reflected shock wave to the buffer plate, use the equation presented by Evans, Hutchinson, and Ashby [1999]:

$$I = BW^{1/3} \exp[-C(R/W^{1/3})],$$

where W is the mass of explosive in TNT equivalent, R is the distance from the charge, $B = 5 \cdot 10^3$ sPa(kg)$^{-1/3}$, and $C = 1.3$ (kg)$^{-1/3}$/m.

References

Adishchev, V.V., and Kornev, V.M. (1979) Calculation of the Shells of Explosion Chambers. *Fizika Goreniya i Vzryva*, **15**, no. 6, pp. 108–114 (in Russian). English translation: *Combustion, Explosion, and Shock Waves*, 1980, May, pp. 780–784.

Afanasenko, S.I. (1990) Application of Porous Media for Shock Damping in Explosive Chambers. In: *Impulse Treatment of Materials* (Edited by A.F. Demchuk, V.F. Nesterenko, V.M. Ogolikhin, A.A. Shtertzer, Y.V. Kolotov, S.A. Pershin, and V.I. Danilevskaya). Special Design Ofice of High-Rate Hydrodynamics and Institute of Theoretical and Applied Mechanics, Novosibirsk, pp. 291–304 (in Russian).

Afanasenko, S.I., Grigoriev, G.S., and Klapovskii, V.E. (1990) Influence on Shock Damping of Structure and Thickness of Porous Layers. In: *Impulse Treatment of Materials* (Edited by A.F. Demchuk, V.F. Nesterenko, V.M. Ogolikhin, A.A. Shtertzer, Y.V. Kolotov, S.A. Pershin, and V.I. Danilevskaya). Special Design Ofice of High-Rate Hydrodynamics and Institute of Theoretical and Applied Mechanics, Novosibirsk, pp. 318–324.

Akhmadeev, N.H., and Bolotnova, R.Kh. (1985) Propagation of Stress Waves in Layered Media under Impact Loading (Acoustical Approximation). *Zhurnal Prikladnoi Mekhaniki i Tehknicheskoi Fiziki*, **26**, no. 1, pp. 125–133 (in Russian). English translation: *Journal of Applied Mechanics and Technical Physics*, July 1985, pp. 114–122.

Akhmadeev, N.H., Akhmadeev, R.H., and Bolotnova, R.H. (1985) About Damping Properties of the Porous Layers under Shock Loading and Spall. *Letters to the Zhurnal Tehknicheskoi Fiziki*, **11**, no. 12, pp. 709–713 (in Russian).

Akhmadeev, N.H., and Bolotnova, R.Kh. (1994) Features of Shock Wave Formation in Soft Materials Generated by Impact of a Plate. *Zhurnal Prikladnoi Mekhaniki i Tehknicheskoi Fiziki*, **35**, no. 5, pp. 19–27 (in Russian). English translation: *Journal of Applied Mechanics and Technical Physics*, 1994, **35**, no. 5, pp. 658–664.

Alekseev, Yu.F., Al'tshuler, L.V., and Krupnikova, V.P. (1971) Shock Compression of Two-Component Paraffin–Tungsten Mixtures. *Zhurnal Prikladnoi Mekhaniki i Tehknicheskoi Fiziki*, **12**, no. 4, pp. 152–155 (in Russian). English translation: *Journal of Applied Mechanics and Technical Physics*, 1973, October, pp. 624–627.

Anikiev, I.I. (1979) Reflection of the Weak Shock Waves in the Air from Plane Obstacles Made from Different Materials. *Applied Mechanics*, **15**, no. 6, pp. 128–131 (in Russian).

Ashby, M.F., Evans, A.G., Fleck, N.A., Gibson, L.J., Hutchinson, J.W., and Wadley, H.N.G. (2000) *Metal Foams: a Design Guide*, Butterworth-Heinemann, London, pp. 166-169.

Bagrovski, Ya., Fruchek, M., and Korzun, N. (1976) Some Problems Connected with Damping of Explosion Effects. In: *Proceedings of III International Symposium on Metal Explosive Working*, Marianske Lazni, Chehoslovakia, Semtin, Pardubice, Chehoslovakia, Vol. 2, pp. 515–522 (in Russian).

Bajenova, G.V., Gvozdeva, L.G. et al. (1986) *Nonstationary Interactions of Shock and Detonation Waves in Gases* (Edited by V.P. Korobeinikov). Nauka, Moscow, p. 205 (in Russian).

Baker, W.E. (1960) The Elastic–Plastic Response of Thin Spherical Shells to Internal Blast Loading. *Transactions of the ASME, J. Applied Mechanics*, March, pp. 139–144.

Barker, L.M. (1971) A Model for Stress Wave Propagation in Composite Materials. *J. Composite Materials*, **5**, pp. 140–162.

Baum, F.A., Orlenko, L.P., Stanykovich, K.P. et al. (1975) *Physics of Explosion.* Nauka, Moscow (in Russian).

Bedford, A., Drumheller, D.S., and Sutherland, H.J. (1976) On Modelling the Dynamics of Composite Materials. In: *Mechanics Today*, Vol. **3**. (Edited by S. Nemat-Nasser). Pergamon Press, New York, pp. 1–54.

Ben-Dor, G., Britan, A., Elperin, T., Igra, O., and Jiang, J.P. (1997) Experimental Investigation of the Interaction Between Weak Shock Waves and Granular Layers. *Experiments in Fluids*, **22**, pp. 432-443.

Benson, D.J., and Nesterenko, V.F. (2001) Anomolous Decay of Shock Impulses in Laminated Composites. *J. Appl. Phys.* **89**, no. 7, pp. 3622–3626.

Bogachev G.A., and Nokolaevskii, V.N. (1976) Shock Waves in the Mixture of Materials. *Mechanics of Liquid and Gas,* no. 4, pp. 113–130 (in Russian).

Buzukov, A.A. (1976) Characteristics of the Behavior of the Walls of Explosion Chambers under the Action of Pulsed Loading. *Fizika Goreniya i Vzryva*, **12**, no. 4, pp. 605–610 (in Russian).
English translation: *Combustion, Explosion, and Shock Waves*, 1977, May, pp. 549–554.

Buzukov, A.A. (2000) Decreasing the Parameters of an Air Shock Wave Using an Air-Water Curtain. *Combustion, Explosion, and Shock Waves*, **36**, no. 3, pp. 395–404.

Checkidov, P.A. (1984) Computer Modeling of Shock Propagation in Heterogeneous Elastic–Plastic Media. *Ph.D. Thesis*, Institute of Theoretical and Applied Mechanics, Novosibirsk, p. 143 (in Russian).

Chelyshev, V.P., Shehter B.I., and Shushko, L.A. (1970) Pressure Variation at the Surface of a Barrier in the Presence of a Contact Explosion. *Fizika Goreniya i Vzryva*, **6**, no. 2, pp. 217–223 (in Russian).
English translation: *Combustion, Explosion, and Shock Waves*, 1973, January, pp. 192–196.

Chhabildas, L.C., Kmetyk, L.N., Reinhard, W.D., and Hall, C.A. (1996) Launch Capabilities to 16 km/s. In: *Proceedings of Conference of the American Physical Society Topical Group on Shock Compression of Condensed Matter*, Seattle, August 13–18, 1995. (Edited by S.C. Schmidt and W.C. Tao). AIP Press, New York, pp. 1197–1200.

Davison, L., and Graham, R.A. (1979) Shock Compression of Solids. *Physics Reports*, **55**, no. 4, pp. 255–379.

Davison, L. (1998) Attenuation of Longitudinal Elastoplastic Pulses, in High-Pressure Shock Compression of Solids III, Chapter 9 (Edited by L. Davison and M. Shahinpoor). Springer-Verlag, New York, pp. 277–327.

Demchuk, A.F. (1968) Method for Designing Explosion Chambers. *Zhurnal Prikladnoi Mekhaniki i Tehknicheskoi Fiziki*, **9**, no. 5, pp. 47–50 (in Russian).
English translation: *Journal of Applied Mechanics and Technical Physics*, 1968, **9**, nos. 4–6, pp. 558–559.

Demchuk, A.F. (1971) Metallic Explosive Chambers. *Candidate of Technical Sciences Dissertation.* Lavrentyev Institute of Hydrodynamics, Novosibirsk, p. 150.

Demchuk, A.F., Kuznetzov, Y.G., and Nesterenko, V.F. (1982) Equipment for Metal Explosive Working. In: *Proceedings of II International Meeting on Materials Explosive Working,* Novosibirsk (Edited by V.F. Nesterenko, G.E. Kuzmin, and I.V. Yakovlev), pp. 202–209 (in Russian).

Deribas, A.A., Nesterenko, V.F., Sapozhnikov, G.A., Teslenko, T.S., and Fomin, V.M. (1979) Investigation of the Shock Damping Process in Metals under Contact Explosion Loading. *Fizika Goreniya i Vzryva,* **15**, no. 2, pp. 126–132 (in Russian).
English translation: *Combustion, Explosion, and Shock Waves,* 1979, September, pp. 220–225.

Dremin, A.N., and Karpuhin, I.A. (1960) Method of Determination of Shock Adiabat of the Disperse Substances. *Zhurnal Prikladnoi Mekhaniki i Tehknicheskoi Fiziki,* **1**, no. 3, pp. 184–188 (in Russian).

Dremin, A.N., and Shvedov, K.K. (1964) Determination of the Chapmen–Juge Pressure and Reaction Time at the Detonation Wave of High Explosives. *Zhurnal Prikladnoi Mekhaniki i Tehknicheskoi Fiziki,* **5**, no. 2, pp. 154–159 (in Russian).

Dremin, A.N., and Adadurov G.A. (1964) The Glass Behavior under Dynamic Loading. *Solid State Physics,* **6**, no. 6, pp. 1757–1764 (in Russian).

Evans, A.G., Hutchinson, J.W., and Ashby, M.F. (1999) Multifunctionality of Cellular Metal Systems. *Progress in Materials Science,* **43**, pp. 171–221.

Fowles G.R. et al. (1979) Acceleration of Flat Plates by Multiple Staging. In *High Pressure in Science and Technology, Proceedings of 6th AIRAPT Conference,* Vol. **2**, pp. 911–916.

Gel'fand, B.E., Gubin, C.A., Kogarko, C.M., and Popov, O.E. (1975) Investigation of the Special Characteristics of the Propagation and Reflection of Pressure Waves in Porous Medium (Foam). *Zhurnal Prikladnoi Mekhaniki i Tehknicheskoi Fiziki,* **16**, no. 6, pp. 74–77 (in Russian).
English translation: *Journal of Applied Mechanics and Technical Physics,* 1975, **16**, no. 6, pp. 897–900.

Gel'fand, B.E., Gubanov, A.V., and Timofeev, V.I. (1983) The Interaction of Shock Waves in Air with Porous Layer. *Izvestiya AN USSR, Mechanics of Liquid and Gas,* no. 4, pp. 79–84 (in Russian).

Gel'fand, B.E., Medvedev, S.P., Polenov, A.N., and Frolov, S.M. (1987) Influence of Porous Compressible Covering on the Nature of Shock-Wave Loading of a Structure. *Zhournal Tekhnicheskoi Fiziki,* **57**, no. 4, pp. 831–833 (in Russian).
English translation: *Sov. Phys. Tech. Phys.,* 1987, **32**, no. 4, pp. 506–508.

Gerasimov, A.V. (1997) Protection of an Explosion Chamber Against Fracture by a Detonation Wave. *Fizika Goreniya i Vzryva,* **33**, no. 1, pp. 131–137 (in Russian).
English translation: *Combustion, Explosion, and Shock Waves,* 1997, **33**, pp. 111–116.

Glenn, L.A. (1986) Buffed Explosions in Steel Pressure Vessels. In: *Proceedings of IX International Conference on High Energy Rate Fabrication* (Edited by V.F. Nesterenko and I.V. Yakovlev). Lavrentyev Institute of Hydrodynamics and Special Design Office of High-Rate Hydrodynamics, Novosibirsk, pp. 350–356.

Grigoriev, G.S., and Klapovskii, V.E. (1987) Chamber for Impulsive Materials Processing. *Fizika Goreniya i Vzryva*, 1987, **23**, no. 1, pp. 104–106 (in Russian).
English translation: *Combustion, Explosion, and Shock Waves*, 1987, July, pp. 9698.

Grigoriev, G.S., Klapovskii, V.E., Koren'kov, V.V., Mineev, V.N., and Shakhidzhanov, E.S. (1987) Estimation of the Efficiency of Perforated Structures in the Design of Chambers for Impulsive Treatment of Materials. *Fizika Goreniya i Vzryva*, **23**, no. 1, pp. 106–110 (in Russian).
English translation: *Combustion, Explosion, and Shock Waves*, 1987, July, pp. 98–101.

Grigoriev, G.S., Vershinin, V.Yu., and Klapovskii, V.E. (1990) The Shielding Structures for Impulse Treatment of Materials. In: *Impulse Treatment of Materials* (Edited by A.F. Demchuk, V.F. Nesterenko, V.M. Ogolikhin, A.A. Shtertzer, Y.V. Kolotov, S.A. Pershin, and V.I. Danilevskaya). Special Design Ofice of High-Rate Hydrodynamics and Institute of Theoretical and Applied Mechanics, Novosibirsk, pp. 336–340.

Guerke, G.H. (1995) Large Plastic Shear Deformation of Thin Steelplates at Impulsive Bloastoading. In: *Metallurgical and Materials Applications of Shock-Wave and High-Strain-Rate Phenomena, Proceedings of the 1995 International Conference EXPLOMET-95,* El Paso, August 6–10 (Edited by L.E. Murr, K.P. Staudhammer, and M.A. Meyers). Elsevier Science, Amsterdam, pp. 847–854.

Gupta, V., Pronin, A., and Anand, K. (1996) Mechanisms and Quantification of Spalling Failure in Laminated Composites under Shock Loading. *J. Composite Materials*, **30**, no. 6, pp. 722–747.

Gvozdeva, L.G., and Faresov, Yu.M. (1984) On Interaction of the Air Shock Wave with the Wall Covered by Porous Compressible Material. *Pis'ma v Zhurnal Tehknicheskoi Fiziki*, **10**, no. 19, pp. 87–90 (in Russian).

Gvozdeva, L.G., and Faresov, Yu.M. (1985) Calculating the Parameters of Steady Shock Waves in Porous Compressible Media. *Zhurnal Tehknicheskoi Fiziki*, **55**, no. 4, pp. 773–775 (in Russian).
English translation: *Soviet Physics—Technical Physics*, 1985, **30**, no. 4, pp. 460–461.

Hofmann, R., Andrews, D.J., and Maxwell, D.E. (1968) Computed Shock Response of Porous Aluminium. *J. Appl. Phys.*, **39**, no. 10, pp. 4555–4562.

Jih, C.J., and Sun, C.T. (1993) Prediction of Delamination in Composite Laminates Subjected to Low Velocity Impact. *J. Composite Materials*, **27**, no. 7, pp. 684–701.

Kompaneetz, A.S. (1956) Shock Waves in Plastic Compressible Media. *Doklady Akademii Nauk SSSR*, **109**, no. 1, pp. 43–52 (in Russian).

Kornev, V.M., Adishchev, V.V., Mitrofanov, A.N., and Grekhov, V.A. (1979) Experimental Investigation and Analysis of the Vibrations of the Shell of an Explosion Chamber. *Fizika Goreniya i Vzryva*, **15**, no. 6, pp. 155–157 (in Russian).
English translation: *Combustion, Explosion, and Shock Waves*, 1980, May, pp. 821–824.

Kostykov, N.A. (1980) Criteria of Tangential Shock Wave Increase with the Help of Porous Materials. *Fizika Goreniya i Vzryva*, **16**, no. 5, pp. 78–80 (in Russian).

English translation: *Combustion, Explosion, and Shock Waves*, 1981, March, pp. 548–550 .

Kozyrev, A.S., Kostyleva, B.E., and Ryazanov, V.T (1969) Cumulation of Shock Waves in Lamellar Media. *Zhurnal Experimentalnoi i Teoreticheskoi Fiziki*, **56**, no. 2, pp. 427–432 (in Russian).
English translation: *Soviet Physics JETP*, 1969, **29**, no. 2, pp. 233–234.

Kravets, V.V., Balanovskii, V.F., Zinchenko, V.V. et al. (1962) Efficiency of Water Curtain for Direct Counteraction to an Air Shock Wave. *Ugol' Ukrainy*, **5**, pp. 38–41 (in Russian).

Kroshko, E.A., and Chubarova, E.V. (1980) Numerical Modeling of High Velocity Impact on Laminate. In: *Proceedings of VI All-Union Conference on Computer Modelling of Elasticity and Plasticity Problems* (Edited by V.M. Fomin). Institute of Theoretical and Applied Mechanics, Novosibirsk, Part 1, pp. 91–104 (in Russian).

Kudinov, V.M., Palamarchuk, B.I., Gel'fand B.E., and Gubin S.A. (1976) The Appication of the Foam for Shock Waves Damping at the Explosive Welding and Cutting. In: *Proceedings of III International Symposium on Metal Explosive Working*, Marianske Lazni, Chehoslovakia, Semtin, Pardubice, Chehoslovakia, Vol. **2**, pp. 511–514 (in Russian).

Kuznetsov V.M., and Shatsukevich, A.F. (1977) On the Interaction of Detonation Products with the Walls of Explosively Created Cavity in Soil and Rocks. *Fizika Goreniya i Vzryva*, **13**, no. 5, pp. 733–737 (in Russian).

Landau, L.D., and Lifshitz E.M. (1976) *Mechanics*, 3d ed. translated from the Russian by J.B. Sykes and J.S. Bell. Pergamon Press, Oxford.

Laptev, V.I., and Trishin, Yu.A. (1974) The Increase of Velocity and Pressure under Impact upon Inhomogeneous Target. *Zhurnal Prikladnoi Mekhaniki i Tehknicheskoi Fiziki*, **15**, no. 6, pp. 128–132 (in Russian).
English translation: *Journal of Applied Mechanics and Technical Physics*, 1976, March, pp. 837–841.

Lee, M.P., Wang, G.M., Sung, P.H. et al. (1986) The Attenuation of Shock Waves in Foam and its Application. In *Shock Waves in Condensed Matter*. Plenum Press, New, pp. 687–692.

Lundergan, C.D., and Drumheller, D.S. (1971) Propagation of Stress Waves in a Laminated Plate Composite. *Journal of Applied Physics*, **42**, no. 2, pp. 669-675.

Neal, N. (1980) Explosive Containment. In: *High Pressure Science and Technology, Proceedings of 7th International AIRAPT Conference,* Le Creusot, France, vol.2, Oxford, p. 1077–1079.

Nesterenko, V.F. (1977) Shock Compression of Multicomponent Materials. In *Dynamics of Continua. Mechanics of Explosive Processes*, **29**, pp. 81–93 (in Russian).

Nesterenko, V.F., and Kuznetsov, Yu.G. et al. (1979) The Testing of the Model Device for the Working Process. Intermediate Report, No. 35–79, Special Design Office of High-Rate Hydrodynamic, Novosibirsk, p. 24 (in Russian).

Nesterenko, V.F., Fomin, V.M., and Cheskidov, P.A. (1983a) Attenuation of Strong Shock Waves in Laminate Materials. In: *Nonlinear Deformation Waves* (Edited by U. Nigul and J. Engelbrecht). Springer-Verlag, Berlin, pp. 191–197.

Nesterenko, V.F., Fomin, V.M., and Cheskidov, P.A. (1983b) Damping of Strong Shocks in Laminar Materials. *Zhurnal Prikladnoi Mekhaniki i Tehknicheskoi Fiziki*, **24**, no. 4, pp. 130–139 (in Russian).

English translation: *Journal of Applied Mechanics and Technical Physics*, January 1984, pp. 567–575.

Nesterenko, V.F. et al. (1984) The Investigation of Possibility to Reduce Mass of Device for Explosion Localization.Testing. Final Report, No. GR 79053160, Special Design Office of High-Rate Hydrodynamic, Novosibirsk, p. 61 (in Russian).

Nesterenko, V.F. et al. (1986) The Development of the Method of Protection of the Concrete Plates from Spall. Final Report, No. 02870009933, Special Design Office of High-Rate Hydrodynamic, Novosibirsk, p. 61 (in Russian).

Nesterenko, V.F. (1988) Nonlinear Phenomena under Impulse Loading of Heterogeneous Condensed Media. Dissertation for the Defense of Doctor of Physics and Mathematics Degree, Novosibirsk, Lavrentyev Institute of Hydrodynamics, p. 369 (in Russian).

Nesterenko, V.F., Afanasenko, S.I., Ipatiev, A.S., Klapovski, V.E., and Grigoriev, G.S. (1988) Shock Impulse Transformation by Laminar Porous Materials. In: *Proceedings X All-Union Conference on Computer Modeling of Elasticity and Plasticity Problems* (Edited by V.M. Fomin). Institute of Theoretical and Applied Mechanics, Novosibirsk, pp. 221–230 (in Russian).

Nesterenko, V.F., Fomin, V.M., and Cheskidov, P.A. (1988) The Limits of Continuum Models for Powder Behavior under Intense Shock-Wave Loading. In: *Proceedings X All-Union Conference on Computer Modeling of Elasticity and Plasticity Problems* (Edited by V.M. Fomin). Institute of Theoretical and Applied Mechanics, Novosibirsk, 1988, pp. 231–236 (in Russian).

Nesterenko, V.F., Afanasenko, S.I., Cheskidov, P.A., Grigoriev, G.S., and Klapovski V.E. (1990) Computer Optimization of Laminar Protective Screens from Metallic and Porous Layers. In: *Impulse Treatment of Materials* (Edited by A.F. Demchuk, V.F. Nesterenko, V.M. Ogolikhin, A.A. Shtertzer, Y.V. Kolotov, S.A. Pershin, and V.I. Danilevskaya). Special Design Ofice of High-Rate Hydrodynamics and Institute of Theoretical and Applied Mechanics, Novosibirsk, pp. 325–335.

Nesterenko, V.F. (1992) *High-Rate Deformation of Heterogeneous Materials*. Nauka, Novosibirsk.

Nigmatullin, R.I., Vainshtein P.B. et al. (1976) Computer Modelling of Physico-Chemical Processes and Shock Wave Propagation in Solids and Composite Materials. *Computer Methods of Mechanics of Continua Media*, **7**, no. 2, pp. 89–108 (in Russian).

Ogarkov, V.A., Purygin, P.P., and Samylov, S.V. (1969) Simple Model of Complex Systems for Obtaining High Velocity. In *Detonation*, Institute of Chemical Physics, Chernogolovka, pp. 155–158 (in Russian).

Pavlovskii, M.N. (1976) The Measurements of the Sound Speed in the Shock Loaded Quartsite, Angidride, NaCl, Paraffine, Plexiglass, Polyethylene, and Teflon. *Zhurnal Prikladnoi Mekhaniki i Tehknicheskoi Fiziki*, **17**, no. 5, pp. 134–139 (in Russian).
English translation: *Journal of Applied Mechanics and Technical Physics*, 1976, **35**, no. 5, pp. 658–664.

Peck, I.C. (1971) Pulse Attenuation in Composites. In: *Shock Waves and the Mechanical Properties of Solids*, Syracuse University Press, New York, pp. 150–165.

Persson, P-A., Holmberg, R., and Lee, J. (1994) *Rock Blasting and Explosives Engineering*. CRC Press, Boca Raton, p. 540.

Raspet, R., Butler, P.B., and Jahani F. (1987) The Reduction of Blast Overpressures from Aqueous Foam in a Rigid Confinement. *Applied Acoustics*, **22**, pp. 35–45.

Raspet, R., Butler, P.B., and Jahani F. (1987) The Effect of Material Properties on Reducing Intermediate Blast Noise. *Applied Acoustics*, **22**, pp. 243–259.

Raspet, R., and Griffiths, S.K. (1983) The Reduction of Blast Noise with Aqueous Foam. *J. Acoust. Soc. Am.*, **74**, no. 6, pp. 1757–1763.

Riney, T. D. (1970) Numerical Evaluation of Hypervelocity Impact Phenomena. In *High Velocity Impact Phenomena* (Edited by R. Kinslow). Academic Press, New York, pp. 157–212.

Sapojnikov, G.A., and Fomin, V.M. (1980) Computer Modeling of Nonrestricted Cumulation in Laminar Materials. In: *Computer Methods of Solving Problems in Elasticity and Plasticity, Proceedings of the All Union Conference* (Edited by V.M. Fomin). Institute of Theoretical and Applied Mechanics, Novosibirsk, Part 1, pp. 91–104 (in Russian).

Steinberg D.J. (1996) Equation of State and Strength Properties of Selected Materials. Preprint, Lawrence Livermore Laboratory, 1996, UCRL-MA-106439, p. 2/96.

Thouvenin, J. (1966) Action d'une onde de choc sur un solide poreux. *J. de Physicue*, **27**, nos. 3/4, pp. 183–189.

Voskoboinikov, I.M., Gogulya, M.F., Voskoboinikova, N.F., and Gel'fand, B.E. (1977) The Possible Scheme for Description of the Shock Compression of the Porous Samples. *Doklady Academii Nauk*, **236**, no. 1, pp. 75–78 (in Russian). English translation: *Soviet Physics–Doklady*, 1977, **22**, no. 9, pp. 512–514.

Whenhui, Z., Honglu, X., Guangquan, Z., and Schleyer, G.K. (1997) Dynamic Response of Cylindrical Explosive Chambers to Internal Blast Loading Produced by a Concentrated Charge. *Int. J. Impact Engng.*, **19**, nos. 9–10, pp. 831–845.

Wierzbicki, T., and Abramowicz, W. (1994) *Blast and Ballistic Loading of Structures*. Butterworth-Heinemann, Oxford.

Wilkins, M. (1964) Calculation of Elastic–Plastic Flow. In: *Methods in Computational Physics, Fundamental Methods in Hydrodynamics*, Vol. **3**. Academic Press, New York, p. 211.

Wilkins, M. (1999) *Computer Simulation of Dynamic Phenomena*. Springer-Verlag, Berlin, p. 246.

Zababakhin, E.I. (1965) Shock Waves in Layered Systems. *Zhurnal Experimentalnoi i Teoreticheskoi Fiziki*, **49**, pp. 642–646 (in Russian). English translation: *Soviet Physics JETP*, 1966, **22**, no. 2, pp. 446–448.

Zababakhin, E.I. (1970) The Unlimited Cumulation Phenomenon. In: *Mechanics in the USSR in Fifty Years*, Vol. **2**. Nauka, Moscow (in Russian).

Zagorel'skii, V.I., Stolovich, N.N., and Fomin, N.A. (1982) Impulse Piezoelectric Gauge with Amplifier for Measurements of Dynamic Pressures. *Ingenerno-Phisicheskii Jurnal*, **2**, pp. 303–306 (in Russian).

Zel'manov, I.L., Kolkov, O.S., Tihomirov, A.M., and Shatsukevich, A.F. (1968) The Motion of Sand under Explosion. *Fizika Goreniya i Vzryva*, **4**, no. 1, pp. 116–121 (in Russian).

4
Shear Localization and Shear Bands Patterning in Heterogeneous Materials

4.1   Introduction

Shear localization plays an important role in the process of high-speed metal cutting, armor resistance to high-velocity impact, and in the break-up of penetrators. The high-strain-rate plastic flow of fractured ceramics during the ballistic impact of ceramic armor plays an important role in governing the depth of penetration, as was shown in the model calculations by Curran, Seaman, Cooper, and Shockey [1993]. Another example is initiation by shear bands of energetic materials under high-strain-rate plastic flow, which is unstable at high strains for most materials.

It is important to establish experimentally and theoretically the main features of the shear localization process (which is inevitable under high strains) in different heterogeneous materials.

Bridgman [1935], Dremin and Breusov [1968], Graham [1993], Batsanov [1993], Enikolopyan [1988] and others experimentally demonstrated the initiation of chemical reactions and phase transformations under quasistatic and dynamic pressure + shear conditions. Winter and Field [1975], Frey [1985], and Kipp [1981] used the shear localization concept as a feasible mechanism of explosive initiation. Frey [1985] emphasized the importance of pressure and shear strain for the initiation of energetic viscous and pressure-dependent materials.

At the same time, there is a lack of experimental information on the details of shear band formation in heterogeneous, porous materials, and on the material behavior within shear bands. The main emphasis of the theoretical and experimental research was concentrated on the behavior of single shear bands. At the same time, the collective behavior of interacting shear bands can be crucial for the material performance. Nesterenko, Meyers et al. [1994] proposed a method providing a strain-controlled reproducible high-strain-rate array of shear bands in heterogeneous porous mixtures. This allows for the investigation of the specific shear band with different strains in a single experiment and also a self-organization of a large number of shear bands (sometimes a few hundred) on different stages of their development. Material structure inside the shear band as a function of the shear strain can also be clarified depending on the precisely tuned overall strain. This prompts a look at the different stages of chemical reactions and the microscopic details of material comminution in high-strain-rate localized shear.

In this chapter the results of the application of the Thick-Walled Cylinder (TWC) method are reviewed for the shear localization and shear band patterning in solid polycrystals, inert ceramics (comminuted and granular armor ceramics Al_2O_3 and SiC with different particle sizes and porosities), and in reactive porous mixtures like Nb + Si, Ti + Si, and Ti + C.

4.2   Thick-Walled Cylinder (TWC) Method

4.2.1   Experimental Set-Up

4.2.1.1   Solid Samples

Most experimental methods are designed to investigate an initiation and development of a single shear band (Bai and Dodd [1992]). At the same time the collective behavior of shear bands is very important for the material response and the appropriate methods must be developed. These methods cannot be one-dimensional and should allow for a relatively simple, initially macroscopically uniform state of strain and stress. The simplest, next to the one-dimensional geometry of experiment, is a cylindrical axisymmetrical geometry of test. It is used in the Thick-Walled Cylinder (TWC) method proposed by Nesterenko et al. [1989], [1994] for solid materials, and later developed for porous inert and reactant mixtures by Nesterenko, Meyers, and coworkers [1994], [1995].

Figure 4.1(a) shows the experimental configuration used to produce the radial collapse of the metallic specimens. The system uses the controlled detonation of an explosive to generate the pressures required for the collapse of a thick-walled cylinder. The metallic specimens are placed within a copper driver tube; the system extremities are composed of steel plugs. The explosive is placed coaxial with the specimen and detonation is initiated at the top, propagating along the cylinder axis.

An internal copper tube establishes the maximum collapse of the sample. In the absence of this tube, total collapse of the sample is obtained and the final radius is ~0.5 mm. The copper tubes with different initial radii were used to arrest the collapse of the sample at different stages. Annealed polycrystalline OFHC copper with grain size 100 μm or less was chosen for driver and stopper tubes because of the high resistance of this material to shear localization.

Instead of copper driver and stopper tubes, tantalum tubes which can dissipate more energy during collapse can also be used. Some advantage with this material may be connected with its high melting temperature and latent heat resulting in lower melting volume in the center of the set up. Higher density, and strength of

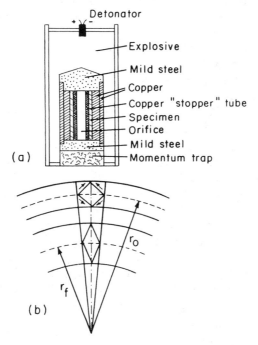

Figure 4.1. (a) Experimental configuration for the collapse of a specimen in the geometry of a TWC method and (b) the pure shear deformation of an element as the tube collapses. (Reprinted from *Acta Materialia*, **46**, Nesterenko, V.F., Meyers, M.A., and Wright T.W., Self-Organization in the Initiation of Adiabatic Shear Bands, pp. 327–340, 1998, with permission from Elsevier Science.)

the tantalum driver and stopper tube can ensure soft recovery at higher initial velocities (resulting in a higher strain rate) and higher dynamic confined pressures, which depend on the density and velocity (Nesterenko [2000]).

One of the very important features of this method is the relatively large size of the sample, allowing it to support uniform boundary conditions on the outside surface of the sample irrespective of the shear band patterning on the inside surface. The number of shear bands can be very large up to a few hundreds and the pattern development can be "frozen" at different stages with a very high accuracy of prescribed final strain.

The use of an inner copper tube allows us to make a "soft" termination of the collapse process and to preserve the details of the shear band geometry, especially the size of displacements on the inner radius of the sample. For example, in experiments with titanium (Nesterenko, Meyers, and Wright [1998]), shear bands were not opened into cracks as was the case in experiments without the inner copper tube (Nesterenko et al. [1995]). This also preserves the details of the shear band structure and ensures that the kinetics of the shear bands development is not replaced by the propagation of the cracks originating from the shear bands during elastic unloading. To arrest the collapse at different radii, central rods with different radii were also used. However, the impact of the titanium inner surface, having a structure developed by shear bands, on the rod destroys details of the shear bands. But the rods inside the inner copper stopper tube were successfully used as an additional method of strain control.

The explosive parameters in the TWC method (detonation velocity, density, and thickness) were carefully selected to provide a "smooth" pore collapse; wave reflection effects are minimized and spalling of the internal cylinder surface is nonexistent. A low detonation explosive ($D = 4000$ m/s) with initial density 1 g/cm^3 was used in the experiments. A description of the method can be found in papers by Nesterenko et al. [1989], [1994], [1995], [2000]. The velocity of the inner wall of the tube was measured by an electromagnetic gauge as described by Nesterenko and Bondar [1994a].

A simple model to calculate the kinematics of implosion driven by explosion can be used (Nesterenko and Benson [1999]). In this model, the hypothesis of instantaneous detonation is applied because the velocity of the collapse is by the order of magnitude less than the detonation speed. The initial conditions in the explosive products correspond to the uniform pressure P_D, the density ρ_D, and the sound speed c_D being equal, correspondingly:

$$P_D = \frac{\rho_0 D^2}{k+1}, \qquad \rho_D = \frac{\rho_0(k+1)}{k}, \qquad c_D = \frac{k}{k+1}D, \tag{4.1}$$

where ρ_0 is the initial density of explosive, D is the detonation speed and k is the coefficient in the polytropic law for detonation products. The dependence of the sound speed c on the current density ρ in expanding detonation products has the following form:

$$c = c_D \left(\frac{\rho_D}{\rho}\right)^{1-k}. \tag{4.2}$$

The material of the cylinder is considered to be incompressible, and elastic–plastic with the strain, strain rate, and temperature-dependent strength.

This approach resulted in very good agreement of the calculations with a measured time of collapse 8 μs for the standard conditions of the TWC test for collapse of a copper cylinder with cavity diameter 11 mm and outside diameter 30 mm. The outside diameter of the explosive was equal to 60 mm, inside diameter 30 mm, $D = 4000$ m/s, and $\rho_0 = 1000$ kg/m^3, and $k = 2.5$. Numerical calculations demonstrated that the time of collapse practically does not depend on the mechanical property of the sample with the inside and outside radii equal to 7 mm and 10.5 mm, the latter being installed between the driver cylinder with an outside radius of 15–16 mm and a stopper cylinder with cavity radius 5.5 mm. In all assemblies the ratio of the explosive mass to the mass of the composite cylinder was constant.

Figure 4.2 shows the (a) initial, and (b), (c) final configurations of the

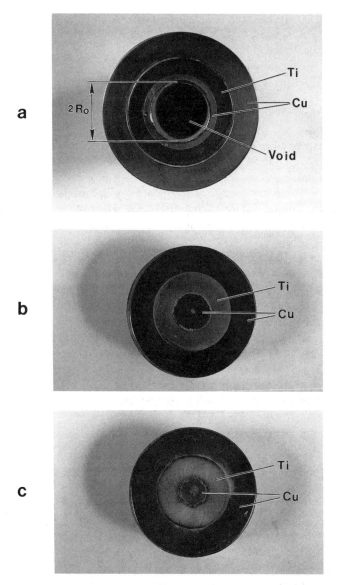

Figure 4.2. (a) Initial and (b),(c) collapsed configurations of titanium specimens: (b) $R_0 = 5.5$ mm; (c) $R_0 = 6.25$ mm. (Reprinted from *Acta Materialia,* **46**, Nesterenko, V.F., Meyers, M.A., and Wright T.W., Self-Organization in the Initiation of Adiabatic Shear Bands, pp. 327–340, 1998, with permission from Elsevier Science.)

specimen cross-sections. Samples of commercial purity titanium (Grade 2, Altemp Alloys, CA) machined from 2 in bar stock, with equiaxed grains having an average size of 20 μm were used in these experiments (Nesterenko, Meyers, and Wright [1998]). The copper "stopper" tubes, which have initial internal radii R_o equal to 5.5 mm and 6.25 mm, are completely collapsed by the implosion, subjecting the titanium tube to two levels of plastic strain. Results for experiments with niobium, aluminum, stainless steel, Ti–6Al–4V alloy, tantalum, Teflon and monocrystale copper are presented in papers by Nesterenko and Bondar [1994], Nesterenko, Meyers et al. [1997], Nemat-Nasser, Okinaka, and Nesterenko [1998], Nemat-Nasser, Okinaka, Nesterenko, and Liu [1998], and Xiu, Nesterenko, and Meyers [2000].

The collapse can also be driven by an impulse magnetic field with detailed measurements of collapse geometry by the impulse X-ray technique. The great advantage of this type of driver is the minimization of axial movement and very well-controlled boundary conditions represented by the magnetic field (pressure) history on the outside surface of the driver tube. The results obtained with this method with thin aluminum alloy shells (Stokes, Oro et al. [1999]) as well as in the geometry of the TWC method for titanium and stainless steel 304 (Stokes, Nesterenko et al. [2001]) are very encouraging from a material science point of view. For example, the dynamic of collapse of titanium and stainless steel 304 was apparently different despite the domination of the total mass of the assembly by the copper driver. This may be associated with the damage induced by earlier shear localization in titanium, as will be shown below.

Another method suitable for the investigation of the collective behavior of shear bands and the subsequent formation of a crack pattern is the "contained fragmented round" or "contained exploding cylinder" (CEC) technique developed by Erlich, Seaman, Shockey, and Curran [1977], and used by Staker [1980], [1981], Semiatin, Staker, and Jonas [1984], Shockey [1986], and others.

4.2.1.2   Prefractured Ceramic

The TWC method was used to investigate the high-strain-rate deformation of prefractured ceramics. For example, the hot-pressed silicon carbide (SiC-I, SiC-II, and SiC-III) specimens manufactured by Cercom, Inc., were tested under identical conditions (Shih, Nesterenko, and Meyers [1998]). Properties of different types of SiC are presented in Table 4.1. AD-998 alumina with specific gravity 3.9 g/cc and crystal grain size 2–50 microns (average 15–30 microns) was manufactured by the Coors Ceramics Company. According to Coors's specifications AD-998 alumina had a flexural strength 331 MPa (temperature 25°C), compressive strength 2071 MPa, and modulus of elasticity 345 GPa.

The modification of the TWC method in this case consists of two explosive events: the first event to fracture the ceramic and the second event to deform the fragmented ceramic (Chen, Meyers, and Nesterenko [1995]; Nesterenko, Meyers, and Chen [1996]).

The experimental steps are outlined in Figure 4.3. A silicon carbide cylinder (16 mm inner diameter and 22 mm outer diameter) was assembled with a copper insert (14.5 mm diameter) and a copper sleeve (23 mm inner diameter and 36 mm outer diameter). A mixture of 3 : 1 volume ratio of ammonite and sand was

Table 4.1. Quasistatic properties of hot-pressed silicon carbide.

	SiC-I	SiC-II	SiC-III
Density (g/cm^3)	3.18	3.20	3.21
Average grain size (μm)	5.6	4.1	4.1
Flexural Strength (MPa)	380	542	477
Weibull modulus	5.9	10.9	11.2
Fracture toughness (MPam$^{1/2}$)	2.5	4.1	4.3
Hardness [Knoop 300g] (GPa)	23.9	22.9	22.6
Young's modulus (GPa)	440	450	450
Poissons' ratio	0.17	0.17	0.17

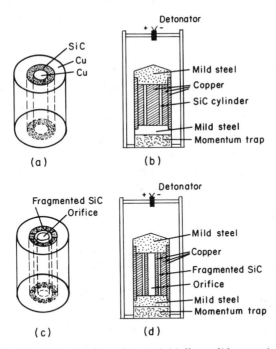

Figure 4.3. Experimental procedures for an initially solid ceramic: (a) specimen assembly; (b) first explosive event-fragmentation of ceramic; (c) orifice drilling; (d) second explosive event-deformation of fragmented ceramic. (Reprinted from *J. Appl. Phys.*, **83**, Shih, C.J., Nesterenko, V.F., and Meyers, M.A., High-Strain-Rate Deformation and Comminution of Silicon Carbide, pp. 4660–4671, 1998.)

used to generate an explosion of low-detonation velocity (3.2 km/s) to fracture the SiC cylinder. Detonation was initiated at the top of the charge and propagated along the cylinder axis. After this explosion, the ceramic remained cylindrical, and contained cracks and incipient shear bands.

A cylindrical orifice (11 mm diameter) was then drilled in the center of the copper insert. The specimen then underwent a second explosive event using 100% ammonite to achieve a detonation velocity of 4.2–4.4 km/s to collapse the center orifice. This explosive event produced large inelastic deformation. The objective was to simulate the flow of the fragmented ceramic under confinement, which in this method was provided by the inertial compressive stresses at the boundary between the ceramic and the copper sleeve, with an amplitude of about 100 MPa.

A similar procedure was applied for two types of solid alumina (9AD-998) tubes with initial inside and outside diameters equal to 14.3 mm (19.05 mm) and 15.875 mm (22.225 mm) (Chen et al. [1996]). The procedure to calculate the global strain is outlined below.

Klopp, Shockey, Curran, and Cooper [1998] and Nesterenko, Klopp et al. [2000] used a different modification of the TWC method. A tapered brass pin was inserted into the central hole, allowing the variation of radial and hoop strains along the length of the assembly. They also did not introduce the preliminary step of prefracturing.

4.2.1.3   Granular Materials (Inert and Reactive)

The TWC method was also applied for granular materials, inert and reactive. For example, silicon carbide powders with three different particle sizes were investigated (Shih, Meyers, and Nesterenko [1997], [1998]). The fine powder was produced by H.C. Starck (Grade UF-15), the medium powder was fabricated by Nagasse (Grade GMF-6S), and the coarse powder was manufactured by Norton (Grade F230). The average particle sizes for the small, medium, and coarse powders were 0.4, 3, and 50 μm, respectively. All powders have a wide range of particle size distribution; nevertheless, these three powders represent particle sizes of two orders of magnitude of difference. These powders all have the α-SiC crystal structure, and "chunky" morphology.

Experiments with granular silicon carbide were conducted on three fine, two medium, and three coarse powder specimens, which were tested under identical conditions. The procedure consists of two explosive events: the first event to densify the ceramic, and the second event to deform the densified ceramic.

The experimental procedures are outlined in Figure 4.4, and are described in detail by Nesterenko et al. [1996]. Silicon carbide powder was loaded into a tubular cavity made up by a central copper rod (14.5 mm diameter) and an outer copper tube (24.6 mm inner diameter and 38 mm outer diameter). A mixture of 3 : 1 volume ratio of ammonite and sand was used to generate an explosion of low-detonation velocity (3.2 km/s) to densify the SiC cylinder. A cylindrical orifice (11 mm diameter) was then drilled in the center of the copper insert. The

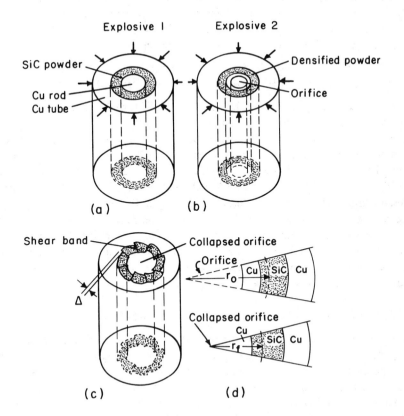

Figure 4.4. Experimental steps for the densification and deformation of granular materials: (a) explosive event 1: densification; (b) explosive event 2: deformation; (c) final configuration; (d) initial and final radii during deformation. (Reprinted from *Acta. Materialia,* **46,** Shih, C.J., Nesterenko, V.F., and Meyers, M.A., High-Strain-Rate Deformation of Granular Silicon Carbide, pp. 4037–4065, 1998, with permission from Elsevier Science.)

specimen then underwent another explosive event using 100% ammonite to achieve a detonation velocity of 4.2–4.4 km/s to collapse the center orifice. This explosive event produced a large inelastic deformation and profuse shear bands. The global strains in the periphery of the sample can be obtained from the strains in the incompressible copper.

After collapse, the composite cylinders were sectioned with a diamond saw and polished by diamond paste. The sections were inserted in a vacuum desiccator and impregnated with epoxy to increase the resistance of the granules and fragments to pull-out. As-received powders were also subjected to the same sectioning and polishing process and the results indicated that the specimen preparation process did not contribute to the observed particle fracturing.

Alumina powders with two particle sizes were investigated in similar manner (Nesterenko et al. [1996]). These powders had a purity level of 99.97% and contained trace amounts of Na, Si, Ca, Ga, Fe, Mg, Ti, and Zn. The powders were donated by CERALOX and have two designations:

(a) APA 3.0—median particle size: 4.17 μm; surface area: 1.5 m^2/g.
(b) APA 0.5—median particle size: 0.37 μm; surface area: 8.9 m^2/g.

These two alumina powders will be denoted "small" (or ~0.4 μm) and "large" (or ~4 μm) in the subsequent discussion.

Reactive mixtures Nb + Si, Ti + Si, Ti + graphite and Ti + ultrafine diamond were investigated with the help of TWC method according to the procedure depicted in Figure 4.4. The niobium, titanium, silicon, and graphite powders were purchased from CERAC and had sizes of 325 mesh (<44 μm) and were of high purity (>99.5%) and irregular shape. These powders were selected because of the previous research in which the reaction between niobium and silicon under shock compression had been thoroughly investigated (Vecchio, Yu, and Meyers [1994]). Ultrafine diamond powder was produced by detonation synthesis and was chemically purified to provide a diamond content ~96% with particle size ~5 nm (Mal'kov and Titov [1995]).

4.2.1.4   Forced Shear Localization in Reactive Materials

Some materials, especially if they are porous or contain "soft" components like graphite, resist "spontaneous" shear localization at relatively small strains. The amorphous foils adjacent to the reactive mixture in the geometry of the TWC method can be used to force the shear localization in the porous reactive materials at very low strains (Nesterenko, Meyers, Chen, and Lasalvia [1994]). This is due to the fact that amorphous materials, because of a lack of the strain hardening effect, are very unstable under shear deformation. This technique can be useful, for example, in the investigation of explosive initiation by localized shear, because dynamic loading can be accompanied by forced localized shear independent of the properties of the explosive.

4.2.2   Estimation of Main Parameters for the TWC Test

4.2.2.1   Strains

The TWC test is strain controlled where high-strain, high-strain-rate deformation may be precisely tuned. The state of strain generated within the collapsing cylinder before shear localization starts is one of pure shear. This is shown in Figure 4.1(b), in which the distortion of an elemental cube of radius r_0 is followed as it moves toward the axis of the cylinder. The strain in the axial direction is zero, and there is no rotation of the elemental cube. The planes of maximum shear lie at 45° to a radius if the material is not pressure sensitive, as indicated by r_0 in Figure 4.1(b).

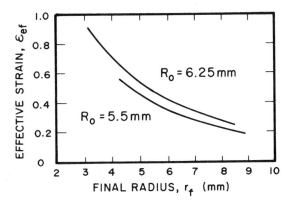

Figure 4.5. Effective strain in titanium as a function of radius in partially collapsed geometry for two configurations: $R_0 = 5.5$ mm (smaller collapse strain) and $R_0 = 6.25$ mm (larger collapse strain). (Reprinted from *Acta Materialia,* **46**, Nesterenko, V.F., Meyers, M.A., and Wright T.W., Self-Organization in the Initiation of Adiabatic Shear Bands, pp. 327–340, 1998, with permission from Elsevier Science.)

As the tube is collapsed inward, its inner surface experiences increasing strains, which tend to infinity as radius of the cavity $R_f \to 0$. The radial, tangential, and axial strains for this geometry are given by

$$\varepsilon_{rr} = \ln\left(\frac{r_0}{r_f}\right), \quad \varepsilon_{\phi\phi} = -\ln\left(\frac{r_0}{r_f}\right) = -\varepsilon_{rr}, \quad \varepsilon_{zz} = 0. \tag{4.3}$$

In many cases, even with samples from initially porous prefractured ceramics, material can be considered as incompressible in the first approximation. Then radii r_0 and r_f (Fig. 4.1(b), Fig. 4.4(d)), representing the initial and final positions of a general point, and the final radii of cavities R_0 and R_f are related by the conservation of mass, if deformation is uniform:

$$r_0^2 - R_0^2 = r_f^2 - R_f^2. \tag{4.4}$$

The effective strains at the points with radius r_f are

$$\varepsilon_{ef} = \frac{2}{\sqrt{3}}\varepsilon_{rr} = \frac{2}{\sqrt{3}} \ln\frac{r_0}{r_f}. \tag{4.5}$$

The strains for the material points being collapsed to different final radii r_f are depicted in Figure 4.5 for the initial cavity radii $R_0 = 5.5$ mm and $R_0 = 6.25$ mm (the initial inner and outer radii of the samples are 7 mm and 10.5 mm, correspondingly, for both cases). Each line designates the strain of a material point as it converges inward. The extremity of the line designates the radius and strain of the inside surface, after the collapse was arrested at R_f equal to 3.1 mm and 4.3 mm, correspondingly.

Equations (4.3)–(4.5) can also be used after the beginning of shear localization outside the area with shear bands. This technique is well suited for the study of both shear band initiation and propagation because it allows precise tuning of the final global strain in the sample.

The local shear strains inside an individual shear band can be evaluated based on the displacement in the shear band Δ (Fig. 4.4(c)) and its thickness.

And, finally, partition of the total plastic strains between uniformly distributed in the volume of the sample and localized inside shear bands can be found, based on the number of shear bands and their corresponding displacements Δ (Nesterenko et al. [1996]). These experiments also reveal the shape, spacing, and configuration of shear bands.

4.2.2.2   Strain Rates

The main step in establishing strain rates is connected with the determination of the velocity field in the sample. If material is incompressible, then the velocity field in the whole assembly can be obtained from the velocity of the wall of the cavity or from the velocity of the outside surface of the copper driver. Differentiating Eq. (4.4) (where R_f and r_f should be replaced by some intermediate radii R and r) with respect to time, we obtain the relationship between the radial velocity υ_R of the inner cavity and the velocity of some point υ_r with current radius r:

$$\upsilon_r = \frac{R}{r}\upsilon_R. \tag{4.6}$$

From this equation the strain-rate field can be obtained for any position inside the assembly. For relatively small strains we obtain the following equations taking into account zero strain along the cylinder axis

$$\varepsilon_{rr} \approx \frac{d\varepsilon_{rr}}{dt} = -\frac{R}{r^2}\upsilon_R = -\frac{\left(R_0 + \int_0^t \upsilon_R dt\right)}{r^2}\upsilon_R, \tag{4.7}$$

$$\varepsilon_{rr} = -\varepsilon_{\phi\phi}.$$

This means that measurement of the inner velocity is enough to provide information for strains and strain rates inside the material. The same is true if the velocity of the outside surface is measured. In this equation, velocity directed toward the center is negative.

The radial velocity of the cylindrical cavity $\upsilon(t)$ can be measured by an electromagnetic technique (Nesterenko et al. [1994]). The insertion of samples, from different materials inside a copper driver tube, does not significantly change the time of collapse in comparison with the uniform copper cylinder having the same geometrical dimensions (Nesterenko and Benson [1999]). This is because the replacement of the central part of the copper by sample material does not essentially change the overall mass which is dominated by the mass of the outside copper of the assembly. That is why the velocity data obtained for a monolithic copper cylinder can be used, as a first approximation, to calculate the strain rate. The calculated strain rate is seen to fluctuate around the corresponding mean values and the variation (±15%) is not significant (Nesterenko et al. [1998]; Nesterenko and Benson [1999]). That is why, to a first approximation, the strain rates for these material points can be considered as constant and equal to $3.5 \cdot 10^4$ s^{-1}.

It should be mentioned that the development of continuous measurements of the kinematics of the inner and outer surfaces of the assembly with large strains is very desirable because it is the simplest two-dimensional test.

4.2.2.3   Stresses

The stresses in this method should be found experimentally or numerically, calculated based on the measured kinematics of implosion and on the corresponding constitutive equations of the materials in the set-up. The situation is the same as in the Taylor test. The advantage of the TWC method is connected with a very simple state of strain (pure shear in the sample bulk during uniform flow and simple shear inside the shear band after instability starts). Also in the TWC method, stress gauges can be used more effectively, due to the same reason.

As was already discussed in this section, material of the sample can be considered incompressible at the investigated level of pressures. The role of the shock wave with amplitude, about 2 GPa, is mainly accomplished by providing initial momentum to the collapsing cylinder, without essentially structural changes in most cases. This is connected with relatively small strains in the shock wave with this small amplitude. The main dynamic input to stresses is connected with inertial effects due to the gradient of the particle velocity in the cylinder.

These "inertia-caused" stresses can be evaluated based on the model of an incompressible ideal liquid with density equal to the density of the material. In this case a pressure distribution along a radius in a collapsing cylinder with zero pressure on its surfaces (this corresponds qualitatively to the stage when the pressure in the detonation products is close to zero due to the rarefaction wave from their interface with air) is expressed by the following equation (Knoepfel [1970]):

$$p(r, R) = \frac{1}{2} \rho_m v_R^2 \left[\left(1 - \frac{R^2}{r^2} \right) - \left(1 - \frac{R^2}{R_1^2} \right) \frac{\ln(r/R)}{\ln(R_1/R)} \right], \tag{4.8}$$

$$\left(\frac{r_{max}}{R} \right)^2 = \left(1 + \frac{R^2}{R_{10}^2 - R_0^2} \right) \ln \left(1 + \frac{R_{10}^2 - R_0^2}{R^2} \right),$$

where ρ_m is the density, r is the radius of one point inside the cylinder, υ_R is the current velocity of the inner surface, r_{max} is the radius corresponding to the pressure maximum when the radius of the internal hole is equal to R (initial R_0), and the outer radius is R_1 (initial R_{10}). For the conditions of the experiment (initial velocity of the inner surface is 200 m/s), the inertial compression stresses, calculated according to Eq. (4.8), are less than 0.1 GPa. Thus, pressure effects can be neglected to a first approximation, at least for solid materials.

The distribution of stresses during implosion will depend on material strength and in general cannot be controlled by this method. Instead, numerical modeling should be used to extract the stress field during the collapse process. Two-dimensional modeling, reflecting plain strain conditions, can be successfully applied predicting the stress field and the relatively complex shape of the cavity (Nemat-Nasser et al. [1998]; Klopp et al. [1998]). At the same time, it is evident that this needs improvements like in situ measurements of stresses during the collapse process, at least on the stage of uniform flow before instability starts.

4.2.2.4 Material Instability Versus Geometrical Instability

The implosion of a relatively thin shell results in flow instability of a geometrical nature (Lindberg [1964]; Florence and Abrahamson [1977]; Serikov [1984]; Ivanov et al. [1992]). This "geometrical" instability of the collapse process, typical for thin shells accelerated toward the center, is not active in the TWC method due to the thick wall of the cylinder.

In conditions of TWC experiments the number of waves, due to the growth of small perturbations on the inner wall caused by this "geometric instability" (any departures from circular form and uniform velocity grow during collapse), can be evaluated and is less than 1. The geometry of the TWC method is selected in such way that material instability breaks the symmetry of the cylindrical collapse resulting in shear localization. Nevertheless, the role of geometric instability may be important as one of the mechanisms for the nucleation sites of shear bands, and needs more careful experimental analysis.

Well-developed geometrical instability during collapse of the thin-walled cylinders may be accompanied by shear localization due to material instability in points where local humps on the outside surface are connected due to local high tensile strength in such places (Stokes, Oro et al. [1999]).

4.3   Experimental Observation of Shear Bands Patterning in Polycrystals

The observations of shear band assemblages under controlled initiation conditions are very important for comparison with theoretical approaches. The role of pressure is not important for localization in solid materials and the strain-controlled TWC method is very suited for this application. An important feature

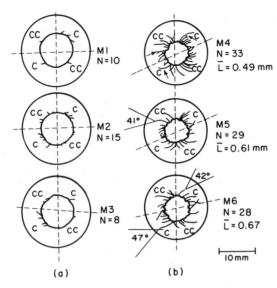

Figure 4.6. Shear band traces for six different experiments: (a) at $R_0 = 5.5$ mm ($\varepsilon_{ef} = 0.554$) and (b) at $R_0 = 6.25$ mm ($\varepsilon_{ef} = 0.923$); C and CC represent regions of clockwise and counterclockwise dominates, respectively. (Reprinted from *Acta Materialia*, **46**, Nesterenko, V.F., Meyers, M.A., and Wright T.W., Self-Organization in the Initiation of Adiabatic Shear Bands, pp. 327–340, 1998, with permission from Elsevier Science.)

of this method is that a relatively large amount of material is in a state of pure shear before the start of instability. This provides conditions for "spontaneous" shear localization which are different than, for example, "forced" localization in hat-shaped specimens (Chen, Meyers, and Nesterenko [1999]).

The modification of the TWC method for this purpose uses an internal tube made from OFHC copper for soft arresting of the collapse with desirable strains at the end of the process (Fig. 4.1). The different initial radii of the cavity and stopper tube allow the strain variations. The final radius of the completely collapsed tubes practically coincides with the calculated ones, based on the incompressibility conditions. This is because only a small amount of material in the center was removed by the central jet, which is easily detected by the cavity on the low plug.

The TWC technique allows for the reproduction of this instability phenomena in strain-controlled conditions. Data for titanium, stainless steel and Ti–6Al–4V alloy (Nesterenko, Meyers, and Wright [1998]; Xue, Nesterenko, and Meyers [2000]) will be considered in this section as an example.

Figure 4.6 shows (a) the initial and (b) final configurations of specimen cross-sections of the titanium sample with equiaxed grains having an average size, $d = 20$ μm (CP–Ti, Grade 2, Altemp Alloys, CA, from 2 in bar, $\sigma_{0.2} = 360$

MPa, $\sigma_t = 470$ MPa, elongation 20%). The copper "stopper" tubes, which have initial internal radii R_0 equal to 5.5 mm and 6.25 mm, are completely collapsed by the implosion, subjecting the titanium tube to two levels of plastic strain corresponding to initiation and well-developed stages of shear banding.

Three different experiments were carried out for each condition. They are labeled M1, M2, and M3 for the first set and M4, M5, and M6 for the second set. The reproducibility of the observed shear band patterns is clearly shown in Figure 4.6(a) and (b). The radii of cavity $R_0 = 5.5$ mm and 6.25 mm were chosen based on the observed strains for shear localization by previous studies. Meyers et al. [1994] obtained shear band propagation in commercial purity titanium (grain size 72 μm; Grade 2), for an effective strain between 0.2 and 0.45.

For simplicity and convenience the effective strains in cylindrical geometry are presented below (with the axial strain equal to zero) to characterize the corresponding stages of shear localization. At the same time, for comparison with experiments performed in the simple shear state of strain, the shear strain that occurred in cylindrical geometry in a plane parallel to the actual plane of localized shear should be taken into account (Staker [1981]; Semiatin, Staker, and Jones [1984]). Criteria for instability in plane strain conditions were developed by Hillier [1963, 1965a,b].

For the two experimental set-ups, the effective strains can be readily calculated (Eq. (4.5)). The strains for the collapsed inner surface of titanium with two initial sizes of inner copper tubes are for $R_0 = 5.5$ mm, $\varepsilon_{ef} = 0.554$ at $r_f = 4.5$ mm, and for $R_0 = 6.25$ mm, $\varepsilon_{ef} = 0.923$ at $r_f = 3.3$ mm.

The value $\varepsilon_{ef} = 0.554$ should provide strains higher than those required for initiation, according to Meyers et al. [1994] and Nesterenko et al. [1995], whereas $\varepsilon_{ef} = 0.923$ should result in a well-developed shear band pattern. This is borne out exactly by the observations shown in Figure 4.6(a) and (b).

The following observations can be made:

(a) The results demonstrate a robust and reproducible pattern of shear localization at the stage of initiation and at the developed stage.

(b) The shear bands initiate at the internal surface of the titanium specimen and propagate outward.

(c) The angle of extremities of the shear bands with radial directions fluctuates around 45°, which is the plane of maximum shear stress (and strain). This is shown for three shear bands in Figure 4.6(b).

(d) Both clockwise (C) and counterclockwise (CC) spirals are observed, in contrast with earlier observations by Nesterenko et al. [1995]. Figure 4.6(b) shows that these spirals organize themselves into four families with approximately equal spacings. The differences between the present results and those of Nesterenko et al. [1995] are not completely understood and may be due to the pre-existing texture in the titanium specimens, which is revealed by the anisotropy of plastic deformation in quasistatic compression tests. This texture may introduce a slight regular change of the cavity shape during collapse, similar to observed in monocrystal copper (Nemat-Nasser et al. [1998]). This will create

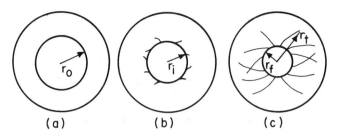

Figure 4.7. Schematic sequence of events during collapse of the cylinder: (a) initial configuration; (b) initiation of shear band formation; and (c) spiral propagation of shear bands during collapse. (Reprinted from *Acta Materialia*, **46**, Nesterenko, V.F., Meyers, M.A., and Wright T.W., Self-Organization in the Initiation of Adiabatic Shear Bands, pp. 327–340, 1998, with permission from Elsevier Science.)

a few segments adjacent to the cavity with a different stress distribution.

(e) The number of shear bands gradually increases with plastic strain. This number (N) counted at the internal surface of the cylinder is marked in Figure 4.6(a) and (b). It was possible to estimate an average spacing between the shear bands for experiments M4, M5, and M6, where the pattern with characteristic periodic spacing was well established. This spacing was calculated considering only shear bands with the same orientation, either clockwise (C) or counterclockwise (CC). The average spacings between the shear bands, also marked in Figure 4.6, are $L = 0.49, 0.61$, and 0.67 mm for specimens M4, M5, and M6, respectively. The averaging procedure was made for a number of shear bands equal to 25, 23, and 23 in Figure 4.6(b). This procedure can cause some decrease in real shear band spacings (up to 27%) as a result of their actually unknown nucleation radius, which in any case should be between radii of the inner titanium tube equal to 4.5 mm and 3.3 mm (Fig. 4.6).

(f) Shear band bifurcation (marked by arrows) is observed in Figure 4.6(b).

The initiation and propagation of shear bands during the collapse is schematically shown in Figure 4.7. When the internal radius of the titanium sample (initial value equal to r_0, Fig. 4.7(a)), reaches a value r_i, the critical shear strain for shear band initiation is attained (Figure 4.7(b)). At this stage the nucleus of the shear band array may be self-organized or each of them can proceed independently.

Upon further collapse, the shear bands grow outward and may go through a second stage of self-organization among well-developed objects (Fig. 4.7(c)).

The experiments cannot track exactly the shear strain at which shear bands initiate on the surface of the cavity. Nevertheless, the small length of the shear bands in Figure 4.6(a) shows clearly that the effective strain of 0.554 is slightly above the minimum strain for the initiation. Global strains at the tips of the bands (at r_t, Fig. 4.7(c)) were also calculated.

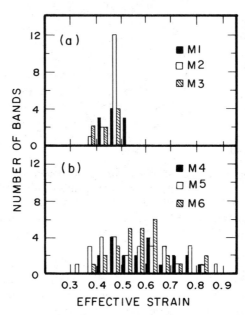

Figure 4.8. Distribution of global effective strains at tips ($r = r_t$) of bands for two different radii: (a) $R_0 = 5.5$ mm; (b) $R_0 = 6.25$ mm. (Reprinted from *Acta Materialia*, **46**, Nesterenko, V.F., Meyers, M.A., and Wright T.W., Self-Organization in the Initiation of Adiabatic Shear Bands, pp. 327–340, 1998, with permission from Elsevier Science.)

Figure 4.8 shows the distribution of strains at the shear band tips. For the earlier stage of collapse ($R_0 = 5.5$ mm, Fig. 4.6(a)), the minimum strain at the shear band extremity is equal to 0.35 and the strains range from 0.35 to 0.5.

For well-developed shear bands (for the later stage of collapse (Fig. 4.6(b), $R_0 = 6.25$ mm), the minimum strain is actually the same (0.3), but a wider distribution of strains 0.3–0.9 is observed. The minimum strain at the tip of the shear band is consistent with the calculated initiation strain, which was estimated to range from 0.2 to 0.45 (Meyers et al. [1994]). It is also consistent with earlier results by Nesterenko et al. [1995]: $\varepsilon_{ef} = 0.22$. The radial span of the shear bands was measured and correlated with the height of the steps at the internal surface, Δ. The results are shown in Figure 4.9. More developed shear bands create a larger step.

The results, shown in Figures 4.6, 4.8, and 4.9 enable the conclusion to be made that shear bands start to form at an effective strain equal to 0.3 along the internal surface of the hollow cylinder. As the collapse proceeds, the number of shear bands increases. The distribution in the radial spans of the shear bands is a direct consequence of their gradual formation: the bands that form first propagate farther outward.

Similar results shown in Figure 4.10, were obtained for austenitic stainless steel AISI 304L (T-304L, solution treated, from 1 in bar, $\sigma_{0.2} = 250$ MPa, $\sigma_t = 640$ MPa, elongation 58%) with a grain size (d) of 30 μm. In this case the

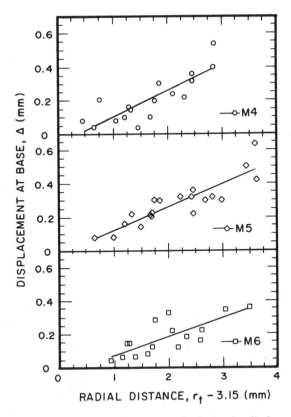

Figure 4.9. Relationship between displacement at the internal cylinder surface, Δ, and the radial range of shear bands, for experiments with $R_0 = 6.25$ mm. (Reprinted from *Acta Materialia*, **46**, Nesterenko, V.F., Meyers, M.A., and Wright T.W., Self-Organization in the Initiation of Adiabatic Shear Bands, pp. 327–340, 1998, with permission from Elsevier Science.)

number of distinguishable shear bands decreases with the increase of the effective strain on the cavity surface, due to the arrest of the growth of some initially created shear bands and distortions of the interface between the copper stopper tube and inner surface of the sample.

The kinetics of shear band initiation and their self-organization affect, to a great extent, the post-localization behavior of material. Multiple shear bands, self-organized into the array with small spacings, did not destroy the symmetry of the collapse process to such an extent as it does fast-growing chaotic shear bands with relatively large spacings. This difference is illustrated by comparing the final configurations of samples from different materials with similar effective strains, as shown in Figure 4.6 for titanium, in Figure 4.10(a),(b) for stainless steel, and in Figure 4.11 for Ti–6Al–4V alloy (MIL-T-9047G from 1 in bar,

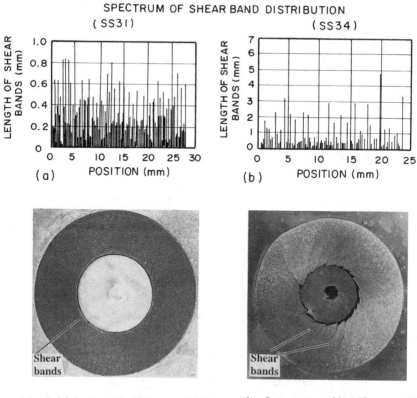

(c)   Initial stage, N=235, ε_{ef} = 0.541   (d)   Later stage, N=165, ε_{ef} = 0.915

_____

10 mm

Figure 4.10. Shear band distribution in stainless steel at different stages: (a), (c) initial stage, ε_{ef} = 0.541; (b), (d) developed stage, ε_{ef} = 0.915. The position of the individual shear band corresponds to some point on the inner surface of the cavity. (Reprinted from *Shock Compression of Condensed Matter—1999,* edited by M.R. Furnish, L.C. Chhabildas, and R.S. Hixson. Xue, Q., Nesterenko, V.F., and Meyers, M.A., Self-Organization of Adiabatic Shear Bands in Ti, Ti–6Al–4V and Stainless Steel, pp. 431–434, 2000.)

$\sigma_{0.2}$ = 980 MPa, σ_t = 1060 MPa, elongation 15%) with d = 5 μm (Xue, Nesterenko, and Meyers [2000]).

The behavior of stainless steel is similar to the behavior of titanium but the shape of Ti–6Al–4V is much more distorted by a smaller amount of fast-growing shear bands.

Data on shear band patterning in copper, tantalum, Teflon, and other materials can be found in papers by Nesterenko and Bondar [1994a,b], and by

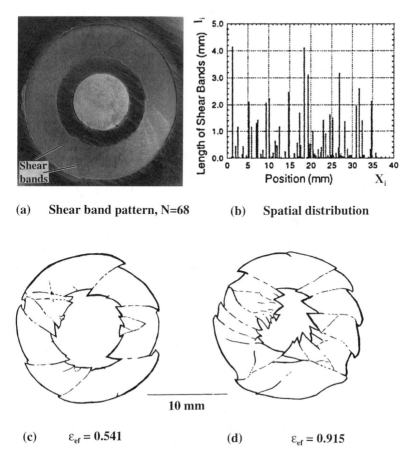

(a)   **Shear band pattern, N=68**   (b)   **Spatial distribution**

(c) $\varepsilon_{ef} = 0.541$ (d) $\varepsilon_{ef} = 0.915$

Figure 4.11. Shear band distribution in Ti–6Al–4V at different stages: (a), (b) initial stage, $\varepsilon_{ef} = 0.264$; (c), (d) developed stage; (c) $\varepsilon_{ef} = 0.541$; (d) $\varepsilon_{ef} = 0.915$. The position of the individual shear band in (b) corresponds to some point on the inner surface of the cavity. (Reprinted from *Shock Compression of Condensed Matter—1999*, edited by M.R. Furnish, L.C. Chhabildas, and R.S. Hixson. Xue, Q., Nesterenko, V.F., and Meyers, M.A., Self-Organization of Adiabatic Shear Bands in Ti, Ti–6Al–4V and Stainless Steel, pp. 431–434, 2000.)

Nesterenko et al. [1995]. In general we can conclude that the collapse of the cylindrical cavity for practically all materials will result in a symmetry break due to the material-induced instability which can be important for some applications (Stokes, Nesterenko et al. [2001]).

The shape of the shear bands assumed after post-localization high-strain plastic flow is different from the shape immediately after initiation (see Problem 12, Nesterenko, Meyers, and Wright [1995]). The most important difference is the decrease of an angle between the shear band and radius.

4.4   Grady–Kipp Model of the Collective Behavior of Shear Bands

In the present experimental configuration corresponding to the condition of the TWC method, the simplest question to be posed is the number of localized shear bands to be expected around the circumference of the inner surface as the cylinder collapses. The following experimental observations allow us to conclude that the shear bands array is self-organized.

Mainly one direction of spiral shear bands—either clockwise or counterclockwise—was observed inside separate segments with multiple shear bands, or even through the entire sample. The second important feature is the existence of some characteristic spacing evident at least in the case of titanium (Fig. 4.6(b)) and stainless steel (Fig. 4.10).

The reasons for this coupled behavior of shear bands may be brought about by two essentially different factors. One may be connected to the influence of an arbitrary originating shear band on the stresses in its vicinity, or by breaking geometrical symmetry through an introduction of the additional nonaxisymmetrical local bending type of deformation.

Another reason may be the interaction of successive nucleating shear bands due to the decrease of shear stress in the first one, even if the geometry of the set-up is not broken but the symmetry of the state of stress is violated. Both mechanisms promote the next shear band to nucleate in a preferential direction ("relay-race"-like coupling, Nesterenko et al. [1994]). The relative input of these mechanisms can be clarified experimentally using different cavity diameters.

If the spacing of shear bands at the same strain and strain-rate conditions does not depend on the cavity radius, then the first mechanism is not active. Probably the first mechanism is not important at the considered radii of cavity because the periodic pattern of shear bands originated at the stage where local bending effects seem small (see Fig. 4.10(c)). At the same time a comparison can be made with existing theories for shear band spacing.

The first theoretical model for shear band spacings was proposed by Grady and Kipp (GK) [1987] for rate-independent materials, and for the one-dimensional shearing of an infinite strip of material without regard to the boundary conditions. Its application in the present case can only be approximate, if the spacing is much smaller than the cavity radius. Nevertheless, it is important, due to dimensional considerations alone, to find the appropriate nondimensional groupings of the physical quantities.

The basic notion, used in the GK analysis, is that the rapid loss of strength or ability to transfer shearing tractions across the developing shear band affects the neighboring material by forcing it to unload. This unloading process is communicated outward by momentum diffusion, rather than by elastic wave propagation. The minimum separation between independently nucleating bands arises from computing the distance traveled by the diffusive unloading front during the time required to unload as localization occurs.

Their analysis also assumes that the width of the shear band adjusts itself so as to achieve a maximum growth rate, the growth rate of narrower bands being limited by thermal diffusion and that of wider bands by inertia.

The predicted spacing, L_{GK}, for the applied shear strain rate ($\dot{\gamma}_0$) is given in their Equation 23. With suitable changes in notation (thermal diffusivity χ has been replaced by thermal conductivity divided by density (ρ) and heat capacity per unit mass (C), $k/\rho C$), it is

$$L_{GK} = 2\left\{ 9kC/\dot{\gamma}_0^3 a^2 \tau_0 \right\}^{\frac{1}{4}}. \tag{4.9}$$

The relation between flow stress and temperature is assumed to be

$$\tau = \tau_0(1 - a\vartheta), \tag{4.10}$$

where τ_0 is the strength at a reference temperature T_0, ϑ is the relative temperature $T - T_0$, and a is a softening term.

It is very interesting that in the frame of the Grady–Kipp model the shear band thickness and spacing are not independent, they adjust to minimize the dissipated energy (Grady and Kipp [1987]).

4.5 Wright–Ockendon Model for Shear Band Spacing

The recently developed analysis by Wright and Ockendon (WO) [1996] is based on the notion that shear bands arise from small but growing disturbances in an otherwise uniform region of constant strain rate. Disturbances do not propagate in perpendicular directions, but simply grow in place, so the most likely minimum spacing is obtained by finding the fastest growing wavelength. The problem is posed by first finding the uniform fields and then finding the differential equations for perturbations with the uniform fields taken as the ground state. Fourier decomposition of the perturbation equations is followed by an asymptotic representation of the solution. Then it is a simple matter of differentiation to find the wavelength that grows the fastest. In outline, the mathematical argument proceeds as follows.

Equations for the balance of momentum, the balance of energy, and a simple flow law may be written as (the subscript implies differentiation with respect to the variable, e.g., $\dfrac{\partial \tau}{\partial y} \equiv \tau_y$):

$$\tau_y = \rho \upsilon_t, \quad \rho C \vartheta_t = k\vartheta_{yy} + s\upsilon_y, \quad \tau = \tau_0\left(1 - a\,\vartheta\right)\left(b\,\upsilon_y\right)^m. \tag{4.11}$$

In these equations y is the coordinate perpendicular to the band and t is the time. The dependent variables τ, υ, and ϑ stand for the shear stress parallel to the band, the particle velocity parallel to the band, and the relative temperature, respectively; the constants ρ, C, k, τ_0, and a are the same as in the description of GK above; and the constants m and b stand for the strain-rate sensitivity and a normalizing time constant, respectively. A homogeneous solution τ_h, V, and Θ that represents uniform shearing may be written

$$V = \dot{\gamma}_0 y, \tag{4.12a}$$

$$\tau_h = \tau_{h0}(1 - a\theta) = \tau_{h0}\exp\left\{-\left(a\tau_{h0}/\rho c\right)\left(\dot{\gamma}_0 t\right)\right\}, \tag{4.12b}$$

where the reference stress is given by

$$\tau_{h0} = \tau_0\left(b\dot{\gamma}_0\right)^m. \tag{4.12c}$$

Perturbation equations are found by considering a solution of the form

$$\tau = \tau_h(t) + \tilde{\tau}(y, t), \qquad \upsilon = V(y) + \tilde{\upsilon}(y, t), \qquad \vartheta = \theta(t) + \tilde{\vartheta}(y, t), \tag{4.13}$$

where parameters with the wave on the top are small quantities. Substituting Eq. (4.13) into Eq. (4.11), and retaining only the linear terms, we obtain

$$\tilde{\tau}_y = \rho\tilde{\upsilon}_t, \quad \rho C\tilde{\vartheta}_t = k\tilde{\vartheta}_{yy} + \tau_h\tilde{\upsilon}_y + \dot{\gamma}_0\tilde{\tau}, \quad \tilde{\tau} = -a\tau_{h0}\tilde{\vartheta} + m\tau_h\tilde{\upsilon}_y/\dot{\gamma}_0, \tag{4.14}$$

where τ_h is a time-dependent term given by Eq. (4.12).

Approximate solutions to Eq. (4.14) may be found by assuming that $\tilde{\tau}$ and $\tilde{\vartheta}$ are even and that $\tilde{\upsilon}$ is odd in y. Then after Fourier decomposition on y, the equations reduce to an infinite set of linear ODEs in time, which may be combined into a single second-order ODE for each Fourier component, ϑ_n. Because these equations have time-dependent coefficients, which arise from the time-dependence of the ground state as shown by Eq. (4.12), and because it is known that the assumption of frozen coefficients can lead to grossly erroneous results, it is necessary to use asymptotic methods to proceed further.

The standard WKB method (e.g., see Bender and Orzag [1978]) may be applied in a straightforward way. The final representation for the nth Fourier component of the strain rate at early times (characteristic only for the development of the shear bands embryos) may be written as

$$\dot{\gamma}_n(t) = \dot{\gamma}_0\left(\frac{a}{m}\right)\vartheta_n(0)e^{\alpha t}. \tag{4.15}$$

Here $\vartheta_n(0)$ is the initial value of the nth Fourier component; see Wright and Ockendon [1996] for details. The coefficient α in the exponent in Eq. (4.15), which determines the characteristic wavelength of the fastest-growing perturbation, is represented by

$$\alpha = \frac{m}{2}\left(\frac{\tau_{h0}}{\rho\dot{\gamma}_0}\right)\left(\frac{n\pi}{l}\right)^2$$

$$\cdot\left\{\left[\left(1 - \frac{k\dot{\gamma}_0}{C\tau_{h0}}\right)^2 + (4+2m)\left(\frac{a\dot{\gamma}_0^2}{Cm^2}\right)\left(\frac{l}{n\pi}\right)^2\right]^{1/2} - 1 - \left(\frac{k\dot{\gamma}_0}{C\tau_{h0}}\right) + \left(\frac{a\dot{\gamma}_0^2}{Cm^2}\right)\left(\frac{l}{n\pi}\right)^2\right\}. \qquad (4.16)$$

Since l/n is the half-wavelength of the nth Fourier component, the maximum of α with respect to l/n may be solved to find the wavelength that grows the fastest. This space scale may be identified with the most probable minimum spacing between the nucleus of shear bands.

The resulting spacing, L_{OW} (to the lowest-order terms) is

$$L_{OW} = 2\pi\left\{\frac{m^3 kC}{\dot{\gamma}_0^3 a^2 \tau_{h0}}\right\}^{1/4}. \qquad (4.17)$$

The approaches taken by GK and OW are completely different. The former concentrates on the stress collapse mechanism when the shear band is a well-developed object. The latter takes into consideration the earliest stages of localization when the shear bands are represented by a nucleus.

Nevertheless, it is a remarkable fact that, except for numerical factors and the rate constants, the two results produce the same functional dependence of spacing on the main material parameters. It should be mentioned that in experiments, the spacings are measured as usual for the well-developed stage when shear bands are distinctive objects qualitatively different from the surrounding material and definitely not just a small perturbation on temperature and strain.

The assumed flow laws (Eqs. (4.10) and the third equation in Eqs. (4.11)) differ only by the rate factor. As a consequence, the major difference between Eqs. (4.9) and (4.17) is that the strain rate sensitivity appears strongly in the numerator and weakly in the denominator of Eq. (4.17), whereas it does not appear at all in Eq. (4.9).

On the other hand, there are only a finite number of ways that the constants may be combined to obtain a length scale. But there are certainly more than one, and combinations such as

$$\left(\tau_{h0}C^2\big/\dot{\gamma}_0^3 ka\right)^{1/2}, \qquad (4.18a)$$

$$\left(k\big/a\rho C\dot{\gamma}_0\right)^{1/2}, \qquad (4.18b)$$

or

$$\left(\tau_{h0}\big/\rho\dot{\gamma}_0^2\right)^{1/2}, \qquad (4.18c)$$

as well as that found in Eq. (4.17), are equally appropriate from this point of view. So it is not quite so surprising that the results are so similar.

4.6   Molinari Model for Shear Band Spacing

Another approach for the analysis of the collective behavior and spacing of adiabatic shear bands in a one-dimensional (simple shearing of an infinitely extended layer of finite thickness) formulation was proposed by Molinari [1997]. The linear perturbation analysis applied was first introduced in the context of adiabatic shear banding by Clifton [1979].

In this model the important step which determines the shear band spacing is also related to the early stages of flow localization, as it is in the WO model. The latter is considered as a small perturbation to the fundamental solution, describing a homogeneous laminar flow with a uniformly increasing temperature as in the WO model. This method allows evaluation of the initial growth rate of perturbation.

The important feature of Molinari's model (MO) is that it includes the strain-hardening effect, and the non-strain-hardening case can be obtained as a partial one. The shear band spacing (L_M) corresponds to the mode with the largest growth rate. The constitutive equation, in general form, is

$$\dot{\gamma} = \psi(\tau, \gamma, T). \tag{4.19}$$

In numerical calculations for a material with characteristics similar to those of the titanium alloy and with the following constitutive equation

$$\tau = \tau_0(\gamma + \gamma_i)^n \dot{\gamma}^m T^\nu, \tag{4.20}$$

where γ_i is a prestrain and n, m, and ν are the exponents describing the strain, the strain-rate hardening and temperature softening demonstrated that the increase of the hardening exponent from 0 (non-strain-hardening case) to 0.3 changes the shear band spacing by two and a half times (Molinari [1997]).

An analytical approach for the partial case of the nonhardening material ($n = 0$ in Eq. (4.19)) results in the shear band spacing being equal to

$$L_M = 2\pi \left(\frac{m^2 kC(T^0)^2}{(1 + m^{-1})\beta^2 (\dot{\gamma}^0)^3 \tau_0 \nu^2} \right)^{1/4}, \tag{4.21}$$

where β is a Taylor–Quinney coefficient of the rate of plastic work transformation into heat (taken as 0.9 for most cases for solid metals) and T^0 and γ^0 are the temperature and the strain at the end of the stable stage of viscoplastic flow.

With thermal softening being a linear function of temperature and with no strain hardening included the spacing is

$$L_{\mathrm{M}} = 2\pi \left(\frac{m^2 kC \left(1 - aT^0 \right)^2}{\left(1 + m^{-1} \right) \beta^2 \left(\dot{\gamma}^0 \right)^3 \tau_0 a^2} \right)^{1/4}.$$

(4.22)

This equation is similar to the WO result (Eq. (4.17)) except the term $\left(1 - aT^0 \right)^2$, creating a difference in calculated values, is of the order of 0.3 (Molinari [1997]). For a general constitutive equation (Eq. (4.19)) including the effect of strain hardening, Molinari [1997] obtained the equation for the spacing

$$L_{\mathrm{M}} = 2\pi \xi_0^{-1} \left(1 + \frac{3}{4} \frac{\rho C \frac{\partial \psi^0}{\partial \gamma}}{\beta \tau^0 \frac{\partial \psi^0}{\partial \gamma}} \right)^{-1},$$

(4.23)

$$\xi_0^2 = \beta \frac{\partial \psi^0}{\partial T} \sqrt{\left(1 + m^{-1} \right) \frac{\tau^0 \dot{\gamma}^0}{C k}}.$$

4.7   Theoretical Predictions Versus Experiments

4.7.1   Shear Bands Patterning and Materials Properties

Most research is concentrated on the threshold conditions of shear localization. Critical conditions for shear band initiation strongly depend on the macroscopic, continuum properties of materials: strength, strain and strain-rate hardening, thermal softening, melting temperature, specific heat, and thermal conductivity (Bai and Dodd [1992]). For example, if the strain rate is high enough then the critical strain γ_{cr} for the shear band initiation (strain-rate sensitivity is neglected) can be introduced which depends on the work (strain) hardening exponent n, thermal softening, density ρ, and heat capacity C (Staker [1981], Semiatin, Staker and Jonas [1984]), is

$$\gamma_{\mathrm{cr}} = -\frac{n\rho C}{\left(\frac{\partial \tau}{\partial \gamma} \right)_{\gamma, \dot{\gamma}}}.$$

(4.24)

Changing macroscopic material properties like strength and microhardness (with the indirect change of work hardening) have a very strong effect on shear band propensity and ballistic performance. For example, an increase of hardness of 4340 steel from HRC 15 to HRC 52 resulted in the change of a bulk-distributed extensive plastic deformation of the target to highly localized shear when impacted by blunt right-angle cylinders (Olson, Mescall, and Azrin [1980]).

At the same time, for the modeling of material behavior at high-strain-rate conditions, predictions of ballistic limits, especially for advanced armor, it is necessary to understand how a number of shear bands and their self-organization depends on material properties. It is clear, from comparison between Eq. (4.24) and Eqs. (4.9), (4.17), that the higher propensity of material to shear localization (lower critical strain) is in accord with smaller spacing between shear bands.

Erlich, Curran, and Seaman [1980], based on surface observations, found a dramatic increase in the number of shear bands (a decrease in shear band spacing) and the extent of propagation in 4340 steel specimens with different hardness HRC 21, 40, and 52 using a contained exploding cylinder (CEC) technique. Unfortunately, in the CEC technique, tensile stresses may replace the kinetics of the shear bands development and two types of shear cracks—cracks along adiabatic shear bands and ductile dimple shear cracking—are competitive modes which are not distinguishable by surface observations (Staker [1980]. Nevertheless, the observed tendency is in the direction equally predicted by different models (Grady–Kipp, Wright–Ockendon, and Molinari).

The nucleation and development of the shear band population in the CEC method was explained, based on the numerical simulation of the plastic flow around a frictionless crack in a plane subjected to pure shear. It was concluded that microstructural imperfections, especially the larger ones, may indeed be nucleation sites, but continuum properties may control which of these imperfections are activated (Shockey [1986]).

Despite a very complex phenomena of shear band initiation and their self-organization, it is important to establish the relationship of the theoretical predictions provided by different models and experimental results. The investigated materials titanium, stainless steel and Ti–6Al–4V (Figs. 4.6 to 4.11) represent a good set of materials with essentially different properties. The patterns of shear bands in stainless steel, titanium, and Ti–6Al–4V alloy exhibit different characteristics during their evolution. For example, in stainless steel the shear bands nucleate at the internal boundary and the distinguishable number at the initial stage is larger than that at later stages (Fig. 4.10). The shear band pattern at the later stage shows that the surviving shear bands compose a new distribution. The number of active shear bands at the later stage significantly decreases with the increase in effective global strain (Meyers, Nesterenko, LaSalvia, and Xue [2001]).

The GK, WO, and MO models were compared by using Eqs. (4.9), (4.17), and (4.22), neglecting the effect of strain hardening. The only difference between them is the coefficient, which relates to the rate sensitivity (strain-rate hardening exponent, m). In the experiment, shear strain rate is equal to $\dot{\gamma}_0 = 6 \cdot 10^4$ s^{-1}. The parameters for titanium can be found in paper by Nesterenko et al. [1998]: $m = 0.052$; $a = 1 \cdot 10^{-3}$ K^{-1}; $k = 19.0$ J/smK; $C = 528$ J/kg K; $\tau_0 = 280$ MPa. For stainless steel found from Stout and Follansbee [1986], $m = 0.012$; $a = 7.2 \cdot 10^{-4}$ K^{-1}; $k = 14.7$ J/smK; $C = 500$ J/kgK; $\tau_0 = 200$ MPa. For Ti–6Al–4V (Nemat-Nasser et al. [1999]; Donachie [1982]), $m = 0.024$; $a = 4.2 \cdot 10^{-3}$K^{-1}; $k =$

Table 4.2. Comparison of theoretical and experimental results for the spacing for stainless steel 304L and CP titanium at the initial stage of patterning and between well-developed shear bands.

Spacing (mm)	Exp. Data Initial (mm)	Exp. Data Developed (mm)	L_{WO} (mm)	L_{GK} (mm)	L_{MO} (mm)
Stainless steel	0.12	3.20	0.17	2.62	0.16
Titanium	0.69	2.57	0.52	3.30	0.36

3.07 J/smK; $C = 564$ J/kgK; $\tau_0 = 490$ MPa. The experimental and calculated spacings are presented in Table 4.2.

It should be emphasized that material is in the extreme state inside a well-developed shear band. There are no experimental data on the material behavior corresponding to the advanced stages of shear localization. A very interesting idea, to use particle velocity and temperature fields in the neighborhood of the shear band to probe into the extreme states of material response inside the shear band, was proposed by Wright and Ravichandran [1997]).

Experimental spacings at the global strain of 0.54, considering all shear bands, is equal to 0.12 mm and 0.69 mm for stainless steel and titanium, respectively (Table 4.2). This stage corresponds to the initiation of shear bands. The data for Ti–6Al–4V are not presented because even at the initial stage it was not possible to identify a periodic pattern. At larger strain there are few shear bands running out of the sample and distorting the geometry and boundary conditions on the outside surface of the sample (Fig. 4.11 (a),(b)).

The experimentally obtained spacings in titanium and stainless steel are in excellent agreement with the predictions of the WO model. Considering the one-dimensional character of the models and the three-dimensional geometry of the experiments, and neglecting the work hardening and averaging of the thermal softening coefficient over a wide temperature range, the agreement of theoretical predictions of the WO model with experimental results should be considered cautiously. Strong dependence of the thermal softening coefficient on the strain at which it is evaluated is discussed by Semiatin, Staker, and Jonas [1984].

It is very important that theoretical models correctly predict the relation between spacing in stainless steel and titanium for both the WO and MO models for the initial stage of shear band patterning. The difference is mainly connected with a different strain-rate sensitivity of these materials.

The spacing of the well-developed shear bands, on the other hand, is 3.2 mm and 2.57 mm (for shear bands with lengths higher than 2.5 mm for stainless steel and 2.62 mm for titanium). The WO and MO models predict a closer estimate for stainless steel and titanium at the initial stage; the GK prediction seems much larger than the experimental results. Since the WO and MO models are based on perturbation analysis, reflecting the initial behavior of shear banding, these may correlate to the initial spacing. If we examine the longer shear bands at the well-developed stage, the GK model gives a closer prediction of spacing; it describes the behavior of shear banding at a later stage, when the shear stress within the shear bands is close to zero.

The experimental results for Ti–6Al–4V alloy reveal different characteristics for shear band evolution. Shear band patterns were examined at four different effective global strains: 0.142, 0.264, 0.55, and 0.92 (Xue, Nesterenko, and Meyers [2000]). At the two higher strains, the patterns become chaotic. Large deformation leads to heavy instability of the material and the shear band patterns do not demonstrate periodicity. Even at the low-strain level of 0.264, the shear bands are well-developed and propagate a large distance though the material bulk (shown in Fig. 4.11(a),(b). The spacing of the shear bands is 0.526 mm. The pattern at a smaller strain (0.142) was monitored to find the initiation state of the shear bands. The average spacing is about 1.12 mm and the general picture does not reveal periodicity. The number of shear bands increases with the increase of effective strain. An interesting comparison can be made considering the initial state of shear banding for stainless steel and Ti–6Al–4V. About 10 shear bands form simultaneously along the internal boundary in stainless steel, while only one shear band appears within the same scale (1 mm) in Ti–6Al–4V (Xue, Nesterenko, and Meyers [2000]).

The large spacing in Ti–6Al–4V, when compared with stainless steel and titanium, cannot be explained by the presented theories for spacing. In fact, they predict the opposite effect. This may suggest that the interaction between the nuclei of shear bands in Ti–6Al–4V is relatively weak. Each shear band grows fast once it nucleates. Unloading during growth of the shear band reduces the new nucleation sites surrounding the developed shear band. This results in a highly distorted shape of the Ti–6Al–4V sample in comparison with the stainless steel case at the same overall strains (compare Figs. 4.10(c),(d) and 4.11(c),(d)). The examination of the ratio of the lengths of shear bands and their edge displacement, L/Δ, also indicates the difference between stainless steel and Ti–6Al–4V. L/Δ is about 9.5 for stainless steel and 26.5 for Ti–6Al–4V at the same Δ.

It can be inferred that Ti–6Al–4V is missing a stage where the nuclei are organized in a periodic pattern with a characteristic spacing. Instead, a stage, with random nucleus spacing, develops into stages with grown shear bands bypassing self-organization of their nucleus. Probably a larger size of sample is necessary to establish a periodic pattern of shear bands in Ti–6Al–4V.

The mentioned difference in the L/Δ ratio reflects the difference between the critical strains for initiation and propagation and between the speed of the shear band development in stainless steel and Ti–6Al–4V. The transition between the different modes of behavior is determined by the relationship between the time for shear propagation through a given size of the sample (d), and the typical time for establishing a self-organized pattern among the shear band nucleii (t_{s-o}). If the velocity of the shear band (V_{sb}) is known, then the typical estimation of time for shear band self-organization may be easily established based on this approach. From this point of view, probably a larger size of Ti–6Al–4V sample is necessary ($d > V_{sb}t_{s-o}$) to observe the self-organization process in shear band population.

It is interesting to note that the number of shear bands observed under the identical conditions of impact on titanium and Ti–6Al–4V was higher in the latter case (Grebe, Pak, and Meyers [1985]).

There is a clear increase in the spacing of the propagating shear bands for stainless steel and titanium as their lengths increase. This phenomenon is analogous to the increasing spacing of the propagating parallel cracks studied by Nemat-Nasser, Keer, and Parihar [1978]. At a certain length of shear band and spacing, the growth becomes unstable and alternate shear bands grow with a new spacing; the other shear bands stop growing.

4.7.2   Shear Bands Patterning and Microstructural Features

The general conclusion made by Shockey [1986], based on results of experiments with the CEC method, is that microstructural features like inclusions, second-phase particles, grain boundaries, triple points, pores, and the like, appear to have little influence on the nucleation, growth, and coalescence of shear bands under dynamic compression in AISI 4340 steel and other materials. This is in striking contrast with tensile failure where the kinetics of the fracture process is determined by the microstructural defects. Shear bands propagate primarily in response to the macrostates of stress and strain, and the microstructural influence on shear band behavior is minor (Shockey [1986]) in the presence of externally induced stress concentrations ("forced" shear localization).

The CEC technique as well as the TWC method are the only techniques where the array of shear bands (a few hundreds in some experiments, see Fig. 4.10) is generated starting from the uniform state of stress and strain without externally induced stress concentration ("spontaneous" shear localization). The difference between these methods is that, in the latter, stresses are compressive on the stage of high strain plastic flow (especially with an inner copper stopper tube), and in the CEC technique the tensile stresses are significant. This results in replacement of the localized shear development by crack propagation in the same plane and in competition between cracks along the shear bands and ductile cracks in the former technique (Staker [1980], [1981]). The TWC method also provides a remarkably soft recovery that is very important for the preservation of the microstructural features developed during dynamic deformation.

It was demonstrated with the CEC and the TWC methods that shear localization closely resembles a nucleation and growth process: shear bands nucleate in the vicinity of the inner surface of the cylindrical samples and propagate into the surrounding material experiencing interaction and self-organization stages (Figs. 4.6, 4.8, 4.10, and 4.11).

Stainless steel AISI 304L has a clear stage of self-organization of shear bands with well-established periodicity (Fig. 4.10). Variation of the grain size in this material from 30 μm to 500 μm was found to have a minor effect on the nucleation and patterning of the shear bands (Meyers, Xue, and Nesterenko [2001]). There is only one publication (Nesterenko, Bondar, and Ershov [1994]) where a decrease in the grain size in copper from 100 μm to 30 μm resulted in an increase in the number of shear bands from 30 to 50 with an attendant decrease of shear band spacing at a very high level of strain of about 1. No change in the number of shear bands was observed when the grain size was decreased from 1000 μm to 100 μm. Amorphous cobalt-based metal alloys (71KNSR) exhibit similar shear band patterning in conditions close to the TWC method with a typical spacing of about 1 mm (Pershin and Nesterenko [1988]; Nesterenko [1992]).

It may be expected that, at a globally uniform state of stress, shear bands should nucleate at stress concentrations existing in the microstructure despite the fact that no clear material heterogeneities were identified as nucleation sites. The model proposed by Shockey, Curran, and Seaman [1985], (Shockey [1986]) uses material parameters like density, melt energy, dynamic yield strength, and the population of nucleating heterogeneities as influential material/microstructural properties to determine the number of nucleating shear bands and the rate of their growth. They emphasized that only large microstructural features (typical scale 0.1 mm) can serve as nucleation sites due to the required adiabaticity.

A qualitatively different approach to the nucleation and self-organization of the shear bands is represented by the models developed by Grady–Kipp [1987], Wright–Ockendon [1996], and Molinari [1997]. In all these models the spacing between the shear bands is determined not by "nucleating heterogeneities" but by the process of shear instability of continuum. Microstructural features influence this process only indirectly through macroproperties. Results of our experiments, especially the evident periodicity in titanium and stainless steel (Figs. 4.6 and 4.10), the correct order of magnitude predicted for spacing by these models strongly support the continuum approach to shear band patterning.

The relation between these two approaches (one is centered on the preexisting nucleation sites and another represents the growth of very small fluctuations in an unstable state) closely resembles the relation between the classical theory of heterogeneous nucleation and spinodal decomposition in phase transformations (Koch [1984]). The realization of the specific mechanism depends partially on the macroscopic conditions (Meyers, Nesterenko, LaSalvia, and Xue [2001]).

Despite the major continuum-based mechanism of the shear band self-organization in titanium and stainless steel grain size scale plays an important role in shear localization for some materials. For example, the width of very diffusive shear bands in tantalum is comparable with grain size and was called "grain scale" localization (Nesterenko, Myers, LaSalvia et al. [1997]). These bands follow the maximal macroscopic shear planes. The number of diffusive shear bands in tantalum was about 70 which results in smaller spacing than in titanium where there are about 30 shear bands at the developed stage of patterning (Fig. 4.6).

But, in this case, mechanism of softening is quite different from thermal softening and may be due to the rotation of the grains toward orientations with a larger Schmidt factor. There are a few important results which demonstrate the importance of microstructural features for the localization process. Asaro [1979] demonstrated that the localization phenomenon can be developed for crystals undergoing multiple shear with positive work hardening. Asaro and Needleman [1985] and Harren and Asaro [1989] in computer modeling predicted shear localization in polycrystals in the presence of work hardening.

Anand and Kalidini [1994], in macroscopically two-dimensional plane strain conditions, numerically observed localized shear bands in agreement with experiments while the macroscopic strain-hardening rate is still positive. They concluded that the localization of plastic flow can occur naturally in ductile polycrystal materials and is "an inherent and natural outcome of a large deformation process, and is closely associated with the concurrent evolution of crystallographic structure."

The rate of softening inside a shear band, when crystal plasticity mechanisms are the major origin of softening, can be strongly microstructurally dependent and microstructural features may effect the shear band spacing and patterning on the macroscale.

The dependence of shear band spacing and patterning in tantalum on the initial grain size will be a very promising subject to probe the influence of the microstructural features on macroscopic dynamic behavior. Chen, Meyers, and Nesterenko [1999] found that shocked tantalum has a greater propensity to shear localization.

The working texture is the microstructural feature that strongly influences the propagation of shear bands in specimens of AISI 4340 steel. It was observed that shear failure developed preferentially on the weaker planes of working (Shockey and Erlich [1980]). This observation suggests that the texture-free methods of materials manufacturing, like powder metallurgy, may be useful to control shear localization at ballistic applications. Ballistic tests of hot isostatically pressed targets from Ti–6Al–4V support this point (Nesterenko, Indrakanti et al. [2000a,b]; [2001].

The TWC method can be used to study shear localization in materials with a large grain size or even in monocrystals (Nemat-Nasser, Okinaka, and Nesterenko [1998]; and Nemat-Nasser, Okinaka, Nesterenko, and Liu [1998]). This important application is based on the weak influence of large distortions of the shape on the inner surface of the boundary conditions on the outside surface of the sample. In this case, the microstructural features have a major impact on the macroscopic behavior, including localization of strains (see Figure 5.8(a)). The final shape of the cavity, specific for a given initial crystal orientation, and the strain localization in monocrystall copper are very well described, based on crystal plasticity without thermal softening despite the high-strain-rate conditions of deformation. The TWC method can be successfully used to probe active slip systems in conditions of high strain, high-strain-rate plastic flow. In general, crystal plasticity effects may dominate the overall plastic flow in conditions of the TWC method when the crystal size becomes comparable with the shear band spacing characteristic for small grain size material.

4.7.3 Scaling of Shear Bands Patterning

An interesting and practically important problem is the scaling of a shear band pattern with an increase of sample size. The nature of this scaling, for example, determines the evolution of the number of fragments during the fracture under dynamic loading for some materials.

Experimental data on the ballistic impact on the 2024-T351 alloy by Backman, Finnegan et al. [1986] suggest that the number of bands in the system scales near the first power of the ratio of the systems dimensions. This is consistent with a constant shear band spacing for different dimensions of the sample under similar strain rate conditions if the size of the sample is much larger than shear band spacing.

The TWC method is a very suitable instrument for controlled experiments on the scaling of a shear band pattern in conditions when this pattern is not obliterated by the fracture process.

4.8   Dependence of Shear Bands Patterning on Initial Structure in Inert (Fractured and Granular) Materials

4.7.3 Large Strain Flow of Fractured Ceramics and Armor Penetration

Ceramic is used as armor material to defeat the projectile through blunting, rupture, erosion, and other dissipation processes including the high-strain-rate flow of comminuted ceramics. The latter mechanism is especially important for a long rod penetrator defeat. In the case of an unconfined ceramic target, the high-strain flow of damaged material is not an important mechanism of energy dissipation, and the main property is resistance of the ceramic to tensile stresses (Wilkins [1978]).

To date, there is no strong correlation between the mechanical properties of the ceramic and the ballistic performance of the armor system. However, all good armor ceramics have high hardness, high Young's modulus, high sonic velocity, low Poissons' ratio, moderate density, and low porosity (Viechnicki et al. [1987]). For example, silicon carbide (SiC) exhibits all of these attributes and is an excellent candidate for armor ceramic. An increase in the critical tensile strength of ceramic is also beneficial for ballistic performance (Wilkins [1978]).

Ceramics have been incorporated into advanced armor systems for over 20 years. The evolution of damage during the ballistic impact of ceramics can be divided into four classes (Viechnicki et al. [1991]; Meyers [1994]):

(1) radial and conical cracks caused by radially expanding stress waves;

(2) formation of a comminuted zone (also known as the Mescall zone), produced by shock waves;

(3) spalling generated by reflected, tensile pulses; and

(4) flow of the comminuted material starting from the front side of the advancing projectile.

In the comminuted zone, the high-amplitude shock waves create stresses that exceed the strength of the ceramic and result in fine ceramic fragments. Recent experiments have demonstrated the presence of the comminuted zone in various ceramics subjected to rod impact (Shockey et al. [1990]; Hauver et al. [1993], [1994]). In order to allow the penetrator to continue moving through the material, the comminuted zone has to flow around the penetrator, under high strain and high-strain rate. The deformation of the comminuted ceramic proceeds in a constrained volumetric condition because the surrounding material imposes a lateral confinement. Curran et al. [1993] developed a microstructural model of a comminuted ceramic subjected to large strain and divergent flow. Inelastic deformation is described by the sliding and ride-up of fragments, and the competition between dilatation and compaction is included. They demonstrated that the most important ceramic properties that determine the penetration resistance of the armor ceramic are:

(1) the friction between comminuted granules;

(2) the unconfined compressive strength of the intact material; and

(3) the strength of the comminuted material.

For example, a threefold increase in intergranular friction decreases the penetration depth by a factor of three.

Different experimental approaches have been developed to investigate the behavior of comminuted materials under high strain, high-strain-rate conditions. Klopp and Shockey [1991] used a laser Doppler velocimeter system to investigate the strength of comminuted SiC at high strain rates (10^5 s^{-1}). They determined that the strength at this condition is essentially lower than at the quasistatic rate. The friction coefficient of 0.23 (in the Mohr–Coulomb model) was reported to be smaller than for quasistatic tests.

Sairam and Clifton [1996], using an inclined plate impact, also reported a lower internal friction at high strain rates in granular alumina. Klopp et al. [1996] performed a spherical cavity expansion experiment with two grades of alumina to provide high-strain-rate data to develop ceramic armor penetration models. They observed that AD-995 alumina is comminuted primarily via a grain boundary fracture and the AD-85 alumina is comminuted by compaction and fracture of the intergranular glassy phase, as well as by the fracture of grains of non-alpha alumina exhibiting multiple slip, and by fracture of the alpha grains. They used a simple model of comminution and were able to predict fragment sizes at various distances from the charge.

Chen and Ravichandran [1997] investigated the behavior of a glass ceramic (Macor) by imposing controlled multiaxial loading on cylindrical ceramic samples using split Hopkinson bars. The confining pressures ranged from 10 MPa to 230 MPa under both quasistatic and dynamic loading conditions. The failure mode changes from complete fragmentation without confinement to localized brittle faulting with lateral confinement.

All results (Klopp and Shockey [1991]; Sairam and Clifton [1996]; Klopp et al. [1996]; Chen and Ravichandran [1997]) demonstrate the necessity of experiments with external stresses of 1 GPa to 10 GPa, effective strains around 1, and strain rates about 10^5 s^{-1}. Dynamic measurements should be accompanied by post-deformation characterization to identify the micromechanical process of deformation, because the micromechanics of granular (comminuted) material deformation is not well established even at a qualitative level. For example, the use of the Mohr–Coulomb approach is problematic under conditions of high strain and high strain rate, where shear localization, particle fracture, and nonequilibrium heat release are involved. Experiments incorporating the dynamic strength measurement, accompanied by post-deformation characterization under controlled conditions, are necessary to develop a micromechanical model of the high-strain-rate flow of comminuted material suitable for the three-dimensional computer code modeling of the penetration phenomena.

The TWC method allows for investigation of the fragmentation and flow of ceramic materials under high strain, high-strain-rate conditions. This was designed to examine the fragmentation and large deformation of ceramics, representing their behavior adjacent to the projectile.

4.8.2   Shear Instability of Prefractured Ceramic

4.8.2.1   Silicon Carbide, Strain Partitioning

The overview of the ceramic specimens after the first and second explosive events in the TWC test, according to Figure 4.3, is shown in Figure 4.12 for SiC III. The qualitative behavior of other materials was the same and can be found in Shih [1998].

After the first explosive event (Fig. 4.12(a)), all the specimens had numerous cracks and incipient shear bands shown by arrows. Shear cracks roughly followed directions 45° to the radial direction; cracks along 90° of the radial direction were also observed. The latter cracks are related to the release stress waves. As shown by Shih, Nesterenko, and Meyers [1998], material comminution can be identified around the cracks and an average fragment size is shown in Table 4.3. Different SiC had different crack surface areas.

SiC-I has the highest crack surface area. Since all specimens are tested under identical conditions, the difference in the crack surface area is directly related to the intrinsic material properties. Consistently, SiC-I has a lower fracture toughness than SiC-II and SiC-III. Thus, the material with a lower fracture toughness has a higher crack surface area during the fragmentation event.

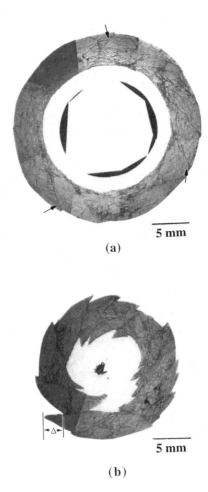

(a)

(b)

Figure 4.12. Overview of initially solid SiC-III ceramic specimens: (a) after the first explosive event-prefragmentation; (b) after the second explosive event-deformation. (Reprinted from *J. Appl. Phys.*, **83**, Shih, C.J., Nesterenko, V.F., and Meyers, M.A., High-Strain-Rate Deformation and Comminution of Silicon Carbide, pp. 4660–4671, 1998).

The second explosive event produced significant shear band displacements (Fig. 4.12(b)) with a relatively small number of the shear bands most probably created during the first stage (see Tables 4.4 and 4.5). This means that faults created during the first low strain event will also be active during the second large strain flow instead of creating brand new faults.

To analyze the material behavior, the total tangential strain (e_t) can be partitioned into a homogeneous strain (e_h) and a shear band strain (e_s):

Table 4.3. Crack surface area after the first explosive event.

Material	S_v (mm²/mm³)
SiC-I	41.2 ± 2.0
SiC-II	29.0 ± 1.6
SiC-III	31.8 ± 1.2

Table 4.4. Deformation during the first explosive event.

	$e_{\varphi\varphi}$ Inner radius	$e_{\varphi\varphi}$ Outer radius	No. of bands	$\Sigma\Delta$ (mm)	e_t	e_s	e_s/e_t
SiC-I	−0.090	−0.04	11	1.24	−0.06	−0.015	0.25
SiC-II	−0.090	−0.03	13	1.27	−0.06	−0.015	0.25
SiC-III	−0.090	−0.04	14	1.83	−0.065	−0.022	0.37

Table 4.5. Deformation during the second explosive event.

Material	$e_{\varphi\varphi}$ Inner radius	$e_{\varphi\varphi}$ Outer radius	No. of bands	$\Sigma\Delta$ (mm)	e_t	e_s	e_s/e_t
SiC-I	0.36	−0.14	10	9.52	0.23	−0.12	0.52
SiC-II	0.38	−0.13	16	10.59	0.23	−0.13	0.57
SiC-III	0.39	−0.14	13	10.00	0.24	−0.13	0.54

$$e_t = e_h + e_s. \tag{4.25}$$

This strain distribution is important because mechanisms of large strain flow inside and outside shear bands may be quite different, as will be shown later.

Since the tangential strain varies from the inner radius to the outer radius, the global tangential strain is approximated at the middle point of the ceramic; i.e.,

$$e_t = \frac{r_{mf}}{r_{m0}} - 1, \tag{4.26}$$

where r_{m0} and r_{mf} are the initial and final mean radii.

Table 4.6. Summary of the shear band configuration of SiC.

	SiC-I	SiC-II	SiC-III
Number of groups	2	2	2
	2	2	2
		2	
Number of shear bands	11	18	15
	10	17	11
		14	
Average shear band spacing	2.81 mm	1.72 mm	2.02 mm
	3.05 mm	1.82 mm	2.75 mm
		2.18 mm	
Average displacement	0.904 mm	0.548 mm	0.669 mm
	0.910 mm	0.561 mm	0.907 mm
		0.884 mm	

Because of the relatively thin layer of ceramic it can be a reasonable representation of the amount of strain for the total specimen. The strain due to shear localization on the mesolevel (intermediate between the sample size and the granular diameter) is obtained by taking the tangential component of the shear band displacement through the following equation (Δ is shown in Fig. 4.12(b):

$$e_s = \frac{\sqrt{2}\Sigma\Delta}{4\pi r_{m0}},$$ (4.27)

where $\Sigma\Delta$ is the summation of all displacements (Nesterenko et al. [1996]).

The tangential strains, after these two explosive events, are listed in Tables 4.4 and 4.5 (Shih, Nesterenko, and Meyers [1998]). The total tangential strain (e_t) for these three different silicon carbides was similar. The first explosive event generated a total tangential strain of –0.06, and the second explosive event provided a total tangential strain of –0.23. During the first explosive event, the ratio between the shear band strain (e_s) and the total strain (e_t) varied from 0.24 to 0.37. As shown in Table 4.5, SiC-III exhibited the highest ratio e_s/e_t in the first explosive event. This did not manifest itself in the second explosive event. During the second explosive event, shear localized strain produced approximately 55% of the total strain for all specimens.

Table 4.6 summarizes the data for the final configurations of all eight specimens after the second explosive event. The average shear band spacing (S) is estimated from the total number of shear bands (N):

$$S = \frac{2\pi r_m}{\sqrt{2}N},$$ (4.28)

where r_m is the mean radius. There are two groups of shear bands: clockwise and counterclockwise. They have roughly the same number of shear bands, and they occupy opposite sides of the cylinder. This grouping of shear bands into clockwise or counterclockwise is indicative of the cooperative material motion and self-organization among the bands. As deformation proceeds beyond the formation of shear bands, the shear band displacement Δ increases by the relative motion of the adjacent blocks. During this process, it is necessary to rotate these fragmented blocks. Therefore, the symmetry of the deformation is broken and the macroscopic state of pure shear is destroyed. As a result, the shear bands do not follow exactly the 45° directions at the end of the deformation. The trajectories of the shear bands at the end of the deformation can be expressed analytically by projecting the original 45° shear bands onto the final dimension. The calculated trajectories are in good agreement with experimental observations (Nesterenko et al. [1995]).

After the second explosive event, shear bands were fully developed. Many shear bands exhibited two adjacent cracks, as shown in Figure 4.13(a). The material microstructure inside and outside the shear bands was quite different.

All shear bands contained numerous fragments, with a bimodal size distribution. The large fragments ranged from 20 μm to 200 μm, and the small fragments were less than, or equal to, the initial grain size (5.6 μm for SiC-I and 4.1 μm for SiC-II and SiC-III). These large fragments did not have sharp corners and contained numerous microcracks, as shown in Figure 4.13(b). Just outside the shear bands, small particles and fragments can be identified, as shown in Figure 4.13(c). These fragments also had the same characteristics as the large fragments inside the shear band: round corners and microcracks.

The localized shear deformation inside the shear band is of simple shear, and the corresponding strain (γ_s) can be estimated from the following equation:

$$\gamma_s = \frac{\Delta}{T}, \tag{4.29}$$

where Δ is the displacement of the shear band, and T is its thickness. This localized shear strain varied from 2 to 14, and the average of the localized shear strains were 7.1, 5.5, and 5.4 for SiC-I, SiC-II, and SiC-III, respectively.

With respect to the interaction between the cracks and microstructure, SiC-II and SiC-III had a predominantly intergranular fracture mode and SiC-I tended to fracture in a transgranular pattern. This is consistent with the results of quasistatic testing (Shih, Nesterenko, and Meyers [1998]). Therefore, it is concluded that the materials with segregated grain boundaries exhibit intergranular fracture, regardless of strain rate. As a result, the cracks in SiC-II and SiC-III were tortuous and the cracks in SiC-I were straight. In SiC-II and SiC-III, material comminution was often observed around the cracks, but SiC-I had less comminution under the same situation. It is obvious that the microstructural differences affect the roughness of the crack surfaces, resulting in differences in comminution, shear band thickness, and localized shear strain. However, the overall macroscopic deformation is not influenced by the microstructure, as shown in Tables 4.3 to 4.5.

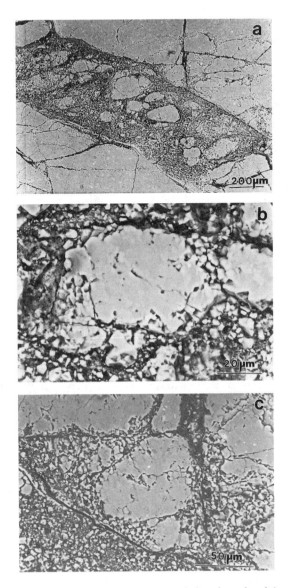

Figure 4.13. Microstructure of a thick segment of the shear band in prefragmented SiC-II: (a) overview; (b) large fragments inside the shear band; (c) fragments along a crack next to the shear band. (Reprinted from *J. Appl. Phys.*, **83**, Shih, C.J., Nesterenko, V.F., and Meyers, M.A., High-Strain-Rate Deformation and Comminution of Silicon Carbide, pp. 4660–4671, 1998.)

4.8.2.2   Mechanisms of Large Strain Flow in Fractured Ceramics

The flow of ceramics under dynamic loading at high strain, high-strain-rate deformation is much less understood than that of metals, which has been extensively investigated. In granular Al_2O_3, Nesterenko et al. [1996] observed shear localization due to a softening mechanism, not directly attributed to thermal effects. Shear localization in granular or fragmented materials can be rationalized as a mechanism to bypass the necessity of dilatation which accompanies large inelastic deformation. Under small or no confinement, the deformation of a fragmented ceramic will exhibit dilatation which is actually the mechanism of a homogeneous large strain flow.

Curran et al. [1993] developed a micromechanical model, using two-dimensional square blocks with voids and dislocation arrays. The deformation of fragmented materials is carried out through the sliding of the blocks. Hegemier [1991] proposed a hexagonal network of fragments to represent a fractured ceramic, as shown in Figure 4.14(a). His approach is used here to model dilatation. Upon deformation, the hexagonal fragments move with respect to each other to open the crack surfaces. As shown in Figure 4.14(b), the volumetric dilatation (ratio of the void volume to the solid volume) can be represented by the areal ratio (δ) between the dashed parallelogram and the hexagon:

$$\delta = \frac{A_{\text{parallelogram}}}{A_{\text{hexagon}}} = \frac{4}{3\sqrt{3}} \frac{\chi}{D}, \tag{4.30}$$

where D is the diagonal length of the hexagonal fragment, and χ is the height of the parallelogram. In the idealized configuration used here, the hexagons move along the direction 60° of the x-axis providing a global simple shear flow of material along the x-axis. The volumetric dilatation can be related to the global shear strain, γ:

$$\delta = \frac{2\sqrt{3}}{3} \frac{\tan \gamma}{1 - \sqrt{3} \tan \gamma}. \tag{4.31}$$

As the deformation continues, the dilatation increases approximately linearly with strain. The dilatation reaches a maximum at

$$\chi_{\max} = \frac{\sqrt{3}}{4} \Delta, \tag{4.32a}$$

corresponding to

$$\tan \gamma = \frac{1}{3\sqrt{3}}, \tag{4.32b}$$

as shown in Figure 4.14(c). After this critical condition, the hexagonal blocks move along the shear direction, leading to a decrease in dilatation, as shown in Figure 4.14(d). The volumetric dilatation is then represented by the following

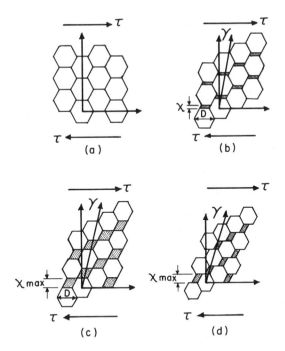

Figure 4.14. Homogeneous deformation through the Hegemier mechanism: (a) formation of hexagonal fragments; (b) crack opening and dilatation; (c) dilatation reaching its maximum; (d) reduction in dilatation. (Reprinted from *J. Appl. Phys.*, **83**, Shih, C.J., Nesterenko, V.F., and Meyers, M.A., High-Strain-Rate Deformation and Comminution of Silicon Carbide, pp. 4660–4671, 1998.)

equation for the corresponding γ

$$\delta = \frac{1}{2} - \frac{\sqrt{3}}{2} \tan \gamma, \tag{4.33a}$$

$$\gamma > \tan \gamma^{-1} \frac{1}{3\sqrt{3}}. \tag{4.33b}$$

Further deformation will return the dilatation to zero because of the rearrangement of the hexagonal fragments and after that the dilatation cycle will be repeated. In reality, not all fragments move along the same direction, and the overall dilatation would not have this simple periodical variation. By integrating the dilatation with respect to the strain, one can find the mean dilatation (δ_m) in this dilatation cycle $\delta_m = 0.16$ (Shih, Nesterenko, and Meyers [1998]) which represents the mean dilatation in the two-dimensional deformation by the motion of hexagonal blocks with the same size.

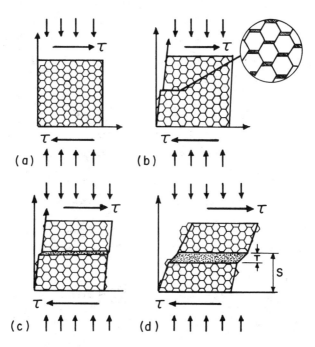

Figure 4.15. Inhomogeneous deformation under high lateral confinement: (a) formation of hexagonal fragments; (b) initiation of shear band; (c) developed shear band; (d) propagation of shear band-thickening. (Reprinted from *J. Appl. Phys.*, **83**, Shih, C.J., Nesterenko, V.F., and Meyers, M.A., High-Strain-Rate Deformation and Comminution of Silicon Carbide, pp. 4660–4671, 1998.)

With lateral confinement (superimposed compressive stresses), the dilatation is constrained, and the homogeneous deformation mechanism is inhibited. The considered homogeneous mechanism due to the block rearrangements cannot be responsible for the large strain, high-strain-rate shear deformation like tan$\gamma > 0.2$ due to kinetic consideration. The sequence of the inhomogeneous deformation is shown in Figure 4.15. A similar array of Hegemier hexagonal fragments is formed first. Then particle fracture occurs preferentially in the direction of maximal shear stresses along a narrow band, which provides a path for future shear localization. This is the softening mechanism in ceramics of a nonthermal nature. Then comminution is initiated in some localized regions and propagates along a shear band. Further macroscopic deformation takes place through the extension and thickening of the shear band.

The total dilatation (δ_t) is determined by the dilatation inside the band (δ_s):

$$\delta_t = \delta_s \frac{T}{S}, \tag{4.34}$$

where T is the shear band thickness and S is the shear band spacing.

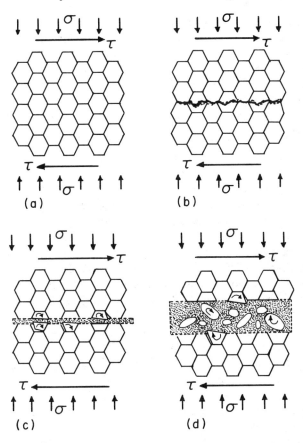

Figure 4.16. Schematic diagram for one of the mechanisms for the shear band formation: (a) fragmentation; (b) comminution along the crack surfaces; (c) incorporation of adjacent fragments; (d) erosion of fragments and shear band thickening. (Reprinted from *J. Appl. Phys.*, **83**, Shih, C.J., Nesterenko, V.F., and Meyers, M.A., High-Strain-Rate Deformation and Comminution of Silicon Carbide, pp. 4660–4671, 1998.)

In this inhomogeneous deformation process, the strain is concentrated in the shear bands, and Hegemier's homogeneous mechanism is applicable inside the shear localization regions (see detail in Fig. 4.15(b)), with a reduction of scale of the unit hexagons. Therefore, one can consider a constant dilatation inside the shear band; i.e., $\delta_s = \delta_m = 0.16$. In experiments, the average shear band thickness is about 150 μm and the shear band spacing is about 2.3 mm. Eq. (4.29) predicts a total volumetric dilatation of 0.01 which is much smaller than the dilatation associated with homogeneous deformation. It is clear that shear localization is a favorable deformation mechanism under constrained dilatation. Based on the microstructural observations, the following mechanism is proposed for shear band formation in fragmented SiC, as depicted in Figure 4.16:

(a) The material is fragmented through the formation of cracks (Fig. 4.16(a)).

(b) Moderate comminution proceeds through the friction of crack surfaces, corresponding to the initiation of shear bands (Fig. 4.16(b)).

(c) When the two interfaces of the shear band move in opposite directions, the adjacent fragments are incorporated into the shear band (Fig. 4.16(c)).

(d) With further shear, these fragments are rotated and eroded in the shear band. The shear band thickens through the continuous incorporation of fragments into the band and their erosion during flow (Fig. 4.16(d)).

Another mechanism resulting in the bimodal particle distribution observed inside shear band (Fig. 4.13) is the incorporation of large fragments between two initial cracks with comminution of materials in the vicinity of this fragment during shear band development. Usually shear bands in comminuted material contain very narrow crack-like segments adjacent to the segments with a structure depicted as in Figure 4.13.

Approximately 55% of the total tangential strain is accommodated by the shear localization which is carried out by comminution, through incorporation and erosion of fragments. The micromechanical model of the inelastic flow of comminuted ceramics proposed by Curran et al. [1993] should be modified to include the effect of shear localization.

Material comminution was also observed outside the shear bands through the advance of a comminution front by localized bending, as shown in Figure 4.17(a) for SiC-II. This comminution front clearly defines a boundary between the comminuted region due to additional compression stresses and the initially fragmented area, as shown in Figure 4.17(b).

A schematic representation is shown in Figure 4.18. A shear band separates two fragmented blocks, and the size of the blocks is determined by the spacing of the shear bands. During shear band propagation, the shear displacement increases accordingly. However, these two blocks cannot keep the same orientation during the large deformation. Their rotation is necessary to accommodate the translation along the shear surface. A localized bending is therefore induced to alter the shape and orientation of these two fragmented blocks. Material comminution is the deformation mechanism that accomplishes the alteration of the geometry of the fragmented blocks. During this localized bending, a comminution front is initiated in the inner radius region and propagates toward the outer radius region.

Comparing Figure 4.17 with Figure 4.13, one can observe that the fraction of material comminution outside the shear bands by bending (Figs. 4.17 and 4.18) is much less than the fraction of comminution inside the shear bands. Therefore, bending comminution is considered as a secondary comminution mechanism. Evidently, this latter mechanism is enhanced by larger spacing between the shear bands.

The observed comminution of ceramic particles due to high-strain plastic flow does not need high shock pressures, but requires only low-detonation velocity industrial energetic materials and can be used for practical applications.

In summary, the shear band spacing introduces a new scale in fragmented

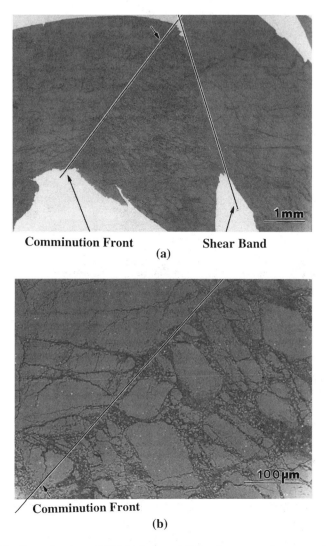

Comminution Front   **Shear Band**

(a)

Comminution Front

(b)

Figure 4.17. Comminution of SiC-II due to additional compression caused by local bending: (a) overview; (b) high magnification at the comminution front. (Reprinted from *J. Appl. Phys.*, **83**, Shih, C.J., Nesterenko, V.F., and Meyers, M.A., High-Strain-Rate Deformation and Comminution of Silicon Carbide, pp. 4660–4671, 1998.)

material: the size of fragmented blocks. The material movement outside the shear localization bands can be considered mainly as the deformation within the fragmented blocks. This deformation is carried out by comminution, through

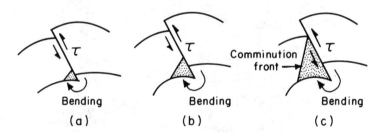

Figure 4.18. Schematic diagram shows the comminution by local bending. (Reprinted from *J. Appl. Phys.*, **83**, Shih, C.J., Nesterenko, V.F., and Meyers, M.A., High-Strain-Rate Deformation and Comminution of Silicon Carbide, pp. 4660–4671, 1998.)

propagating comminution fronts. Similar results were obtained for prefractured alumina (Chen et al. [1995], Chen [1997]).

4.8.3   Granular Ceramics

4.8.3.1   Granular SiC, Partitioning of Strain and Spacing Between Shear Bands

Granular ceramic materials are different from prefractured initially solid ceramic mainly because of higher porosity. Particle size in granular materials is the main structural parameter. For example, in tapped granular materials the thickness of a "weak" shear localization band without particle fracture is 10 to 15 times the mean particle diameter (Vardoulakis [1980]).

Shih, Meyers, and Nesterenko [1998] tested fine, medium, and coarse powder specimens of SiC under identical conditions of the TWC method (see Fig. 4.4). The compact density (after powder was loaded into the cavity), densified density (after the first explosive event), and deformed density (after the second explosive event), are listed in Table 4.7. The density of a powder compact is determined by the particle characteristics, such as size distribution and morphology. As shown in Table 4.7, the initial powder packing density varied from 33% to 56% of the theoretical density. After the first explosive event, approximately 85% of theoretical density was achieved. The second explosive event reduced the average density to about 71% of the theoretical density.

The objective of the first explosive event was to densify the powder compact through dynamic compressive loading. Scanning electron microscope (SEM) analysis indicated that the microstructure was uniform from the inner radius area to the outer radius region. All densified ceramics had a bimodal particle distribution, consisting of large particles surrounded by small particles. Particle break-up can be clearly identified in the coarse and medium powder specimens

Table 4.7. Density of granular SiC after different stages.

Powder type Average particle size		Fine (0.4 μm)	Medium (3 μm)	Coarse (50 μm)
Compact density	1	39 %	34 %	59 %
	2	37 %	33 %	54 %
	3	38 %		56 %
Average		38 %	33 %	56 %
Density	1	73 %	81 %	92 %
After the first explosive	2	87 %	94 %	84 %
	3	88 %		88 %
Average		83 %	87 %	88 %
Density	1	67 %	63 %	78 %
After the second explosive	2	77 %	63 %	78 %
	3	66 %		78 %
Average		70 %	63 %	78 %

(Shih [1998]; Shih, Meyers, and Nesterenko [1998]). All powders had a wide range of particle size distribution, and the densification is mainly through: (1) break-up of large particles; and (2) rearrangement of the comminuted particles to achieve a high packing density. These three different SiC powders represent particle sizes of two orders of magnitude difference. The largest particles in the fine powder are around 2 μm, and are much smaller than the smallest particles in the coarse powder. No marks of shear localization were observed at this stage.

The second explosive event provided high-strain-rate deformation of the densified material. The overall view of samples with three ceramic particle sizes is shown in Figure 4.19. Profuse shear localization can be seen for all of them.

In the inner radius region, all shear bands followed roughly an angle of 45° with the radial direction, corresponding to the maximum shear direction. As shown in Figure 4.19(d), the angle (θ) of the shear bands with respect to the radial direction increases as the trajectory moves toward the outer radius regions. The angle is determined by the magnitude of the principal stresses and shear strength, and can be predicted from the Mohr–Coulomb flow criterion. This angle (θ) could provide a way to determine the friction coefficient in the Mohr–Coulomb criterion determining the resistance to flow in conditions when particle fracture is involved (Nesterenko and Lam [1999]).

The shear band configurations for these eight granular specimens are summarized in Table 4.8. The spacing between the shear bands is fairly regular, indicating a self-organization phenomenon with minor influence on the initial particle size on spacing which varies from 0.5 mm to 1 mm in different materials. This behavior indicates that continuum-based theory of shear band patterning can be developed for granular materials as is a case with the solids considered in Sections 4.4 to 4.7. Shear patterning in granular ceramics is in

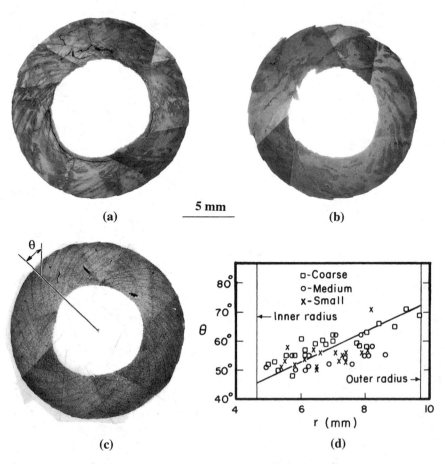

Figure 4.19. Overall configuration after densification and deformation: (a) fine powder; (b) medium powder; (c) coarse powder, (d) relationship between the tangential angle and radius. (Reprinted from *Acta Materialia*, **46,** Shih, C.J., Nesterenko, V.F., and Meyers, M.A., High-Strain-Rate Deformation of Granular Silicon Carbide, pp. 4037–4065, 1998, with permission from Elsevier Science.)

striking contrast to the behavior of the prefractured (more dense) ceramic, compare Figure 4.12(b) and Figure 4.19(b)–(c).

The shear band spacing is significantly lower than the one for the prefractured SiC, described by Shih, Nesterenko, and Meyers [1998] (Table 4.5): the spacings vary between 1.7 mm and 3.1 mm. The same differences were observed between granular and prefractured Al_2O_3 (Chen, Meyers, and Nesterenko [1995]; Nesterenko, Meyers, and Chen [1996]). The average shear band spacing for the prefractured Al_2O_3 is 2.0 mm, and the spacings are 0.49 mm and 0.61 mm for

Table 4.8. Summary of shear band configurations for granular SiC in different experiments (3 for fine, 2 for medium, and 3 for coarse particles).

Particle size	0.4 µm	3 µm	50 µm
Number of groups	4	2	4
	2	2	4
	4	—	4
Number of shear bands	72	34	52
	51	24	54
	52	—	48
Average shear band spacing	0.42 mm	0.86 mm	0.62 mm
	0.58 mm	1.20 mm	0.57 mm
	0.58 mm	—	0.66 mm
	0.53 ± 0.07	1.03 ± 0.17	0.62 ± 0.03
Average shear band displacement	26 µm	72 µm	—
	41 µm	133 µm	—
	37 µm	—	—
	35 ± 6 µm	102 ± 30 µm	—

Table 4.9. Deformation during the second explosive event.

Powder	$e_{\varphi\varphi}$ Inner radius	$e_{\varphi\varphi}$ Outer radius	No. of bands	$\Sigma\Delta$ mm	ε_{eff}	e_t	e_s	e_s/e_t
0.4 µm	−0.35	−0.12	58	1.95	−0.28	−0.22	−0.03	0.12
3 µm	−0.37	−0.1	29	2.82	−0.28	−0.21	−0.04	0.18
50 µm	−0.36	−0.1	51	—	−0.27	−0.21	—	—

the granular Al_2O_3 with 0.4 µm and 4 µm particle size, respectively. In general, a change of particle size more than two orders of magnitude did not essentially effect spacing between the shear bands. This means that this parameter is determined by some macroscopic characteristics of granular material which are not very sensitive to particle size. It is also possible to conclude that slightly less dense ceramics may provide more stability for structures collapsing under high strain deformation.

Both clockwise and counterclockwise shear bands can be identified. There were either two or four groups of the clockwise and counterclockwise shear bands. But mainly shear bands have the same orientation. This preferential orientation and grouping is indicative of cooperative material motion and self-organization among the bands.

The partitioning of global strain (e_t) into homogeneous deformation (e_h) and deformation localized in shear bands (e_s) can be calculated as described in the previous section (Nesterenko, Meyers, and Chen [1996]; Shih, Meyers, and Nesterenko [1998], see Eqs. (4.24)–(4.26). The average results, after the second explosive events for each particle size, are listed in Table 4.9. The global tangential strains (e_t) for these three SiC powders were almost identical (–0.21). However, the shear band strain (e_s) varied from 0 to –0.038. The coarse powders did not exhibit any distinguishable shear displacement, although their shear bands can be clearly identified. The medium-sized powders had the highest amount of deformation localized in shear bands. The behavior of different materials will be analyzed in the following sections.

4.8.3.2   Deformation of Coarse Granular SiC

Even though shear bands were clearly identified in the coarse granular SiC, the shear band displacement (Δ) is negligible, as shown in Figure 4.19(c) and Table 4.8. A bimodal particle size distribution is observed inside the shear bands, as shown in Figure 4.20(a). The large particles range from 10 μm to 30 μm, and the small particles vary from 1.5 μm to 5 μm. The large particles have round corners and contain numerous microcracks, as shown in Figure 4.20(b). Outside the shear band, small and large particles can also be identified, as shown in Figure 4.20(c). The large particles exhibit the same characteristics as the large particles inside the shear band—round corners and microcracks. They contain characteristic crack patterns, emanating from their contact points (marked by arrows in Fig. 4.20(c)).

Figure 4.20 illustrates the main mechanism of softening in granular materials—microcracking in the individual particles in the preferential direction of maximal shear stress. The preferential microcracking outside the shear band along its direction is clear in Figure 4.20(a),(c). The merger of these microcracks into one macrocrack determines the shear band formation. Shih, Meyers, and Nesterenko [1998] presented an analysis of the stresses within particles subjected to compressive tractions.

An interesting picture of the shear bands intersection is presented in Figure 4.21. These sites can be especially effective as local hot spots in granular energetic materials for reaction initiation.

The existence of these two shear bands can be caused by a preferential particles fracture in two directions (compare Figs. 4.20 and 4.21). The microcracking of particles in one of the directions is evident. The kinking is connected with different times of shear band origination. Also it is clear that large fragments incorporated into the shear band are composed from small particles and stick together. Only coming through subsequent large strains will break these fragments into small separated pieces and mix them. This phenomenon of multiple fracture and the mixing of fragments is a very

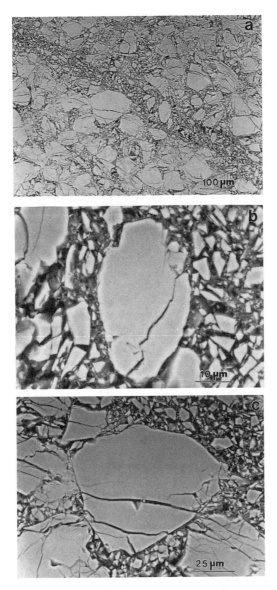

Figure 4.20. A shear band in coarse powder: (a) overview; (b) particle inside the shear band; (c) particle at the interface of the shear band. (Reprinted from *Acta Materialia,* **46,** Shih, C.J., Nesterenko, V.F., and Meyers, M.A., High-Strain-Rate Deformation of Granular Silicon Carbide, pp. 4037–4065, 1998, with permission from Elsevier Science.)

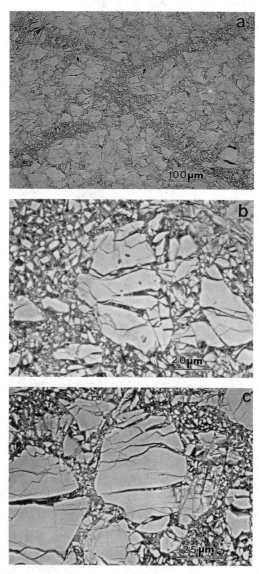

Figure 4.21. Intersection of shear bands in coarse powder: (a) overview; (b) particles inside the shear band, close to intersection; (c) particles outside the shear band. (Reprinted from *Acta Materialia,* **46,** Shih, C.J., Nesterenko, V.F., and Meyers, M.A., High-Strain-Rate Deformation of Granular Silicon Carbide, pp. 4037–4065, 1998, with permission from Elsevier Science.)

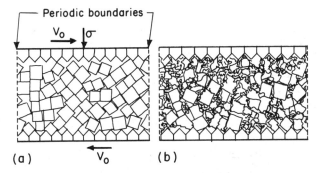

Figure 4.22. Computer modeling of particle comminution inside a sheared area: (a) simulated shear cell; (b) fragmented particles inside the shear band. (Reprinted from *Powder Technology*, **94**, Potapov, A.V., and Campbell, C.S., Computer Simulation of Shear-Induced Particle Attrition, pp. 109–122, 1997, with permission from Elsevier Science.)

important feature making shear bands very effective for reaction initiation in a mixture of solid materials.

Shih, Nesterenko, and Meyers [1998] observed and described in detail the same mechanism in fragmented SiC, subject to similar deformation. Shear band formation in fragmented SiC is propitiated by comminution of fragments. The process proceeds through the incorporation of nearby fragments into the shear band and their erosion inside the shear band (Fig. 4.16(d)). During deformation, large particles are crushed and broken down, and the resulting distribution of smaller particles can be rearranged to pack more efficiently. The comminution and rearrangement are localized in the shear bands.

Bimodal distribution of particles inside the shear band (Figs. 4.20 and 4.21) can be caused not only by the incorporation of adjacent large particles into a uniformly comminuted region of the shear band as shown in Figure 4.16(d), but also by the retarded kinetics of fracture of some individual particles inside the shear band. These particles practically did not change their position but were not comminuted to the same extent as surrounding particles. This may be the natural result of the highly heterogeneous conditions of particle loading in granular materials. The preferentially loaded particles are fractured first creating small fragments, which retard the fracture of the remnant particles due to their more uniform loading in the matrix of a small fragments.

Computer simulation, of brittle particle arrays subjected to simple shear, predicts the same mechanism for particle break-up (Potapov and Campbell [1997]; Fig. 4.22(b)). Since the granular material is porous, the localized comminution and rearrangement are sufficient to accommodate the large strain without the extension of the shear bands.

Recently, Schwarz and Horie [1998] used the two-dimensional discrete meso-dynamic method (DM2) to study a behavior of granular material at shear strains up to $2/3 \cdot 10^4$ s^{-1}. They found that the stress field in granular flow is highly heterogeneous and that stress fluctuations exceed the average normal stress by an order of magnitude. This can cause a local fracture of particles inside the force chains.

Although the measured shear band strain (e_s in Table 4.9) is negligible, the global deformation is still inhomogeneous in nature. The small displacements in the shear bands indicate that they were created during the final stage of the collapse process; this means that their initiation strain is higher than the other two granular materials.

The reason why the large grain-size granular material is more stable with respect to shear localization may be due to some specific for the granular materials hardening mechanism. This mechanism can be connected to the fracture of large particles, with a subsequent rearrangement and increased density. The decrease in initial particle size makes this mechanism less efficient due to the higher strength of the smaller particles.

4.8.3.3   Deformation of Medium-Size Granular SiC, "Sintering" Within Shear Bands

This group of specimens had the largest shear band displacement (Δ), as shown in Table 4.8. Figure 4.23 shows a shear band in which comminution occurs. Inside the shear band, there is no particle larger than 1.5 µm, because all the large particles have been comminuted.

This type of shear band is similar to the shear bands in the coarse powder specimens and in fragmented SiC. However, this comminution phenomenon was observed only in the shear bands with small displacement (<150 µm).

For shear band displacements larger than 150 µm, a layer of dense SiC along the shear band can be identified, as shown in Figure 4.24. The density of this SiC layer decreases gradually away from the crack, as shown in Figure 4.23(b).

It is expected that the large displacement generates local heating. The calculated temperature rise is presented in Shih, Meyers, and Nesterenko [1998].

Figure 4.24(b) shows the micro-indentation markings that were made on the dense SiC layer in the shear bands. The material did not crumble; on the contrary, the indentation markings are clear and confirm that the powder is fully bonded inside the shear localization regions. The hardness value obtained is 26 GPa, which compares favorably with a reported hardness of 23 GPa for fully dense SiC (Shih, Meyers, and Nesterenko [1998]). The difference could be due to the small load (10 gf) used in the current experiments.

It is thought that the strong bonding between the particles, under the influence of high superimposed pressure, is related to the combined (and coupled) effects of intense plastic deformation and heating, during the shear localization

Figure 4.23. A shear band of medium powder with small displacement (Δ < 200µm, γ_s < 12): (a) overview; (b) high magnification. (Reprinted from *Acta Materialia*, **46**, Shih, C.J., Nesterenko, V.F., and Meyers, M.A., High-Strain-Rate Deformation of Granular Silicon Carbide, pp. 4037–4065, 1998, with permission from Elsevier Science.)

event. This is indeed surprising and is a novel phenomenon, since conventional sintering requires times that are many orders of magnitude higher. The possibility of accelerated diffusion due to the generation of large dislocation densities in the deforming particles should be considered, as well as phase transformation of SiC under high shear strains.

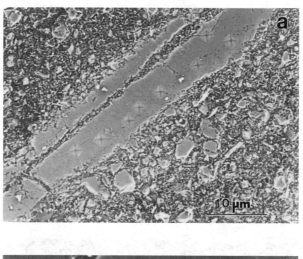

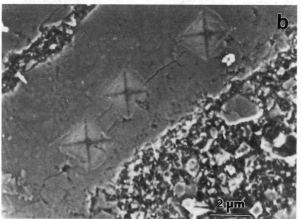

Figure 4.24. A shear band of medium powder with large displacement ($\Delta > 200\mu m$, $\gamma_s >$ 12): (a) overview; (b) high magnification and microhardness indentation on the dense SiC layer within the shear band. (Reprinted from *Acta Materialia*, **46**, Shih, C.J., Nesterenko, V.F., and Meyers, M.A., High-Strain-Rate Deformation of Granular Silicon Carbide, pp. 4037–4065, 1998, with permission from Elsevier Science.)

4.8.3.4   Deformation of Fine Granular SiC

In fine granular SiC, a solid sintered layer of SiC was also identified in shear bands with large displacement ($\Delta > 150\ \mu m$), as shown in Figure 4.25. This sintered SiC layer was similar to the one observed in the medium powder specimens (Fig. 4.24).

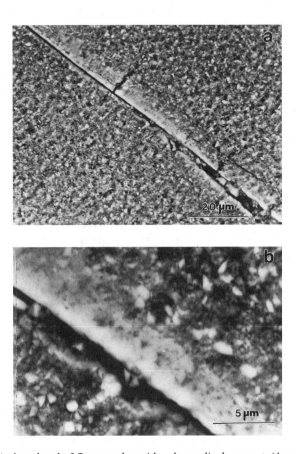

Figure 4.25. A shear band of fine powder with a large displacement ($\Delta > 150\mu$m, $\gamma_s >$ 15): (a) overview; (b) high magnification. (Reprinted from *Acta Materialia*, **46**, Shih, C.J., Nesterenko, V.F., and Meyers, M.A., High-Strain-Rate Deformation of Granular Silicon Carbide, pp. 4037–4065, 1998, with permission from Elsevier Science.)

If the shear-band displacement was between 100 and 150 μm, only rounded SiC nodules were formed. These nodules represent the onset of bonding between the powders, and were also observed in Al_2O_3 granules with 0.4 μm particle size (Nesterenko, Meyers, and Chen [1996]). It is obvious that there is a critical displacement necessary to generate sufficient heat to effect the particle bonding.

Particle break-up was not activated in fine SiC powders, but pore compaction was identified. This is particularly evident for shear bands with low displacement ($\Delta < 100$ μm). The particle size inside the shear band is nearly identical to the particle size outside the shear band. However, the packing density is higher inside the shear band; this is evident by the dark color in the secondary electron SEM images (Shih, Meyers, and Nesterenko [1998]).

4.8.3.5   Particle Sizes and Thicknesses of Shear Bands in Granular Materials

The most interesting phenomenon is that shear band thickness depends only slightly on initial particle size, and increases with a decrease in initial porosity. The average shear band thickness for the prefractured SiC is about 150 μm (Shih, Nesterenko, and Meyers [1998]); the shear band thicknesses are 40, 25, and 15 μm for the granular SiC with 50, 3, and 0.4 μm particle sizes, respectively. These results show that the shear band thickness is dependent on the type of materials: prefractured SiC has a much larger shear band thickness than the granular SiC.

However, the particle size in the granular SiC does not effect the thickness significantly. This is quite different from the well-established fact of proportionality of shear band thickness to particle size under quasistatic deformation without particle fracture (see the review by Oda and Kazama [1998]). This difference is caused by the apparent fact that in confined dense ceramics, particle size does not represent the structural element responsible for formation of the shear band as in granular materials under normal conditions.

4.8.3.6   Fracture and Heat Release in Granular Materials Under High Shear Strain

It is clear from previous considerations that if the shear strain inside the shear bands is about 10 and the initial particle size is less than 3 μm, the granular material can be heated up enough to initiate plastic flow and the subsequent bonding process between SiC particles. This phenomenon was not observed for larger particles and in comminuted ceramic. It is important to understand the reason for this difference in behavior.

One aspect of ceramic flow a under high-strain-rate condition is connected with the particle fracture which depends on the initial particle size. The variation in initial particle size results in an essential change of the overall behavior of the granular material (Chen [1997]; Shih, Nesterenko, and Meyers [1997], [1998]).

One of the important causes of the observed difference can be the dependence of the energy fraction converted into heat during the flow of granular material if the particle fracture is involved. Relatively small particles will not break and the whole energy of plastic deformation will be converted into heat, whereas the large particles are able to convert an essential part of the plastic work into the creation of new surfaces thus leaving less energy for heat generation.

To evaluate the possibility of this qualitative dependence of granular material behavior on particle size, let us find the ratio K_{ph} of thermal energy to the work of plastic deformation W_p for granular material undergoing fracture (Nesterenko [1998]).

The difference between these energies is connected with the creation of new particle surfaces and also with internal damage of the newly created particles W_f:

$$K_{ph} = \frac{W_p - W_f}{W_p} = 1 - \frac{\sigma \pi (n-1) S(d)}{\tau \gamma_s d}, \tag{4.35}$$

where σ is the specific energy associated with the creation of a new surface area of fragments, τ is the global shear strength of granular material, γ_s is the shear strain inside the shear band, d is the initial particle size, n is the ratio of a newly created fragment diameter to the initial particle diameter d, and $S(d)$ is the factor responsible for the difference between the surface area of newly created fragments and the total new surface created during the fragmentation.

The coefficient S can be of the order of magnitude 10^3 that reflects the fact that the fragment size does not represent the real amount of damage and total created surface, as emphasized by Meyers [1994]. For numerical evaluations, the following parameters for SiC were selected: $\sigma = 20$ J/m^2 (Chiang et al. [1997]), $\tau = 2$ GPa, and $n = 10$ (Shih, Nesterenko, and Meyers [1997], [1998]), $d = 50$ μm. At these material parameters and shear strain $\gamma_s = 0.1$ the value of $S(d) = 100$ will result in $K_{ph} = 0.4$. For the shear strain $\gamma_s = 10$ the same K_{ph} will be obtained with $S(d) = 10^4$. So it is possible to conclude that particle fragmentation can be an important mechanism of energy dissipation inside the shear bands.

It is interesting to mention the nonmonotonous dependence of K_{ph} on the initial particle size. If the particle size is getting smaller with constant n and $S(d)$, it will result in less heat generation for smaller particle sizes. On the contrary, for sufficiently small particles with d less than 0.3 μm, it is evident that $L \rightarrow 1$ as well as $S(d) \rightarrow 1$ because the fracture process is not developing. For this limit the whole energy of plastic work will be converted into heat and $K_{ph} = 1$.

4.9   Behavior of Reactant Granular Materials Under Conditions of Controlled Localized Shear

The large strain localized shear exhibits very important features like the comminution of particles and their mixing, combined with the heat effect, as was shown in the previous section. It was observed by Nesterenko, Meyers et al. [1994], [1995] that the shear localization of relatively high amplitude can result in chemical reaction in a mixture of the solid materials. The different stages of the shear localization corresponding to the various overall and local shear strains are presented in Figure 4.26 for the Ti + Si mixture.

It is evident that a slight increase of the overall strain from $\varepsilon_{eff} = 0.33$

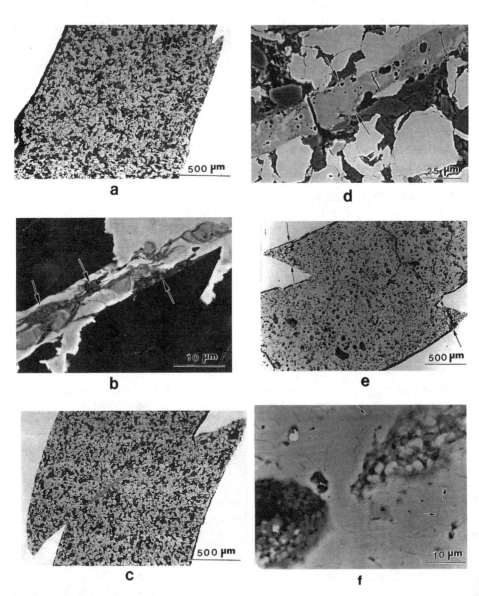

Figure 4.26. Overall views and shear band structures in the Ti + Si mixture for an increasing amount of global strain: (a),(b) ε_{eff} = 0.33; (c),(d) ε_{eff} = 0.35; (e),(f) ε_{eff} = 0.38. (Reprinted from *Shock Compression of Condensed Matter*—1997, *AIP Conference Proceedings,* edited by S.C. Schmidt, D.P. Dandekar, and J.W. Forbes. Nesterenko, V.F., Controlled High-Strain-Rate Shear Bands in Inert and Reactant Porous Materials, pp. 609–614, 1998.)

(Fig. 4.26(a),(b)) to $\varepsilon_{eff} = 0.38$ (Fig. 4.26 (e),(f)) results in a complete reaction in the material bulk being initiated inside the shear band. Small unreacted parts (shown by arrows in Fig. 4.26(e)) were used to evaluate the reaction rate in a dense, plastically deformed mixture outside the shear bands (Chen, LaSalvia, and Nesterenko [1998]). On the basis of the observations, a mechanism of *shear-assisted* chemical reaction can be proposed (Nesterenko, Meyers, Chen, and LaSalvia [1995]). The main stages are:

(1) Niobium and titanium particles, being inside the localized macroshear zone, are split into foils with a thickness on the order of magnitude of 0.1–1 μm by localized mesoshear (a "cascade" of the shear bands). They are heated as a result of the intense shear deformation which precedes their formation and have "fresh" surfaces. The spacing between these small scale shear bands in metal particles can be evaluated based on the models developed by Grady and Kipp [1987], Wright and Ockendon [1996], and Molinari [1997], which provide the numbers of the same order of magnitude. The possibility of fragmenting particles inside the shear zone is very important for the reaction mechanism.

For example, the spacing between the shear bands in Wright–Ockendon (L_{WO}) can be calculated using Eq. (4.17). Nesterenko, Meyers, and Wright [1998] demonstrated that at a strain rate of $3.5 \cdot 10^4$ s^{-1} the Wright-Ockendon model provides very good agreement with experiments on titanium. The strain rate inside the shear band in a mixture can be evaluated based on the thickness of the shear band δ, its displacement Δ, and the time of collapse t_c:

$$\dot{\gamma} \approx \frac{\Delta}{\delta t_c}. \qquad (4.36)$$

The magnitude of the averaged (over time of experiment and thickness of shear band) macroshear strain rates in (Eq. 4.36) can reach a value up to 10^7 s^{-1}. Due to this averaging procedure, the evaluation of shear spacing will provide only the upper boundary. For titanium, using characteristic values of its parameters will result in spacings $L_{WO} = 7$ μm. This value is larger than the thickness of the titanium particles created due to the mesoshear which are of the order of magnitude 1 μm or smaller (Fig. 4.26(b)). Taking into account the average evaluation of $\dot{\gamma}$ over the total time of collapse and thickness of the shear band, it is possible to conclude that the shear localization is a reasonable mechanism of particle fracture inside the macroshear band.

(2) Reaction begins due to the extensive relative flow of niobium (titanium) particles and silicon inside the shear band, which is accompanied by the melting of silicon. The reaction proceeds along the Nb(sliver)–Si interfaces, with a thickness l_r up to 1 μm as mentioned by Thadhani [1993]. Also the relative motion inside the macroshear band will promote the fracture of the reaction product between the metals and silicon providing the mechanism of reaction acceleration.

(3) Vorticity inside the shear band, originating from the instability of the gradient flow, facilitates the mixing of materials and results in the coiling of

partially reacted niobium foils (as snow balls) together with adjacent silicon into rounded particles with diameters up to 8 μm (Nesterenko, Meyers et al. [1994], [1995]; Fig. 4.26(b)). The development of a spatial pattern inside the shear band was speculated by Leroy and Molinari [1993].

(4) Reaction continues in places where temperatures are sufficiently high and is then quenched by the surrounding material (Nb–Si, Ti–Si, Ti–C) or can proceed through the surrounding material as in some experiments with Ti–Si (Fig. 4.26(e),(f)).

The shear localization in a Ti–graphite mixture was suppressed, presumably due to the lubricant properties of graphite allowing a uniform plastic flow of the mixture at higher strains in comparison with Ti–Si or Nb–Si. Only initiation of the chemical reaction in separate localized areas was observed in the Ti–graphite mixture. The replacement of the graphite by ultrafine diamond at the same chemical content of phases drastically changed the mechanical behavior: intense shear localization was observed in the latter case (Chen [1997]; Nesterenko [1998]; Chen, Nesterenko, and Meyers [1998b]).

The propagation of the reaction into the material bulk in the Ti–Si mixture and its extinction inside the shear bands in Nb–Si was explained (Chen, Nesterenko, and Meyers [1998a]) on the basis of thermochemical analysis. The difference is due to the smaller heat release from the reaction in Nb–Si and the higher thermal conductivity of this mixture. Shear band spacing is independent of the scale of the experiment; for the TWC assemblies with mean diameters of 18 mm and 34 mm, the shear band spacing is the same: $L \sim 0.6$–1 mm.

From the advance of the reaction front for the Ti–Si mixture, parallel to the shear band interface, it is possible to estimate a reaction time, by integrating the velocity (variable) over the shear band semispacing. The reaction time is estimated to be, for Ti–Si, 10 ms, and does not depend on the sample size being a function of the shear band spacing. This is in stark contrast with combustion synthesis, where the reaction time depends on the sample size in the wave propagation regime and is of the order of seconds for the given sample size and shock compression synthesis, where the reaction time can be as low as 1 μs. Thus, an intermediate reaction regime has been obtained.

Tamura and Horie [1998] performed a mesodynamic simulation of chemical reactions in shear band developing in a porous mixture of niobium and silicon. They found that a cluster of brittle material could become the initiation point of reaction and a higher void ratio inside the shear band enhances heat generation and mixing of materials.

The phenomenon of reaction-induced plasticity (RIP) was considered in papers by Levitas, Nesterenko, and Meyers [1998a,b]. In the frame of the continuum mechanics of large strain plasticity, it was shown that shear stresses may provide a positive effect on the reaction in the solid state, due to the necessity of fulfilment of the yield condition for the transforming material and also as a result of the additional heating due to the RIP.

The observed phenomena have interesting similarities to reactions in geological conditions on a completely different time scale. In naturally and experimentally deformed rocks at much smaller strain rates, the reactions between the constituent mineral, due to heat release, can occur inside the shear zones.

The grain size of the reaction products in the narrow shear zones was only 0.1–0.2 µm. In some cases it may enhance deformability (Rubie [1990]).

Howe [2000] presented extensive experimentally based analysis of the influence of microstructure upon shear band formation in explosives. He concluded that many important questions are still unanswered and this area is particularly rich for investigation.

Reactive shear bands, where heating due to viscoplastic work is combined with reactive heating, were considered analytically with applications to the initiation of HMX explosive by Dienes [1986]. The analysis of Frank-Kamenetskii is modified to account for the mechanical effect of heat generation due to shear flow. It was concluded that shear bands are capable of igniting HMX at very high strain rates when the stresses are also high.

Guirguis [2000] considered the possibility of ignition by macroscopic shear in the frame of a viscous fluid and perfectly plastic models. The viscous model predicts the localization of energy dissipation at the edges, where ignition is usually observed in experiments.

The shear-induced ignition of solid energetic materials was analyzed by Powers [1999]. It was concluded that for the HMX-based explosive LX-14, the enhanced plastic work in the localized area accelerates the reaction, but at this time the plastically deformed surrounding material is also at the threshold of reaction. This means that shear localization in such materials is not a cause of ignition because the latter is an imminent process based on the conditions in the material bulk. The author mentioned that similar conclusions may be valid for other energetic materials, which react at a temperature where the material strength is already lost, but a different scenario may be expected for materials without this property.

This predicted behavior is different from that, for example, demonstrated by the Ti + Si (Fig. 4.26(d)) mixture where reaction inside the shear band most probably started due to the melting of silicon with a subsequent lost of strength but in some cases is quenched inside the shear band (Chen, Nesterenko, and Meyers [1998a]; Chen, LaSalvia, Nesterenko, and Meyers [1998]). Also, ignition by the array of shear bands (Chen, Nesterenko, and Meyers [1998a]) should be considered instead of ignition by a single shear band. At relatively close spacing between the shear bands, the result may be quite different than in the case of one shear band.

4.10 Conclusion

The presented results of applying the TWC method to a variety of new problems demonstrate that this is a valuable technique. It allows for the investigation of complex phenomena starting from the state of pure shear and later in the state of plain strain. It was shown that material instability due to shear localization is a very common phenomenon in high strain, high-strain-rate deformation polycrystal solids, prefractured and granular inert, and reactive materials. This is mainly a strain-controlled method, but allows stress measurements during the process and the testing of constitutive equations, due to a simple state of strains.

Spacing developed during self-organization of the shear bands in titanium, and stainless steel at the range of macrostrains close to the initiation level can be correctly described by the WO and MO models, while the GK model shows good agreement with the behavior at the developed stage of the shear bands. The different behavior exhibited by Ti–6Al–4V is attributed to the fast growth of chaotically initiated shear bands bypassing the self-organized stage of their nucleus.

Shear instability accompanies the high strain, high-strain-rate flow of granular ceramics. It is very sensitive to the initial porosity and particle size affects it less extensively. The origin of the softening mechanism is the preferential microcracking of particles in the direction of the maximal shear. Heat release inside the shear band strongly depends on initial particle size as a result of competition with energy dissipation due to multiple fracture. Sintering of micron and submicron SiC particles was observed inside shear bands with a strain more than 10. Despite the dramatically different behavior of granular materials with fine (0.4 μm) and coarse particles (50 μm) inside the large strain shear band the shear band patterns in both cases are very similar. This may justify efforts toward development of a continuum approach to shear band patterning in granular media.

Multiple fracture of particles and their mixing together with heat release may initiate reaction inside the shear band. The main differences between shear- and shock-assisted chemical reactions are due to the essential differences in the shear strain, resulting in a different amount of displacements of the neighboring particles. Under shock the relative displacements of particles are comparable to or less than the particle sizes ($\gamma < 1$) and cannot induce essential shear localization and fracture inside particles, as is observed inside the shear band. The flow instabilities inherent for large strain deformation result in vorticities, thus enhancing the mixing and subsequent reactivity of mixtures inside the shear bands. The array of shear bands can ignite energetic materials at appropriate conditions of global deformation.

Problems

1. Derive Eq. (4.5) using the equation of motion for an incompressible liquid in cylindrical geometry.

2. How does particle size in granular material affect the shear band thickness in tapped density, unconfined and in dense, confined samples?

3. How will work hardening change shear band spacings according to the WO model?

4. Explain why "dynamic sintering" was not observed in experiments with prefractured ceramics in given conditions of experiments.

5. Derive Eq. (4.20). Why is the size of the fragment not important for dilation?

6. Compare energy density under shear and shock loading.

7. Explain how particle size can affect the rate of heating inside a shear localization band in granular ceramics.

8. What are the main differences between shock- and shear-assisted reactions?

9. Explain why shear localization can increase the rate of energy release in energetic materials.

10. Explain the difference between shear-induced reactions in Ti + Si and Nb + Si mixtures.

11. Derive an equation for the strain rate in incompressible material for TWC geometry for large strains.

12. Find a shape of shear bands assumed after the post-localization large strain flow during a cylinder collapse (Nesterenko, Meyers, and Wright [1995]):

$$\Theta_i = \exp\left(\frac{\gamma_c}{2}\right)\left\{\frac{\sqrt{2}Li}{R_0} + \frac{1}{2}\ln\left(1 + \frac{r^2}{R_0^2}\right)\exp\left(\frac{\gamma_c}{2}\right)\right\},$$

where i is the number of shear band, L is the spacing between shear bands, γ_c is the critical strain for shear band initiation and propagation, Θ_i, r are the coordinates of the point of the i-th shear band in the polar reference system, and R_0 is the initial radius of the cavity.

13. Find the ratio of shear band spacings in two 304 stainless steels with different grain sizes (10 μm and 100 μm) using the Hall–Petch equation for the dependence of τ_0 on grain size in the frame of the WO model. Ignore the strain rate dependence of the yield strength on grain size.

14. In the GK model (Grady and Kipp [1987]) there is a dependence of the localization time t_c on the shear band thickness δ:

$$\frac{t_c}{t_{c0}} = \frac{1}{4^{\frac{1}{3}}}\left(\frac{a}{a_0} + \frac{t_c a_0}{t_{c0} a}\right)^{2/3},$$

where δ_0 is the normalized length and t_{c0} is the normalized time. Find an optimum shear band width corresponding to a maximum thermal softening (minimum t_c).

15. Parameters δ_0 and t_{c0} are equal (Grady and Kipp [1987]):

$$t_{c0} = \left(\frac{36\rho^2 Ck}{\tau_0^3 a^2 \dot\gamma}\right)^{1/2}, \qquad \delta_0 = \left(\frac{9k^3}{\tau_0^3 a^2 \gamma C}\right)^{1/2}.$$

Show, using the result of Problem 14, that the optimum shear band width is controlled by thermal diffusion over the localization time t_{c0}.

16. The shear band spacing L_{GK} is given by Eq. (4.9). Show that L_{GK} is controlled by the propagation of stress release over the localization time t_{c0}. Use equation for t_{c0} from the previous problem.

17. Using equation from the Problem 15 for the optimum shear band width and Eq. (4.9) for the spacing between shear bands, find the fraction of shear banded material in the GK model (Grady and Kipp [1987]). How does this fraction depend on localization time?

References

Aifantis, E.C. (1992) On the Role of Gradients in the Localization of Deformation and Fracture. *Int. J. Engng. Sci.*, **30**, no. 10, pp. 1279–1299.

Anand, L., and Kalidini, S.R. (1994) The Process of Shear band Formation in Plane Strain Comprssion of fcc Metals: Effects of Crystallographic Texture. *Mechanics of Materials*, **17**, pp. 223–243.

Andrade, U., Meyers, M.A., Vecchio, K.S., and Chokshi A.H. (1994) Dynamic Recrystallization in High-Strain, High-Strain-Rate Plastic Deformation of Copper. *Acta metall. mater.*, **42**, no. 9, pp. 3183–3195.

Asaro, R.J. (1979) Geometric Effects in the Inhomogeneous Deformation of Ductile Single Crystals. *Acta metall.*, **27**, pp. 445–453.

Asaro, R.J., and Needleman, A. (1985) Texture Development and Strain Hardening in Rate Dependent Polycrystals. *Acta Metall.*, **33**, no.6, pp. 923–953.

Backman, M.E., Finnegan, S.A., Schulz, J.C., and Pringle, J.K. (1986) Scaling Rules for Adiabatic Shear. In: *Metallurgical Applications of Shock-Wave and High-Strain-Rate Phenomena* (Edited by L.E. Murr, K.P. Staudhammer, and M.A. Meyers). Marcell Dekker, New York, pp. 675–687.

Bai, Y. (1981) A Criterion for Thermo-Plastic Shear Instability. In: *Shock Waves and High-Strain-Rate Phenomena in Metals: Concepts and Applications* (Edited by M.A. Meyers and L.E. Murr). Plenum, Press New York, pp. 277–284.

Bai, Y., and Dodd, B. (1992) *Adiabatic Shear Localization*. Pergamon Press, Oxford, pp. 302, 303, 305, 332, 333, 352.

Batsanov, S.S. (1993) *Effects of Explosions on Materials,* Springer-Verlag, New York.

Bondar, M.P., and Nesterenko, V.F. (1991) Strain Correlation at Different Structural Levels for Dynamically Loaded Hollow Copper Cylinders. *Journal de Physique IV (Colloque C3)*, **1**, supplement au *Journal de Physique III*, pp. 163–170.

Bondar, M.P., Pervuhina, O.L., Nesterenko, V.F., and Luk'yanov, Ya.L. (1998) The Development of Deformation Structure During Explosive Collapse of Thick-Walled Cylinders. *Combustion, Explosion, and Shock Waves*, **34**, no. 5, pp. 122–129 (in Russian).
English translation: *Combustion, Explosion, and Shock Waves*, 1999, **34**, no. 5, pp. 590–597.

Bridgman, P.W. (1935) Effects of High Shearing Stress Combined with High Hydrostatic Pressure. *Phys. Rev.*, **48**, pp. 825–847

Brockenbough, J.R., Suresh, S., and Duffy, J. (1988) An Analysis of Dynamic Fracture in Microcracking Brittle Solids. *Phil. Mag.* **58**, pp. 619–634.

Chen, H.C. (1997) Shear Localization in High-Strain-Rate Deformation of Inert and Reactive Porous Materials. PhD. Thesis, University of California, San Diego, p. 193.

Chen, H.C., Meyers, M.A., and Nesterenko, V.F. (1995) Chemical Reaction in TiSi Mixture under Controlled High-Strain-Rate Plastic Deformation. In: *Metallurgical and Materials Applications of Shock-Wave and High-Strain-Rate Phenomena, Proceedings of the 1995 International Conference EXPLOMET-95*, El Paso, August 6-10 (Edited by L.E. Murr, K.P. Staudhammer, and M.A. Meyers). Elsevier Science, Amsterdam, pp. 723–729.

Chen, H.C., Meyers, M.A., and Nesterenko, V.F. (1996) Shear Localization in Granular and Comminuted Alumina. In: *Shock Compression of Condensed Matter—1995. Proceedings of the Conference of the American Physical Society Topical Group on Shock Compression of Condensed Matter*, Seattle, August 13–18, 1995 (Edited by S.C. Schmidt and W.C. Tao). AIP Press, New York, pp. 607–610.

Chen, H.C., Nesterenko, V.F., LaSalvia, J.C., and Meyers, M.A. (1997) Shear-Induced Exothermic Chemical Reactions. *Journal de Physique IV* (Colloque C3), supplement au *Journal de Physique III*, 7, no. 7, pp. C3-27–C3-32.

Chen, Y.J., LaSalvia, J.C., Nesterenko, V.F., Meyers, M.A., Bondar, M.P., and Lukyanov, Y.L. (1997) High-Strain, High-Strain-Rate Deformation, Shear Localization and Recrystallization in Tantalum. *Journal de Physique IV* (Colloque C3), supplement au *Journal de Physique III*, 7, no. 7, pp. C3-435–C3-440.

Chen, H.C., LaSalvia, J.C., Nesterenko, V.F., and Meyers, M.A. (1998) Shear Localization and Chemical Reaction in High-Strain, High-Strain-Rate Deformation of Ti–Si Powder Mixtures", *Acta Materialia*, 46, no. 9, pp. 3033–3046.

Chen, H.C., Nesterenko, V.F., and Meyers, M.A. (1998a) Shear Localization and Chemical Reaction in Ti–Si and Nb–Si Powder Mixtures: Thermochemical Analysis. *J. Appl. Phys.*, 84, no. 6, pp. 3098–3106.

Chen, H.C., Nesterenko, V.F., and Meyers, M.A. (1998b) Controlled High-Rate Deformation of Ti–Graphite and Ti–Ultrafine Diamond Mixtures. Unpublished results.

Chen, W., and Ravichandran, G. ((1997) Dynamic Compressive Failure of a Glass-Ceramic under Lateral Confinement. *J. Mech. Phys. Solids*, 45, pp. 1303–1315.

Chen, Y.-J., Meyers, M.A., and Nesterenko, V.F. (1999) Spontaneous and Forced Shear Localization in High-Strain-Rate Deformation of Tantalum. *Material Science and Engineering*, A268, pp. 70–82.

Chiang, Y-M, Birnie, D., and Kingery, W.D. (1997) *Physical Ceramics*. Wiley, New York, Ch. 5, p. 359.

Clifton, R.J. (1979) Material Response to Ultra High Loading Rates. Rep. NMAB—356, NMAB, NAS, Washington, DC, Chapter 8.

Curran, D.R., Seaman, L., Cooper, T., and Shockey, D.A. (1993) Micromechanical Model for Comminution and Granular Flow of Brittle Material under High Strain Rate Application to Penetration of Ceramic Targets. *Int. J. Impact Eng.*, 13, pp. 53–83.

Curran, D.R., Seaman, L., Klopp, R.W., de Resseguier, T., and Kanazawa, C. (1994) A Granulated Material for Quasibrittle Solids. In: *Fracture and Damage in Quasibrittle Structures* (Edited by Z.P. Bazant, Z. Bittnar, M. Jirasek, and J. Mazars). E&FN SPON, London.

Dienes, J.K. (1986) On Reactive Shear Bands. *Phys. Lett. A*, **118**, no. 9, pp. 433–438.

Donachie, M.J. Jr. (1982) Introduction to Titanium and Titanium Alloys. In: *Titanium and Titanium Alloys* (Edited by Donachie, M.J. Jr.). ASM Press, Metals Park, Ohio, pp. 3–19.

Dremin, A.N., and Breusov, O.N. (1968) Processes Occuring in Solids Under the Action of Powerful Shock Waves. *Russ. Chem. Rev.*, **37**, no. 5, 392–402.

Enikolopyan, N.S. (1988) Detonation, Solid-Phase Chemical Reaction. *Doklady Akademii Nauk SSSR*, **302**, pp. 630–634 (in Russian).

Erlich, D., Seaman, L., Shockey, D., and Curran, D. (1977) Development and Application of a Computational Shear Band Model. Final Report, Contract DAAD05-76-C-0762, May 1977. U.S. Army Ballistic Research Laboratories, Aberdeen Proving Ground, MD.

Erlich, D., Curran, D., and Seaman, L. (1980) Further Development of a Computational Shear Band Model. Final Report, Contract DAAG 46-77-C-0043, March 1980. U.S. Army Materials and Mechanics Research Center, Watertown, MA.

Epshtein, G.N. (1988) *The Structure of Explosively Deformed Metals*. Moscow, Metallurgy, p. 66 (in Russian).

Ezis, A. (1994) Monolithic, Fully Dense Silicon Carbide Mirror and Method of Manufacturing. U.S. Patent, no. 5,302,561.

Florence, A.L., and Abrahamson, G.R. (1977) Critical Velocity for Collapse of Viscoplastic Cylindrical Shells Without Buckling. *Trans. of the ASME, J. Appl. Mech.*, March, pp. 89–94.

Follansbee, P.S. (1986) High-Strain-Rate Deformation of FCC Metals and Alloys. In: *Metallurgical Applications of Shock-Wave and High-Strain-Rate Phenomena* (Edited by L.E. Murr, K.P. Staudhammer, and M.A. Meyers). Marcel Dekker, New York, pp. 451–479.

Freund, L.B. (1990) *Dynamic Fracture Mechanics*. Cambridge University Press, Cambridge.

Frey R.B. (1985) The Initiation of Explosive Charges by rapid Shear. In: *Proceedings of the Seventh Symposium on Detonation*, U.S. Naval Academy, pp. 35–41.

Grady, D.E., and Kipp, M.E. (1980) Continuum Modeling of Explosive Fracture in Oil Shale. *Int. J. Rock Mech. Min. Sci.* **17**, pp. 147–157.

Grady, D.E., and Kipp, M.E. (1987) The Growth of Unstable Thermoplastic Shear with Application to Steady-Wave Shock Compression in Solids. *J. Mech. Phys. Solids*, **35**, pp. 95–119.

Graham., R.A. (1993) *Solids Under High Pressure Shock Compression: Mechanics, Physics and Chemistry*, Springer-Verlag, New York.

Grebe, H.A., Pak, H-R., and Meyers, M.A. (1985) Adiabatic Shear Localization in Titanium and Ti–6PctAl–4PctV Alloy, *Metallurgical Transactions*, **16A**, pp. 761–775.

Guirguis, R.H. (2000) Ignition Due to Macroscopic Shear. In: *Shock Compression of Condensed Matter—1999*, CP505 (Edited by M.D. Furnish, L.C. Chhabildas, and R.C. Hixson). American Institute of Physics, Melville, NY, pp. 647–650.

Harren, S.V., and Asaro, R.J. (1989) Nonuniform Deformation in Polycrystals and Aspects of the Validity of Taylor Model. *J. Mech. Phys. Solids*, **37**, no. 2, pp. 191–232.

Hauver, G.E., Netherwood, P.H., Benck, R.F., and Kecskes, L.J. (1993) Ballistic Performance of Ceramic Targets. In *Proceedings of 13th Army Symposium on Solid Mechanics*, pp. 23–34

Hauver, G.E., Netherwood, P.H., Benck, R.F., and Kecskes, L.J. (1994) Enhanced Ballistic Performance of Ceramics. In: *Proceedings of 19th Army Science Conference*, pp. 1–8.

Hegemier, G. (1991) UCSD, private communication.

Hillier, M.J. (1965a) Tensile Plastic Instability under Complex Stress. *Int. J. Mech. Sci.*, **5**, pp. 57–67.

Hillier, M.J. (1965b) Tensile Plastic Instability of Thin Tubes—I. *Int. J. Mech. Sci.*, **7**, pp. 531–538.

Hillier, M.J. (1965c) Tensile Plastic Instability of Thin Tubes—II. *Int. J. Mech. Sci.*, **7**, pp. 539–549.

Howe, P.M. (2000) Explosive Behaviors and the Effects of Microstructure. In: *Solid Propellant Chemistry, Combustion, and Motor Interior Ballistics,* Chapter 1.6 (Edited by V. Yang, T. B. Brill, and W. Z. Ren). Progress in Astronautics and Aeronautics, American Institute of Aeronautics and Astronautics, Vol. 185, Reston, VA.

Ivanov, A.G., Ogorodnikov, V.A., and Tyun'kin, E.S. (1992) The Behavior of Shells under Impulsive Loading. Small Perturbations. *Zhurnal Prikladnoi Mekhaniki i Tekhnicheskoi Fiziki*, **33**, no. 6, pp. 112–115.

English translation: *Journal of Applied Mechanics and Technical Physics*, 1993, May, pp. 871–874.

Johnson, C.A., and Tucker, W.T. (1991) Advanced Statistical Concepts of Fracture in Brittle Materials. In *Engineered Materials Handbook*, Vol. **4**, Ceramics and Glasses, ASM International, pp. 709–715.

Jones, R.H., Schilling, C.H., and Schoenlein, L.H. (1989) Grain Boundary and Interfacial Chemistry and Structure in Ceramics and Ceramic Composites. *Materials Science Forum*, **46**, pp. 277–308.

Kipp, M.E. (1981) Modeling Granular Explosive Detonation with Shear Band Concepts. In: *Proceedings of the Eigth Detonation Symposium*, U.S. Naval Academy, Annapolis, MD, pp. 36–42.

Klopp, R.W., and Shockey, D.A. (1991) The Strength Behavior of Granulated Silicon Carbide at High Strain Rates and Confining Pressure. *J. Appl. Phys.* **70**, pp. 7318–7326.

Klopp, R.W., Shockey, D.A., Seaman, L., Curran, D.R., McGinn, J.T., and Resseguier, T. (1996) A Spherical Cavity Expansion Experiment for Characterizing Penetration Resistance of Armor Ceramics. In *Mechanical Testing of Ceramics and Ceramic Composites*, AMD, Vol. **197**, ASME, pp. 41–55.

Klopp, R.W., Shockey, D.A., Curran D.R., and Cooper, T. (1998) A Granular Flow Model for Developing Smart Armor Ceramics. Final Technical Report on Contract DAAH04-94-K-0001, January 31, 1998, pp. 85.

Knoepfel, H. (1970) *Pulsed High Magnetic Fields*, North-Holland, London, pp. 206–209.

Koch, S.W. (1984) *Dynamics of First-Order Phase Transitions in Equilibrium and Nonequilibrium Systems*. Springer-Verlag, Berlin, p. 148.

Kubin, L.P. (1993) Dislocation patterning during multiple slip of FCC crystals. A Simulation Approach. *Physica Status Solidi A*, **135**, no. 2, pp. 433–443.

LaSalvia, J.C., Chen, Y.J., Meyers, M.A., Nesterenko, V.F., Bondar, M.P., and Lukyanov, Y.L. (1996) High-Strain, High-Strain-Rate Response of Annealed and Shocked Tantalum. In: *Tantalum* (Edited by E. Chen, A. Crowson, E. Lavernia, W. Ebihara, P. Kumar). The Minerals, Metals & Materials Society, Warrendale, PA, pp. 139–144.

Leroy, Y.M., and Molinari, A. (1993) Spatial Pattern and Size Effects in Shear Zones: a Hyperelastic Model with Higher-Order Gradients. *J. Mech. Phys. Solids*, **41**, pp. 631–663.

Levitas, V.I., Nesterenko, V.F., and Meyers, M.A. (1998a) Strain-Induced Structural Changes and Chemical Reactions. Part I. *Acta Materialia*, **46**, no. 16, pp. 5929–5945.

Levitas, V.I., Nesterenko, V.F., and Meyers, M.A. (1998b) Strain-Induced Structural Changes and Chemical Reactions. Part II. *Acta Materialia*, **46**, no. 16, pp. 5947–5963.

Lindberg, H.E. (1964) Buckling of a Very Thin Cylindrical Shell Due to an Impulsive Pressure. *Trans. ASME, E, J. Appl. Mech*, **31**, pp. 267–273.

Mal'kov, I. Yu., and Titov, V.M. (1995) The Regimes of Diamond Formation under Detonation. In: *Metall. and Mater. Applications of Shock-Wave and High-Strain-Rate Phenomena* (Edited by L.E. Murr, K.P. Staudhammer, and M.A. Meyers). Elsevier, Amsterdam, pp. 669–676.

Mason, J.J., Rosakis, A.J., and Ravichandran, G. (1994) On the Strain and Strain Rate Dependence of the Fraction of Plastic Work Converted to Heat: An Experimental Study Using High Speed Infrared Detectors and the Kolsky Bar. *Mechanics of Materials*, **17**, pp. 135–145.

Mechanics of Materials (1994) (Special Issue on Shear Instabilities and Viscoplasticity Theories) **17**.

Meyers, M.A., and Chawla, K.K. (1984) *Mechanical Metallurgy*, Prentice-Hall, Englewood Cliffs, NJ, p. 497.

Meyers, M.A., Subhash, G., Kad, B.K., and Prasad, L. (1994) Evolution of Microstructure and Shear-Band Formation in α-hcp Titanium. *Mechanics of Materials,* **17**, pp. 175–183.

Meyers, M.A. (1994) *Dynamic Behavior of Materials*, Wiley, New York, p. 597.

Meyers, M.A., Nesterenko, V.F., Chen, Y.J., LaSalvia, J.C., Bondar, M.P., and Lukyanov, Y.L. (1995) High-Strain, High-Strain-Rate Deformation of Tantalum: the Thick-Walled Cylinder Method. In: *Metallurgical and Materials Applications of Shock-Wave and High-Strain-Rate Phenomena, Proceedings of the 1995 International Conference EXPLOMET-95,* El Paso, August 6–10 (Edited by L.E. Murr, K.P. Staudhammer, and M.A. Meyers). Elsevier, Amsterdam, pp. 487–494.

Meyers, M.A., Nesterenko, V.F., Vecchio, K.S., and Batsanov, S.S. (1998) Shock and Shear Induced Chemical Reactions in Mo-Si, Nb-Si, and Ti-Si Systems. In: *Molybdenum and Molybdenum Alloys* (Edited by A. Crowson, E.S. Chen, J.A. Shields, and P.R. Subramanian). The Minerals, Metals and Materials Society, Warrendale, PA, pp. 221–239.

Meyers, M.A., Nesterenko, V.F., Chen, Y.J., LaSalvia, and Qing Xue, Q. (2001) Shear Localization in Dynamic Deformation of Materials: Microstructural Evolution and Self-Organization. *Materials Science and Engineering A* (in press).

Meyers, M.A., Xue, Q., and Nesterenko, V.F. (2001) Evolution of Self-Organization of Adiabatic Shear Bands. Presentation at *APS Conference on Shock Compression of Condensed Matter*, Atlanta, June 24–29.

Molinari, A., and Clifton, R.J. (1983) Localization de la Deformation Viscoplastique en Cisaillement Simple: Resultats Exacts en Theorie Non Lineaire. *C.R. Acad. Sci. Paris*, **296**, pp. 1–4.

Molinari, A. (1997) Collective Behavior and Spacing of Adiabatic Shear Bands. *J. Mech. Phys. Sol.* **45**, no. 9, pp. 1551–1575.

Munz, D.G., Shannon, J.L., and Bubsey, R.T. (1980) Fracture Toughness Calculation from Maximum Load in Four Point Bend Tests of Chevron Notch Specimens, *Int. Journal of Fracture* **16**, pp. R137–141.

Nakano, M., Kishida, K., Yamauchi, Y., and Sogabe, Y. (1994) Dynamic Fracture Initiation in Brittle Materials under Combined Mode I/II Loading. *J. De Physique* IV pp. C8 695–700.

Nemat-Nasser, S., Keer, L.M., and Parihar, K.S. (1978) Unstable Growth of Thermally Induced Interacting Cracks in Brittle Solids, *Int. Journal Solid Structures*, **14**, pp. 409–430.

Nemat-Nasser, S., Okinaka, T., Nesterenko, V.F., and Liu, M. (1998) Dynamic Void Collapse in Crystals: Computational Modeling and Experiments. *Phil Mag.*, **78**, no. 5, pp. 1151–1174.

Nemat-Nasser, S., Okinaka, T., and Nesterenko, V.F. (1998) Experimental Observation and Computational Simulation of Dynamic Void Collapse in Single Crystal Copper. *Materials Science and Engineering*, **A249**, nos. 1–2, pp. 22–29.

Nemat-Nasser, S., Guo, W.-G., Nesterenko, V.F., Indrakanti, S.S., and Gu, Y.-B. (2001) Dynamic Response of Conventional and Hot-Isostatically Pressed Ti–6Al–4V Alloys: Experements and Modeling. *Mechanics of Materials* (submitted).

Nesterenko, V.F. (2000) Thick Walled Cylinder Testing. In *ASM Handbook, Mechanical Testing and Evaluation,* vol. 8, ASM International, pp. 455–457.

Nesterenko, V.F., Lazaridi, A.N., and Pershin, S.A. (1989) Localization of Deformation in Copper by Explosive Compression of Hollow Cylinders. *Fizika Goreniya i Vzryva,* **25**, no. 4, pp. 154–155 (in Russian).

Nesterenko, V.F., and Pershin, S.A. (1989) The Shear Localization at Explosive Compaction of Rapidly Solidified Metal Powders. In: *High Pressure in Science and Technology: Proceedings of XI AIRAPT International High Pressure Conference,* Naukova Dumka, Kiev, Vol. 4, pp. 166–169.

Nesterenko, V.F. (1992) *High-Rate Deformation of Heterogeneous Materials.* Nauka, Novosibirsk.

Nesterenko, V.F., Bondar, M.P., and Ershov, I.V. (1994) Instability of Plastic Flow at Dynamic Pore Collapse. In: *High-Pressure Science and Technology—1993. Joint International Association for Research and Advancement of High Pressure Science and Technology and American Physical Society Topical Group on Shock Compression of Condensed Matter Conference,* Colorado Springs, CO, 28 June–2 July 1993. (Edited by S.C. Schmidt, J.W. Shaner, G.A. Samara, and M. Ross). AIP Press, New York, pp. 1173–1176.

Nesterenko, V.F., and Bondar, M.P. (1994a) Localization of Deformation in Collapse of a Thick Walled Cylinder. *Fizika Goreniyai Vzryva,* **30**, no. 4, pp. 99–111 (in Russian).

English translation: *Combustion, Explosion, and Shock Waves*, 1994, **30**, no. 4, pp. 500–509.

Nesterenko, V.F., and Bondar, M.P. (1994b) Investigation of Deformation Localization by the "Thick-Walled Cylinder" Method. *DYMAT Journal*, **1**, no. 3, pp. 245–251.

Nesterenko, V.F., Meyers, M.A., Chen, H.C., and LaSalvia, J.C. (1994) Controlled High-Rate Localized Shear in Porous Reactive Media. *Applied Physics Letters*, **65**, no. 24, pp. 3069–3071.

Nesterenko, V.F., Meyers, M.A., Chen, C.H., and LaSalvia, J. (1995) The Structure of Controlled Shear Bands in Dynamically Deformed Reactive Mixtures. *Metallurgical and Materials Transactions A*, **26A**, pp. 2511–2519.

Nesterenko, V.F., Meyers, M.A., and Wright, T.W. (1995) Collective Behavior of Shear Bands. In: *Metallurgical and Materials Applications of Shock-Wave and High-Strain-Rate Phenomena, Proceedings of the 1995 International Conference EXPLOMET-95*, El Paso, August 6–10 (Edited by L.E. Murr, K.P. Staudhammer, and M.A. Meyers). Elsevier, Amsterdam, pp. 397–404.

Nesterenko, V.F., Meyers, M.A., and Chen, H.C. (1996) Shear Localization in High-Strain-Rate Deformation of Granular Alumina. *Acta Materialia*, **44**, no. 5, pp. 2017–2026.

Nesterenko, V.F., Meyers, M.A., Chen, Y.J., and LaSalvia, J.C. (1996) Chemical Reactions in Controlled High-Strain-Rate Shear Bands. In: *Shock Compression of Condensed Matter—1995, Proceedings of the Conference of the American Physical Society Topical Group on Shock Compression of Condensed Matter*, Seattle, August 13–18, 1995 (Edited by S.C. Schmidt and W.C. Tao). AIP Press, New York, pp. 713–716.

Nesterenko, V.F., Meyers, M.A., and Wright, T. W. (1997) Characteristic Spacing in a System of Adiabatic Shear Bands. In: *Proceedings of Plasticity' 97: The Sixth International Symposium on Plasticity and its Current Applications*, Juneau, Alaska, July 14–18 (Edited by Akhtar S. Khan), pp. 131–132.

Nesterenko, V.F., Meyers, M.A., LaSalvia, J., Bondar, M.P., Chen, Y.J., and Lukyanov, Y.L. (1997) Shear Localization and Recrystallization in High-Strain, High-Strain-Rate Deformation of Tantalum. *Materials Science and Engineering*, **A229**, pp. 23–41.

Nesterenko, V.F., Meyers, M.A., and Wright, T.W. (1998) Self-Organization in the Initiation of Adiabatic Shear Bands. *Acta Materialia*, **46**, no. 1, pp. 327–340.

Nesterenko, V.F. (1998) Controlled High-Strain-Rate Shear Bands in Inert and Reactant Porous Materials. In: *Shock Compression of Condensed Matter—1997, Proceedings of the Conference of the American Physical Society Topical Group on Shock Compression of Condensed Matter*, Amherst, 27 July–1 August 1997 (Edited by S.C. Schmidt, D.P. Dandekar, and J.W. Forbes). *AIP Conference Proceedings* Vol. **429**. American Institute of Physics, Woodbury, NY, pp. 609–614.

Nesterenko, V.F. and Benson, D.J. (1999) Unpublished results.

Nesterenko, V.F. and Lam, H. (1999) Unpublished results.

Nesterenko, V.F., Indrakanti, S.S., Brar, S. and Gu, Y. (2000a) Long Rod Penetration Test of Hot Isostatically Pressed Ti-Based Targets. In: *Shock Compression of Condensed Matter—1999. AIP Conference Proceedings* (Edited by M.R. Furnish, L.C. Chhabildas, and R.S. Hixson). Vol. **505**, American Institute of Physics, Melville, NY, pp. 419–422.

Nesterenko, V.F., Indrakanti, S.S., Brar, S. and Gu, Y. (2000b) Ballistic Performance of Hot Isostatically Pressed (Hiped) Ti-Based Targets. In: *Key Engineering Materials*, Vols. **177–180**. Trans Tech Publications, Switzerland, pp. 243–248.

Nesterenko, V.F., Klopp, R., Shockey, D., Curran, D., and Cooper, T. (2000) Experimental Study of Instability of Large Strain Flow in Dense Granular Material. In *Abstracts*, MRS Spring Meeting, San Francisco, CA, pp. 480.

Nesterenko, V.F. (2000) Thick Walled Cylinder Testing. In: *ASM Handbook, Mechanical Testing and Evaluation*, Vol. **8**, ASM International, Materials Park, OH, pp. 455–457.

Nesterenko, V.F., Indrakanti. S.S., Goldsmith, W., and Gu, Y. (2001) Plug Formation and Fracture of Hot Isostatically Pressed (HIPed) Ti-6Al-4V Targets. In: *Fundamental Issues and Applications of Shock-Wave and High-Strain-rate Phenomena*. (Edited by K.P. Staudhammer, L.E. Murr, and M.A. Meyers), Elsevier, Amsterdam, pp. 593–600.

Oda, M., and Kazama, H. (1998) Microstructure of Shear Bands and its Relation to the Mechanism of Dilatancy and Failure of Dense Granular Soils. *Geotechnique*, **48**, no. 4, pp. 465–481.

Olson, G.B., Mescall, J.F., and Azrin, M. (1980) Adiabatic Deformation and Strain Localization. In: *Shock Waves and High-Strain-Rate Phenomena in Metals* (Edited by M.A. Meyers and L.E. Murr). Plenum Press, New York, pp. 221–247.

Pershin, S.A., and Nesterenko, V.F. (1988) Shear Strain Localization with Pulsed Compaction of Rapidly Quenched Alloy Foils. *Fizika Goreniya i Vzryva*, **24**, no. 6, pp. 120–123 (in Russian).
English translation: *Combustion, Explosion, and Shock Waves*, May 1989, pp. 752–755.

Potapov, A.V., and Campbell, C.S. (1997) Computer Simulation of Shear-Induced Particle Attrition. *Powder Technology*, **94**, pp. 109–122.

Powers, J.M. (1999) Thermal Explosion Theory for Shear Localizing Energetic Solids. *Combust. Theory Modelling*, 3, pp. 103–122.

Recht, R.F. (1964) Catastrophic Thermoplastic Shear. *J. Appl. Mech., Transactions of the ASME,* **31**, June, pp. 189–193.

Rubie D.C. (1990) Mechanisms of Reaction-Enhanced Deformability in Minerals and Rocks. In: *Deformation Processes in Minerals, Ceramics and Rocks* (Edited by D.J. Barber and P.G. Meredith). Unwin Hyman, London, pp. 262–295.

Sairam, S., and Clifton, R.J. (1996) Pressure-Shear Impact Investigation of Dynamic Fragmentation and Flow of Ceramics. In: *Mechanical Testing of Ceramics and Ceramic Composites*, AMD Vol. **197**. ASME, pp. 23–34.

Schwarz, O.J., and Horie, Y. (1998) Stress Fluctuation and Order Generation in Shearing of Granular Materials. In: *Shock Compression of Condensed Matter—1997. Proceedings of the Conference of the American Physical Society Topical Group on Shock Compression of Condensed Matter*, Amherst, 27 July–1 August 1997 (Edited by S.C. Schmidt, D.P. Dandekar, and J.W. Forbes). AIP Conference Proceedings vol. **429**. American Institute of Physics, Woodbury, NY, pp. 263–266.

Semiatin, S.L., Staker, M.R., and Jonas, J.J. (1984) Plastic Instability and Flow Localization in Shear at High Rates of Deformartion. *Acta Metallurgica*, **32**, no. 9, pp. 1347–1354.

Serikov, S.V. (1984) Stability of a Viscoplastic Ring. *Zhurnal Prikladnoi Mekhaniki i Tekhnicheskoi Fiziki,* **25**, no. 1, pp. 157–168

English translation: *Journal of Applied Mechanics and Technical Physics,* 1984, **25**, July, pp. 142–153.

Shetty, D.K., Rosenfield, A.R., and Duckworth, W.H. (1986) Mixed-Mode Fracture of Ceramics in Diametral Compression. *J. Am. Ceram. Soc.* **69**, pp. 437–443.

Shetty, D.K. (1987) Mixed-Mode Fracture Criteria for Reliability Analysis and Design with Structural Ceramics. *J. Eng. Gas Turbines and Power* **109**, pp. 282–289.

Shih, C.J. (1998) Dynamic Deformation of Silicon Carbide, PhD Dissertation, University of California, San Diego, 331 pp.

Shih, C.J., Nesterenko, V.F., and Meyers, M.A. (1997) Shear Localization and Comminution of Granular and Fragmented Silicon Carbide. *Journal de Physique IV,* Colloque C3, supplement au *Journal de Physique* III, **7**, no. 7, pp. C3-577C–3-582.

Shih, C.J., Meyers, M.A., and Nesterenko, V.F. (1998) High-Strain-Rate Deformation of Granular Silicon Carbide. *Acta Materialia,* **46**, no. 11, pp. 4037–4065.

Shih, C.J., Nesterenko, V.F., and Meyers, M.A. (1998) High-Strain-Rate Deformation and Comminution of Silicon Carbide. *J. Appl. Phys.* **83**, no. 9, pp. 4660–4671.

Shockey, D.A. and Erlich, D.C. (1980) Metallurgical Influences on Shear Band Activity. In: *Shock Waves and High-Strain-Rate Phenomena in Metals* (edited by M.A. Meyers and L.E. Murr). Plenum Press, New York, pp. 249–261.

Shockey, D., Curran, D., and Seaman, L., (1985) Development of Improved Dynamic Failure Models. Final Report, Contract DAAG 29-81-K-0125, February 1985, U.S. Army Research Office, Durham, NC.

Shockey, D.A. (1986) Materials Aspects of the Adiabatic Shear Phenomenon. In: *Metallurgical Applications of Shock-Wave and High-Strain-Rate Phenomena* (Edited by L.E. Murr, K.P. Staudhammer, and M.A. Meyers). Marcel Dekker, New York, pp. 633–656.

Shockey, D.A., Marchaud, A.K., Skaggs, S.R., Corte, G.E., Burkett, M.W., and Parker, R. (1990) Failure Phenomenology of Confined Ceramic Targets and Impacting Rods. *Int. J. Impact Eng.* **9**, pp. 263–275.

Singh, D., and Shetty, D.K. (1989) Fracture Toughness of Polycrystalline Ceramics in Combined Mode I and Mode II Loading. *J. Am. Ceram. Soc.* **72**, pp. 78–84.

Srinivasan, M. (1989) The Silicon Carbide Family of Structural Ceramics. *Structural Ceramics, Treatise on Materials Science and Technology,* **29**, pp. 99–160.

Staker, M.A. (1980) On Adiabatic Shear Band Determinations by Surface Observations (of AISI 4340 Steel). *Scripta Metallurgica,* **14**, no. 6, pp. 677–680.

Staker, M.R. (1981) The Relation Between Adiabatic Shear Instability Strain and Material Properties. *Acta Metallurgica,* **29**, no. 4, pp. 683–689.

Stokes, J.L, Oro, D., Fulton, R.D., Morgan, D., Obst, A., Oona, H., Anderson, W., Chandler, E., and Egan, P. (1999) Material Failure and Pattern Growth in Shock-Driven Aluminium Cylinders at the Pegasus Facility. *Bulletin of the American Physical Society,* **44**, no. 2, p. 33.

Stokes, J. L., Nesterenko, V.F., Shlachter, J. S., Fulton, R. D., Indrakanti, S.S., and Gu, Y. (2001) Comparative Behavior of Ti and 304 Stainless Steel in Magnetically-Driven Implosion at the Pegasus-II Facility. In: *Fundamental Issues and Applications of Shock-Wave and High-Strain-Rate Phenomena* (edited by K.P. Staudhammer, L.E. Murr, and M.A. Meyers). Elsevier, Amsterdam, pp. 585–592.

Stout, M.G., Follansbee, P.S. (1986) Strain Rate Sensitivity, Strain Hardening, and Yield Behavior of 304L Stainless Steel. *Transactions of the ASME. Journal of Engineering Materials and Technology*, **108**, no. 4, pp. 344–353.

Suresh, S., Nakamura, T., Yeshurun, Y., Yangv, K.-H., and Duffy, J. (1990) Tensile Fracture Toughness of Ceramic Materials: Effects of Dynamic Loading and Elevated Temperatures. *J. Am. Ceram. Soc.*, **73**, pp. 2457–2466.

Tamura, S., and Horie, Y. (1998) Discrete Meso-Element Simulation of Chemical Reactions in Shear Bands. In: *Shock Compression of Condensed Matter—1997. Proceedings of the Conference of the American Physical Society Topical Group on Shock Compression of Condensed Matter*, Amherst, 27 July–1 August 1997 (Edited by S.C. Schmidt, D.P. Dandekar, and J.W. Forbes). *AIP Conference Proceedings*, Vol. **429**. American Institute of Physics, Woodbury, NY, pp. 377–380.

Taylor, G.I., and Quinney, M.A. (1934) The Latent Energy Remaining in a Metal after Cold Working. *Proc. Royal Soc. London,* **A143**, p. 307.

Telle, R. (1994) Boride and Carbide Ceramics. In *Materials Science and Technology*, Vol. **11**. *Structure and Properties of Ceramics*, VCH, Weinheim, pp. 173–266.

Thadhani N.N. (1993) Shock-Induced Chemical Reactions and Synthesis of Materials. *Prog. Mat. Sci.*, **37**, pp. 117–126.

Tikare, V., and Choi, S.R. (1993) Combined Mode I and Mode II Fracture of Monolithic Ceramics. *J. Am. Ceram. Soc.* **76**, pp. 2265–2272.

Tresca, M.H. (1878) On Further Applications of the Flow of Solids. *Proc. Inst. Mech. Engrs.* **30**, p. 301.

Vecchio, K.S., Yu, L.-H., and Meyers, M.A. (1994) Shock Synthesis of Silicides–I. Experimentation and Microstructural Evaluation. *Acta Metallurgica et Materiallia*, **42**, no. 3, pp. 701–714.

Viechnicki, D., Blumenthal, W., Slavin, M., Tracy, C., and Skeele, H. (1987) Armor Ceramics—1987. In: *Proceedings of the Third TACOM Armor Coordinating Conference*, 17–19 February 1987, Monterey, CA.

Viechnicki, D.J., Slavin, M.J., and Kliman, M.I. (1991) Development and Current Status of Armor Ceramics. *Cer. Bull.* **70**, pp. 1035–1039.

Watanabe, T. (1994) The Impact of Grain Boundary Character Distribution on Fracture in Polycrystals. *Mat. Sci. and Eng.* **A176**, pp. 39–49.

Wilkins, M.L. (1978) Computer Simulation of Penetration Phenomena. Preprint Lawrence Livermore Laboratory, UCRL-81262, July 5, 1978, p. 35.

Winter, R.E., and Field, J.E. (1975) The Role of Localized Plastic Flow in the Impact Initiation of Explosives. *Proc. R. Soc. Lond.* A, **343**, pp. 399–413.

Wright, T.W., and Ockendon, H. (1996) Research Note: A Scaling Law for the Effect of Inertia on the Formation of Adiabatic Shear Bands. *International Journal of Plasticity*, **12**, no. 7, pp. 927–934.

Wright, T.W., and Ravichandran, G. (1997) Canonical Aspects of Adiabatic Shear Bands. *International Journal of Plasticity*, **123**, no. 4, pp. 309–325.

Xue, Q., Nesterenko, V.F., and Meyers, M.A. (2000) Self-Organization of Adiabatic Shear Bands in Ti, Ti-6Al-4V and Stainless Steel. In: *Shock Compression of Condensed Matter—1999. AIP Conference Proceedings* (Edited by M.R. Furnish, L.C. Chhabildas, and R.S. Hixson). Vol. **505**, American Institute of Physics, Melville, NY, pp. 431–434.

Zener, C., and Hollomon, J.H. (1944) Effect of Strain Rate upon Plastic Flow of Steel. *J. Appl. Phys.* **15**, pp. 22–32.

5
Nonequilibrium Heating of
Powders Under Shock Loading

5.1   Introduction

Nonequilibrium heat release in powders under shock loading is strongly coupled
with viscoplastic nonuniform material flow during the densification process (see
Chapter 2). At the same time, the essential difference in time scales for
mechanical and thermodynamical equilibrium for relatively large particles makes
possible the decoupling of mechanical processes on the stage of compaction and
the subsequent phenomena connected with heat diffusion. This is also possible
due to the weak dependence of shock macroparameters on the details of energy
release.

The mesoscopic level of the porous material structure, characterized by
particle size, morphology, and particle size distribution, has a strong influence
on its mechanical behavior as demonstrated in Chapter 2. In this chapter, the
thermodynamical aspects of the shock deformation of powders will be
emphasized. The measurements of the thermal characteristics of the powders
under dynamic loading are of principal importance. The reason for this is the
multiscale nature of the porous material, including the development of the new
scales during the deformation process, which does not exist in the initial state,
and which is difficult to model.

Granular materials under strong shock can be considered as an example of
highly heterogeneous materials, in the sense that the typical scale of parameter
variation is comparable with particle size and their spread from the mean value is
comparable with the value itself. Graham [1997] uses the term "highly porous"
materials, which emphasizes the fact that interacting void configurations are not
described by models based on an isolated single pore, as was considered in detail
in Chapter 2.

It is probable that the characteristic scales for the equilibrium establishment
of various parameters like temperature, density, or pressure can be essentially
different for granular materials (powders). This feature is perfectly demonstrated
by the laminar model of the powder (Hofmann, Andrews, and Maxwell [1968]).
In the case of relatively weak shock, complete compaction is accomplished when

the leading wave closes the corresponding gap. But the establishment of thermal equilibrium is delayed and is provided by the train of five to ten shock waves in already compacted material.

5.2   Experimental Results

5.2.1   Single Shock Loading

5.2.1.1   Results of Optical Measurements

The very high heating rate of heavily deformed parts of particles resulting from high-strain-rate deformation of the material in the process of the pore collapse can naturally produce highly nonequilibrium energy release in the densified material. This is explained by the lack of sufficient time to establish equilibrium between nonuniformly deformed parts of the porous material. The first authors who were able to investigate this basic phenomenon experimentally, in the powder during shock loading with the help of the optical method, were Blackburn and Seely [1962], [1964]. Their results demonstrated the existence of areas with high temperatures in the shocked powder, despite the difficulty of identifying which part of the powder (or gas trapped in the pores) is the source of the radiation.

Belyakov, Rodionov, and Samosadnui [1977] observed very high temperatures in metallic powders after shock exit from the free surface. They also observed the dependence of this temperature on the particle size.

Matytsin and Popov [1987] used an optical method to measure the brightness temperature in shocked ceramic powders on their contact with a transparent plate. The difference between the calculated equilibrium temperature and the experimental values at pressures 15–25 GPa is observed to be three to five times.

5.2.1.2   Microstructural Observations

Metallographical analysis of the compacted material after shock loading allows the identification of zones with locally high temperatures due to the structural changes, as was first demonstrated in the case of ceramics (Carlson, Porembka, and Simons [1966]). Local melts and other characteristic features of nonuniform heat release in the shocked powders of many materials were described later in numerous papers (Belyakov et al. [1974]; Meyers et al. [1981], [1994]; Staver [1981]; Raybould [1981]; Morris [1981]; Morris [1982]; Gourdin [1984a,b], [1986]; Molotkov et al. [1991]; and others).

The microstructures of rapidly solidified granules of Ti-based alloy BT3-1 (in vacuum $6.55 \cdot 10^{-3}$ Pa) and stainless steel after explosive densification at normal temperatures are presented in Figure 5.1. The local zones (white) indicate structural changes due to extensive heat releases that are apparently caused by the sliding of one particle relative to another. These areas with a high concentration

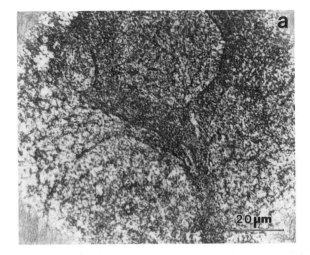

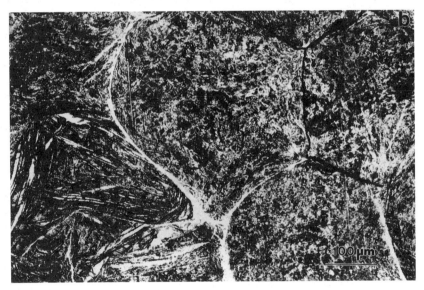

Figure 5.1. The adjacent areas of different plastic deformation of particle contacts for (a) stainless steel EP-450 (shock pressure 20 GPa) and (b) Ti-based spherical particles (shock pressure 2 GPa) after explosive loading.

of thermal energy leading to structural peculiarities are located alongside the particle boundaries and in places that can be considered the center of the pore collapse at the final stage of densification.

The interesting feature in Figure 5.1 is that different parts of the same particle can have a qualitatively different type of deformation, resulting in a large temperature gradient inside the particle. The parts that underwent the "quasistatic" type of contact deformation (simple blunting of the contact area) had a lower temperature than those which suffered a localized shear flow—manifestation of the "dynamic" type of deformation (Chapter 2).

Based on the thickness of the melted layers on the particle contacts in granular Fe-based, Ni-based, and Al-based alloy powders, Morris [1982] concluded that the quenching rate of the contact area is of the order 10^5–10^{10} K/s.

Now it is a well-established fact that the nonequilibrium character of powder heating is a typical feature under conditions of shock-wave loading. At the same time, the extent of thermal nonequilibrium is not a constant parameter. It depends on the shock pressure, particle size, and other parameters like initial temperature. For example, experiments with initial preheating of the powder (1000 °C, vacuum 10^{-3} Pa) gave rise to the uniform structural changes throughout all the contact surfaces (Molotkov et al. [1991]).

This testifies that heat release during hot compaction was more uniform than under cold explosive compaction. It was caused by thermal softening of the material before the densification that decreases the amount of energy to be dissipated for complete densification in the "quasistatic" deformation mode. Then the available excess of energy is dissipated throughout the plastic deformation localized along the contact surfaces, more or less uniformly.

Staudhammer [1999] emphasized the dependence of the amount of melt on particle size. The melt volume per particle (tungsten rod) for larger rods (3.2 mm diameter) is about twice as large as for the smaller rods (0.8 mm diameter) at the same conditions of loading, with shock pressures of 22 GPa.

5.2.1.3   Thermoelectric Power Measurements

There is one experimental method that allows us to measure the average interface temperature on the contact of particles inside metal powders. It is based on the use of the natural thermoelectric couple made by the contact of two different metal powders (Nesterenko [1974], [1975], [1976]; Nesterenko and Staver [1974]; Schwarz, Kasiraj, and Vreeland [1986]). The experimental set-up is presented in Figure 2.1. Not every pair of metals could be used for this purpose. The necessary condition is the mechanical similarity of the powder materials, with an essential difference of their thermoelectric power, which must be independent of shock pressure.

The analysis of the electric effects developing on the metal interfaces of solid metals identified Cu–Ni (Nesterenko [1975]) as well as Cu–constantan pairs (Ishutkin, Kuzmin, and Pai [1986]) as one of the possible candidates. Usually, experiments with different particle sizes were conducted with the same initial density and loading conditions. This makes conclusions based on the comparison between experiments more reliable, because in this case, both materials are under the same dynamic pressure. That is why differences between the thermoelectric powers are based only on differences in temperature.

This method provided the first experimental data for the shock increase time τ for the contact temperature of the particles. For particle size 0.1–0.5 mm (copper and nickel powders) τ is in the interval 10^{-6}–10^{-7} s (Table 2.1) and depends on the pressure of the shock. At relatively high pressures ($P > 3$ GPa), the temperature shock rise time can be estimated according to the equation

$$t \approx \frac{a}{D}, \tag{5.1}$$

where a is the average size of the particles in the powder and D is the shock speed (Nesterenko [1975], [1992]). This result was confirmed by Schwarz, Kasiraj, and Vreeland [1986] in experiments using the same method with the Cu–constantan pair. From this data the rate of heating in the shock front can be estimated as high as 10^9–10^{10} K/s at a temperature increase of 10^2–10^3 K. This heating rate is very high. For example, it is higher than the quenching rate typical for the manufacturing of some metallic amorphous materials, 10^4–10^6 K/s. By the way, that is why the shock heating does not result in crystallization on the shock front of such materials.

The most important fact for modeling of the dissipation processes in the shocked powders is that the nonequilibrium temperature–time increase in the shock is of the same order of magnitude as the particle velocity rise time $\tau_M \sim a/D$ (Nesterenko [1975], [1992], see also Chapter 2). This means that processes of heat release are mainly completed on the same space and time scales as the compaction stage (during and maybe shortly after it), and no long time scale processes in the compacted material contribute to a temperature increase. This is in complete accord with the two-dimensional computer simulations by Benson, Nesterenko et al. [1997] which demonstrate that microkinetic energy exists on the same scale as shock thickness (see Chapter 2).

This method, based on the measurements of the thermoelectric power between two powders in contact (typical oscillograms are shown in Fig. 5.2), was used to determine the dependence of the contact temperature on particle size and shock pressure (Nesterenko [1975], [1977], [1986], [1992]). It produced the first experimental data to demonstrate that at the shock pressure range less than 3 GPa the contact temperature of the large particles can exceed the temperature for small particles by two to three times for the given materials (see Table 5.1).

This dependence of the contact temperature on particle size allows selection of particle size as one of the parameters to achieve a better bonding between particles (Roman, Nesterenko, and Pikus [1979]).

The average particle contact temperatures calculated from measurements of the thermoelectric power are presented in Table 5.1. The experimental spread reflects the data spread of the thermoelectric power values.

The experimental fact of the difference between contact temperatures in particles with different particle sizes is explained by the dependence of the specific particle surface on particle size (Nesterenko [1976]; [1977]; Roman, Nesterenko, and Pikus [1979]; Nesterenko [1986], [1992]; Belyakov, Rodionov,

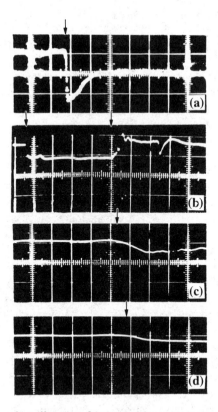

Figure 5.2. Examples of oscillograms for temperature measurements on the contact of porous Cu–porous Ni at the same pressures and for different particle sizes. Shock wave arrivals on the first and second contacts are shown by arrows: (a), (b) $P = 2.4$ GPa, (a) particle size 0.5–1 mm, $H = 40$ mm, 10 mV/div, 3 μs/div; (b) particle size 1–20 μm, $H = 6$ mm, 11 mV/div, 0.625 μs/div; (c), (d) $P = 0.9$ GPa, (c) particle size 0.1–0.5 mm, $H = 40$ mm, 10 mV/div, 0.6 μs/div; (d) particle size 1–20 μm, $H = 5.9$ mm, 10 mV/div, 0.6 μs/div.

and Samosadnui [1977]; Gourdin [1986]). Indeed the internal shock energy (per unit mass) being independent of particle size at the same initial density and shock pressure after distribution on the smaller specific surface s results in higher temperatures. For spherical particles the parameter $s = 6/\rho_0 a$ (a is the particle diameter and ρ_0 is the density of the particle material). This means that the larger particle size leads to the smaller specific surface s and, as a consequence, to the larger contact temperature. The contact temperature dependence on particle size is not linearly proportional to the particle size. This was connected with the dependence of the heat release zone on the particle contacts on the particle size (Nesterenko [1976], [1992]).

Table 5.1. The data for contact (porous Cu–porous Ni) temperature increase for particles with different particle sizes (T_1 corresponds to small particles, 1–20 μm; T_2 to large particles, 0.1–0.5 mm and 0.51.0 mm (data in parentheses), and T_{eq} is the equilibrium temperature) in dependence on shock pressure.

P (GPa)	T_1 (°C)	T_{eq} (°C)	T_2 (°C)
0.9 ± 0.2	120 ± 40		280 ± 40
1.6 ± 0.1	200 ± 40		320 ± 40
2.4 ± 0.5	240 ± 40		740 ± 40 (770 ± 80)
3.0 ± 0.3	360 ± 40		800 ± 80
4.5 ± 0.5	800 ± 80		1070 ± 160
6.5 ± 0.5	980 ± 80		1100 ± 80
11.6 ± 0.6		1120 ± 70	1300 ± 160
13.0 ± 1.0		1182 ± 120	1580 ± 120
14.3 ± 1.0		1370 ± 50	1750 ± 160
25.5 ± 2.0		2550 ± 100	2700 ± 200

Despite the fact that the relation between specific surfaces for particles of different sizes obviously does not essentially depend on the pressure, the differences between temperatures in powders with different particle sizes decrease with the pressure increase (for $P > 3$ GPa, Table 5.1). For pressures more than 6.5 GPa the differences are inside the experimental data spread. This behavior can be explained by the increase of the thickness of the heat release zone (zone of intensive plastic deformation) on the particle interfaces with shock pressure, and by the increase of energy dissipating in the particle body from shock reverberations. One of the dissipation mechanisms under high-strain-rate deformation of the particle material can be shear localization. This process results in the creation of the hot spots, not only on the particle interfaces but also inside the particles.

The fact that the time of temperature equilibration is much longer ($\tau_T \sim 1$–5 μs) than the particle velocity rise time in the shock wave ($\tau_M \sim a/D \sim 0.1$ μs) can raise a question about the stationarity of the shock waves in powders with large grain sizes in real experimental conditions when the duration of the loading impulse is 1–10 μs. The independence of the mechanical shock parameters of the powders with different particle sizes (Dianov, Zlatin et al. [1976]; Nesterenko [1976], [1977]; Dianov, Zlatin et al. [1979]) testifies that the nonstationarity caused by thermodynamic nonequilibrium does not influence the mechanics of shock loading. That is why the process of thermal relaxation can be considered in the first approximation as developing under constant pressure. The result of this process will be the redistribution in the total pressure between the cold and thermal parts under a small change in the total volume, in comparison with volume change during densification. Nevertheless, it is necessary to mention that this process can result in additional plastic deformation behind the shock front.

For temperatures less than the copper melting temperature (1083 °C), the standard dependence of thermoelectric power on temperature for the Cu–Ni

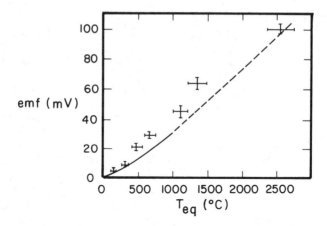

Figure 5.3. The dependence of the experimentally measured thermoelectric power (+) for small particles on the calculated equilibrium temperature T_{eq} and the initial density of copper powder $\rho_{00} = 4.67$ g/cm^3. The solid line corresponds to the calibration curve for the Cu–Ni thermocouple at normal conditions and the broken line is extrapolation of this curve.

thermocouple was used to calculate the values of temperatures from the emf data.

For small particles at $T > 1083$ °C only the calculated equilibrium values of the temperatures T_{eq} are presented in Table 5.1, because there is no data on the emf for the contact of nickel and liquid copper at normal conditions. These equilibrium values were calculated on the basis of the total specific internal energy of the powder and heat capacity of the copper being constant and equal to 0.545 kJ/kgK.

For temperatures higher than 1083 °C, the experimental data for the emf versus T_{eq} were used as a calibration curve instead of using the emf data at normal conditions. The justification for this step is based on a small difference between the observed thermoelectric power for small and large particles at this temperature range (see Table 5.1). This suggests the establishment of thermal equilibrium even for powders with large particles as a result of intensive plastic deformation in their bodies at high pressures. The hypothesis about establishing thermal equilibrium in the powder with small particles makes possible the use of this experimental data for calibration of the thermoelectric power in a temperature range higher than the melting temperature (Fig. 5.3, Nesterenko [1976]). This calibration data were used to calculate the contact temperatures T_2 for large particles above the melting temperature. The temperatures for small particles at pressures <3 GPa are in good agreement with the calculated equilibrium values T_{eq}.

It is worth emphasizing that to calculate the temperature T_{eq} for a calibration

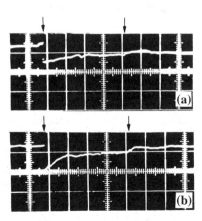

Figure 5.4. Asymmetric character of the contact heating under shock loading: a fast increase in temperature accompanied by relatively slow relaxation: (a) small particles (1–20 μm); (b) large particles (0.1–0.5 mm). Shock pressure for both experiments is 11.6 GPa, scales 45 mV/div and 3 μs/div for both oscillograms. Shock arrivals on the first and second contacts are shown by arrows.

curve at high temperatures, only the total internal energy of the shocked copper and heat capacity were used. The amount of cold energy is small under these conditions and can be neglected as well as the latent heat. This is equivalent to the hypothesis of the superheated solid on the shock front during the time 10^{-7} s. If the latent heat of melting is taken into account together with the extrapolated pressure-dependent melting temperatures of nickel (Tonkov [1979]) and copper (Moriarty [1986]), then even the top point of the "calibration" curve in Figure 5.3 (emf = 100 mV) would be shifted to the left only at approximately 240 °C. This shift is comparable to the experimental spread caused by the error for pressure measurements and, for simplicity the melting effects were not taken into account.

This technique can be used to provide calibration curves up to the evaporation temperature for other thermocouples at very high temperatures if the pressure effect is negligible. A simple check on the absence of the strong pressure dependence of the thermoelectric signal can be made based on its behavior in the controlled rarefaction wave. Then these pairs can be used for dynamic temperature measurements for the different impulse processes connected with large energy releases. It is hard to imagine another way of making this calibration procedure for contact involving melting in at least one of the metals.

The essential differences between the thermoelectric and optical methods should be mentioned because they reflect the temperature values for different parts of the shocked powder. The thermoelectric data represent the average temperature throughout the particle interfaces. The data from the optical measurements are connected with the maximal temperatures which are characteristic for a small portion of material, probably for hot gases and vapors in the pores.

The high-quenching rate is also the characteristic phenomenon of the shock compression of porous materials. For example, at the shock pressure 6.5 GPa in the powder with particle size 0.1 mm $< a <$ 0.5 mm the relaxation time of the nonequilibrium contact temperature is ~5 μs (Nesterenko [1976], [1977]; Fig. 5.4). This value corresponds to the quenching rate ~10^8 K/s, which is higher than the one for traditional quenching methods (Davies [1978]).

Another principal peculiarity of the shock compression of the powders is the asymmetrical character of the processes of heating and cooling (Fig. 5.4; see also Fig. 5.2(a)) also shown by Schwarz, Kasiraj, and Vreeland [1986]. This means that the time of contact heating is essentially less than the time of temperature drop. It also points out the importance of the finite thickness of the heat source on the interfaces—this thickness should be essentially larger than the typical scale of the heat diffusion process during a temperature increase on the shock front. Only under this condition is the asymmetrical temperature profile possible.

5.2.2 Secondary Shock Loading of Powders

The heterogeneous behavior of the powder heating in shock waves is also characteristic for secondary shock loading (Nesterenko [1992]). Only the thermoelectric method of shock temperature measurements enables us to get the secondary electrical signal (see Fig. 5.5) from the interface of the different powders caused by the secondary shock.

This secondary shock arises when a rarefaction wave, previously originated from the contact copper plate–copper powder, arrives on the contact detonation products–copper plate. At that moment, if the pressure in the detonation products is still high, the secondary shock emerges from this contact. Suppose that the Hugoniot of the shocked powder is close to the solid copper. Using the known Hugoniot and isentrope of the detonation products (Baum, Orlenko, and Stanyukovich [1975]), find the estimation of the secondary shock amplitude to be close to 15 GPa for the explosive TNT/RDX 50/50. The corresponding increase in temperature at this shock pressure at the solid copper (~70 K) should result in a thermoelectric signal ~1 mV. This response can be expected, if the contact of the shocked powders is ideal and does not produce additional heating.

In experiments (Fig. 5.5), the amplitude caused by the secondary shock loading was 10–16 mV (Nesterenko [1974], [1975], [1988]). The electrical signal from the secondary shock loading is more than an order of magnitude larger than is expected from the ideal interface. This can be interpreted due to the presence of the residual porosity in the shocked powder even at the pressure level in the first shock >10 GPa where the density of the porous copper should be larger than the solid density under normal conditions (Bakanova et al. [1974]). The important physical consequences of this observation are connected with

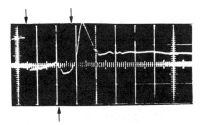

Figure 5.5. The explosive loading by generator 2 (the thickness and diameter of the explosive charge are equal to 70 mm) causes the secondary shock loading and the thermoelectric signal (shown by the arrow at the bottom of the picture). First shock arrivals on the first and second contacts are shown by the arrows at the top. Shock pressure 14.3 GPa, particle size 1–20 μm, $h = 2.1$ mm, $H = 5.8$ mm, scales 40 mV/div, 0.6 μs/div.

the possibility of activating hot spots even in the shocked granular (powder) materials under secondary shock loading.

The absence of such a signal from the secondary shock loading in Figure 5.4 (despite the same shock generator as for the experiment in Fig. 5.5) is connected with the increase in thickness of the buffer plate up to 12 mm in experiments corresponding to Figure 5.4, in comparison with 6 mm for Figure 5.5. As a result, prior to the arrival of the rarefaction wave on the contact with explosive, pressure in the detonation products was essentially reduced resulting in a low amplitude of the reflected secondary shock.

5.2.3   Contact Between Solid and Powder

Thermoelectric measurements (Fig. 5.6(a)) were also used for evaluations of the contact temperature between solid material and powder (solid copper–nickel powder of different particle sizes). One very interesting and unexpected result is that this contact temperature has the same value as the contact temperature in the powders (contact of the copper powder and nickel powder).

The result was confirmed in a later paper by Zagarin, Kuz'min, and Yakovlev [1989]. Also notice the faster temperature relaxation on contact with small particles in comparison with larger ones. A larger amplitude of the thermoelectric signal corresponding to shock loading of the second contact in later cases, despite the same geometry of samples and type of loading (heigth of nickel powder is $H = 40$ mm), is characteristic for these experiments. In addition, the previously discussed thermoelectric signal caused by secondary shock loading of the contact is evident for both experiments.

The reason that the initial contact temperature is equal to the temperature

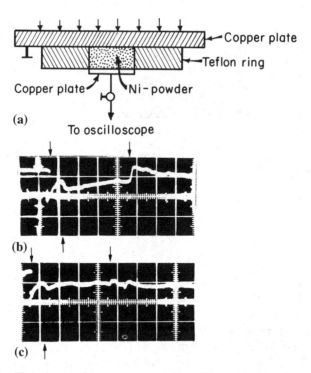

Figure 5.6. (a) The scheme of measurement of the solid–powder interface temperature and corresponding oscillograms: (b) for large (0.1–0.5 mm); and (c) for small (1–20 μm) particles. Arrows at the top mark the arrival of the first shocks on the first and second contacts, arrows on the bottom correspond to the arrival of the secondary shock on the first contact, shock pressure in the first shock 14.3 GPa, scales 40 mV/div, 3 μs/div.

inside the powder can be connected with the plastic deformation of the surface layer of the solid copper plate caused by the pores in the powder. The scale of this heavily deformed area should be comparable with the pore size that is of the same order as the particle size in the investigated granular material. This explains a faster relaxation of the contact temperature for small particles in powder (compare Figs. 5.6(b) and (c)).

This phenomenon is illustrated in Figure 5.7 where the microstructural picture of the interface deformation for the contact of stainless steel plate and stainless steel particles together with a schematic representation of this phenomenon are presented. From Figure 5.8(a) it is clear that in the solid–powder interface the essential plastic flow of solid material takes place. This phenomenon makes physically incorrect any attempts to obtain contact temperatures from the Hugoniot curves for solid and powder by solving the

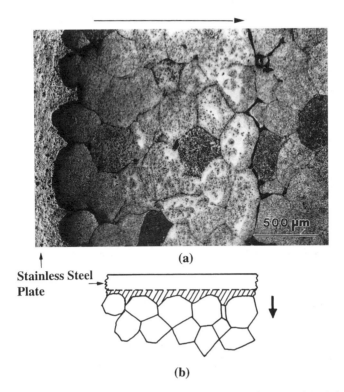

(a)

**Stainless Steel
Plate**

(b)

Figure 5.7. (a) The structure of the interface of solid stainless steel–stainless steel powder and (b) the schematic representation of plastic flow in the vicinity of the solid–powder interface. The plastically deformed area of solid plate adjacent to the powder is shown by hatches, the direction of the shock wave is marked by arrows.

corresponding heat conductivity contact problem for two bodies with initially different temperatures.

So, the surface layers of the solid material take an equal part in the void's filling. This results in an increase in the local temperature of the solid adjacent to the contact close to the temperature of the powder. The resulting interface temperature approaches the same value. The solid–powder contact also demonstrates heterogeneous behavior under secondary shock loading similar to the behavior of the powders contact (Fig. 5.5) under the secondary shock: the thermoelectric signal corresponding to the secondary shock is shown by an arrow (Fig. 5.6(b),(c)).

5.3   Thermodynamic Models for Heterogeneous Heating Under Shock-Wave Loading

5.3.1   Contact Thermal Energy Source Approach

5.3.1.1   Main Assumptions and Results

The heterogeneous structure of porous material and the rapid change in pressure in the shock front are the main reasons for essentially nonequilibrium and heterogeneous types of thermal energy release under shock loading. The first attempt to model this process was made by Belyakov, Rodionov, and Samosadnui [1977]. Their approach is based on the following assumptions:

The thermal energy source is concentrated on the particle interfaces in the area with zero thickness.

The source of the thermal energy has a constant intensity F_0 providing heat release equal to the shock specific internal energy E during the shock rise time τ. From these two assumptions we obtain

$$F_0 = \frac{E}{\tau s_k}, \qquad s_k = \frac{3}{\rho_0 a}, \tag{5.2}$$

where s_k is the specific surface of the particle contacts per unit mass, ρ_0 is the solid density of the particle's material, and a is the diameter of the particles.

The zone of the subsequent redistribution of the thermal energy is determined by the diffusion-type heat conductivity process. The size of the area involved in this process is much smaller than the characteristic particle size (diameter for spherical particles). For relatively large particles this is a realistic hypothesis.

Shock rise time τ was calculated based on the hypothesis that the shock thickness is equal to the particle size $\tau = a/D$, where D is the shock speed. This hypothesis is in agreement with the thermoelectric measurements of the shock front rise time for strong shocks (see Table 2.2) and with Eq. (5.1) which reflects the thermal energy release in the powder mainly during shock loading.

In this approach, a complete decoupling of the thermal processes and densification is introduced. The problem is described by the heat conductivity equation for a semi-infinite solid:

$$\frac{\partial T}{\partial t} = \kappa \frac{\partial^2 T}{\partial x^2}, \qquad 0 < x < +\infty, \quad 0 < t < +\infty, \tag{5.3}$$

$$T(x, t = 0) = T_0, \qquad 0 < x < +\infty,$$

$$\rho c \kappa \frac{\partial T(x = 0, t)}{\partial x} = \frac{F_0}{2}, \qquad 0 < t \leq \tau,$$

$$\rho c \kappa \frac{\partial T(x = 0, t)}{\partial x} = 0, \qquad \tau < t \leq \infty,$$

where T is the temperature, x is the coordinate, t is the time; κ is the thermal diffusivity coefficient, c is the specific heat coefficient, and t is the initial temperature $T_0 = 0$ °C. In Eq. (5.3) the heat flux at $x = 0$ (contact of the particles) is assumed to be constant and equal to $F_0/2$ for $t < \tau$ and zero at $t > \tau$.

The solution of Eq. (5.3) can be found in the book by Carslaw and Jaeger [1988]. The equation for temperature at $x = 0$ is

$$T(x = 0, t) = \frac{F_0}{\kappa \rho c} \left(\frac{\kappa t}{\pi}\right)^{1/2}, \qquad 0 < t < \tau, \tag{5.4}$$

$$T(x = 0, t) = \frac{F_0 \kappa^{1/2}}{\kappa \rho c} \left\{ t^{1/2} \, ierfc \frac{x}{2\kappa^{1/2}t^{1/2}} - (t - \tau)^{1/2} \, ierfc \frac{x}{2\kappa^{1/2}(t - \tau)^{1/2}} \right\}, \qquad t > \tau.$$

The maximal temperature at the contact point between particles ($x = 0$) corresponds to $t = \tau$ and can be obtained from Eq. (5.4):

$$T_{c,\,max} = \frac{E}{3c}\left(\frac{aD}{\kappa\pi}\right)^{1/2}. \tag{5.5}$$

The important feature of Eq. (5.5) is the explicit dependence of the maximal contact temperature on the initial particle size. It is interesting to understand how the particle size affected the value of the maximal temperature, which was not caused by the dependence of F_0 on the particle size (Eq. (5.2)). Substitution of Eq. (5.1) into Eq. (5.2) demonstrates that the intensity F_0 does not depend on the particle size. The reason why particle size appears in the final equation for the maximal temperature is hidden in the boundary conditions of the heat conductivity equation which prescribes the time τ responsible for the heat generation by this source.

To some extent, the inverted problem is considered by Raybould [1980] within the framework of the same basic approximations as in the work by Belyakov, Rodionov, and Samosadnui [1977]. He obtained the relation between the time of energy supply by the source on the interface t, the increase of the interface temperature T_c, and the total energy of the source (transformed into a final equilibrium temperature increase ΔT):

$$t = \frac{1}{12\kappa}\left(\frac{a\Delta T}{2T_c}\right)^2. \tag{5.6}$$

Eq. (5.6) was used to find the time t_m that is necessary to reach the melt temperature on the interface. It can be converted to obtain an expression for T_c similar to Eq. (5.5), if time t is known. These relations differ only by a factor of $3/4$.

To connect Eq. (5.6) with the nonequilibrium heating inside the shock front Raybould uses the value $t = \tau = 3.6a/D$. Such evaluation of the shock rise time τ results in overestimation of the time of energy release for relatively strong shocks. In our experiments for powder with particle sizes 0.5 mm $< a < 1.0$ mm at the shock pressure 2.4 GPa, the value $\tau = 3.6a/D = 1.64$ μs. That number is 2.7 times longer than the experimental rise time (see Table 2.2).

Gourdin [1984a,b], [1986] used the same assumptions mentioned at the beginning of this section. But he additionally introduced the difference between the geometrical macrosize of the particles determining the shock front thickness, and the size of the surface roughness, which are responsible for the specific surface area of the powder. Gourdin solved the problem for spherical geometry and took into account the melting.

The main advantages of the approach proposed by Belyakov et al., Raybold, and Gourdin (the approach deserves to be called the BRG model according to the authors of this approach) should be emphasized. First of all, the model allows the estimation of the upper time for the establishment of thermal equilibrium. The introduction of additional factors such as heat release in the particle volume or final size of the energy source on the contacts will result in a more uniform initial energy distribution, decreasing the characteristic time for establishing equilibrium.

Second, the maximal possible temperatures of the particle contacts can be correctly estimated in the frame of this model. This is explained by the fact that only diffusion heat transfer is taken into account and the whole energy is deposited into a zero thickness heat source.

Finally, this model is the closed-type approach which uses only main initial parameters and does not need additional hypotheses.

As in any model approach, it also has some shortcomings which should be taken into account when interpretating the experimental results.

For example, the BRG model predicts the following contact temperature dependence on particle size $T_c \sim a^{1/2}$ (see Eq. (5.5)). According to the model this dependence should be valid for any shock pressure. This does not agree with the decrease in the differences in contact temperatures between particles of different sizes with the pressure increase observed in the experiments (see Table 5.1). This experimental behavior can be explained by an effective increase of the surface on which the heat release has taken place or, more likely, by an increase of the thickness of the energy release zone on the particle contacts with pressure (Nesterenko [1976], [1977], [1986]; Roman, Nesterenko, and Pikus [1979]). Evidently, no contact temperature dependence on particle size exists for powder with the same density and under the same loading conditions, when this zone is comparable with the particle size. The dependence of the energy release zone on particle interfaces on pressure should be incorporated in the model.

The account of the initial real surface of the powder particles in the BRG model is not justified as a physically necessary addition because of the absence of "right" scaling estimations. This looks like a good step only if the thickness of the heavily deformed interface layer, resulting in heat release on the interface, is much less than the typical scale of the surface roughness. Another scale that should be taken into account is the thickness of the area involved in the heat conductivity process. If this scale is much larger than the typical roughness height, then an account of the roughness is also problematic. During the plastic flow of the particle, the initial real surface of the particle can be essentially changed.

The hypothesis of the zero thickness of the source of thermal energy should provide an approximately symmetrical curve of the interface temperature during its increase and decrease stages. This behavior is in contradiction with the experimental results obtained by Nesterenko [1975] (see Figs. 5.3 and 5.5) and Schwarz, Kasiraj, and Vreeland [1986].

Nesterenko [1977], [1986], [1992a] and Staver, Kuzmin, and Nesterenko [1982] emphasized that the energy released on the particle contacts in copper powder, with large particles and porosity $\alpha_0 = 2$ at the shock front at pressure 3 GPa (this pressure corresponds to the maximal deviation from equilibrium), is of the same order of magnitude as the total specific internal shock energy E. This is not a trivial conclusion because particles deformed in the bulk require some additional energy. This means that the viscoplastic flow of the material in the vicinity of the particle interfaces is an important factor of energy dissipation and can provide an essential input for the establishment of thermal equilibrium behind the shock front. It was also concluded that the mechanism of energy dissipation connected with the shock reverberations alone is inadequate to describe the energy dissipation processes behind shock waves in powders (Nesterenko [1976], [1977], [1992]).

The hypothesis of a constant intensity heat source on the particle contacts is very simple and useful for an analytical treatment of the problem, but this is not proved either. For example, it is reasonable to expect that the thermal softening of the material during plastic flow in the vicinity of the interface should make the work of the source less efficient at higher temperatures toward the end of the compaction process. Moreover, this hypothesis results in an unusual shock front for powders.

In fact, if the shock front is the stationary one and the internal energy at each point of the shock equals $1/2u^2$ (u is the particle velocity at this point), then at condition $F = F_0 = $ constant we get (neglecting the "cold" energy):

$$\frac{1}{2}u^2 = F_0 s_c t \quad \Rightarrow \quad u = u_f \sqrt{\frac{t}{\tau}}. \tag{5.7}$$

The time t is calculated from the start of the front and u_f is the final particle velocity behind the shock. This shape of the shock front (Eq. (5.7)) does not agree with the experimental results for the powders (Roman, Nesterenko, and Pikus [1979]).

Instead, postulating $F_0 = $ constant, the real shock profiles in powders can be used to get more realistic expressions for the dependence of the heat release intensity F on time. It is interesting to estimate how different approximations for $F(t)$ can change the maximal temperatures of the contacts. The hypothesis F_0 = constant should be considered as robust if switching to different time dependencies of F doesn't essentially change the maximum contact temperature.

For simplicity let us consider a one-dimensional case. Under the assumptions of the BRG model (except F is now a function of time) the relation between the contact temperature and the heat source intensity $F(t)$ can be expressed as follows (Carslaw and Jaeger [1988]):

$$T_c(\tau) = \frac{\kappa^{\frac{3}{2}}}{k\pi^{\frac{1}{2}}} \int_0^\tau F(\tau - \xi) \frac{d\xi}{\xi^{\frac{1}{2}}}, \tag{5.8}$$

where k is the heat conductivity coefficient. The liner, quadratic, and cubic dependencies for $F(t)$, under a condition of the same total deposited energy and shock rise time τ, result in a subsequent increase in the maximal contact temperature in 1.33, 1.6, and 1.83 times (Problem 5.1). This means that deviations of the heat source intensity from the hypothesis $F_0 =$ constant can provide essential differences in the estimations of the maximal contact temperature. At the same time, these calculations demonstrate that a considered hypothesis can be reliable at least inside the order of the magnitude evaluation of the maximal contact temperature.

5.3.1.2   Comparison with Experiments

The optical measurements of the temperature demonstrated its dependence on particle size, in qualitative agreement with the BRG model (Belyakov, Rodionov, and Samosadnui [1977]). But this dependence was essentially weaker than predicted by Eq. (5.5). For example, the increase in the particle size by five times in their experiments resulted only in the increase in maximal temperature by 25%, rather than more than twice, according to Eq. (5.5). The discrepancy between the measured values of the temperature and evaluated from Eq. (5.5) is about by an order of magnitude. Additionally, the weaker dependence of the temperature on shock pressure in comparison with the dependence expected from Eq. (5.5) was observed in experiments. The difference between measured temperatures in powders with different grain sizes slightly increased with pressure.

It is worthwhile also mentioning that the average temperature evaluated on the basis of heat equilibrium has a linear dependence on shock pressure.

The discussed experiments testified in favor of the high nonequilibrium temperature probably connected with the hot gas expanding from the free surface of the powder. Definitely the temperature of the hot gas on the free surface of the powder is different from the temperature on the particles contacts. The observed discrepancy can reflect the effects of the initial gas compression in the pores which screen the temperature of the interfaces being much lower. Also an additional problem of the experiment's interpretation is connected with the difference in the pore collapse processes in the powder bulk and in the powder layer adjacent to the free surface.

Gourdin [1984a,b], [1986] checked the model predictions by the comparison of the amount of melt in theoretical consideration and after the shock loading of three different powders (steel 4330V, copper and Al–6%Si alloy). The best agreement was obtained for the last alloy.

Nesterenko [1986], [1992] proposed using assumptions similar to the BRG model for an upper estimation of the contact energy deposition from experimental dependencies of the contact temperature on time. It is supposed that the heat source thickness is equal to zero and that the particle size is much larger than the area involved in the thermal conductivity process $(\kappa\tau)^{1/2} \ll a$. This condition is valid for copper powder with an average size of the particles $a \sim 0.3$ mm and $\tau \sim 3 \cdot 10^{-7}$ s, $\kappa \sim 1$ cm^2/s as well as it is valid for nickel powder. This proves the validity of the one-dimensional approximation for the heat conductivity process.

In this case, the relation between $F(t)$ and $T_c(t)$ is the following (Carslaw and Jaeger [1988]):

$$F(t) = \frac{k}{\pi^{1/2}\kappa^{1/2}} \frac{d}{dt} \int_0^t \frac{T_c(\xi)}{\sqrt{t-\xi}} \, d\xi. \tag{5.9}$$

The temperature in the shock front can be approximated by a linear function (see Fig. 5.2):

$$T_c(t) = \frac{T_{c,\max}}{\tau} t, \quad 0 < t < \tau. \tag{5.10}$$

Substitution of Eq. (5.10) into Eq. (5.9) gives the following result:

$$E_c = s_c \int_0^\tau F(t) \, dt = \frac{4}{3} s_c \rho_0 c T_{c,\max} \left(\frac{\kappa\tau}{\pi}\right)^{1/2}. \tag{5.11}$$

The contact surface of the particles is changing during the compaction process. The constant value $s_c = (3/\rho_0 a)$ is used in calculations providing the higher value of E_c because in reality the s_c is lower. An additional increase in E_c is due to the artificial increase in rise time, because of the nonplanar wavy interface between copper and nickel powders, resulting in a longer rise time detected in experiments in comparison with real shock rise time. The aforementioned expression for s_c will be exact if the main processes of energy release are developed at the stage of complete densification. The calculated values of E_c for copper powder (large particles) are presented in Table 5.2 as well as the values of the specific internal energy E and their ratios. The averaged values of $a = 0.3$ mm (for particle sizes 0.1 mm $< a <$ 0.5 mm) and $a = 0.75$ mm (for particle sizes in the interval 0.5 mm $< a <$ 1.0 mm) with the following values for $\kappa = 10^{-4}$ m^2/s and $c = 0.4$ kJ/kgK were used in the calculations. For $T_{c,\max}$ the corresponding values T_2 were taken from Table 5.1. At the pressure 2.4 GPa value in parentheses corresponds to the average particle size 0.75 mm.

The values of E_c calculated from experimental temperature profiles according to Eq. (5.11) definitely provide lower values than the values of the total internal specific shock energy E, and their discrepancy is more than the order of magnitude for high pressures. The main reason for this is connected to the

Table 5.2. The calculated values of energy deposited on the contacts E_c, and the internal specific energy E, and their ratios for different shock pressures.

P (GPa)	E_c (kJ/kg)	E (kJ/kg)	E/E_c
0.9 ± 0.2	9.25	28.8	3.2
1.6 ± 0.1	8.35	61.3	7.4
2.4 ± 0.5	17.25 (7.2)	105.8	6.2 (14.6)
3.0 ± 0.3	20.05	135.2	6.8
4.5 ± 0.5	17.65	217.8	12.4
6.5 ± 0.5	17.2	312.1	18.2
11.6 ± 0.6	≤ 21.45	561.8	≥ 26
13.0 ± 1.0	16.5	644	39
14.3 ± 1.0	28.85	720	25
25.5 ± 2.0	≤ 40.65	1394	≥ 34

hypothesis of zero thickness for the heat source—the main feature in the BRG model.

The differences between the calculated values of E_c and E are less in the lower range of pressures. These results can be interpreted as follows. The mechanism of the energy dissipation operating on the interfaces according to the rules of the BRG model can be responsible, only partially, for energy release in the range $(0.1–0.3)E$ (Nesterenko [1986], [1992]). This mechanism cannot be responsible for higher values because of the naturally small shock rise time and moderate average contact temperatures. The rest of the internal energy dissipates in the finite thickness heavily deformed contact layer (Nesterenko [1977]) and into the bulk of the particles. This is in agreement with the calculations by Morris [1981], based on the microstructural measurements of melt in the shocked powders.

If the energy dissipated in the BRG model can be separated from the total energy and equals $E/(3 \div 6)$, then it can be in good agreement with the experimental temperature profiles at the pressures $1 < P < 4$ GPa. It is not clear how such a separation can be justified physically.

Mutz [1992] arrived at a similar conclusion based on his experiments: a 70% bulk and 30% interface energy release from the total shock energy input are in good agreement with experimental data on the amount of melt.

It should be emphasized that large discrepancies between the calculated values of E_c and E do not mean that the main part of the internal energy cannot be deposited near the interfaces, but manifests only that the real thickness of the heat source on the interfaces is much larger than the characteristic scale of the area ($\sim(\kappa\tau)^{1/2}$) involved in the diffusive thermal conductivity process during shock loading in the front.

Thus the BRG model does not include the right scaling of the area involved in the process of energy dissipation during intense viscoplastic flow of the interface area and the subsequent heat conductivity process. The necessity of the introduction of a new scale, the thickness of the plastically deformed zone on the interfaces, was emphasized by Nesterenko [1976], [1977], [1986], [1992]. The different character of energy dissipation in shocked powders, rather than in the BRG model, was described by Nesterenko and Muzykantov [1985], in the skin model of energy release where the hot layer with a final thickness is instantaneously introduced into the particle surface.

Trebinski and Wlodarczyk [1987a,b] also recognized that the BRG model is based on the condition that a heavily plastically deformed layer on the particle surface (the actual origin of heat) is less than the thermal relaxation layer. Otherwise, the volume character of the heat source will dominate the process. They proposed to estimate the thickness of this heat source based on the volume of pores, because in the course of the compression process, pores are filled with the material of the grain. So, it was postulated that the volume of the surface layer subjected to intense deformation is equal to the volume removed by the compaction process.

This hypothesis definitely reflects the necessity of introducing the final scale for the heat release source, but does not take into account that the plastic flow during filling of the pore space is highly localized. This localization develops a new space scale that is not directly connected with the scale derived from the volume of the filled pores.

5.3.2   Modeling a Local Heating by Pore Collapse

5.3.2.1   Symmetrical Pore Collapse

There were numerous attempts to use the Carroll–Holt model (Carroll and Holt [1972]) for a description of the thermodynamics of shocked powders (Khasainov, Borisov, Ermolaev, and Korotkov [1980]; Dunin and Surkov [1982]; Attetkov, Vlasova, Selivanov, and Solovev [1984a,b]; and others). Some interesting results of this modeling can be mentioned, such as the dependence of the maximal temperature on the initial pore size (Attetkov et al. [1984a]). It is worth mentioning here that the continuum approach based on the energy consideration and the hypothesis of equilibrium results in temperature dependence only on porosity!

It should be emphasized that the Carroll–Holt model was originally proposed for the kinematics of powder deformation, addressing mainly the question of how material moves into the empty space of pores. The main disadvantage of the expansion of this approach to the thermodynamic applications is connected with the fact that symmetrical pore collapse results in any unrestricted values of temperatures at the center. As the result of initial spherical geometry, the specific dissipated energy on pore boundaries has a logarithmic-type singularity due to plasticity. Power-law-type singularity $(\alpha - 1)^{-5/6}$ is caused by viscous dissipation (Dunin and Surkov [1982]). Also the area of high temperature is concentrated near the internal surface of the pore (also as the result of specific model geometry), in contradiction with the energy dissipation near the particle interfaces (Figs. 5.1). The modified model (Chapter 2, Section 2.2) is able only partially to improve that shortcoming. It can be used to get an estimate of the energy dissipation at the stage of pore collapse. But the modified model is unable to describe the dissipation of microkinetic energy into thermal energy after pore closure.

Kusubov, Nesterenko, Wilkins, Reaugh, and Cline [1989] made an attempt to explain, based on the modified Carroll–Holt model, the establishment of heat equilibrium behind the shock front in the powder with large particles under a high shock pressure $P > 10$ GPa during the shock rise time. The necessity for such an explanation is connected with the fact that at this range of pressure, no essential difference exists between the measured temperature in powders composed from large or small particles (Table 5.1, Fig. 5.4). It suggests that the heat equilibrium is established by some mechanically based process in both powders, because the thermal diffusion is unable to monitor effectively the equilibrium in large particles for a short shock rise time $((\kappa \tau)^{1/2} \ll a)$.

The main idea was that thermal softening (which is included in the modified Carroll–Holt model) of the inner layer of the collapsing pore would result in more intensive deformation and a subsequent thermal softening of the outer layers. It was expected that heavily deformed and hot areas would mechanically propagate into material bulk from the inner layers into the outside. But it turned out in the computer calculations that this mechanism did not work and could not explain the establishment of the thermal equilibrium in large particles. For example, at the shock pressure $P > 10$ GPa, only 20% of the material volume of large particles with $a = 0.3$ mm was heated by viscoplastic flow up to the melting temperature during the pore collapse and the rest of the material had a temperature in the range 250–300°C.

This apparently is in disagreement with experimental results, when the whole material should be more or less uniformly heated at this shock pressure level. The reason for this behavior of the temperature distribution in the modified pore collapse model is that the inertial resistance is the main one that determines the collapse process time. At this level of pressure, the spherical geometry of the flow automatically localizes the hot area in the small volume near the center. So, it is possible to conclude that even the modified Carroll–Holt model is unable to describe the experimental fact of the thermal equilibrium establishment during the shock rise time for high pressures. One result which can be extracted from this model is that for high-shock pressures material in the vicinity of the pore center accumulates an essential amount of microkinetic energy which can only be dissipated during the stage when the pores are closed.

5.3.2.2   Instability of Pore Collapse and Localized Shear

There is one very important feature of pore collapse, which restricts application of the single-pore model for prediction of the nonequilibrium heating of powders. This is connected with the intrinsic property of material instability, which destroys the spherically or cylindrically symmetrical process of collapse, even for

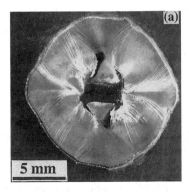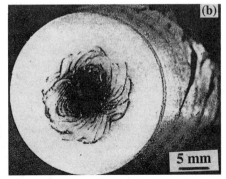

Figure 5.8. Localization of the deformation during pore collapse in (a) monocrystal copper and (b) teflon cylinders at the same loading conditions. (Fig. 5.8(a) reprinted from *Phil. Mag.*, **78**, no. 5, Nemat-Nasser, S., Okinaka, T., Nesterenko, V.F., and Liu, M., Dynamic Void Collapse in Crystals: Computational Modeling and Experiments, pp. 1151–1174, 1998, with permission from Taylor & Francis Ltd. (http://www.tandf.co.uk))

perfect initial symmetry (Nesterenko, Bondar, and Ershov [1994]; Nesterenko, Meyers, and Wright [1998]). A typical picture of the shear localization process that breaks the symmetry of collapse in different materials (e.g., monocrystal copper, Teflon) is shown in Figure 5.8. Figures 4.10(d) and 4.11(c),(d) from the previous chapter also illustrate this point very clearly.

Symmetry of the atomic lattice destroys the initial axial macrosymmetry of the collapse process in monocrystals at large strains due to the crystal plasticity effects (Fig. 5.8(a); Nemat-Nasser, Okinaka, Nesterenko, and Liu [1998]).

The picturesque molecular dynamic simulations of an imploding monocrystal cylinder losing its symmetry at the end of the collapse with 0.4 million atoms interacting with the Lennard–Jones potential can be found on the web site (http://linux.lanl.gov/bifrost/MD/implode_0.4m.html). The calculations were made by Peter Lomdahl (LANL).

This phenomenon demonstrates that shear localization at the late stage of collapse is the main source of energy dissipation, practically in any material with an effective strain more than 0.5–1. This is connected with the concentration of deformation inside the shear bands and with the arrest of plastic flow in the areas between them. It also means that the geometry of hot spots even in initially perfect pores may be one-dimensional following the plane geometry of shear bands, rather than three-dimensional due to the initial geometry.

5.3.3 Laminar Model of Powders

Another alternative for the modeling of shock thermodynamics is the laminar model (Thouvenin [1966]). The kinematics of powder densification in the frame of this model was discussed in Chapter 3. The mechanism of energy dissipation here is connected with differences in the loading (shock) and unloading (rarefaction) paths which alternate in the material point before it comes into an equilibrium state (Fig. 5.9).

The great advantage of this model is that paths in the P-V-E space are completely described by the constitutive equations of solid materials. This can be applied to the porous mixture of different materials with the known constitutive equations of components. This feature eliminates the problem of energy distribution between components (Nigmatullin, Vainshtein et al. [1976]) because it naturally selects the path from the initial to the final equilibrium compressed state.

This model is able to predict the final state behind the shock wave, but is unable to describe the right scale for the shock front (Figs. 5.2 and 5.4, and Section 3.3) in powders. As was discussed earlier, the space scale of the shock front in the granular material, where the material achieves its final state, is about the particle size for a strong shock. From Figure 5.9 it is evident that the thickness of the shock in the laminar model is an order of magnitude longer.

It should be emphasized that the ability of some model to describe the final state in the shocked powder does not mean that this model describes correctly the path that is followed by the material to get into this final state. Actually, any dissipation process eventually brings material into the same final state if the material is loaded by a stationary impulse.

Despite that the laminar model definitely fails to explain the main path which real porous materials follow to get into the final state, the nonperiodical three-dimensional shock wave on the scale of the particle size can be the important mechanism of the energy dissipation. This is especially true for strong shocks in the dynamic regime of particle deformation or when the viscoplastic flow does not efficiently dissipate the kinetic energy (Benson, Nesterenko et al. [1997]).

5.3.4 Two-Dimensional Numerical Modeling
of Heterogeneous Heat Release

5.3.4.1 Ordered Packing of Particles

The first attempt to move from simple one-dimensional approaches to two-dimensional numerical calculations of an ordered array of a small number of particles was made by Williamson and Berry [1986]. There are few qualitatively important results obtained by these authors and their group, based on computer simulation of the two-dimensional process of the viscoplastic flow of cylindrical particles during shock loading.

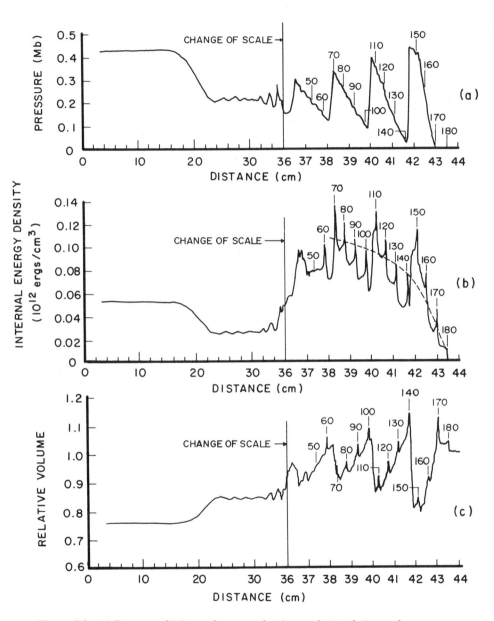

Figure 5.9. (a) Pressure, (b) internal energy density, and (c) relative volume versus distance for a laminar model of the powder. (Reprinted from *J. Appl. Phys.*, **39**, no. 10, Hofmann, R., Andrews, D.J., and Maxwell, D.E., Computed Shock Response of Porous Aluminum, pp. 4555–4562, 1968, with permission from the author.)

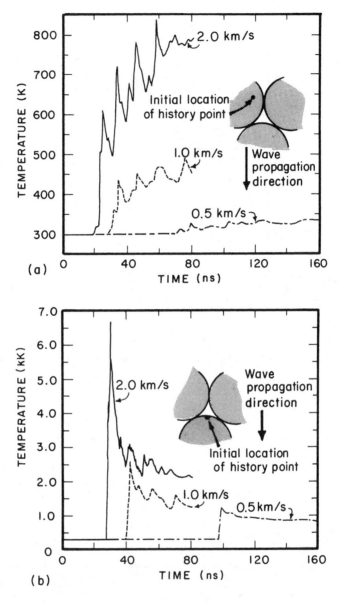

Figure 5.10. Powder model with ordered particles. The predicted temperature histories at two different tracer points on the interior of a particle for three impact velocities. (Reprinted from *J. Appl. Phys.*, **68**, no. 3, Williamson, R.L., Parametric Studies of Dynamic Powder Consolidation Using a Particle-Level Numerical Model, pp. 1287–1296, 1990, with permission from the author.)

One of them is the demonstration (Williamson and Berry [1986]) that the work of viscoplastic deformation creates the source of heat near the interface area even when friction between the adjacent particles was absent. The size of this intensively deformed zone was comparable with the particle sizes (diameter 75 μm) and was essentially larger than the size of the heat conductivity zone $(\kappa\tau)^{1/2}$ ~ 0.5 μm at κ ~ 0.1 cm²/s and $\tau = 28 \cdot 10^{-9}$ s, where κ is the thermal conductivity coefficient for steel and τ is the compaction time. This behavior demonstrates that contact friction (a very difficult problem by itself) can be neglected at least for high-shock amplitudes resulting in the densification of granular materials. It also clearly demonstrates the nature of a hot layer originating on particle contacts.

Another example is connected with the clarification of the role of shocks inside particles on the dissipation process together with viscoplastic flow (Williamson [1990]; Fig. 5.10). It is interesting that different tracer points are affected by microshocks in a qualitatively different manner. The influence of a high-velocity jet (case (b)) resulting in much higher temperatures is accompanied by a relatively slow influence of the microshocks. At the same time, the point which does not experience a jet impact (case (a)) was essentially effected by heat release during the train of microshocks. This again emphasizes the highly heterogeneous state of granular materials under strong shock loading.

Also, this two-dimensional approach was used to investigate the influence of bimodal particle distribution, hollow particles, the role of interstitial gases, and shock densification of composite materials (Williamson [1990]; Williamson, Wright, Korth, and Rabin [1989]). The experiments with an explosive compaction of wires by Gao, Shao, and Zhang [1991] as well as by Williamson, Wright, Korth, and Rabin [1989] are in good agreement with computer modeling.

The shortcomings of the approach based on the regular packing of a small amount of cylinders include the impossibility of introducing vorticity (due to symmetry). It is also impossible to explain how microflow in the frame of this model is related to a stationary shock regime. Actually, a demonstration of shock stationarity was never attempted with this approach.

5.3.4.2   Random Packing of Particles

Two-dimensional computer simulation with random particle sizes is the only way to avoid constraints arising from the specific initial geometry of the pores or particles. The two-dimensional computer model developed by Benson [1994], [1995] includes the complicated geometry of the material flow and is the most advanced method of treating the micromechanical and coupled thermochemical problems of the shock dynamics of powders. This modeling naturally includes thermal softening and what is most important—a geometry of material flow close to the real one. In this approach the geometry of the nonequilibrium heat release is in strong coupling with the viscoplastic flow of material during the compaction stage. The main equations for this process are presented in Chapter

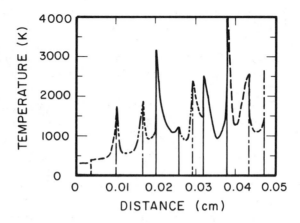

Figure 5.11. Highly heterogeneous temperature profile through an IN718 powder with a 1.10 km/s particle velocity, each particle is traced with a distinct type of line for clarity. (Reprinted from *Net Shape Processing of Powder Materials*, AMD-**216**, Benson, D.J., Nesterenko, V.F., and Jonsdottir, F., Numerical Simulation of Dynamic Compaction, pp. 1–8, 1995, with permission from ASME International.)

2. In this section, only the main results relevant to the thermodynamic aspects of shock loading will be emphasized.

Figure 5.11 presents the temperature distribution for cylindrical particles with different sizes ranging from 7.4 μm to 8.8 μm which were initially packed to a density of approximately 40% of the solid density. The final geometry of particles is scaled by the particle size and results of the calculations do not depend on this scaling. The temperature calculations don't account for the heat of melting, therefore the temperature above melting is overpredicted. The results of the two-dimensional computer simulation of the resulting particle geometry, in comparison with the experimental structure obtained from shock loading of the spherical particles are shown in Figure 5.12 (Benson, Nesterenko, and Jonsdottir [1995]). The following most important observations can be made from Figures 5.11 and 5.12:

1. The characteristic size of the hot zones concentrated on the particle's boundaries is comparable with particle size, only a few times less than the initial size (no heat conductivity is included in the model!). This is in qualitative agreement with the previous discussion of the BRG model and experimental results.

2. The different parts of particles can have different temperatures. Their surfaces are heated very nonuniformly, as the result of nonuniform deformation. Some areas were actually not deformed and preserved their initial spherical shape. This is in complete agreement with experimental results. For example, in Figure 5.1 different sides of the same particle underwent a completely different type of

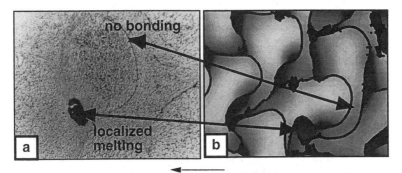

Figure 5.12. The structure of compacted material (stainless steel EP-450) after (a) shock consolidation and (b) contours of temperature obtained in computer calculations for a shock wave with a 1.1 km/s particle velocity. Darker regions have a higher temperature, with $T > 2500$ K being black. The direction of the shock propagation is shown by the arrow at the bottom. (Reprinted from *Net Shape Processing of Powder Materials*, AMD-**216**, Benson, D.J., Nesterenko, V.F., and Jonsdottir, F., Numerical Simulation of Dynamic Compaction, pp. 1–8, 1995, with permission from ASME International.)

deformation that resulted in their different temperatures.

3. The difference between maximal temperatures in the contact areas ($T_{c,max} = 3000$–4000 K) and the equilibrium temperature ($T_{eq} = 1500$ K) is two to three times in accord with the maximal difference between equilibrium and the contact temperatures in experiments with copper and nickel powders (Table 5.1).

4. The local melting in the experiment and the concentration of the internal energy in the simulation is definitely the result of a jet-like flow. It leads to melting after collision with the neighboring particles. This type of material flow, where only part of the particle is heavily involved in the pore collapse process, is qualitatively different from a single-pore model approach.

5. The computer model also correctly predicts the heat release along part of the particle surfaces as the result of shear localization—the feature of qualitative importance for bonding between particles in dynamic experiments.

6. The vorticity inside the melting pools is predicted by two-dimensional computer simulations if particles are initially packed nonuniformly. This feature is a typical one for Benson's model and is not included in the case with the symmetrical packing of particles.

It is possible to conclude that the two-dimensional assembly of irregularly packed particles is a reasonable model to explain the three-dimensional granular material behavior with spherical particles during dynamic densification, and qualitatively correctly describes the experimental results at least on the scale of the particle size.

On the contrary, the phenomena developing on the scale of the jets thicknesses colliding with neighboring particles are sensitive to the geometry

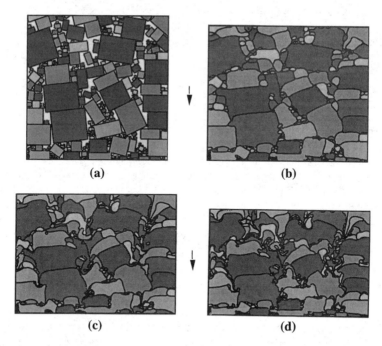

Figure 5.13. (a) Initial and (b), (c), (d) final configurations for the rectangular particles with an initial porosity of 19% and particle velocities of 0.25, 1.00, and 2.00 km/s, correspondingly. The direction of shock propagation is shown by the arrows. Notice transition from (b) quasistatic to (c), (d) dynamic deformation. (Reprinted from *Modelling Simul. Mater. Sci. Eng.*, 2, Benson, D.J., An Analysis by Direct Numerical Simulation of the Effects of Particle Morphology on the Shock Compaction of Copper Powder, pp. 535–550, 1994, with permission from the author.)

changes from the axisymmetric to the cylindrically symmetric ones (Cooper, Benson, and Nesterenko [2000]). At the same loading conditions and geometrical sizes, the axisymmetric model exhibited much higher temperatures, in part due to jet formation in the void with a higher velocity.

Even powders with particle shapes different from the cylindrical symmetry can be treated straightforwardly (Benson [1994]). The case of particles with a square shape deforming under shock loading with a different amplitude of particle velocity from Benson's paper is presented in Figure 5.13. The physical properties of rectangular particles were the same as for the cylindrical ones. Notice the transition from quasistatic deformation (Fig. 5.13(a)) to dynamic deformation (Fig. 5.13(c),(d)) that was considered in detail in Section 2.4.

An interesting comparison was made between particles of different shapes with the same loading conditions and initial porosity. In the case of rectangular particles, areas of the energy release were concentrated mainly along the particle boundaries parallel to the direction of shock-wave propagation (Fig. 5.13). In the assembly of cylindrical particles, the distribution of thermal energy along the particle's boundaries was more uniform (Fig. 5.12(b)). In comparison with cylindrical particles, the rectangular ones exhibit a much less turbulent flow, which should result in the weaker bonding between particles. From this observation it is possible to expect that the strength of compact can be very anisotropic in the case of rectangular particles (Benson [1994]).

5.4 "Skin" Model and Thermal Relaxation in Shocked Powders

In the previous section the different models of heterogeneous heating were discussed. The process of thermal relaxation in the shocked powder is equally important. The approach to analyzing this process is represented by the "skin" model developed by Nesterenko [1977], and Nesterenko and Muzykantov [1985]. It is based on the hypothesis that during the characteristic shock rise time τ, the thickness of the layer in which the heat is generated is characterized by a final size δ and the influence of the heat conductivity is negligible.

Evidently, this is true if

$$(\kappa\tau)^{1/2} \ll \delta. \tag{5.12}$$

Because of the relatively strong shock wave $\tau \sim a/D$ (Eq. (5.1) and Table 2.1), Eq. (5.12) can be written as

$$a\kappa/\delta^2 D \ll 1. \tag{5.13}$$

The typical parameters for the amorphous powders are $a \sim 10^{-3}$ cm, $\kappa \sim 0.1$ cm²/s, and $D \sim 10^5$ cm/s. Eq. (5.13) is valid if $\delta \gg 3 \cdot 10^{-5}$ cm = 0.3 μm. This restriction is reliable from an experimental point of view and also agrees with the natural condition $a > \delta$. A similar approach was developed by Schwarz, Kasiraj, and Vreeland [1986] and Kondo, Soga, Sawaoka, and Araki [1985].

The "skin" model is supported by the experimental fact of a nonsymmetric interface temperature profile: a relatively fast heating time $\tau \sim a/D$ and an essentially slower time of their cooling (see Figs. 5.2 and 5.4). If the temperature profile is symmetric, the "skin" model has no rights to exist. The asymmetry of the temperature profile is evidence of the finite thickness of the layer where heat is released during the viscoplastic flow along the particle contact.

In the evaluation of the cooling rate for particle contact surfaces, their mutual thermal interaction should be taken into account. This can be simplified by consideration of the thermal relaxation process in a nonuniform one-dimensional

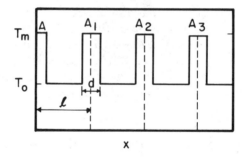

Figure 5.14. Initial distribution of temperature for the "skin" model.

system of cold and hot regions (Fig. 5.14). The former corresponds to the volume of particles, and the latter corresponds to their surface layers.

A one-dimensional system is a reasonable description of powder consolidation of the fragmented amorphous foils or consolidation of the ribbons without their fragmentation (Nesterenko et al. [1987]; Zolotarev, Denisova, Nesterenko et al. [1989]; Friend and Mackenzie [1987]; Nesterenko [1992]; Heczko and Ruuskanen [1993], [1994]; Rakhimov [1993]).

Heat flow at points with the coordinates x = 0, $l/2$, l, $l/2$, $3l/2$, $2l$, ... equals zero as a result of the periodicity of the initial temperature distribution in this model. Therefore, cells with length l (or $l/2$) shown in Figure 5.14 are thermally insulated from each other. The whole problem can be solved in the region $0 \leq x \leq l$. An equation for thermal conductivity, initial and boundary conditions for the cell with length l, can be written in the dimensionless form

$$\frac{\partial u}{\partial \tau} = \frac{\partial^2 u}{\partial \xi^2}, \quad 0 < \xi < 1, \quad 0 < \tau < +\infty, \tag{5.14}$$

$$u(\xi, 0) = \begin{cases} 1, & 0 \leq \xi < \delta/2; \\ 0, & \delta/2 \leq \xi \leq 1 - \delta/2; \\ 1, & 1 - \delta/2 < \xi \leq 1; \end{cases}$$

$$u_\xi(0, \tau) = 0, \quad u_\xi(1, \tau) = 0, \quad 0 < \tau < +\infty,$$

where $u = T/T_m$, $\xi = x/l$, $\tau = \kappa t/l^2$, $\delta = d/l$, T is the temperature, x is the coordinate, t is the time, κ is the thermal conductivity, and T_m is the temperature of the hot areas. It is assumed that the initial temperature is $T_0 = 0\,°C$.

The solution to this problem is found in the form of a Fourier series (Carslaw and Jaeger [1988]):

$$u(\xi, \tau) = \delta + \frac{2}{\pi} \sum_{n=1}^{\infty} \frac{\sin(n\pi\delta)}{n} \cos(2\pi n\xi) e^{-4\pi^2 n^2 \tau}. \tag{5.15}$$

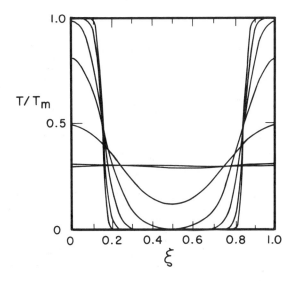

Figure 5.15. The temperature distributions in the particle ($\delta = 0.3$) during relaxation process for different moments $\tau_T = 10^{-4}$, $4 \cdot 10^{-4}$, $16 \cdot 10^{-4}$, $64 \cdot 10^{-4}$, $256 \cdot 10^{-4}$, $1024 \cdot 10^{-4}$ and the equilibrium curve.

The exponential factor provides absolute convergence of the series with $\tau > 0$. In order to obtain the sum of the series with sufficient accuracy in the stages of thermal relaxation ($\tau \geq 10^{-4}$) the summing was limited by 100 terms.

Evolution of the temperature distribution obtained is shown in Figure 5.15 for different instants of time τ inside one cell. It can be seen that the approximate time for establishing thermal equilibrium in the cell with the given $\delta = 0.3$ is $\tau_{eq} \approx 0.1 t_{eq} \approx 0.1 l^2/k$. Temperature is at a maximum, and the material cooling rate is at a minimum in the centers of the hot zones, i.e., at the particle contact surface (in the model they correspond to points $\xi = 0$ and $\xi = l$ in Fig. 5.14). These points are considered to be the most critical for retaining the amorphous initial state.

The dependence of temperature on time at the points $\xi = 0$ ($\xi = 1$) is shown in Figure 5.16 for different values of the initial hot layer thickness δ. In can be seen that the cooling time to the equilibrium temperature is practically independent of the initial size of the heated zone, i.e., on the value of δ and equal to $\tau_{eq} = 0.1$. This leads to the average cooling rate $(T_m - T_{eq})/t_{eq}$ falling proportionally to δ:

$$\frac{T_m - T_{eq}}{t_p} = 10 \frac{T_m(1 - \delta)k}{l^2}. \tag{5.16}$$

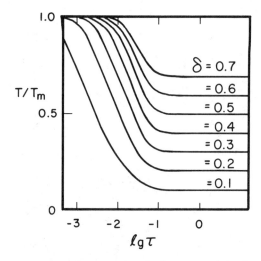

Figure 5.16. Temperature dependence on time in the center of the heated areas, points $\xi = 0$ ($\xi = 1$) for different values of δ.

Eq. (5.16), obtained for the cooling rate in the frame of the "skin" model, may be used to evaluate conditions for quenching of the contact surface areas to the amorphous state during shock consolidation and also in other dynamic processes of materials densification.

It is necessary to note that there are two typical scales of cooling times corresponding to two space dimensions: d and l, for the process with the initial temperature distribution given in Figure 5.14. The characteristic time t_{eq} determines the time for thermal equilibrium to be established in the cell, and $t_1 = d^2/\kappa$ characterizes the average rate of the temperature drop of the contact surfaces in the early stages of thermal relaxation, until the size of the heated region is comparable with d. The existence of the relaxation time t_1 reflects the emergence of the new structural level in the powder system, due to the preferential heat release on the particles contacts.

5.5   Nonequilibrium Thermodynamics of Powder Mixtures

5.5.1   Analysis of Energy Distribution Between Components

The analysis of the shock loading of porous mixtures of the condensed materials and the calculations of their Hugoniots demands an introduction of the principles of the thermal energy distribution between components. The different approaches to this problem can be illustrated by the phenomenological principle of the additivity (Dremin and Karpuhin [1960]) or by micromechanical modeling like the one-dimensional laminar model considered in Section 5.3. It was shown analytically that the component with less density in the mixtures is underheated and the component with a larger density is overheated in comparison with their temperatures according to the single shock Hugoniot at the same pressure (Nikolaevskii [1969]; Bogachev [1973]; Bogachev and Nikolaevskii [1976]). The opposite effect of overheating the less dense component over the larger density component was discussed by Nesterenko [1976], [1977]. Drumheller [1987a,b] studied analytically the behavior of shocked mixtures using different approaches to the energy distribution between components in porous mixtures.

Krueger and Vreeland [1991] developed a model for the Hugoniot for solid and porous two-component mixtures, which is based only on the static thermodynamic properties of the components. The model does not presuppose either the relative magnitude of the thermal and elastic energies or the temperature equilibrium between two components. It is shown that for a mixture, the conservation equations define a Hugoniot surface and that the ratio of the thermal energy of the components determines where the shocked state of the mixture lies on this surface. This ratio, which may strongly affect the shock-initiated chemical reactions and the properties of consolidated powder mixtures, is found to have only a minor effect on the Hugoniot of a mixture.

Let us try to analyze the different types of energy distribution which are possible in porous mixtures based on the distribution of the pore space between components. The total specific volume of the porous mixture at corresponding pressures can be written as follows:

$$V_{mix} = \alpha_1 V_1 + \alpha_2 V_2 + V_p, \qquad \alpha_1 + \alpha_2 = 1, \tag{5.17}$$

where V_{mix}, V_1, and V_2 are specific volumes of the mixture, and of the components in solid state, α_1 and α_2 are the mass fractions of the components in the mixture and V_p is the volume of pores per unit mass of mixture.

For shock loading of the mixture the Hugoniot equation of energy connecting the initial and final (equilibrium or "frozen") states behind the shock front is supposed to be valid. If the mechanical equilibrium with respect to particle velocity and pressure is established, then the following equation for the Hugoniot can be used:

$$E = \tfrac{1}{2}P\big(V_{00,mix} - V_{f,mix}\big), \tag{5.18}$$

where $V_{00,mix}$ are the initial and $V_{f,mix}$ final specific volumes of the mixture. It is considered that pressure equilibrium is attained in the mixture. The temperature equilibrium is not the prerequisite to use Eq. (5.18) because of the relatively long time of establishment of the thermal equilibrium. The "frozen" shock Hugoniot can be introduced in this case. In porous mixtures, the volume change, due to the establishment of the thermal equilibrium, is essentially less than the volume change due to compaction on the shock front, and the "frozen" Hugoniot in the P-V coordinates should not be too different from the equilibrium one.

For simplicity it is supposed that the material is completely compacted behind the shock, and that the individual densities of the components (ρ_1 and ρ_2) are equal to their solid densities in the normal conditions (ρ_{01} and ρ_{02}) despite the pressure. This is a reliable approximation for the porous materials with a porosity of about 2. It is based on the experimental fact that under pressures of about 10 GPa the volume change of metals is ~ 5%, which is small in comparison with the volume change resulting from pore closure. Then the initial ($V_{00,mix}$) and final ($V_{f,mix}$) specific volumes of the mixture can be written as follows:

$$V_{00,mix} = \alpha_1 V_{01} + \alpha_2 V_{02} + V_p, \qquad V_{f,mix} = \alpha_1 V_{01} + \alpha_2 V_{02}, \tag{5.19}$$

where V_{01} and V_{02} are the specific intrinsic volumes of the components at normal conditions.

Let us consider three different cases of pore distribution and find how they will effect the internal energy of the components in the final state.

Case 1. Initial pores in a mixture distributed between components proportionally to their mass fractions.

This means that the pore volumes $\alpha_1 V_p$ and $\alpha_2 V_p$ ($\alpha_1 V_p + \alpha_2 V_p = V_p$) are attached to the volumes of the corresponding solids $\alpha_1 V_{01}$ and $\alpha_2 V_{02}$. The total specific energy of shock compression can be found from the global Hugoniot for the mixture

$$E = \frac{1}{2} P\left(V_{00,mix} - V_{f,mix}\right) = \alpha_1 \frac{P}{2} V_p + \alpha_2 \frac{P}{2} V_p = \alpha_1 \varepsilon_1 + \alpha_2 \varepsilon_2. \tag{5.20}$$

In Eq. (5.20) the energies ε_1 and ε_2 can be interpreted as the specific energy of each component in the mixture behind the shock, according to the fact that the mass fractions of the components do not change during compaction. From this consideration, it is evident that each component can be assigned the specific energies which are equal in this particular case

$$\varepsilon_1 = \varepsilon_2 = \frac{P}{2} V_p. \tag{5.21}$$

It is worth mentioning that in this case the energies of the components in the unit mass of mixture have a ratio equal to the ratio of mass fractions (ε_1 and ε_2 are energies per unit mass of the corresponding components).

The possible justification for this distribution of initial pores and the resulting energies can be the fact that the ratio of the initial kinetic energies of the components in the mixture before the shock front is proportional to their mass fractions. So, if each component dissipates the kinetic energy which it carried before the shock front (in the reference system connected with the shock), then it results in the described energy distribution. The proposed derivation of energy distribution does not exhaust all the possible variants. Actually, the infinite amount of the pores distribution between components can be proposed, which results in a different energy distribution between components.

Case 2. Porosities of the components are equal.

This case is likely to be a reliable hypothesis if there is no preferential mechanism allowing one component to use the pores more efficiently (or more quickly) than another. At least this mechanism looks more "democratic" in the distribution of the initial open space between components, in comparison with Case 1. The initial pore volume can be represented as

$$V_p = V_{p1} + V_{p2}, \tag{5.22}$$

where V_{p1} and V_{p2} are the pore volumes attached to each of the components. Their values can be determined based on our hypothesis that "porosities" x_1 and x_2 (the ratio of powder volume to the volume of solid in the powder) of the components are equal. Because each component occupies the volume $\alpha_1 V_1 + V_{p1}$ and $\alpha_2 V_2 + V_{p2}$, the condition for equal porosities is

$$\frac{\rho_{01}\left(\alpha_1 V_{01} + V_{p1}\right)}{\alpha_1} = \frac{\rho_{02}\left(\alpha_2 V_{02} + V_{p2}\right)}{\alpha_2}, \tag{5.23}$$

where ρ_{01} and ρ_{02} are the intrinsic densities of the solid material for each component. Using Eq. (5.23) together with Eq. (5.22) we obtain values for V_{p1} and V_{p2}:

$$V_{p1} = \frac{\rho_2 \alpha_1}{\rho_2 \alpha_1 + \rho_1 \alpha_2} V_p, \tag{5.24a}$$

$$V_{p2} = \frac{\rho_1 \alpha_2}{\rho_2 \alpha_1 + \rho_1 \alpha_2} V_p, \tag{5.24b}$$

Finally, from Eq. (5.17) and the expressions for V_{p1} and V_{p2} (Eq. (5.24)), the porosities for components $x_1 = x_2$ are derived

$$x_1 = \frac{\rho_1\left(\alpha_1 V_1 + V_{p1}\right)}{\alpha_1} = \frac{V_{00,mix}}{V_{00,mix} - V_p} = x_2. \tag{5.25}$$

It is easy to show (Problem 6) that condition $x_1 = x_2$ results in the equality of the porosities x_1, x_2 to the global porosity of the mixture. Finally, for the specific internal energy in this case, we get

$$
\begin{aligned}
E &= \frac{1}{2}P\left(V_{00,mix} - V_{f,mix}\right) = \frac{1}{2}P\left[V_{p1} + V_{p2}\right] \\
&= \alpha_1 \frac{P}{2}\frac{V_{p1}}{\alpha_1} + \alpha_2 \frac{P}{2}\frac{V_{p2}}{\alpha_2} = \alpha_1 \varepsilon_1' + \alpha_2 \varepsilon_2'.
\end{aligned}
\tag{5.26}
$$

The natural interpretation of Eq. (5.26) is that the specific internal energies of each component ε'_1 and ε'_2 in the mixture are equal, correspondingly,

$$\varepsilon'_1 = \frac{P}{2}\frac{V_{p1}}{\alpha_1},$$
(5.27a)

$$\varepsilon'_2 = \frac{P}{2}\frac{V_{p2}}{\alpha_2}.$$
(5.27b)

The ratio of the specific internal energies from Eqs. (5.27) and (5.24) is

$$\frac{\varepsilon'_1}{\varepsilon'_2} = \frac{\rho_2}{\rho_1}.$$
(5.28)

So, in contrast to Case 1, the specific energies of the components are not equal but inversely proportional to their densities. If $\rho_1 \ll \rho_2$, then $\varepsilon'_1 \gg \varepsilon'_2$. This means that the considered method of internal energy distribution provides a preferential energy release into the light component, which can only be fulfilled by transformation of the part of the initial kinetic energy of the heavy particles into the specific energy of the light one. It is possible to expect that this case is true if both components are represented by particles of the same size, and with nearly equal strength.

Case 3. Finally, let us consider an extreme situation where the total internal energy is dissipated within only one component. This means that all the pores in the mixture are initially attached to this component. For example, consider the mixture W + boron nitride or W + graphite. Tungsten represents the large particles with high strength and hexagonal boron nitride (BN) or graphite represents easy deformable particles with lubricant properties. The dynamic loading is provided by a plate moving with velocity u. For simplicity of the consideration, imagine that this is manufactured from tungsten. Then the total specific internal energy of the mixture (the internal energy of the unit mass of the mixture) will be not less than $u^2/8$. This conclusion is based on the natural fact that the Hugoniot of the porous mixture W + BN is disposed lower than the Hugoniot of the solid tungsten in the pressure–particle velocity plane. According to this reason, the particle velocity of the mixture will be larger than $1/2u$, resulting in the aforementioned value of the internal energy.

It is considered that in this mixture all the specific internal energy will be released in the light particles of BN as a result of their plastic flow between large nondeformable tungsten particles. If this is true the specific internal energy of BN (energy per one unit mass of BN) will be equal to $u^2/8m$, where m is the weight content of BN in the mixture.

For real mixtures of tapped density the value of m can be close to ~0.33, providing the complete surrounding of the large tungsten particles by small BN particles. Under these conditions, the specific internal energy of BN will be as much as $8u^2/2$. If, for comparison, the same plate were used to load porous BN with any density, its internal energy would not exceed $u^2/2$, because the particle velocity of the powder cannot be larger than the velocity of the impacting plate. Even in these calculations we can see the difference in the order of magnitude in the internal energies of the same material loaded in the mixture and in a pure state by the same method. For the considered mixture W + BN under loading by the tungsten plate with a velocity of 5 km/s and taking the heat capacity of BN equal to 2 kJ/kgK we get a BN temperature equal to $5 \cdot 10^4$ K. It is clear that using less dense materials than BN in the mixture with tungsten, higher temperatures can be obtained.

From an energy point of view in the shock system of reference, large heavy particles in the mixture transform their kinetic energy, which they had before the shock front, into the thermal energy of small particles behind the shock front. Thus, a mixture under shock loading works like a natural transformer of the kinetic energy of one component into the internal energy of another.

5.5.2 Experimental Results

Despite the obvious practical significance of the nonequilibrium thermodynamics of porous mixtures (e.g., for material synthesis) there is a shortage of experimental research in this area. Nesterenko [1976], [1977] conducted experimental measurements of the thermoelectric force (emf) on the contact of the powder mixtures Ni + BN and Cu + BN with the use of the electric scheme presented in Figure 5.17.

The idea of these experiments is based on two main approximations:

- BN is dielectric and does not produce an input at the electrical signal of the composite thermocouple; and
- The mechanical behavior of copper and nickel in the mixture can be considered as similar one.

In this case the electrical signal from the thermocouple represents the average

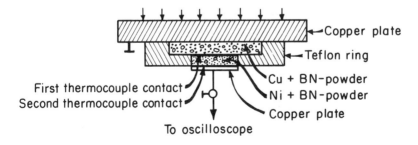

Figure 5.17. The scheme of measurement of the metallic particle's contact temperature in a mixture with dielectric particles.

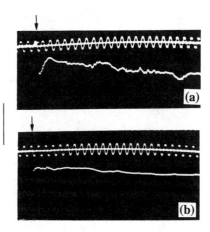

Figure 5.18. The oscillograms of the emf from the contact of powders Cu + BN and Ni + BN for (a) small (1–20 μm) and (b) large (0.1–0.5 mm) particles of copper and nickel in the mixture with BN. Shock pressure, 24 GPa. Arrows at the top mark the arrival of the first shocks on the first contact, the vertical scale is equal to 150 mV, the frequency of the calibrating signal is 1 MHz. The height of the Ni + BN powder layer (H = 45 mm) ensures a long time for the uninterrupted signal.

temperature of one of the mixture components—the metallic one.

The examples of recorded signals for the small and large particles of metals are given in Figure 5.18. From the comparison of the oscillograms, it is clear that the temperature on the contact of the large particles on the shock front and behind it is essentially less than the contact temperature of small particles. The amplitudes and time behavior of emf signals are perfectly reproducible (Nesterenko [1988]). The origin of the reproducible drop of the emf signal for small particles (Fig. 5.18(a)) is unknown.

So, the temperature of the metallic component on the shock front decreased with the increase in the particle size of this component at the same macroscopic conditions (mixture content, initial density, and loading conditions). This is opposite to the increase in the particle temperature with increasing particle size in a porous one-component powder. The Hugoniot of these mixtures did not depend on the particle size of the components. Thus, these results point to the dependence of the thermal energy (the main part of the internal energy for a relatively small initial density of the powders) distribution on the relative particle sizes of the components.

One of the explanations of this result could be connected with the lubricant effect of the small BN particles, during shock densification of the mixture that can be facilitated with the increase in size of the metallic particles. This decreases the extent of plastic deformation of the metallic particles resulting in the decrease of their temperatures. According to the evaluations, the value of the BN temperature can be as high as 10^4 K in a mixture with a copper for weight ratio

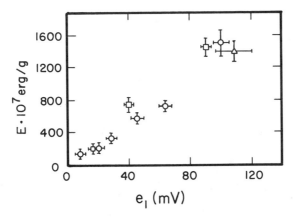

Figure 5.19. Experimental data of the dependence of the emf signal on the internal energy of the one-component powder and mixtures for small particle sizes of the metallic component: circles, contact of copper and nickel powders, $\rho_{00} = 4.67$ g/cm^3; triangles, contact of Cu + BN and Ni + BN powders, weight ratio 15:1, $\rho_{00} = 4.67$ g/cm^3; squares, contact of Cu + BN and Ni + BN powders, weight ratio 9:1, $\rho_{00} = 4.15$ g/cm^3.

of 15 : 1, initial density 4.67 g/cm^3, at shock pressure 25 GPa (Nesterenko [1977]). This behavior of the porous mixtures can be used for the melting of the component with small particles, with a subsequent quenching of the melt due to heat conductivity in the bulk of the large particles. This can be accomplished during the high-pressure stage and after shock loading.

The experimental data (Fig. 5.19) demonstrate that the transition through the region of existence of the high density and strength modifications of BN (cubic boron nitride) did not effect the heat release in the metallic part of the mixture.

Points for a mixture of 9:1 ($\rho_{00} = 4.15$ g/cm^3) on the graph of the dependence of thermoelectric power on the internal energy are situated around the straight line. This could prove that the main part of the hexagonal BN is transformed into dense phases behind the shock front after the end of the compaction process.

The high temperatures of the light component in the mixture are able to evaporate it and induce growth of new phases from the gaseous state on the relatively cold metallic particles. This mechanism of creating high temperatures can be responsible for the diamond formation in mixtures metal + graphite under shock loading (Kleiman, Heinmann, Hawken, and Salansky [1984]).

Even the mixture of particles of the same material has an interesting behavior under shock loading. The densification of the mixture in the quasistatic regime is provided mainly by the plastic deformations of the small particles, as is demonstrated in Figure 5.20 (Nesterenko [1985]). Raybould [1987] obtained the same result for the mixture of large and small spherical particles of Ti–6Al–4V alloy where the large particles remain spherical while the smaller ones undergo

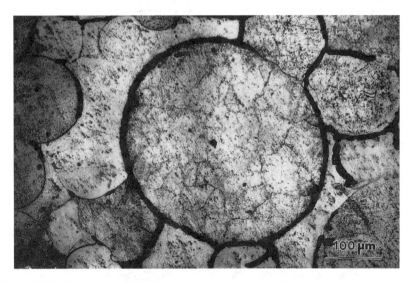

Figure 5.20. Shock densified mixture of particles Ti–Al–Nb–Zr alloy with different particle sizes (quasistatic regime).

extensive plastic deformation during shock loading. Vukcevic, Glisic, and Uskokovic [1993] also mentioned that most of the energy released owing to plastic deformation is accumulated on the smaller particles of Al–Li–X alloy.

Because of the plastic deformation of the particles being the main source of energy dissipation, the average temperature of the small particles will be higher than the temperature of the larger ones. The mechanism of this phenomenon can be explained by the higher local elastic stresses on the contacts of small particles, in comparison with the contact of large particles at the beginning of shock loading. This results in the development of plastic flow initially on the contacts of the small particles. It increases the contact's temperature and produces thermal softening with a further development of plastic flow in these regions.

This type of heterogeneous heating can result in the melting of small copper particles (10–20 μm) in mixture with large particles (100–400 μm) at a shock pressure of 32 GPa as is illustrated in Figure 5.21. The structure of this material, compacted without melting, can be seen in Figure 2.30(a) close to the central rod.

Using a method similar to one proposed by Nesterenko [1974], Mutz and Vreeland [1993] experimentally demonstrated that preferential deposition of the shock energy is observed in small particles being in contact with larger powder.

As was already mentioned, the mixture of large heavy particles and small light particles is a natural transformer of kinetic energy of the former into the plastic deformation and flow of the latter. This can be used for the ultrafine crushing of small ceramic particles during shock loading, as was demonstrated by

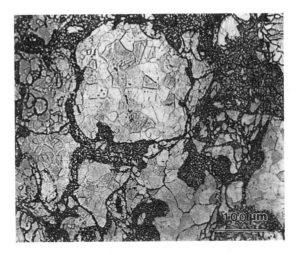

Figure 5.21. The melting of small (10–20 μm) particles of copper in a mixture with large (100–400 μm) particles. Shock pressure, 32 GPa.

Nesterenko, Panin, Kulkov, and Melnikov [1992] for a mixture of steel particles with sizes of 100–500 μm and submicron zirkonia powder.

Another reason for heterogeneous plastic deformation in the porous mixtures is connected to a different strength of the constituent particles. Deribas, Staver, and Stertser et al. [1977] demonstrated that, in the mixture of spherical steel and copper particles with equal volume ratio, the densification is provided by plastic flow of the weak material—copper particles. This produces nonuniform heating of the mixture with overheating of the soft component. The nonuniform plastic deformation of the powder mixtures, depending on the component's particle-size relations, can effect the final strength of the densified materials, as was shown by Raybould [1982] for mixture of aluminum and lead powders.

5.6   Shock and Chemical Reactions in Powders

The chemical reactions in powder materials were extensive areas of research during the last 10 years. A detailed analysis of different aspects of this phenomenon can be found in the books by Graham [1994], Batsanov [1993], and Meyers [1994]. The existing approaches are represented in papers by Krueger, Mutz, and Vreeland [1991], [1992], Thadhani [1993], [1994], Bennett, Horie, and Hwang [1994], Iyer, Bennett, Sorrell, and Horie [1994], Horie [1995], Batsanov [1996], Thadhani, Graham et al. [1997], Vreeland, Montila, and Mutz [1997],

Baer [1997], Thadhani and Aizawa [1997], Bennet, Tanaka, and Horie [1997], and others.

There are different interpretations of the results on chemical reactions under shock loading. For example, Thadhani, Graham et al. [1997] concluded that "the initiation of chemical reactions in the shock compression of powders is controlled by solid-state mechanochemical processes and cannot be qualitatively or quantitatively described by thermomechanical models."

On the other hand, Batsanov, Marquis, and Meyers [1995] arrived at the conclusion that "the mechanism of the reaction was by the formation of a liquid Nb–Si or Mo–Si layer surrounding the solid niobium and molibdenum particles respectively."

Vreeland, Montilla, and Mutz [1997] stated that in Ti–Si porous mixtures "reaction initiation occurs at melt pools of silicon formed in localized sites of excess porosity, with no evidence of solid-state reaction." It is interesting that, despite the essential differences in the interpretation of the results of chemical reactions under shock loading, different authors agree that the micromechanics and nonequilibrium thermodynamics of powders play a paramount role in the reaction initiation. There is no common point of view on the mechanism of chemical transformations under shock wave loading, and actually a single mechanism for all cases can hardly exist.

This section will briefly outline only the main features of powder micromechanics and thermodynamics which can be crucial in reactive porous materials. There is no doubt that porosity and grain structure are very important for chemical reactions as is mentioned by almost all researchers. It is evident that the shock-wave mechanics of powder deformation in their inert stage is very important for establishing the necessary conditions for chemical reactions. It is possible that a strong coupling between plastic flow and reactions takes place.

One of the prerequisites of the initiation of a chemical reaction can be the transition from a quasistatic to a dynamic mode of particles deformation described in Chapter 2. This is due to the fact that high pressure in the shock wave, as well as a small time of shock loading, should not result in the promotion of a chemical reaction by itself. The only thing distinguishing shock loading from quasistatic pressing is the possibility of qualitatively new effects, such as shear localization on the particle's boundaries, local concentration of microkinetic energy, jet flows, mutual localized plastic flow of components, vorticity resulting from instability of shear flow, and hot spots. All these effects are noticeable only in the dynamic mode of powder deformation. So, transition to the dynamic mode of powder deformation should be considered as the natural necessary condition for any fast chemical reactions.

The two-dimensional computer calculations (Chapter 2) clearly demonstrate that microkinetic energy, even in a developed dynamic regime for typical condensed porous materials, exists only on the time and space scale comparable to the shock front itself. This means that the typical time scale for some peculiar interaction between components is comparable with the shock rise time. After this period, it is hardly probable to expect that there is some intense movement between components, allowing them to give some results different from quasistatic compression. This restricts the initiation period of truly shock-induced chemical reactions to the order of magnitude about the shock front rise time.

The nonequilibrium heating of powder and its dependence on particle size is very important for the initiation of chemical reactions. At the same time, the role of nonequilibrium heating should be most important at the intermediate level of pressures, because the difference between the temperatures of particle contacts and the equilibrium temperatures is negligible at developed dynamic regimes of particle deformation. Heterogeneous heating of the powders is of special importance for porous explosives. For example, the influence of the particle sizes on detonation parameters is observed by Attetkov and Solov'ev [1987] and Howe, Frey, and Boyle [1976].

A recent review on this subject can be found in the paper by Sheffield, Gustavsen, and Anderson [1997]. It is interesting to mention that Schilperoord [1976] used the coefficient of energy concentration in the hot spots equal to three to get agreement between experimental data and model predictions. This is in accord with the thermoelectric data on the maximal difference of contact temperature in powder with large particles in comparison with equilibrium temperature (Table 5.1), despite the different properties of the explosive and metal particles. At the same time, it should be emphasized that, according to our data, no one pressure-independent parameter, like the coefficient of energy concentration in the hot spots, exists.

The behavior of condensed reactive mixtures with a nonexplosive type of reaction is also sensitive to the particle size. Thadhani, Graham et al. [1997] demonstrated the dependence of the reaction in a mixture of Ti–Si on particle size with a maximum for intermediate particle sizes. This was connected with the change of deformation mode between fracture and plastic deformation in silicon powders. Fine morphology particles showed particle agglomeration, while coarse Si powders underwent extensive fracture and entrapment within the plastically deformed titanium. Both features were detrimental for chemical reaction. At the same time, the simultaneous plastic deformation of both titanium and silicon particles played a crucial role to initiate reaction in this mixture. This example clearly demonstrates that the certain mechanical mode of the plastic flow of the powder mixture is really the prerequisite for chemical reaction. It is very important to establish some critical parameters of this mutual flow which will ensure the beginning of the reaction.

The shock behavior of mixtures of a few materials with essentially different particle sizes and properties is practically an unexplored area in chemical reactions. For example, it is evident that in such mixtures, the shock pressure cannot be a reliable parameter to describe the initiation of a chemical reaction between components. If three materials are present in the mixture with two of them being able to react, then the properties of the third material can be very crucial to direct the kinetic energy into these reactant components and cause their mutual plastic flow and heating. This can be one of the reasons why detonation in explosives with heavy additives depends on their particle size. Shvedov et al. [1980] described two different detonation regimes in explosive mixtures highly diluted by inert material. The specific role of heavy additives in an explosive mixture can be connected not only with additional inertial confinement of the detonation products but also with superheating of the fine explosive component by transformation of the kinetic energy of heavy particles into internal energy of the explosive. This will result in a more effective explosive transformation and different detonation regimes (Nesterenko [1992]).

The residual porosity left even after strong shock loading of the porous material can be important in evaluating the sensitivity of powder explosives to secondary shock loading (Boyle and Pilarski [1981]).

Another important feature of the mixtures is the ability to cause melting and even evaporation in one component (or components) in a mixture with relatively cold particles, if their size is large enough. This can provide a unique capability for the quenching of high-pressure/high-temperature phases due to heat transport into the cold body of large particles on a time scale comparable with shock duration.

The dependence of jet formation on the pore morphology (Cooper, Benson, and Nesterenko [2000]) definitely demonstrates that this characteristic can be used to design the most efficient granular materials for the initiation of chemical reactions or phase transformations. Owing to this dependence, it is possible to design such materials in which a chemical reaction will not be initiated up to some extent of impact loading.

Howe [2000] provided a detailed analysis of the effects of microstructures upon explosive sensitivity and performance, and outlined further desirable activity in this area. It was mentioned that the influence of microstructural features on the sensitivity is typically much greater than its influence upon performance. Such behavior may be connected with the initiation process starting at the beginning of the developed dynamic regime of deformation when the initial structural features are destroyed to a large extent by shock wave compression.

One more relatively unexplored area is connected with the use of experimental methods such as the method of natural thermocouples to test the three-dimensional computer programs for multiphase, porous reactant materials. The comparison of results of such programs, with experimental and computed shock velocity–particle velocity or even shock front thickness, has very limited value because the good fit can be obtained with a wide variety of equations of state or using practically any rules describing the energy redistribution between components in the porous mixtures. Only testing of the internal energy distribution between components, even in model ones, can be a solid justification of a numerical model.

5.7   Conclusion

As a result of the "highly" heterogeneous nature of thermal energy release on essentially different space scales during shock densification, a few characteristic temperatures can be introduced into the powder. For example, the average temperature of the particle contacts, T_c, average temperature of the gases into the pores, T_g, average temperature over the volume of the compacted material, T_v, average temperature over the volume of the relatively cold, bulk part of the particle's material, T_b and, finally, equilibrium temperature T_{eq}, when thermal equilibrium is established in the material bulk. The existing experimental data demonstrate the difference between these values and in some cases it can be more than the order of magnitude. Thus, shock loading of granular materials (powders) results in a highly heterogeneous thermodynamical state behind the shock front.

The unique feature of this state is that the differences between neighboring portions of the same particle can be very large. Moreover, this difference can be monitored by the initial particle size. Its increase facilitates the temperature differences between the contact area and the bulk of the particle if the shock pressures are not very high. After exceeding the critical pressure depending on particle strength the temperature can be "mechanically" equilibrated on the shock front.

The contact area between the granular material and solid is highly affected by the nonequilibrium heat release.

A reasonable description of this highly heterogeneous state can be provided by two-dimensional numerical modeling where the heat release is coupled with a complex geometry of viscoplastic flow. Existing phenomenological models are also useful instruments for the qualitative understanding of shock thermodynamics and the processes of thermal relaxation behind the shock front.

The shock compression of porous mixtures represents a unique possibility of obtaining the high-temperature states of small light particles in a mixture with large and heavy ones. In combination with relatively cold neighboring large particles, it may help to obtain rapidly solidified states of materials under high pressure.

Problems

1. Prove that the linear, quadratic, and cubic dependencies for $F(t)$ under conditions of the same total deposited energy and shock rise time τ in the frame of the BRG model assumptions results in a subsequent increase in the maximal contact temperature by 33%, 60%, and 83% correspondingly in comparison with $F_0 = $ constant.

2. What is the maximal amount of nonequilibrium melt which can be obtained at the shock loading of porous copper with initial density 4.45 g/cm^3 and shock pressure 150 kbar. The heat capacity of copper can be considered in the first approximation constant and is equal to 0.4 kJ/kgK, the heat of fusion does not depend on pressure in the first approximation and is equal to 13 kJ/mol.

3. Find the maximal interface temperature for copper powder loaded by a shock wave with a pressure of 20 GPa, initial density 4.45 g/cm^3, and particle size 0.5 mm according to the BRG model.

4. Estimate the ratio of the maximal interface temperatures for copper and nickel powders, according to the BRG model.

5. Calculate the dependence of the inner surface temperature of the collapsing pore in the rigid–plastic material for the following dependence of the yield strength on temperature:

$$Y = \begin{cases} Y_1\left(1 - [T/T_m]^2\right), & T \leq T_m, \\ 0, & T > T_m. \end{cases}$$

6. Show that a condition of equal porosities of components $x_1 = x_2$ in two-component porous mixture results in their equality to the global porosity of the mixture.

7. Find the shock pressure which will result in a complete melting of fine-grained copper mixed with large tungsten particles. The volume content of copper is 80%, the initial density of the mixture is half of the solid density, and the heat of copper fusion is 13 kJ/mol.

8. What is the main reason for the decoupling of the mechanical and thermal processes in shock-loaded powders? Make simple calculations.

9. Propose the possible tailoring of the initiation in a mixture of explosive particles by adding inert particles depending on their properties.

10. What is the main difference between heat release in porous solids (with spherical pores) and in granular materials?

11. Why do events inside a shock front in porous mixtures play the most important role in the initiation of chemical reactions?

References

Attetkov, A.V., Vlasova, L.N., Selivanov, V.V., and Solov'ev, V.S. (1984a) Local Heating of a Material in the Vicinity of a Pore upon its Collapse. *Zhurnal Prikladnoi Mekhaniki i Tehknicheskoi Fiziki*, **25**, no. 2, pp. 128–132 (in Russian).
   English translation: *Journal of Applied Mechanics and Technical Physics*, September 1984, pp. 286–291.
Attetkov, A.V., Vlasova, L.N., Selivanov, V.V., and Solov'ev, V.S. (1984b) Effect of Nonequilibrium Heating on the Behavior of a Porous Material in Shock Compression. *Zhurnal Prikladnoi Mekhaniki i Tehknicheskoi Fiziki*, **25**, no. 6, pp. 120–127 (in Russian).

English translation: *Journal of Applied Mechanics and Technical Physics*, May 1985, pp. 914–921.

Attetkov, A.V., Vlasova, L.N., and Solov'ev, V.S. (1986) The Influence of the Structural Peculiarities on the Shock Compressibility of Porous Media. In: *Procedeengs of IX International Conference on High Energy Rate Fabrication* (Edited by V.F. Nesterenko and I.V. Yakovlev). Lavrentyev Institute of Hydrodynamics, Special Design Office of High-Rate Hydrodynamics, Novosibirsk, pp. 136–140 (in Russian).

Attetkov, A.V., and Solov'ev, V.S. (1987) Heterogeneous Explosive Decomposition in a Weak Shock Wave. *Fizika Goreniya i Vzryva*, July–August, **23**, no. 4, pp. 113–125 (in Russian).

English translation: *Combustion, Explosion, and Shock Waves*, 1988, **23**, January, pp. 482–491.

Attetkov, A.V., and Solov'ev, V.S. (1989) On Mechanism of Thermal Dissipation under Shock Compression of Porous Material. In: *Procedeengs of International Seminar on High-Energy Working of Rapidly Solidified Materials and High-T_C Ceramics* (Edited by V.F. Nesterenko and A.A. Shtertzer). Special Design Office of High-Rate Hydrodynamics, Institute of Theoretical and Applied Mechanics, Novosibirsk, pp. 191–195 (in Russian).

Baer, M.R. (1997) Continuum Mixture Modeling of Reactive Porous Media. In: *High-Pressure Shock Compression of Solids IV. Response of Highly Porous Solids to Shock Loading* (Edited by L. Davison, Y. Horie, and M. Shahinpoor). Springer-Verlag, New York, pp. 63–82.

Bakanova, A.A., Dudoladov, I.P., and Sutulov, Yu.N. (1974) The Shock Compressibility of Porous Tungsten, Molybdenum, Copper and Aluminum in the Range of Low Shock Pressures. *Zhurnal Prikladnoi Mekhaniki i Tehknicheskoi Fiziki*, **15**, no. 2, pp. 117–122 (in Russian).

English translation: *Journal of Applied Mechanics and Technical Physics*, 1975, October, pp. 241–245.

Batsanov, S.S. (1994) *Effect of Explosions on Materials*. Springer-Verlag, New York.

Batsanov, S.S., Marquis, F.D.S., and Meyers, M.A. (1995) Shock Induced Synthesis of Silicides. In *Metallurgical and Materials Applications of Shock-Wave and High-Strain-Rate Phenomena* (Edited by Murr L.E., Staudhammer K.P., and Meyers M.A.). Elsevier Science, Amsterdam, pp. 715–722.

Batsanov, S.S. (1996) Solid-Phase Reactions in Shock Waves: Kinetic Studies and Mechanisms. *Fizika Goreniya i Vzryva*, January–February, **32**, no. 1, pp. 115–128 (in Russian).

English translation: *Combustion, Explosion, and Shock Waves*, 1996, **32**, no. 1, pp. 102–113.

Baum, F.A., Orlenko, L.P., and Stanyukovich, K.P. et al. (1975) *Physics of Explosion*. Nauka, Moscow (in Russian).

Belyakov, G.V., Livshits, l.D., and Rodionov, V.N. (1974) Shock Deformation of Heterogeneous Media Modeled by the Set of Steel Spheres. *Izvestiya Akademii Nauk SSSR, Fizika Zemli*, no. 10, pp. 92–95 (in Russian).

Belyakov, G.V., Rodionov, V.N., and Samosadnui, V.P. (1977) Heating of Porous Material under Impact Compression. *Fizika Goreniya i Vzryva*, **13**, no. 4, pp. 614–619 (in Russian).

English translation: *Combustion, Explosion, and Shock Waves,* 1978, February, pp. 524–528.

Bennett, L.S., Horie, Y., and Hwang, M.M. (1994) Constitutive Model of Shock-Induced Chemical Reactions in Inorganic Powder Mixtures. *Journal of Applied Physics,* **76**, no. 6, pp. 3394–4402.

Bennett, L.S., Tanaka, K., and Horie, Y. (1997) Developments in Constitutive Modeling of Shock-Induced Reactions in Powder Mixtures. In: *High-Pressure Shock Compression of Solids IV. Response of Highly Porous Solids to Shock Loading* (Edited by L. Davison, Y. Horie, and M. Shahinpoor). Springer-Verlag, New York, pp. 105–142.

Benson, D.J. (1994) An Analysis by Direct Numerical Simulation of the Effects of Particle Morphology on the Shock Compaction of Copper Powder. *Modelling Simul. Mater. Sci. Eng.,* **2**, pp. 535–550.

Benson, D.J. (1995) A Multi-Material Eulerian Formulation for the Efficient Solution of Impact and Penetration Problems. *Computational Mechanics,* **15**, pp. 558–571.

Benson, D.J., Tong, W., and Ravichandran, G. (1995) Particle-Level Modeling of Dynamic Consolidation of Ti–SiC Powders. *Modelling and Simulation in Materials Science and Engineering,* **3**, no. 6, pp. 771–796.

Benson, D.J., Nesterenko, V.F., and Jonsdottir, F. (1995) Numerical Simulations of Dynamic Compaction. In *Net Shape Processing of Powder Materials,* AMD-**216** (Edited by S. Krishnaswami, R.M. McMeeking, and J.R.L. Trasorras). ASME, New York, pp. 1–8.

Benson, D.J., Nesterenko, V.F., Jonsdottir, F., and Meyers, M.A. (1997) Quasistatic and Dynamic Regimes of Granular Material Deformation Under Impulse Loading. *Journal of the Mechanics and Physics of Solids,* **45**, nos. 11/12, pp. 1955–1999.

Blackburn, J.H., and Seely, L.B. (1962) Source of the Light Recorded in Photographs of Shocked Granular Pressing. *Nature,* **194**, April 28, pp. 370–371.

Blackburn, J.H., and Seely, L.B. (1964) Light Emitted from Shocked Granular Sodium Chlorides in a Vacuum. *Nature,* **202**, April 18, pp. 276–277.

Bogachev, G.A. (1973) Calculation of the Shock-Wave Adiabats for Some Heterogeneous Mixtures. *Zhurnal Prikladnoi Mekhaniki i Tehknicheskoi Fiziki,* no. 4, pp. 130–137 (in Russian).
English translation: *Journal of Applied Mechanics and Technical Physics,* 1975, February, pp. 546–552.

Bogachev, G.A., and Nikolaevskii, V.N. (1976) Shock Waves in Materials Mixture. Hydrodynamic Approximation. *Mechanics of Liquid and Gas,* no. 4, pp. 113–130 (in Russian).

Bondar, M.P., and Nesterenko, V.F. (1991) Strain Correlation at Different Structural Levels for Dynamically Loaded Hollow Copper Cylinders. *Journal de Physique IV (Colloque C3),* **1**, supplement au *Journal de Physique III,* pp. 163–170.

Boyle, V.M., and Pilarski, D.L. (1981) Shock Ignition Sensitivity of Multiply-Shocked Pressed TNT. In: *Proceedings of the 7th Symposium (International) on Detonation.* Office of Naval Research, NSWC MP 82-334, pp. 906–913.

Carlson, R.J., Porembka, S.W., and Simons, C.C. (1966) Explosive Compaction of Ceramic Materials. *Ceramic Bull.,* **45**, no. 3, pp. 266–270.

Carroll, M.M., and Holt, A.C. (1972) Static and Dynamic Pore-Collapse Relations for Ductile Porous Materials. *J. Appl. Phys.* **43**, pp. 1626–1635.

Carroll, M.M., Kim, K.T., and Nesterenko, V.F. (1986) The Effect of Temperature on Viscoplastic Pore Collapse. *Journal of Applied Physics*, **59**, no. 6, pp. 1962–1967.

Carslaw, H.S., and Jaeger, J.C. (1988) *Conduction of Heat in Solids*, Clarendon Press, Oxford.

Cooper, S.R., Benson, D.J., and Nesterenko, V.F. (2000) The Role of Void Geometry on the Mechanics of Void Collapse and Hot Spot Formation in Ductile Materials. *Int. Journal of Plasticity*, **16**, pp. 525–540.

Davies, H.A. (1978) Rapid Quenching Techniques and Formation of Metallic Glasses. In: *Rapidly Quenched Metals III: Proceedings of the Third International Conference on Rapidly Quenched Metals*, Vol. **1** (Edited by B. Cantor). Metals Society, London, pp. 1–21.

Deribas, A.A., Staver, A.M., Stertser, A.A. et al. (1977) Explosive Compression of Steel and Copper Powder Mixtures. *Izvestiya Sibirskogo Otdeleniya Akademii Nauk SSSR, Seriya Tekhn Nauk*, **1**, no. 3, pp. 45–50 (in Russian).

Dianov, M.D., Zlatin, N.A., Mochalov, S.M. et al. (1976) The Shock Compressibility of Dry and Water Saturated Sand. *Letters to Journal of Technical Physics*, **2**, no. 12, pp. 529–532 (in Russian).

Dianov, M.D., Zlatin, N.A., Pugachev, G.S., and Rosomaho, L.H. (1979) The Shock Compressibility of Finely Dispersed Media. *Letters to Journal of Technical Physics*, **5**, no. 11, pp. 692–694 (in Russian).

Dremin, A.N., and Karpuhin, I.A. (1960) Method of Determination of Shock Adiabats of Disperse Materials. *Zhurnal Prikladnoi Mekhaniki i Tehknicheskoi Fiziki*, no. 3, pp. 184–188 (in Russian).

Drumheller, D.S. (1987) Hypervelocity Impact of Mixtures. *Int. J. Impact Engng.*, **5**, pp. 261–268.

Drumheller, D.S. (1987) A Theory for Dynamic Compaction of Wet Porous Solids. *Int. J. Solids Structures,* **5**, pp. 261–268.

Dunin, S.Z., and Surkov, V.V. (1982) Effects of Energy Dissipation and Melting on Shock Compression of Porous Bodies. *Zhurnal Prikladnoi Mekhaniki i Tehknicheskoi Fiziki*, **23**, no. 1, pp. 131–142 (in Russian). English translation: *J. Appl. Mech. and Tech. Phys.*, July, 1982, pp. 123–134.

Flinn, J.E., Williamson, R.L., Berry, R.A., Wright, R.N. et al. (1988) Dynamic Consolidation of Type 304 Stainless-Steel Powders in Gas Gun Experiments. *Journal of Applied Physics,* **64**, no. 3, pp. 1446–1456.

Friend, C.M., and MacKenzie, P.J. (1987) Fabrication of Multi-Laminar Metallic Glass/Aluminium Composites by Explosive Compaction. *J. Mater. Sci.*, **6**, no. 1, pp. 103–105.

Gao, J., Shao, B., and Zhang, K. (1991) A Study of the Mechanism of Consolidating Metal Powder Under Explosive–Implosive Shock Waves. *J. Appl. Phys.* **69**, no. 11, pp. 7547–7555.

Gourdin, W.H. (1984a) Microstructure and Deformation in a Dynamically Compacted Copper Powder. *Mater. Sci. Eng.*, **67**, pp. 179–184.

Gourdin, W.H. (1984b) Energy Deposition and Microstructural Changes in Dynamically Consolidated Metal Powders. *J. Appl. Phys.*, **55**, no. 1, pp. 172–181.

Gourdin, W.H. (1986) Dynamic Consolidation of Metal Powders. *Progress in Materials Science*, **30**, pp. 39–80.

Graham, R.A. (1993) *Solids Under High-Pressure Shock Compression*, Springer-Verlag, New York.

Graham, R.A. (1997) Comments on Shock Compression Science in Highly Porous Solids. In: *High-Pressure Shock Compression of Solids IV. Response of Highly Porous Solids to Shock Loading* (Edited by L. Davison, Y. Horie, and M. Shahinpoor). Springer-Verlag, New York, pp. 1–21.

Heczko, O., and Ruuskanen, P. (1993) Magnetic Properties of Compacted Alloy Fe73.5Cu1Nb3Si13.5B9 in Amorphous and Nanocrystalline State. *IEEE Transactions on Magnetics,* **29**, no. 6, pt. 1, pp. 2670–2672.

Heczko, O., Ruuskanen, P., Kraus, L., and Haslar, V. (1994) Study of Magnetization in Compacted Amorphous and Nanocrystalline Alloy Fe73.5Cu1Nb3Si13.5B9. *IEEE Transactions on Magnetics,* **30**, no. 2, pt. 2, pp. 513–515.

Hofmann, R., Andrews, D.J., and Maxwell, D.E. (1968) Computed Shock Response of Porous Aluminum. *J. Appl. Phys.,* **39**, no. 10, pp. 4555–4562.

Horie, Y. (1995) Mass Mixing and Nucleation and Growth of Chemical Reactions in Shock Compression of Powder Mixtures. In: *Metallurgical and Materials Applications of Shock-Wave and High-Strain-Rate Phenomena, Proceedings of the 1995 International Conference* (Edited by L.E. Murr, K.P. Staudhammer, and M.A. Meyers). Elsevier, Amsterdam, pp. 603–614.

Howe, P., Frey, R., and Boyle, V. (1976) Shock Initiation and the Critical Energy Concept. In: *Proceedings of the 6th Symposium (International) on Detonation.* Office of Naval Research ACR-221, pp. 11–20.

Howe, P.M. (2000) Explosive Behaviors and the Effects of Microstructure. In: *Solid Propellant Chemistry, Combustion, and Motor Interior Ballistics* (Edited by V. Yang, T.B. Brill, and W.Z. Ren). Progress in Astronautics and Aeronautics. American Institute of Aeronautics and Astronautics. Reston, VA, Vol. 185, Chapter 1.6.

Ishutkin, S.N., Kuzmin, G.E., and Pai, V.V. (1986) Thermocouple Measurements of Temperature in the Shock Compression of Metals. *Fizika Goreniya i Vzryva,* **22**, no. 5, pp. 96–104 (in Russian).
English translation: *Combustion, Explosion, and Shock Waves.* 1987, March, pp. 582–589.

Iyer, K.R., Bennett, L.S., Sorrell, F.Y., and Horie, Y. (1994) Solid State Chemical Reactions at the Shock Front. In: *High-Pressure Science and Technology—1993. Joint International Association for Research and Advancement of High Pressure Science and Technology and American Physical Society Topical Group on Shock Compression of Condensed Matter   Conference* (Edited by S.C. Schmidt, J.W. Shaner, G.A. Samara, and M. Ross). AIP Press, New York, no. 309, pt. 2, pp. 1337–1340.

Kasiraj, P., Vreeland, T. Jr., Schwarz, R.B., and Ahrens T.J. (1984) Shock Consolidation of a Rapidly Solidified Steel Powder. *Acta Metall.,* **32**, no. 8, pp. 1235–1241.

Khasainov, B.A., Borisov, A.A., Ermolaev, B.S., and Korotkov, A.I. (1980) The Model of Shock-Wave Initiation of Detonation in High Density Explosives. In: *Chemical Physics of Combustion and Explosion Processes: Detonation,* Institute of Chemical Physics, Chernogolovka, pp. 52–55.

Kleiman, J., Heinmann, R.B., Hawken, D., and Salansky, N.M. (1984) Shock Compression and Flash Heating of Graphite/Metal Mixture at Temperatures up to 3200 K and Pressures up to 25 GPa. *J. Appl. Phys.,* **56**, no. 5, pp. 1440–1454.

Kondo, Ken-ichi, Soga, S., Sawaoka, A., and Araki, M. (1985) Shock Compaction of Silicon Carbide Powder. *J. Mater. Sci.*, **20**, pp. 1033–1048.

Krueger, B.R., Mutz, A.H., and Vreeland, T. (1992) Shock-Induced and Self-Propagating High-Temperature Synthesis Reactions in Two Powder Mixtures: 5:3 Atomic Ratio Ti/Si and 1:1 Atomic Ratio Ni/S. *Metallurgical Transactions A*, **23A**, no. 1, pp. 55–58.

Krueger, B.R., Mutz, A.H., and Vreeland, T. Jr. (1991) Correlation of Shock Initiated and Thermally Initiated Chemical Reactions in a 1:1 Atomic Ratio Nickel–Silicon Mixture. *Journal of Applied Physics, 70*, no. 10, pt. 1, pp. 5362–5368.

Krueger, B.R., and Vreeland, T. Jr. (1991) A Hugoniot Theory for Solid and Powder Mixtures. *Journal of Applied Physics, 69*, no. 2, pp. 710–716.

Kusubov, A.S., Nesterenko, V.F., Wilkins, M.L., Reaugh, J.E., and Cline, C.F. (1989) Dynamic Deformation of Powdered Materials as a Function of Particle Size. In: *Proc. International Seminar on High-Energy Working of Rapidly Solidified Materials and High-T_C Ceramics* (Edited by V.F. Nesterenko and A.A. Shtertzer). Special Design Office of High-Rate Hydrodynamics, Institute of Theoretical and Applied Mechanics, Novosibirsk, pp. 139–156 (in Russian).

Matytsin, A.I., and Popov, S.T. (1987) Determination of the Brightness Temperature with Emergence of a Shock Wave from Powder onto the Boundary with Transparent Barrier. *Fizika Goreniya i Vzryva*, May–June, **23**, no. 3, pp. 126–132 (in Russian).
English translation: *Combustion, Explosion, and Shock Waves*, 1987, November, pp. 364–369.

Meyers, M.A., Gupta, B.B., and Murr, L.E. (1981) Shock-Wave Consolidation of Rapidly Solidified Superalloy Powders. *Journal of Metals*, October, pp. 21–26.

Meyers, M.A. (1994) *Dynamic Behavior of Materials*. Wiley, New York, Ch. 17, pp. 616–636.

Meyers, M.A., Benson, D.J., and Olevsky, E.A. (1999) Shock Consolidation: Microstructurally-Based Analysis and Computational Modeling. *Acta Mater.*, **47**, no. 7, pp. 2089–2108.

Molotkov, A.V., Notkin, A.B., Elagin, D.V., Nesterenko, V.F., and Lazaridi, A.N. (1991) Microstructure after Heat Treatment for Explosive Compacts Made from Granules of Rapidly Quenched Titanium Alloys. *Fizika Goreniya i Vzryva*, **27**, no. 3, pp. 117–126 (in Russian).
English translation: *Combustion, Explosion, and Shock Waves*, November 1991, pp. 377–384.

Moriarty, J.A. (1986) High-Pressure Ion-Thermal Properties of Metals from AB Initio Interatomic Potentials. In: *Proceedings of the American Physical Society Topical Conference "Shock Waves of Condensed Matter"* (Edited by Y.M. Gupta). Plenum Press, New York, pp. 101–106.

Morris, D.G. (1981) Melting and Solidification During Dynamic Compaction of Tool Steel. *Metal Sci.*, March, pp. 116–124.

Morris, D.G. (1982) Rapid-Solidification Phenomena. *Metal Sci.*, October, pp. 457–466.

Mutz, A.H., and Vreeland, T., Jr., (1993) Thermoelectric Measurements of Energy Deposition During Shock-Wave Consolidation of Metal Powders of Several Sizes. *J.Appl. Phys.*, **73**, no. 10, pp. 4862–4868.

Mutz, A.H. (1992) PhD thesis, California Institute of Technology, Pasadena, CA.

Nemat-Nasser, S., Okinaka, T., Nesterenko, V.F., and Liu, M. (1998) Dynamic Void Collapse in Crystals: Computational Modeling and Experiments, *Phil Mag.*, **78**, no. 5, pp. 1151–1174.

Nesterenko, V.F. (1974) Electrical Phenomena Under Shock Loading of Metals and Their Connection With Shock Wave Parameters. PhD thesis, Lavrentyev Institute of Hydrodynamics, Russian Academy of Sciences, Siberian Branch, Novosobirsk.

Nesterenko, V.F., and Staver, A.M. (1974) Temperature Determination for Shock Loading of a Metal Interface. *Fizika Goreniya i Vzryva*, **10**, no. 6, pp. 904–907 (in Russian).
English translation: *Combustion, Explosion, and Shock Waves*. March 1976, pp. 811–813.

Nesterenko, V.F. (1975) Electrical Effects in Shock Loading of Metal Contacts. *Fizika Goreniya i Vzryva*, **11**, no. 3, pp. 444–456 (in Russian).
English translation: *Combustion, Explosion, and Shock Waves*. July 1976, pp. 376–385.

Nesterenko, V.F. (1976) The Thermodynamics of Shock Compaction of Powders. In: *Proceedings of III International Symposium on Metal Explosive Working*, Marianske Lazni, Chehoslovakia, Semtin, Pardubice, Chehoslovakia, Vol. **2**, pp. 419–432 (in Russian).

Nesterenko, V.F. (1977) Shock Compression of Multicomponent Materials. *Dynamics of Continua. Mechanics of Explosive Processes*, **29**, pp. 81–93 (in Russian).

Nesterenko, V.F., and Muzykantov, A.V. (1985) Evaluation of Conditions for Retaining an Amorphous Material Structure During Consolidation by Explosion. *Fizika Goreniya i Vzryva*, **21**, no. 2, pp. 120–126 (in Russian).
English translation: *Combustion, Explosion, and Shock Waves*, September 1985, pp. 240–245.

Nesterenko, V.F. (1985) Potential of Shock-Wave Methods for Preparing and Compacting Rapidly Quenched Materials. *Fizika Goreniya i Vzryva*, **21**, no. 6, 85–98 (in Russian).
English translation: *Physics of Explosion, Combustion and Shock Waves*, 1986, May, pp. 730–740.

Nesterenko, V.F. (1986) Heterogeneous Heating of Porous Materials at Shock Wave Loading and Criteria of Strong Compacts. In: *Procedeengs of IX International Conference on High Energy Rate Fabrication* (Edited by V.F. Nesterenko and I.V. Yakovlev). Lavrentyev Institute of Hydrodynamics, Special Design Office of High-Rate Hydrodynamics, Novosibirsk, pp. 157–163 (in Russian).

Nesterenko, V.F. et al., (1987) Explosive Consolidation of Rapidly Solidified Materials for Electromagnetic Applications. Final Report #GR 01860022255, Special Design Office of High-Rate Hydrodynamics. Siberian Branch of the USSR Academy of Sciences, Novosibirsk, 62 pp.

Nesterenko, V.F. (1988) Non-linear Phenomena Under Impulse Loading of Heterogeneous Media. *Dissertation for Doctor's Degree in Physics and Mathematics*. Lavrentyev Institute of Hydrodynamics, Russian Academy of Sciences, Siberian Branch, Novosibirsk, 370 pp.

Nesterenko, V.F., Panin, V.E., Kulkov, S.N., and Melnikov, A.G. (1992) Modification of Submicron Ceramics under Pulse Loading. *High Pressure Research*, **10**, pp. 791–795.

Nesterenko, V.F. (1992) *High-Rate Deformation of Heterogeneous Materials*. Nauka, Novosibirsk.

Nesterenko, V.F., Bondar, M.P., and Ershov, I.V. (1994) Instability of Plastic Flow at Dynamic Pore Collapse. In: *High-Pressure Science and Technology—1993. Joint International Association for Research and Advancement of High Pressure Science and Technology and American Physical Society Topical Group on Shock Compression of Condensed Matter Conference* (Edited by S.C. Schmidt, J.W. Shaner, G.A. Samara, and M. Ross). AIP Press, New York, no. 309, pt. 2, pp. 1173–1176.

Nesterenko, V.F., Meyers, M.A., and Wright, T.W. (1998) Self-Organization in the Initiation of Adiabatic Shear Bands. *Acta Materialia*, **46**, no. 1, pp. 327–340.

Nigmatullin, R.I., Vainshtein, P.B. et al. (1976) Computer Modeling of Physico-Chemical Processes and Shock Wave Propagation in Solids and Composite Materials. *Computer Methods in Mechanics of Continuum Media*, **7**, no. 2, pp. 89–108 (in Russian).

Nikolaevskii, V.N. (1969) Hydrodynamic Analysis of Shock Adiabats of Heterogeneous Mixtures of Substances. *Zhurnal Prikladnoi Mekhaniki i Tehknicheskoi Fiziki*, no. 3, pp. 82–88 (in Russian).
English translation: *Journal of Applied Mechanics and Technical Physics*, 1969, May–June, pp. 406–411.

Rakhimov, A.E. (1993) Optical Microstructure of Explosively Compacted Ribbon Toroids from Fe-based Amorphous Alloy. *J. Mater. Sci. Lett.*, **12**, pp. 1891–1893.

Raybould, D. (1980) The Cold Welding of Powders by Dynamic Compaction. *Int. J. Powder Metallurgy and Powder Technology*, **16**, no. 1, pp. 1–10.

Raybould, D. (1981) The Properties of Stainless Steel Compacted Dynamically to Produce Cold Interparticle Welding. *J. Mat. Sci.*, **16**, pp. 589–598.

Raybould, D. (1982) Wear-resistant Al–Steel–Pb Admixed Alloys Produced by Dynamic Compaction. *Powder Metallurgy*, **25**, no. 1, pp. 35–41.

Raybould, D. (1987) Cold Dynamic Compaction of Pre-Alloyed Titanium and Activated Sintering. In: *New Perspective in Powder Metallurgy (Fundamentals, Methods, and Applications)*, vol. **8**, Powder Metallurgy for Full Density Products. Metal Powder Industries Federation, Princeton, NJ, pp. 575–589.

Roman, O.V., Nesterenko, V.F., and Pikus, I.M. (1979) Influence of the Powder Particle Size on the Explosive Pressing. *Fizika Goreniya i Vzryva*, **15**, no. 5, pp. 102–107 (in Russian).
English translation: *Combustion, Explosion, and Shock Waves*, March 1980, pp. 644–649.

Schilperoord, A.A. (1976) A Simple Model for the Simulation of the Initiation of Detonation by a Shock Wave in Heterogeneous Explosive. In *Procedeengs of the 6th Symposium (International) on Detonation*. Office of Naval Research ACR-221, pp. 371–380.

Schwarz, R.B., Kasiraj, P., Vreeland, T. Jr., and Ahrens T.J. (1984) A Theory for the Shock-Wave Consolidation of Powders. *Acta Metall.*, 32, no. 8, pp. 1243–1252.

Schwarz, R.B., Kasiraj, P., Vreeland, T. Jr. (1986) Temperature Kinetics During Shock-Wave Consolidation of Metallic Powders. In: *Metallurgical Applications of Shock-Wave and High-Strain-Rate Phenomena* (Edited by L.E. Murr, K.S. Staudhammer, and M.A. Meyers). Marcel Dekker, New York, pp. 313–327.

Sheffield, S.A., Gustavsen, R.L., and Anderson, M.U. (1997) Shock Loading of Porous High Explosives. In: *High-Pressure Shock Compression of Solids IV. Response of Highly Porous Solids to Shock Loading* (Edited by L. Davison, Y. Horie, and M. Shahinpoor). Springer-Verlag, New York, pp. 23–61.

Shvedov, K.K, Aniskin, A.I., Il'in, A.N., and Dremin, A.N. (1980) Detonation of Highly Diluted Porous Explosives. I. Effect of Inert Additives on Detonation Parameters. *Fizika Goreniya i Vzryva*, **16**, no. 3, pp. 92–101 (in Russian). English translation: *Combustion, Explosion, and Shock Waves*, 1980, November, pp. 324–331.

Staudhammer, K.P. (1999) Fundamental Temperature Aspects in Shock Consolidation of Metal Powders. In: *Powder Materials: Current Research and Industrial Practices*, (Edited by F.D.S. Marquis). The Minerals, Metals and Materials Society, Warrendale, PA, pp. 317–325.

Staver, A.M. (1981) Metallurgical Effects Under Shock Compression of Powder Materials. In: *Shock Waves and High-Strain-Rate Phenomena in Metals, Concepts and Applications* (Edited by M.A. Meyers and L.E. Murr). Plenum Press, New York, pp. 865–880.

Staver, A.M., Kuzmin, G.E., and Nesterenko, V.F. (1982) Experimental Investigation of Shock Waves in Porous Media. In: *Proceedings of II International Meeting on Materials Explosive Working* (Edited by V.F. Nesterenko, G.E. Kuzmin, and I.V. Yakovlev). Lavrentyev Institute of Hydrodynamics, Novosibirsk pp. 150–156 (in Russian).

Thadhani, N.N. (1993) Shock-Induced Chemical Reactions and Synthesis of Materials. *Progress in Materials Science*, **37**, no. 2, pp. 117–226.

Thadhani, N.N. (1994) Shock-Induced and Shock-Assisted Solid-State Chemical Reactions in Powder Mixtures. *Journal of Applied Physics*, **76**, no. 4, pp. 2129–2138.

Thadhani, N.N., and Aizawa, T. (1997) Materials Issues in Shock-Compression-Induced Chemical Reactions in Porous Solids. In: *High-Pressure Shock Compression of Solids IV. Response of Highly Porous Solids to Shock Loading* (Edited by L. Davison, Y. Horie, and M. Shahinpoor). Springer-Verlag, New York, pp. 257–287.

Thadhani, N.N., Graham, R.A., Royal, T., Dunbar, E., Anderson, M.U., and Holman, G.T. (1997) Shock-Induced Chemical Reactions in Titanium–Silicon Mixtures of Different Morphologies: Time-Resolved Pressure Measurements and Materials Analysis. *J. Appl. Phys.* **82**, no. 3, pp. 1113–1128.

Thouvenin, J. ((1966) Action d'une onde de choc sur un solide poreux. *J. Physics*, **27**, nos. 3/4, pp. 183–189.

Tonkov, E.Yu. (1979) *Phase Diagrams of Elements at High Pressures*. Nauka, Moscow, (in Russian).

Trebinski, R., and Wlodarczyk, E. (1987a) Estimation of the Local Value of the Temperature in a Shock Loaded Porous Body. *J. Technical Physics*, **28**, no. 3, pp. 327–346.

Trebinski, R., and Wlodarczyk, E. (1987b) A Method for Estimating the Local Temperature in a Shock Loaded Porous Medium. *J. Technical Physics*, **28**, no. 4, pp. 431–450.

Vreeland, T. Jr., Montilla, K., and Mutz, A.H. (1997) Shock Wave Initiation of the Ti_5Si_3 Reaction in Elemental Powders. *Journal of Applied Physics*, **82**, no. 6, pp. 2840–2847.

Vukcevic, M., Glisic, S., and Uskokovic, D. (1993) Dynamic Compaction of Al–Li–X Powder Obtained by a Rotating Electrode Process. *Materials Science and Engineering*, **A168**, pp. 5–10.

Williamson, R.L. and Berry, R.A. (1986) Microlevel Numerical Modeling of the Shock Wave Induced Consolidation of Metal Powders. In: *Proceedings of the Fourth American Physical Society Topical Conference on Shock Waves in Condensed Matter* (Edited by Y.M. Gupta). Plenum Press, New York, pp. 341–346.

Williamson, R.L., Wright, R.N., Korth, G.T., and Rabin, B.H. (1989) Numerical Simulation of Dynamic Consolidation of SiC Fiber-Reinforced Aluminum Composite. *J. Appl. Phys.* **66**, no. 4, pp. 1826–1831.

Williamson, R.L., and Wright, R.N. (1990) A Particle-Level Numerical Simulation of the Dynamic Consolidation of a Metal Matrix Composite Material. In: *Proceedings of the American Physical Society Topical Conference "Shock Waves of Condensed Matter—1989* (Edited by S.C. Schmidt J.N. Johnson, and L.W. Davison). Elsevier Science, Amsterdam, pp. 487–490.

Williamson, R.L. (1990) Parametric Studies of Dynamic Powder Consolidation Using a Particle-Level Numerical Model. *J. Appl. Phys.* **68**, no. 3, pp. 1287–1296.

Wlodarczyk, E., and Trebinski, R. (1989) On Estimating of the Local Value of Temperature in Shocked Granular Medium. In: *Procedeengs of International Seminar on High-Energy Working of Rapidly Solidified Materials and High-T_C Ceramics* (Edited by V.F. Nesterenko and A.A. Shtertzer). Special Design Office of High-Rate Hydrodynamics, Institute of Theoretical and Applied Mechanics, Novosibirsk, pp. 186–190 (in Russian).

Zagarin, Yu.V., Kuz'min, G.E., and Yakovlev, I.V. (1989) Measurement of Pressure and Temperature with Shock Loading for Porous Composite Materials. *Fizika Goreniya i Vzryva*, **25**, no. 2, pp. 129–133 (in Russian).
English translation: *Combustion, Explosion, and Shock Waves*, 1989, September, pp. 248–251.

Zolotarev, S.N., Denisova, O.N., Nesterenko, V.F., Pershin, S.A., and Lazaridi, A.N. (1989) Explosive Compacts from Amorphous Soft-Magnetic Alloys. In: *Procedeengs of International Seminar on High-Energy Working of Rapidly Solidified Materials and High-T_C Ceramics* (Edited by V.F. Nesterenko and A.A. Shtertzer). Special Design Office of High-Rate Hydrodynamics, Institute of Theoretical and Applied Mechanics, Novosibirsk, pp. 97–112 (in Russian).

6
Advanced Materials Treatment by Shock Waves

6.1 Introduction

The shock treatment of powders, for example, to increase defect density and activate sintering, or to densify them to solid density has a relatively long history of application attempts in the area of advanced materials. Nevertheless, the scale of industrial fabrication is very small in comparison, not only with the traditional powder technology methods, but even in comparison with the explosive forming, welding, or explosive hardening, which use the same type of loading. There are two main reasons for this.

The first one is connected with the fact that, for example, explosive welding can provide commercial products with traditional materials, widely available and with well-known properties. The positive effect in this case results from combinations of the very different properties of components (steel + aluminum, copper, or titanium, etc.) and is very difficult to achieve by other means. The benefits of the shock treatment of powders (explosive compaction) can be ensured only with advanced materials, often with metastable properties not very well characterized, with limited sources of supply for research, and often with an unknown market.

The second reason has roots in the process itself. The simplicity of the process, so attractive at the initial stage, turned out to be just a mirage. In reality, the process of dynamic densification is complex and multiscale. For example, the quality of bonding between particles, initiation of the phase transitions, and chemical reactions are determined by mesoscopic phenomena, like shear localization and hot spot formation on the particle interfaces. Other scales are introduced by the quenching processes at interfaces which can be very small (or large) in comparison with other characteristic times like duration of the pressure pulse, and can depend on particle size.

That is why it is very important to understand, at least qualitatively, the interconnections between these thermomechanical phenomena. They will enable us to predict the final properties of compacts, and to establish the main restrictions of the method as well as to recognize its real potential. Future progress in this area is impossible without at least a qualitative insight into the micromechanics of powder deformation, which provides some basis for material treatment (see Chapters 2 and 5).

6.2   Powder Densification and Consolidation by Shock Waves

6.2.1   Densification Versus Consolidation

It is useful to outline the difference between two physically different processes, densification and consolidation, which can be accomplished through the shock loading of powders. Explosive (shock) compaction or pressing are used as general terms identical to shock loading without relation to any specific process inside the powder.

Densification is characterized by the increased density due to the reduction of porosity as a result of plastic flow or repacking of particles. In many cases, this process is not accompanied by any physical bonding between neighboring particles. The latter requires activation of bonding between particles on an atomic or molecular level that is not a necessary condition for densification. The overall low strength of the densified material can be caused by the simple mechanical hitching of particles.

If dynamic loading is accompanied, not only by porosity reduction, but also by the creation of physical bonding between contacting particles, then it is considered as a consolidation process. Usually, densification precedes consolidation with an increase in shock pressure. As a rule, strong bonding between particles, resulting in tensile strength comparable to the strength of the corresponding solid material, can be accomplished at shock pressures essentially exceeding pressures ensuring complete densification.

It is clear from this preliminary consideration that the criteria for densification (reduction of porosity) and consolidation (atomic/molecular bonding between particles) must be qualitatively different.

6.2.2   Densification Criterion

Dependence of the density behind a shock wave on pressure is described by the Hugoniot relation, established based on the experimental data for stationary shocks. In the case of an initial porosity about two and a final density comparable with the solid density, the volume change during pressure release can be neglected in practically all cases. That is why the final density after shock loading may be considered as closely approximated by density during the shock process represented by the Hugoniot curve. The criterion on shock pressure to achieve the necessary level of densification can be based on the experimental Hugoniot data.

In the range of shock pressures where the density of shocked material is below the solid density, the kinematics of particle deformation is close to the one caused by the application of static loading (see Section 2.1). This means that particles are deformed mainly by simple blunting of the contacts and this type of deformation is enough to ensure the dissipation of the total shock energy.

The main differences between this quasistatic shock deformation and a truly static one can be connected with the strain rate sensitivity of the material, and with thermal softening. In the former case it can be localized as a result of adiabatic deformation due to the short time of deformation in the shock front.

A good estimation of shock pressures, ensuring the desirable level of final density, can be obtained based on experimental data for static powder densification or on theoretical consideration. A review of the related models of static densification, and relations between pressures and density, can be found in a paper authored by Olevsky [1998].

Prummer [1983] introduced a very useful engineering criterion for achieving fully and uniform densification

$$P_t = HV, \tag{6.1}$$

where P_t is the detonation pressure in cylindrical geometry and HV is the initial microhardness in GPa. Of course, the additional condition on the amount of explosive should be added to ensure a cone-shaped shock-wave configuration resulting in the uniform densification. Usually this condition is met by selecting a thickness of explosive comparable to the radius of the powder.

Meyers [1994] applied a similar criterion for a variety of materials and associated the P_t with pressure in the shock wave. There are no reasons to expect that the quasistatic deformation of particles, even under high pressure, will induce physical bonding between particles. The diffusion process during application of shock pressures with a typical duration of 10–100 μs cannot be considered as an efficient mechanism to provide bonding between particles.

6.2.3   Consolidation Criterion Based on Critical Amount of Melt

It is clear from previous considerations that physical bonding between particles and subsequent consolidation can be accomplished only by an excess of shock energy over the amount that is necessary to densify powder. This excess of shock energy can change the geometry of particle deformation, especially in the vicinity of particle contacts.

Even more important is that it will introduce qualitatively new processes like intensive shear flow and local melting which are absent under static loading. It is a common point that bonding between particles is the result of local energy depositions on the contacts. Different approaches were developed for the prediction of strong bonding between particles. Some of them use the criteria which are based on a critical amount of melt.

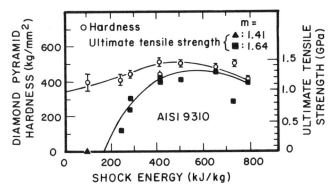

Figure 6.1. Microhardness and ultimate tensile strength of shock consolidated AISI 9310 as a function of specific shock energy. Note the high value of microhardness at zero tensile strength. (Reprinted from *Shock Waves in Condensed Matter—1983*, edited by J.R. Asay, R.A. Graham and G.K. Straub. Kasiraj, P., Vreeland, T. Jr., Schwarz, R.B., and Ahrens, T.J., Mechanical Properties of a Shock Consolidated Steel Powder, pp. 439–442, 1984, Elsevier Science, with permission from the author.)

For example, to address the bonding condition, Schwarz, Kasiraj, Vreeland, and Ahrens [1984a,b] proposed the following criterion:

$$L = \frac{E}{c_v(T_m - T_0) + \lambda} > L_{min},$$ (6.2)

where c_v is the average value of the specific heat, λ is the latent heat of fusion, T_0 and T_m are the initial and melting temperatures, E is the specific shock energy, and L_{min} is constant. They have shown that a tensile strength $\sigma > \sigma_s/2$, where σ_s is the strength of the solid material, is reached when $L_{min} = 0.22$ for AISI 9310 alloy powder with a particle size near 60 μm and an initial porosity of $\alpha_0 = 1.7$. The measured properties of compacts are shown in Figure 6.1.

This criterion is based on the hypothesis that melting is a necessary condition for bonding. It does not predict the observed dependence of the transition pressure on material strength for the same melting temperature (Nesterenko and Lazaridi [1990]; Nesterenko [1992]).

Schwarz et al. [1984] considered the additional requirement $t_r > t_s$ where t_r is the time during which the specimen is under compression before the arrival of the expansion wave and t_s is the crystallization time of the melt along the particle boundaries.

Ahrens, Kostka, Vreeland, Schwarz, and Kasiraj [1984] proposed a pressure criterion for the formation of a molten zone of thickness δ along the particle boundaries, responsible for bonding, as a result of friction:

$$P \geq \frac{3\delta\left[(T_m - T_0)c_v + \lambda\right]\rho_0}{\mu a(\alpha - 1)},$$ (6.3)

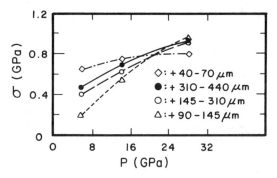

Figure 6.2. Ultimate tensile strength of shock consolidated stainless steel particles EP-450 (atomized, annealed powder, initial microhardness 2.90 GPa) as a function of shock pressure for different particle sizes. Each point is averaged over four independent measurements.

where μ is the interparticle friction coefficient. The dependence of the critical pressure on the particle size in Eq. (6.3) contradicts the results of experiments by Nesterenko and Lazaridi [1990] and Lazaridi [1990] with the atomized annealed powder EP-450 having initial microhardness $HV = 2.90 \pm 0.40$ GPa (Fig. 6.2). In those experiments, under the same shock loading, the strength of the compacts did not noticeably vary as the particle sizes changed from 40–70 μm to 310–440 μm at pressures higher than 14 GPa. For particle sizes less than 40 μm, the ultimate tensile strength was equal to 0.8 GPa after consolidation at a shock pressure of 28 GPa (Lazaridi [1990], not shown in Fig. 6.2). More importantly, for a shock pressure of 6 GPa the strength of the compact (0.6 GPa) with the smallest particle size 40–70 μm is the highest one.

Meyers, Benson, and Olevsky [1999] analyzed various energy dissipation mechanisms like interparticle melting, plastic deformation, and particle fracture for different materials. They calculated the total energy of shock wave required to consolidate a given material. The key point in their approach is also a prescribed thickness of molten material on the particle interfaces needed for good bonding.

As a result they obtained a scaling of shock pressure with respect to the flow stress of the material. For different materials depending on particle size they predicted threshold pressures for good bonding in the range of shock pressures $\overline{Y} \div 4.5\,\overline{Y}$, where $\overline{Y}$ is the characteristic initial yield strength of the material.

This prediction is close to the transition pressure from quasistatic to dynamic regimes ($P \approx HV \approx 3\,\overline{Y}$), but below a pressure for strong bonding ($P \approx 2HV \approx 6\,\overline{Y}$) for the corresponding initial porosity proposed by Nesterenko [1992], [1995]. The important difference between this approach and that by Meyers et al. [1999] is the dependence of the transition pressure on particle size in the latter one. The available data (Fig. 6.2) does not support the dependence of critical shock pressure for achieving a strong bonding on particle size.

Additional data on the dependence of the ultimate tensile strength on particle size for different materials is certainly needed to reach a final conclusion on the validity of approach, based on the prescribed thickness of the molten layer on the particle contacts. The absence of such dependence will testify against it and a noticeable effect of particle size will support it. Particle size is certainly important for contact temperatures at low-shock pressures (see Chapter 5) and can be instrumental in preventing the cracking of particles under shock loading (Meyers et al. [1999]). On the other hand, the combination of a relatively large amount of melt on particle interfaces with a short duration of compression pulse is detrimental for the consolidation process (Schwarz et al. [1984b]).

6.2.4 Criterion Based on Microkinetic Energy

A key hypothesis proposed by Nesterenko [1988a,b,c] (see also Nesterenko and Lazaridi [1990]), for establishing the bonding criteria for powders, is the idea that strong bonding is the result of a qualitative change of the kinematics of interface deformation—the transition from a quasistatic to a dynamic regime with the accumulation of microkinetic energy (see Chapter 2, Section 2.4). A criterion for obtaining strong compacts, on the basis of a comparison of the specific energy dissipated during the spherical pore collapse and the specific shock energy, was proposed (Nesterenko [1992]). The difference between these parameters provides the microkinetic energy required to strip the surface contact layers and to create on them the necessary degree of mutual plastic deformation to ensure a strong bond between the particles (Bondar' and Nesterenko [1991]).

Based on this idea, the critical pressure for strong bonding (consolidation) can be found for the case where the initial density is one-half of the solid density (the initial and final porosities are $\alpha_0 = 2$ and $\alpha_1 = 1$, correspondingly):

$$P_{cb} = \frac{2\rho_s E_d}{K_1(\alpha_0 - 1)} \approx \frac{1.85}{K_1}Y. \tag{6.4}$$

The value proposed for the coefficient K_1, to ensure the developed dynamic deformation regime resulting in sound compact formation, was 0.25; thus, P_{cb} can be established. The most important feature of the criterion expressed by Eq. (6.4) is that it scales the threshold pressure based on a single material parameter—its initial strength. The sources of energy dissipation that are involved to dissipate the microkinetic energy accumulated during the collapse stage are actually taken into account by coefficient K_1. None of them (like melting of the interface layer) is specified as a necessary condition for particle bonding.

Figure 6.3 illustrates the highly heterogeneous character of the particle interface deformation, typical for a developed dynamic regime (atomized EP-450 powder in "as is" conditions, the initial microhardness is 6.60 ± 0.40 GPa, also see Fig. 2.19). Quite different types of contact deformation can coexist in the vicinity of the same particle, or even on different parts of the contact surface of

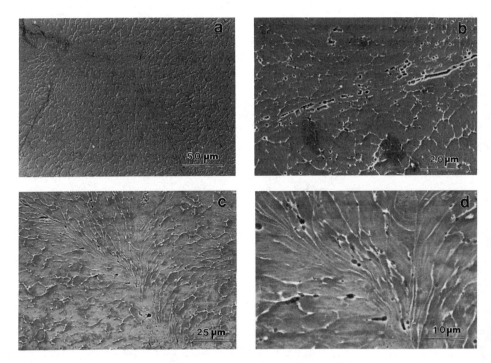

Figure 6.3. Different types of plastic flow on particle interfaces (atomized "as is" stainless steel powder EP-450, initial microhardness 6.60 GPa), shock pressure 28 GPa, particle size 310–440 μm (a), (c), (d) and 145–310 μm (b); (a) shear localization, local melting and simple blunting of particle contacts; (b), (c), (d) localized plastic deformation accompanied by concentration of globular carbides on the particle interfaces; (d) corresponds to the central part of (c).

the same particle. For example, the intensive mutual plastic flow of neighboring particles due to shear localization coexist with simple blunting (Figs. 6.3(a) and 2.19(a),(c),(e)).

Shear localization resulting in strong bonding sometimes is accompanied by local melting (Fig. 2.19(a),(b)). But it is clear (see Fig. 6.3(b),(c),(d)) that the strength of these compacts is not associated with this localized melting, as is usual in triple points of contacts (Nesterenko [1992], [1995]).

This behavior of contact deformation was successfully modeled (Benson, Nesterenko, and Jonsdottir [1995], [1996]) with the help of a two-dimensional computer code developed by Benson [1994] (see also Fig. 2.20).

Experiments with atomized EP-450 powder in "as is" conditions (initial microhardness is 6.60 ± 0.40 GPa) at a shock pressure of 28 GPa did not demonstrate correlation of the ultimate tensile strength (σ_t) with the initial particle size. For example, for particle size 90–145 μm, $\sigma_t = 0.35$ GPa, for particle size 145–310 μm, $\sigma_t = 0.5$ GPa, and for particle size 310–440 μm $\sigma_t = 0.4$ GPa; the data spread for σ_t is 20–30%, being largest for the largest particle sizes. Sheet tensile specimens had a total length of 50 mm (gauge length 30 mm) and a rectangular cross section $3 \cdot 5$ mm^2. The presented values of the ultimate tensile strength is about two times lower than for the shock consolidated annealed atomized powder EP-450 at the same shock pressure (see Fig. 6.2), and is also less than the tensile strength of conventionally manufactured steel EP-450 (0.64 GPa). This may be attributed to the heterogeneity connected with different types of particle deformation with no bonding for simple blunting, or with local melting and cracking (Fig. 6.3(a)).

The interesting phenomenon of the elongation and coalescence of carbide particles was observed inside the shear localization zone adjacent to the particle interfaces (Fig. 6.3(b),(c),(d)). It is especially noticeable in Figure 6.3(b). The mechanism of this phenomenon is connected with the contraction of grain sizes in the direction perpendicular to the shear direction. The critical strain inside the shear zone for this effect can be evaluated based on the condition that the thickness of sheared grains in the direction perpendicular to shear should be comparable to the initial diameter of the carbide inclusions (see Problem 13). Some additional carbidization in the area with high shear strains can also be expected. On some contacts this phenomenon even resulted in the cracks being detrimental for the compact strength.

In a certain sense, bond formation in shock compaction is, above all, a mechanical process of the intense mutual plastic deformation of neighboring particles. The existence of local melts in the developed dynamic regime (Figs. 6.3(c) and 2.19(a),(b)) can to some extent be considered as a side effect and not as a prerequisite for strong bonding. The dynamic character of deformation, needed to produce good bonding, must be accompanied by local concentrations of microkinetic energy, which often cannot be dissipated without melting due to highly localized flow. This is in accord with Staudhammer's point of view that melting is not necessary for efficient consolidation (Staudhammer [1999]).

The data presented in Figure 6.1 are in agreement with the transition to a well-developed dynamic regime of deformation. Indeed, the microhardness of the rapidly solidified steel powder in experiments performed by Schwarz et al. [1984] was 3.44 GPa. Therefore, the use of the criterion $P > HV$ (Eq. (2.50b), $K = 0.5$, $\overline{Y} = HV/3$, HV is the microhardness in GPa) leads to a pressure of transition from the quasistatic to the dynamic regime of about 3.4 GPa. This agrees with the beginning of the bonding process in this material where the noticeable strength was measured only at a shock pressure of 5.9 GPa (see Fig. 6.1, shock energy 280 kJ/kg corresponds to a shock pressure equal to 7 GPa).

The developed dynamic regime ($P > 2HV$, $K = 0.25$) resulting in a strong compact, needs a pressure of 6.9 GPa. This value is in good agreement with the results of Kasiraj et al. [1984a,b] where the threshold nature of the development of intergranular bonds was observed (see dependence of ultimate strength on shock energy, Fig. 6.1). The beginning of this threshold can be interpreted as the onset of the dynamic regime, the pressure being in the interval between 3.6 GPa and 5.9 GPa. The attainment of the compact strength $\sigma > \sigma_s/2$, when $P = 8$ GPa, coincides with the establishment of the developed dynamic deformation regime.

An important feature of the compact's ultimate strength dependence on shock energy is that the energy interval where bonding achieves the saturation is comparable with the threshold of the shock energy to start the bonding process. It is clear from Figure 6.1 that the compact's hardness does not reflect the strength of densified powder at low pressures.

Shtertser [1993] also followed this approach based on the idea that microkinetic energy is the main parameter determining the possibility for bonding between particles. He proposed an expression for minimum pressure P_b to obtain bonding, taking into account an oxide layer on the particle surfaces and based on the analogy with explosive welding:

$$P_b = \frac{k^2 HV}{(\alpha_0 - 1)}, \tag{6.5}$$

where the coefficient k depends on the oxide layer thickness. This approach is similar to the one based on the geometrically necessary energy and specific shock energy proposed by Nesterenko [1992] with a difference only in the coefficient depending on the oxide layer thickness. Batsanov [1994] considered other alternative approaches to the bonding. In general, scaling of the critical shock pressure to achieve good bonding with initial microhardness can be currently considered the most established approach.

6.2.5   Criteria Based on Surface Temperature

Other attempts at the bonding process are based on the specification of the contact temperature, or on other conditions on the particle interfaces. For example, Raybould [1980] assumed that a strong bonding between particles is formed when the material temperature at the contact approaches the melting temperature ($T = T_m$). The necessary condition for this is

$$\tau \leq t_m, \qquad t_m = \frac{1}{12\kappa}\left(\frac{a\Delta T}{2\Delta T_m}\right)^2, \qquad \tau = \frac{3.6a}{D}, \tag{6.6}$$

where ΔT_m and ΔT are the differences between the melting point and the average steady state temperature behind the front and initial temperature, respectively, D is the shock velocity, τ is the width of the shock front, t_m is the characteristic time, and κ is the thermal diffusivity. The physical sense of this criterion (Eq. (6.6)) is that energy release at the particle contact should be fast enough to overcome heat diffusion into the cold bulk of the particle and to achieve the melting temperature. This can easily be converted into a criterion on shock pressure (see Problem 14).

Nesterenko et al. [1990], [1991], [1992] investigated the strength of the compacts made of the annealed powders of ferrite–martensite steel EP-450 (microhardness 2.90 GPa) having the same porosity and with different particle sizes 40–70 μm, 90–145 μm, 145–310 μm, and 310–440 μm under the same loading conditions (Fig. 6.2). For large particles the shock pressure of 6 GPa was near the lower experimental limit required for obtaining the strong compacts. From Eq. (6.6) it follows that, for the finest particle sizes, the critical pressure to obtain bonding must be approximately two times higher. In other words, the compact from a fine particle should not have any tensile strength. However, the strength of the fine particle compacts is even higher than that of the largest particle sizes. Thus, this experimental result is in clear disagreement with the criterion described by Eq. (6.6).

6.2.6   Particles Bonding and Mach Reflection in the Cylindrical Geometry of Explosive Consolidation

The cylindrical geometry of explosive loading is widely used for shock compaction (Lenon, Williams, and Bhalla [1978]; Prummer [1987]; Meyers [1994]; and others). The main restriction of this geometry is the Mach reflection of the shock wave in the center of the cylinder, which often results in the "central zone" inhomogeneity (Kostyukov and Kuz'min [1986]) where local melting and a subsequent central cavity can be developed. The reason for melting is the high shock pressure behind the Mach stem because of its wave speed being equal to the detonation speed is relatively high (usually in the interval 4–6 km/s, see Problem 15). Another complication of Mach reflection is the tangential jump of the particle velocity on its circumference that creates a high-velocity "jet" in the central part of the sample. These features may result in a central cavity and in cracks in the explosively compacted samples (Kostyukov and Kuz'min [1986]). To avoid such complications and obtain uniformly densified material, the regular reflection of the shock wave in the center (cone-shaped shock-wave configuration) is required (Deribas and Staver [1974]; Staver [1981]; Prummer [1987]).

It is very instructive to compare the criterion for obtaining a strong consolidated compact under plane shock loading following from Eq. (6.4)

$$P > 2HV, \tag{6.7}$$

where P is the pressure in the first shock (initial porosity is about 2), with the phenomenological criterion (Prummer [1983]) for obtaining a uniform and dense cylindrical compact $P_t = HV$, where P_t is the detonation pressure (see Eq. (6.1)). The increase of detonation pressure beyond that determined by the latter criterion results in a Mach reflection in the center with subsequent macrodefects (hole, cracks). A comparison of Eqs. (6.1) and (6.7) (in cylindrical geometry usually $P = 0.4P_t$, as shown by Stertser [1988]) enables us to conclude that these criteria are incompatible (Nesterenko [1992], [1995]).

The practically important consequence of this is that the shock wave necessary for the developed dynamic deformation regime of powder deformation to provide strong bonding between the particles in the cylindrical geometry will lead to Mach reflection at the center. This means that it is impossible to obtain a uniform compact with high strength, because high-velocity flow behind the Mach wave results in melting and cracking in the center. One may also conclude that uniform densification in cylindrical geometry will always be accompanied by a quasistatic deformation of particles.

The use of a double-tube geometry enables higher pressures for a lower detonation velocity, so the Mach wave can be minimized at high pressures and shock parameters can be better tuned (Meyers and Wang [1988]).

Another important feature of cylindrical geometry is connected to the energy which should be spent on the plastic deformation of the container tube in addition to the energy supplied to the powder for its densification and bonding (Lenon, Williams, and Bhalla [1978]).

6.2.7   Explosive Welding Versus Explosive Consolidation

It is useful to compare the conditions and criteria for obtaining a strong bond between the particles in the shock compaction of powders and in explosive welding. Some authors (e.g., Raybould [1987]; Staver [1981]) pointed to the similarity between the mechanisms of shock consolidation and explosive welding. Wang, Meyers, and Szecket [1988] even conducted a large scale model experiment with a collision of a flat plate with a semicylinder surface to get a better understanding of the bonding processes during shock consolidation. The resultant morphologies of the interface regions were strongly dependent on the initial angle of collision. The normal impact did not result in bonding, and only some critical initial angles ensured the bonding process. This is in complete accord with the explosive welding criterion. The authors arrived at the conclusion that only an inclined impact between particle surfaces in powder generating highly localized deformation and, often, melting and jetting, is the process responsible for bonding. The important conclusion was made that even well-consolidated powders have local regions with particles that are not well bonded.

This is in agreement with the observation by Nesterenko [1985], [1986] that the most unfavorable situation for bonding is the quasistatic deformation of a contact with the resulting plane geometry and bonding as the result of localized plastic deformation on the particles contacts (Fig. 6.3). It is highly improbable to expect the development of localized plastic flow on all contact surfaces. This is because the beginning of the localized plastic flow on some contact will result in the arrest of plastic flow in the neighboring particles (Nesterenko [1986]; Fig. 6.3). As a result it is impossible to achieve uniform bonding over all interfaces of contacting particles. The compacted sample will always have poor bonded contacts causing a low plasticity of material.

Shtertser [1993] used the explosive welding analogy to introduce his criterion of bonding. The analogy with explosive welding cannot be straightforwardly applied to the bonding in shock consolidated powders, due to the qualitatively different regimes of collision in explosive welding and in powder consolidation processes. Whereas the welding regimes can be described in the coordinates, collision angle/collision point velocity, it is impossible to introduce such quantities for compaction. During shock consolidation there are no steady particle collision regimes and in a number of cases no collision at all. In many cases a wavy interface was not observed in the deformation of powders where good bonding was achieved, a typical example is Figure 6.3. Only in rare cases with relatively large particles were local wave formation and jets noticeable (Staver [1981]; Raybould [1987]).

Another important difference is that jet formation in explosive welding, that efficiently removes the "contaminated" surface layer, cannot play the same role inside the powder body. Common to both processes is the importance of the local mutual plastic deformation of the contacting bodies—more or less uniform in explosive welding and highly heterogeneous in explosive consolidation (Bondar' and Nesterenko [1991].

6.2.8 Macroshear Localization as a Bonding Mechanism

A special type of bonding mechanism for the practically important class of porous materials composed from amorphous foils (Nesterenko [1992]) should be mentioned. In this case the relative displacement cannot provide conditions for mutual plastic deformation and the subsequent bonding between foils. But the shear localization (Fig. 6.4(a)), usually a precursor of fracture in amorphous materials under normal conditions, plays a critical role in the bonding process, ensuring compact strength. The development of this type of shear localization needs a critical strain and compressive stresses which could allow the local heating and quenching of amorphous material inside the shear zone.

The structure inside the shear bands can be different from the surrounding amorphous material. For example, Chen et al. [1994] found nanocrystalline formation (face-centered cubic aluminum grains, 7–10 nm in diameter) in aluminum-based amorphous alloys in shear bands induced by bending. The authors even suggested that the mechanical deformation with high plastic strain might be used to form high-strength amorphous-nanocrystalline composites. Figure 6.4(a) corresponds to the structure of explosively compacted cylinders from foils of amorphous alloy (Co-based alloy, $Co_{58}Ni_{16}Fe_5Si_5B_{16}$ (71 KNSR)) with inner and outside diameters of 8 mm and 22 mm. Their global compressive strength was 0.2 GPa in the direction parallel to the foil planes and 4.2 GPa in the perpendicular direction (Tabachnikova et al. [1990]) despite fracture of the foils in some places of the sample. Samples can be cut mechanically and machined without fracture.

Another interesting feature of this compact is the wavy shape (wavelength about 100–200 μm or five to ten times larger than the foil thickness) of the

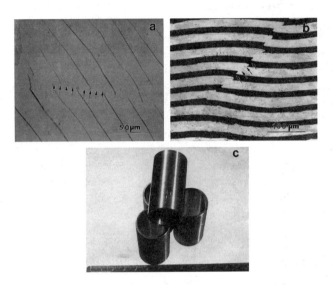

Figure 6.4. Shear localization as a mechanism of bonding (a) in compact from amorphous Co-based ribbons; (b) in laminar composite from amorphous (white) and copper ribbons; (c) the cylinders manufactured from the composite. Note that there is no fracture in shear bands, shown by the arrows.

amorphous foils assumed after the compaction which preceded the shear localization (Zolotarev, Denisova, Nesterenko, Pershin, and Lazaridi [1989]). This is the result of the accommodation of compressive strain in the tangential direction by the amorphous foils, which cannot be plastically deformed.

Localized macroshear can also be used as a bonding mechanism in composite laminar materials (Pershin and Nesterenko [1988]). Figures 6.4(b),(c) correspond to the material fabricated by the explosive densification of an alternating sequence of copper and amorphous alloy (Co-based alloy, $Co_{58}Ni_{16}Fe_5Si_5B_{16}$, 71 KNSR) foils with a thickness of 15 μm and 26 μm, correspondingly. They were coiled with an overall density of 60% of solid density about the central copper rod. The final strength of the compact allowed its machining, and it can be used as a structural element for magnetic shielding (Fig. 6.4(c)) with good electromagnetic properties (Nesterenko et al. [1991], [1992]). The addition of plastic copper foils reduced the total amount of localized deformation by three to four times in comparison with an explosive compaction of foils made from the same amorphous alloy in the same geometry and loading conditions. Figure 6.4(b) demonstrates the possibility to use this geometry for "forced" shear localization in copper and other ductile materials like tantalum.

Zolotarev, Denisova, Nesterenko, Pershin, and Lazaridi [1989] also explosively compacted samples from amorphous Fe-based foils with a dielectric coating (using a suspension of MgO powder in alcohol) on one or both sides of the foil. A wavy pattern of foil deformation and shear localization was also characteristic for samples of this material. In general, the dielectric coating was intact after compaction.

It should be mentioned that shear band patterning is typical for dynamic densification of amorphous materials (powders and foils) in a cylindrical geometry with characteristic shear band spacing of about 1 mm (Pershin and Nesterenko [1988], Nesterenko [1992]).

6.3   Preservation of the Amorphous State Under Dynamic Loading

A series of metal alloys exhibiting unique properties, strength, corrosion, and radiation resistance, etc., is prepared in an amorphous condition (Otooni [1998]). At the same time, it is difficult to use these alloys as structural materials because, in most cases, the methods to obtain them are based on the rapid cooling of the melt. As a result they can be used only for very small sizes of samples because the fast rate of diffusion cooling is ensured only by the small size of the melt, usually less than 100 μm. As a result only foils, wires, and powders are prepared with rapid quenching.

For some metallic alloys, the cooling rates which are required for glass formation can be relatively low allowing us to reach a thickness up to 40 mm (Ti–Zr–TM, Zr–Ti–TM–Be and Pd–Ni–Cu–P systems, Inoe [1996]) or rod diameters up to 25 mm (Pd–Ni–P, Pd–Cu–P, He, and Schwarz [1996]). Some of these alloys remain free of crystal nucleation up to 100 K in excess of the glass transition temperature during a reasonable time scale that ensures a good possibility for processing. A pentiary ($Zr_{41.2}Ti_{13.8}Cu_{12.5}Ni_{10.0}Be_{22.5}$) alloy, highly processable in the undercooled liquid regime, was developed by Peker and Johnson [1993]. This needs a cooling rate of about 10 K/s, which allows preparation of amorphous rods up to 14 mm in diameter.

Nevertheless, as a rule, only high cooling rates (10^3–10^6 K/s) can ensure the amorphization of a broad class of metallic alloys.

The method of creating the bulk articles of amorphous powders (foils) related to the topic of this chapter is consolidation by dynamic loading. The gas (powder) gun technique or explosive loading can be used for this purpose. Heterogeneous heat release is typical for the shock compaction, leading to a high temperature of particle contact surfaces with a comparatively low bulk temperature. There is an important dependence of the contact surface temperature T_c on the particles size at the same pressure and initial density (see Chapter 5). The time for the existence of a high nonequilibrium temperature for the contact surfaces is governed by the size of the heated zone. As shown experimentally for copper/nickel powders, the relaxation time for the nonequilibrium temperature is ~5 ·10^{-6} s with a particle size of 0.1–0.5 mm (Nesterenko [1975], [1992]). The corresponding rate of cooling is 10^7–10^8 K/s, which is higher than the quenching rate (10^5–10^6 K/s) for traditional methods of amorphous materials fabrication.

The distinguishing feature of shock-wave compression is the high heating rate of contact surfaces experimentally estimated at a value of 10^9–10^{10} K/s (Nesterenko [1975], see Section 5.1). This may lead to high-density compacts and to the formation of strong bonds between particles with retention of the initial metastable state.

It is important to establish the criteria that determine the possibility of retaining the initial amorphous structure during and after dynamic densification. Morris [1980] and Roman et al. [1983] concluded that an amorphous structure may only be retained under the condition $T_{th} < T_1$, where T_{th} is the calculated equilibrium theoretical temperature after dynamic densification, and T_1 is a temperature about 200 K higher than the crystallization temperature. Taking into account the uncertainty in estimating T_{th} and the possibility of its increase, the condition for an amorphous structure retaining may be approximately presented in the form $T_{th} < T_{cr}$, where T_{cr} is the crystallization temperature (Roman, Gorobtsov, and Mitin [1982]).

The nonequilibrium temperature of the contact surfaces during dynamic compression may exceed, by several times, the residual equilibrium temperature in the technologically important pressure range (Section 5.1). If the rate of cooling in a certain temperature range is lower than that in the process of manufacturing an amorphous material, then crystallization may commence at the stage of temperature equilibration. In this case, the required condition $T_{th} < T_{cr}$ does not guarantee retention of an amorphous structure at the stage of temperature leveling between the cold and hot areas.

The effect of nonequilibrium heat release in a porous material on the probable crystallization process was emphasized by Roman et al. [1983]. They suggested that the contact areas being melted during explosive loading are quenched into the amorphous condition as a result of the anomalously high values of thermal conductivity resulting from turbulent flow inside the contact layers. If this hypothesis is accepted, then conditions for turbulent movement in molten contact zones, in addition to the criterion of retaining an amorphous state ($T_{th} < T_{cr}$), should be provided. This cannot be accurately described by a simple equation.

Negishi et al. [1985] reported the essential difference of the final properties of the dynamically compacted amorphous $Pd_{78}Cu_6Si_{16}$ depending on initial particle size, due to the different kinetics of quenching of the heterogeneously heated powder in the shock wave. These authors also proposed a model of nonequilibrium heat release based on the adiabatic conditions for plastic deformation of the interface layers of adjacent particles. Unfortunately, there are no systematic data on the crystallization kinetics of dynamically compacted amorphous materials. It is possible to expect that shock loading is able to qualitatively change the recrystalization kinetics of amorphous alloys, as is the case for rapidly solidified Ti–Al alloy (Section 6.5).

The experimental results for the shock-compacted amorphous alloy $Ni_{78}P_{22}$ (Sucic, Minelic, and Uskokovic [1990]) demonstrated the essential decrease of the activation energy of crystallization (from an initial value of 212 kJ/mol to 150 kJ/mol) under the shock pressures 50 kbar, 70 kbar, and 90 kbar. No significant change in the crystallization temperature of the compacted amorphous material was observed. Deformation-induced microstructural changes in $Fe_{40}Ni_{40}P_{14}B_6$ metallic glass, causing the modification of the thermal crystallization sequence, were observed after high-energy ball milling by Fan et al. [1999].

Ishakov et al. [1984] evaluated the cooling time (and rate) of locally hot areas, using the relationship $t \sim R^2/\kappa$, where R is particle radius and κ is material thermal diffusivity. However, it is clear that calculations of this type imply an error at the level of an order of magnitude. They may not serve as a reliable basis for conclusions about the possibility of retaining an amorphous structure in the stage of temperature equilibration.

The cooling rate is estimated by Roman et al. [1982] and Schwarz et al. [1984a] in a similar way, with the difference that for its calculation they did not take particle radius R, but the melted layer thickness at its surface. As will be seen later this approach may give overestimated cooling rates when the thickness of the melt layer is comparable with the particle size.

The thermal effects at the particle contact surfaces were studied by Gourdin [1984a,b,c] on the basis of analyses of the thermal conductivity process with a given energy flux at the particle boundary due to mechanical work. The latter is governed by the overall energy of shock compression and the shock front time as well as by the specific particle surface area. This model (Belyakov–Reybould–Gourdin, BRG model) may be used to estimate possibilities for the crystallization process taking account the temperature transition kinetics. However, the basic assumption of the BRG model is that heat-release processes operate along the particle boundary during dynamic consolidation on a space scale much smaller than the region embraced by the thermal conductivity process. This is not valid for metal powders, at least for weak (<1 GPa) and high (>10 GPa) shock pressures, as was discussed in Section 5.2.

Generally speaking, it is necessary to know the kinetic crystallization parameters, the size of heated zones, the temperature close to the contact surfaces, materials thermophysical properties, and the critical cooling rates, in order to solve the problem of the retention of an amorphous structure during the process of establishing thermal equilibrium. In the majority of cases, it is not possible to determine these parameters.

The "skin" model proposed by Nesterenko and Muzykantov [1985] (see Section 5.3) addresses this problem. It enables us to establish the criterion for preventing an amorphous phase from crystallization, by comparing quenching processes resulting in the initial amorphous state with a cooling rate at the stage of the establishment of thermal equilibrium in a heterogeneously heated material after shock densification.

As was already discussed, there are different points of view on bonding conditions under the shock compaction of powders. For amorphous powders, welding with local melting of contact areas was also proposed (Roman et al. [1982]). A strong bonding associated with melting was observed in shock consolidated rapidly solidified Al–Li–X powder (Vukcevic, Glisic, and Uskokovic [1993]). A sufficient thermophysical condition for forming bonds at the contact surfaces of amorphous particles may be $T_c = T_m$, where T_m is the melting temperature. This condition does not seem to be a necessary one and strong bonding, for example, in explosive welding, was obtained without noticeable melting (Deribas [1980]) as was the case with the consolidation of stainless steel powder (see Fig. 6.3). The condition of melting increases the probability of the crystallization process during consolidation, in comparison with the case of bonding due to "cold" welding. Since in regimes of explosive consolidation $P \leq 10$ GPa, the dependence $T_m(P)$ may be ignored in the first approximation.

In accordance with the Nesterenko and Muzykantov approach, it is necessary to compare the cooling kinetics in a layered system, representing the heterogeneously heated shocked powder (see Fig. 5.15), with cooling conditions during preparation of the initial amorphous material. Determination of the actual cooling conditions during quenching presents significant difficulties and they are normally evaluated on the basis of the Newton rule with an empirically selected heat-transfer interface coefficient. The latter depends on the quality of the contact surfaces, and may be markedly different in relation to the type of quenching process. In addition, application of the Newton rule makes it possible to estimate only the average cooling rate through ribbon volume. As a matter of fact, the time dependence of temperature for different regions in the original amorphous material may differ significantly.

In order to estimate the cooling rate, during preparation of the amorphous material, the following two models can be used (Fig. 6.5).

For a one-sided quenching process (Fig. 6.5(a)), it is assumed that a melted layer of finite thickness l (1), initially uniformly heated to T_m, is brought into ideal thermal contact with a massive substrate (2) exhibiting infinite thermal conductivity, or with a forcible cooled substrate. The temperature in the interface may vary in different processes within the limits of hundreds of degrees. For simplicity, its value is chosen constant and equal to the initial temperature $T_0 = 0$ °C. At the free boundary of the melt ($\xi = 0$) a condition of thermal insulation is assumed.

For a two-sided geometry of cooling (Fig. 6.5(b)) the same boundary conditions are used on the symmetric interfaces melt–substrates. The first case can be converted into the second one, decreasing by two times a scale of the system. The symmetrical central point $\xi = l/2$ for two-side quenching corresponds to the condition of thermal insulation.

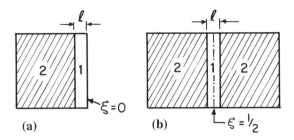

Figure 6.5. The geometry of an amorphous material preparation for (a) one-sided and (b) two-sided geometry of quenching; 1, layer of melt; 2, cold substrate with constant temperature T_0.

It is clear that these cooling regimes provide the greatest rate for temperature drop compared with all of the procedures for obtaining amorphous materials involving diffusive heat conductivity processes with one-sided and two-sided cooling. The thermal conductivity equation in the region width l and the initial and boundary conditions are written in the following way for one-sided cooling (Fig. 6.5):

$$\frac{\partial u}{\partial \tau} = \frac{\partial^2 u}{\partial \xi^2}, \quad 0 < \xi < 1, \quad 0 < \tau < +\infty; \tag{6.8}$$

$$u(\xi, 0) = \begin{cases} 1, & 0 \le \xi < 1; \\ 0, & \xi = 1; \end{cases}$$

$$u_\xi(0, \tau) = 0, \quad u(1, \tau) = 0, \quad 0 < \tau < +\infty.$$

In these equations, dimensionless parameters are equal to, correspondingly, $u = T/T_m$, $\xi = x/l$, $\tau = \kappa t/l^2$, where T is temperature, x is coordinate, t is time; κ is thermal diffusivity, and l is the thickness of the melted layer.

The solution to this problem is found in the form of a Fourier series (Carslaw and Jaeger [1988]):

$$u(\xi, \tau) = \frac{2}{\pi} \sum_{n=1}^{\infty} \frac{(-1)^{n+1}}{\left(n - \frac{1}{2}\right)} \cos\left(\left(n - \frac{1}{2}\right)\pi\xi\right) e^{-\pi^2\left(n - \frac{1}{2}\right)^2 \tau}. \tag{6.9}$$

The exponential factor in Eq. (6.9) provides an absolute convergence of the series with $\tau > 0$. In order to obtain the sum of the series, with sufficient accuracy in the stages of thermal relaxation, ($\tau \ge 10^{-4}$) 100 terms were used. The most characteristic points are those where crystallization can occur first of all. These points are evidently at $\xi = 0$ for one-side quenching (Fig. 6.5(a)) and at $\xi = 1/2$ for two-side cooling (Fig. 6.5(b)).

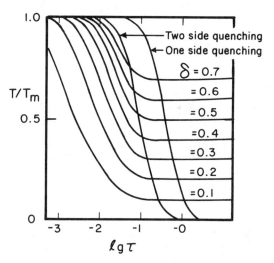

Figure 6.6. Time dependence of the contact temperature for various thicknesses of the hot layer $\delta = d/l$ and cooling curves for the one-sided and two-sided methods of quenching.

The dependence of temperature at point $\xi = 0$ on dimensionless time for one side quenching is given in Figure 6.6. The solution to this problem for two-sided cooling may be obtained from Eq. (6.9) by an appropriate change in scale ($l/2$ instead of l) and is also represented in Figure 6.6. The temperature in this case corresponds to the point with the coordinate $\xi = 1/2$ (Fig. 6.5(b)). The temperature histories in these points are taken as the "critical quenching schedules" which can ensure obtaining the amorphous state throughout the bulk of the specimen. It is assumed that the original material is amorphous throughout the whole volume. The temperature dependence, at the most dangerous points for crystallization after dynamic densification, should be compared with this "critical quenching schedule."

As has been noted, the cooling rates for different regions of the heterogeneously heated shocked material differ from each other. During consolidation it is at a minimum in areas of particle contact (Fig. 5.14, points A_1). The corresponding temperature profiles calculated, based on Eq. (5.15) for a periodic system of "cold" and "hot " layers in the "skin" model, are presented in Figure 6.6 for different values of the dimensionless thickness of the hot layer δ.

It can be seen by comparing the positions of the "critical quenching schedules" and cooling curves for the "skin" model in Figure 6.6 that the cooling rate in the shocked material in the temperature region from T_m to the equilibrium temperature T_{th} is greater than that in the two-sided method with $\delta \leq$ 0.4. For the one-sided quenching method, this statement is valid even at a larger thickness of the hot layer in the "skin" model $\delta \leq 0.7$. These values of δ can be

Table 6.1. Melting, crystallization temperatures, and critical sizes of the melted zone on the particle contacts for the "skin" model.

Alloy	T_m, °C	T_{cr}, °C	$\delta'_{cr} = T_{cr}/T_m$
$Fe_{91}B_9$	1355	327	0.24
$Fe_{83}B_{17}$	1175	487	0.41
$Co_{75}Si_{15}B_{10}$	1120	512	0.46
$Fe_{79}Si_{10}B_{11}$	1146	545	0.48
$Ni_{15}Si_8B_{17}$	1067	509	0.48
$Fe_{80}P_{13}C_7$	985	463	0.47
$Pd_{82}Si_{18}$	798	384	0.48
$Ni_{62.4}Nb_{37.6}$	1169	672	0.57
$Pd_{77.5}Cu_6Si_{16.5}$	742	380	0.54
$Fe_{41.5}Ni_{41.5}B_{17}$	1079	447	0.41
$Au_{77.8}Ge_{13.8}Si_{18.4}$	356	20	0.06

considered as a critical width for the melted zone on the particle contacts δ_c for preservation of the amorphous state.

Thus, if $\delta \leq \delta_c = 0.4$–$0.7$, partially melted material can be solidified into an amorphous state during shock consolidation. If this condition is valid, then crystallization of amorphous material after shock compression may only proceed at the stage of thermal equilibrium in a specimen with the condition $T_{th} > T_{cr}$.

The second limitation on the width of the melting zone, to preserve the amorphous state during shock consolidation, arises as a result of the requirement $T_{th} < T_{cr}$. In the model considered find $T_{th} = \delta T_m$. Therefore, this requirement leads to a limitation on the width of the melted zone, expressed in the following way: $\delta \leq \delta'_c = T_{cr}/T_m$. This criterion is not connected with the cooling kinetics for contact surfaces. The maximum permissible values of the melted zone (δ'_c) are given in Table 6.1. It can be seen that the possible size of the melted zones is of the order of particle sizes if the criterion $\delta \leq \delta'_c$ is observed.

It is worthwhile mentioning that in the modified Carroll–Holt model (see Section 2.3) it is supposed that the whole internal energy (microkinetic plus thermal) up to the moment of pore collapse is concentrated in the outer spherical part. We can suppose that after pore collapse, the microkinetic energy mainly dissipates in the outer shell, at least at the regimes where the microkinetic energy is not too high. In this case, the relation d/l for the one-dimensional "skin" model can be obtained from the relation of the mass of the outer shell to the whole mass of the model cell. A similar condition should determine the transition of the results from the one-dimensional case to the three-dimensional case. The ratio of the masses and not the ratio d_r/R (where d_r is the thickness of the hot layer on the surface of the spherical particle with radius R) should establish the correlation between the one-dimensional model and the three-dimensional system of spherical particles. It is clear from the fact that the establishment of thermal equilibrium is determined by the mass ratio between the hot and cold parts, which only in the one-dimensional case is equal to d/l.

Since T_{cr} and T_m are calculated from the initial temperature $T_0 = 0$ °C, the size of the permissible heated zones along boundaries may be increased by prior specimen cooling. For alloys with $T_{cr} \approx 0.5T_m$, preheating with the retention of quite high critical values of the relative size of heated zones (δ) during consolidation is possible (Nesterenko and Muzykantov [1985]).

Comparison of the two found limitations on δ (δ_c and δ'_c) indicate that there are materials for which observing the conditions $\delta \leq \delta'_c$ ($T_{th} < T_{cr}$) also leads to observing the criterion $\delta \leq 0.4$, which follows by comparing the cooling kinetics (see, e.g., $Fe_{83}B_{17}$, $Fe_{41.5}Ni_{41.5}B_{17}$). At the same time, there are materials (e.g., $Ni_{62.4}Nb_{37.6}$) for which observing only the condition $\delta \leq \delta'_c$ may lead to lower cooling rates after consolidation than in the preparation process. However, considering the approximate nature of the used model of ideal quenching, it is possible to suppose that in the majority of cases the requirement $\delta \leq \delta'_c$ ($T_{th} \leq T_{cr}$) also leads to observation of the kinetics criterion ($\delta_c \leq 0.4$–0.7) for retaining an amorphous state. It is apparent (see Fig. 6.6) that this is especially valid for materials obtained by one-sided quenching.

It was proposed by Schwarz et al. [1984a] that the proportion of melt required, in order to obtain a strong compact, is 0.22. There is a considerable number of materials which tolerate molten zones of this size with the retention of an amorphous structure (see Table 6.1).

The considered problem has a straightforward relation to the explosive consolidation of amorphous ribbons (or ribbon-like powder particles), important for practical applications (Nesterenko et al. [1987]; Friend and Mackenzie [1987]; Zolotarev, Denisova, Nesterenko et al. [1989]; Heczko and Ruuskanen [1993]; Rakhimov [1993]). Under the dynamic compaction of amorphous ribbons, the specific shock energy corresponding to the stationary shock condition in the system of parallel layers (ribbons) is ensured by the circulation of the shock and rarefaction waves in the bulk of material of the layers (Section 5.2). It creates a less concentrated energy release than in the powder and can be less efficient for creating a strong bonding between particles (Prummer [1988]). At the same time this type of dissipation process is more reliable in preventing crystallization.

The heat transfer processes in heterogeneously heated consolidated amorphous materials are also considered by Roman, Gorobtsov, Pikus et al. [1986]. Their conclusion is that for a particle size of 100 μm, the thickness of melt can reach up to 10 μm and the rate of cooling is enough to requench the amorphous state. This is similar to the results of the "skin" model.

Prummer and Klemm [1986] concluded, based on the calculation of the critical rate of cooling of the melted sphere, that the rate of cooling (10^6 K/s), which is sufficient to preserve the initially amorphous state, is ensured only for a melted particle size with diameter less than 250 μm. From this consideration, they introduced a critical size of powder particle, 250 μm, which can be consolidated with preservation of the amorphous concluded state. It should be mentioned that melting usually occurs on the periphery of the spherical particles and heat transfer is better described by the "skin" model where the critical size of the melt should be scaled with the initial particle thickness (diameter).

Friend and MacKenzie [1987] and Hare, Murr, and Carlson [1984] mentioned that for the retention of the initially amorphous structure it is necessary to avoid excess heating and melting of the particle interfaces. As is justified in our calculations it is possible to melt up to 40% of the amorphous particles and still be able to preserve the amorphous state.

A specific class of amorphous (or nanocrystalline) materials can be synthesized by mechanical alloying in a ball mill or in attritors. In this case no critical cooling rate can be identified to ensure the initial amorphous or nanocrystalline state. Nesterenko, Avvakumov et al. [1989] successfully used explosive densification in plane geometry to obtain dense compacts from mechanically activated Fe–Nd–B alloy with retention of the initial substructure.

Explosive consolidation (double-tube geometry (Meyers and Wang [1988])) of nanocrystalline ball-milled powders from Ti-based (Ti–55Al and Ti–24Al–11Nb) and Fe-based (Fe–N) alloys produced fully dense compacts while retaining the initial scale of the substructure (5–25 nm) (Korth and Williamson [1995]). Pressure in the first shock wave evaluated by numerical one-dimensional modeling was about 10 GPa. A comparison of samples obtained by explosive consolidation, HIP, and Cercom methods demonstrated that the former technique retained the nanostructure much better than the other two methods. As a typical feature of explosive compaction the fully densified samples contained microcracking.

Jain and Christman [1994] used a plane shock wave to consolidate the nanophase Fe–28Al–2Cr powders obtained by ball milling. Fully dense compacts with a grain size of about 80 nm were produced. The compressive flow stress (2.1 GPa) of the compact was much higher than the yield strength of the solid coarse-grained material (0.25 GPa). In tension, the nanophase compact has a failure strength of 0.65 GPa which is comparable to a coarse-grained material. It should be mentioned that a grain refinement from 100 nm to 7 nm, and in the bulk of the same Fe–28Al–2Cr alloy, can be obtained without powder precursors after room temperature compressive deformation (Jain and Christman [1996]).

In conclusion, the high rate of cooling of the contact areas in the shock consolidated material can guarantee the requenching of the amorphous state due to the diffusion-type mechanisms of heat transfer, even for a relatively large amount of melt on the interfaces. The necessary and sufficient condition, for retention of the amorphous state under dynamic consolidation for the known alloys, is the inequality $T_{th} \leq T_{cr}$. This condition takes into account the possibility of crystallization at the stage of fast temperature relaxation and at the stage of relatively slow heat removal into the surrounding capsule. To avoid the sample destruction in the rarefaction wave, other conditions should also be satisfied (Wilkins [1984]). Very often, the amorphous state is preserved, and most of the sample demonstrates a good bonding between particles; and still the sample is saturated by the transparticle cracks.

6.4   Obtaining Supercooled States Under Shock-Wave Loading

6.4.1   Amorphization of Solids Under Shock Loading

The heterogeneous high-strain-rate deformation of solids in shock waves connected with nonequilibrium heating throughout the bulk promotes the generation of rapidly solidified structures, including the amorphous ones. This is due to the small sizes of locally "hot" areas embedded into a relatively "cold" matrix.

Decarli and Jamieson [1959] were the first to discover an amorphous form of quartz under shock conditions. Ahrens and Rosenberg [1968], Anan'in, Breusov et al. [1974], and Arndt [1984] observed the amorphous state in solid quartz after shock loading. One of the reasons for amorphization is melting in the shear localization zones and the subsequent rapid quenching under compression before the pressure is dropped in the rarefaction wave. Evidently this mechanism can be active if the average equilibrium temperature is less than the melting temperature of quartz. Gratz, Nellis et al. [1992] performed shock experiments with quartz at various initial temperatures (-117 to $+1000$ °C) and showed that the amount of glass recovered increases with both pressure and initial temperature. They found that shock produces two types of disorded material: quenched melt resulting from melting along the microfaults and diaplectic glass (amorphous material with no signs of melting behavior) in transformation lamellae. The main difference between these two structures is the density difference: the quenched melt expands during a release of pressure and temperature creating a density lower than quartz, while recovered diaplectic glass has a density closer to that of quartz. At pressures lower than 15 GPa, quartz transforms mostly due to melting connected with shear localization. At higher pressures, it converts mostly along the transformation lamellae.

It is worth mentioning that even the static compression of quartz (or coesite) produces gradual amorpization, which becomes complete by 35 GPa, as demonstrated by Hemley et al. [1988] in experiments with a diamond anvil cell. Molecular dynamic simulation by Tse and Klug [1991] and by Somayazulu, Sharma et al. [1994] have reproduced the creation of "pressure glass" as well as the first-principle calculations by Binggeli and Chelikowsky [1991].

The process of pressure amorphization is now recognized as a common phenomenon in materials with a relatively complex structure like SnI_4 (Fujii et al. [1985]), $AlPO_4$ (Kruger [1990]; Chaplot and Sikka [1993]), $Ca(OH)_2$ (Sikka, Sharma, and Chidambaram [1994]), and others. This pressure-induced amorphization is the result of a kinetically favorable intermediate amorphous state on the way to transformation to another denser stable phase. There is growing evidence that these materials could also be amorphized under shock-loading conditions probably with the resulting microstructures being different (see, e.g., the discussion in Gratz, Nellis et al. [1992]), due to the different strain and strain-rate history (Somayazulu, Sharma et al. [1994]). The importance of shear stresses in the amorphization of SiO_2 was experimentally demonstrated under static pressures by Kingma et al. [1993]. Sikka and Sharma [1992] showed that many crystalline-to-amorphous transitions under pressure occur when the Van der Waals contacts in these materials reach the limiting values found at normal pressures.

The analogous structures for normal metals were not found. The specific possibility of a shock wave creating localized amorphous states as a result of the deformation defects was demonstrated by Psahie, Korostelev, and Panin [1988] in computer calculations.

6.4.2 Local Amorphization in Powders Under Shock Loading

New possibilities for forming rapidly solidified (amorphous) structures are connected with heterogeneous heat release in shocked powders. Raybould [1981] and Meyers, Gupta and Murr [1981] observed a microcrystalline phase with characteristic grain size, ~1 μm, on the contact boundaries of particles of stainless steel 304L, Ni-based alloy MAR M-200, and Al–11%Si. In the case of 304L stainless steel these rapidly solidified structures were relatively ductile, connecting together particles with high microhardness.

The amorphous phase was also observed on the particle's boundaries in ceramic AlN powder by Gourdin, Echer, Cline, and Tanner [1981] with the help of transparent electron microscopy. This phase was connected with an enrichment of the surface layers by silicon which promoted amorphization.

The detailed description and analysis of microcrystalline structures created by rapid quenching of the surface layers in Fe-, Ni-, and Al-based powders are provided by Morris [1981], [1982a]. The quenching rate was determined based on the cell size of rapidly solidified alloys. The found values were in the range 10^6–10^{10} K/s. There are some advantages to this shock-assisted method of quenching:

(1) In one macrosample, there are many microsamples quenched with a different rate depending on the melt thickness but at the same pressure. We may use this technique to establish the amorphization criteria for different alloys.

(2) The contact between the melt and "substrate" is not contaminated by surface oxidation and can be considered as ideal.

Morris [1982a] used the shock-loaded powders to find the dependence between the quenching rate and the characteristic sizes of the created structures (cells or secondary dendrite arms). It was shown that the corresponding curves can be extrapolated from relatively low quenching rates (10^5 K/s) to quenching rates up to 10^{10} K/s. Despite this, the fast-quenching amorphous phase was observed only for Inconel 718. This particular behavior can be connected with initial contamination of the surface layers by oxides and carbides in addition to rapid solidification.

Additionally, this method allows for the investigation of quenching under high-dynamic pressures for relatively thin-melt layers, which can produce new properties of the rapidly solidified alloys. The experimental possibility of rapid quenching under the action of high pressure is confirmed by observation of the high density of dislocation in the melted zone for Ni-based alloy APK-1 and Inconel 718 by Morris [1982a].

Vreeland, Kasiraj, Ahrens, and Schwarz [1983] performed interesting experiments on the shock amorphization of the initially microcrystalline (with a small amorphous part) alloy Markomet 3.11. At the ratio $E/e_m = 0.38$ (E is the total specific shock energy and e_m is the energy which is necessary to increase the temperature up to the melting point) the amount of the amorphous phase increased in relation to the initial state. The further increase in internal energy decreased the initial amount of the amorphous phase. These experiments clearly demonstrated the importance of the fast-quenching processes, as well as the importance of the post-shock recrystallization. The results of this paper are in very good agreement with the critical thicknesses of the melt obtained in the frame of the "skin" model by Nesterenko and Muzykantov [1985] (see Section 6.2).

Ferreira, Meyers, and Thadhani [1992] obtained the interparticle amorphous Ti–Al alloy in the Ti_3Al powders, and the mixture of amorphous and crystalline regions in the Ti_3Al-based alloy mechanically blended with niobium. These rapidly solidified materials resulted from interparticle melting and quenching under explosive consolidation.

High cooling rates are attained in all the aforementioned materials by using small thicknesses of molten metal in contact with a cold substrate. The cold bulk of the particles serves as a substrate. In these examples, the role of the dynamic pressures was limited to a rapid generation of the thin layers of melts by converting mechanical energy into heat. The subsequent rapid solidification was ensured by the small melt thickness, as it was in traditional approaches.

Frenkel [1955] made very useful comments on the mechanism of fusion (amorphization). He emphasized that the volume is a main factor which determines the transition from a crystalline to an amorphous structure of a body and that "... the increase of volume associated with fusion ("free" volume) is simultaneously the direct cause and effect" of the loss of order in the crystal structure. Frenkel called the smallest volume compatible with the disappearance of long-range order in a body of the "amorphization volume" V^*, in the limit $V \rightarrow V^*$ the "amorphization" temperature must tend to the absolute zero. This means that if the volume of the solid crystalline body is uniformly increased to V^* ("free" volume is distributed in the body in a continuous way), then an amorphous state may be achieved at the temperature below melting temperature.

One of the ways to achieve this crystal/amorphous transformation is severe plastic deformation, which may provide a volume increase, and ensure condition $V > V^*$ due to the creation of various lattice defects. Refining the microstructure of a solid body to a nanometer range by severe plastic deformation (SPD) was experimentally demonstrated for a few metallic materials (Valiev, Lowe, and Mukherjee [2000]). SPD can be applied to the consolidation and refinement of the initial structure of powders. High-pressure torsional straining of the micrometer-size titanium powders or its mixture with TiO_2 results in nanocrystalline titanium and $Ti–TiO_2$ composites (Stolyarov, Zhu, Lowe et al. [2000]).

Dynamic densification of powders in the shock wave, due to the large porosity, results in severe plastic deformation of the layers of material adjacent to the surface of particles (see Figs. 2.3, 2.19, 2.20, 5.12, and 6.3). Another example is the large-strain deformation inside the shear bands (Figs. 4.10, 4.11, and 4.23 to 4.26). This may result in nanocrystalline (or even amorphous) states similar to that observed under quasistatic severe plastic deformation (SPD).

6.4.3 Pressure-Induced Supercooling Under Shock Loading

A qualitatively different way of obtaining supercooled states is the melting of material in a shock wave and cooling it in the rarefaction wave. Breusov [1978] pointed out that an effective cooling rate in this case can be as high as 10^{10} K/s (temperature drop $\sim 10^3$ K during the typical time of 10^{-7} s). This is above the record levels attained by traditional methods. It is worth mentioning that if a substance is quickly transferred to a temperature range well below the melting point T_m, then crystallization is retarded by the high viscosity. By increasing the quenching rate and magnitude of supercooling, the microcrystalline or even amorphous states can be obtained for any materials (Salli [1972]). However, when the substance is shock melted, it usually remains liquid after the pressure drop. This is due to the relative position of isentropes and melting curves on the $P–T$ plane for normal materials (McQueen and Marsh [1960]; Fig. 6.7(a)).

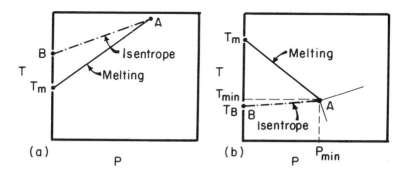

Figure 6.7. The schematic phase diagram of materials with (a) normal and (b) anomalous melting curves; isentropes describe the unloading path after shock loading.

Table 6.2. The melting temperatures in normal conditions, minimal melting tempratures under pressure P_{min}, and undercooling $T_m - T_{min}$ possible under a quick pressure release.

Material	T_m (K)	T_{min} (K)	P_{min} (kbar)	$T_m - T_{min}$ (K)
Ge	1210	723	95	487
Si	1688	1000	115	688
Ce	1077	933	33	144
Ba	1002	773	65	229
InSb	798	593	25	205
InAs	1173	773	100	400
InP	1373	973	125	400

But there exists a class of materials where the properties of dynamic loading can be used more fully under certain conditions (Nesterenko [1983a,b,c], [1984]). These are the substances in which the shape of the melting curve is anomalous, i.e., the melting temperature decreases with a pressure increase in a certain pressure range (Fig. 6.7(b)).

It is clear from Figure 6.7(b) that if the material is melted by the shock wave near triple point A, then during the rarefaction process along the curve AB it will be in the supercooled state at zero pressure. This is mainly because the melting temperature will increase in the course of a pressure decrease. This happens even without cooling, due to the rarefaction process or heat conductivity into the adjacent cold material. It is assumed that crystallization does not occur during the pressure drop. The attained supercoolings $(T_m - T_{min})$ thereby may be as much as 150–700 K (Table 6.2).

The data in Table 6.2 are taken from Tonkov [1979] and Stishov [1968]. For liquid germanium purified of impurities, a maximum supercooling of 227 K has been attained by the traditional method (Salli [1972]), i.e., less than by about a factor of 2 than the magnitude attained in the proposed process.

If the width of the rarefaction wave is about 10^{-7} s, then the effective cooling rate is in the range $10^9–10^{10}$ K/s, according to the supercooling magnitude in Table 6.2. Moreover, high cooling rates are observed below T_{min}, i.e., material is transferred to the supercooled state without passing through the high-temperature range $T_m \div T_B$, where the crystallization may rapidly occur. Therefore, in the proposed method, the cooling from T_m to T_B by heat transfer is replaced by a pressure drop in the rarefaction wave.

The possibility of melting a material in bulk by a shock wave with a pressure close to P_{min} is far from being obvious, since pressures substantially exceeding P_{min} may be needed to produce a liquid. In the shock front the material passes through all pressures from $P = 0$ to P_{min} and may be partially melted at any intermediate P, which corresponds to a latent heat of melting λ_p. Therefore, to produce a complete melting at the final pressure P_{min}, it is necessary to have the specific internal shock energy in the range between ε_1 and ε_2:

$$\varepsilon_2 = \lambda_p + c\left(T_{\min} - T_0\right) + \varepsilon_c,$$
$$\varepsilon_1 = \lambda_0 + c\left(T_{\min} - T_0\right) + \varepsilon_c,$$

(6.10)

where λ_p and λ_0 are the latent heats of fusion at $P = P_{\min}$ and $P = 0$, $T_{\min}$ is a minimal melting point under pressure $P_{\min}$, T_0 is the initial temperature, and ε_c is the energy of "cold" compression at $P = P_{\min}$. It is assumed that the specific heat c is independent of temperature, while the latent heat of melting varies monotonically with pressure, and λ_p corresponds to the minimum value. If the relationship is not monotone, it is necessary to take λ_p and λ_0 as the maximum and minimum values of λ in the given pressure range.

If the energy required for melting is provided, and the duration of the shock wave is greater than the melting time at the given pressure, then a solid trasforms into a liquid behind the shock front. The experiments by Mineev and Savinov [1967], Kormer, Sinitsyn, Kirillov, and Urlin [1965], Belyakov, Valitskii, Zlatin, and Mochalov [1966], and Asay and Hayes [1975] give an estimate for the melting time of less than 10^{-6} s. The melting can be promoted if the material is preheated to a temperature closer to T_m while remaining crystalline.

Let us consider the possibility of performing this process for silicon, which has a high melting point, a high latent heat, and the maximum difference ($T_m - T_{\min}$) combined with a high value of latent heat. For this purpose, we elucidate how the latent heat λ is dependent on the pressure. One can estimate its change from Clapeyron's equation

$$\frac{dT_m}{dP} = \frac{T_m \Delta V}{\lambda},$$

(6.11)

where ΔV is the volume change on melting. The derivative of the melting temperature with respect to pressure is approximately constant up to 115 kbar (Tonkov [1979]). Therefore,

$$\lambda_p = \lambda_0 \frac{T_{\min}}{T_m} \frac{\Delta V}{\Delta V_0}.$$

(6.12)

Here ΔV and ΔV_0 are the volume changes on melting at $T_{\min}$ and T_m. The volume change on melting for a material with anomalous behavior is related mainly to the change in the mutual disposition of the atoms. Application of the pressure makes the atoms more closely packed, and a natural assumption is that $\Delta V < \Delta V_0$. Therefore

$$\lambda_p \le \lambda_0 \frac{T_{\min}}{T_m} \approx 0.6\lambda_0.$$

(6.13)

One can estimate ε_c from the Hugoniot for monolithic silicon (Altshuler [1965]):

$$\varepsilon_c \le \frac{1}{2} P \Delta V_a, \tag{6.14}$$

where ΔV_a is the change in specific volume on shock compression at $P = 115$ kbar and $\Delta V_a \approx 0.05$ cm^3/g. If $\lambda_0 = 1.78 \cdot 10^{10}$ erg/g (Tonkov [1979]) and $c = 0.89$ erg/gK, we get $\varepsilon_1 = 2.72 \cdot 10^{10}$ erg/g and $\varepsilon_2 < 2 \cdot 10^{10}$ erg/g.

Let us now estimate the specific energy ε for the shock compression of porous silicon of which the initial density is half of the density of solid silicon. If it is assumed that the density at the front alters by about a factor of 2, obtain for $P = 115$ kbar that

$$\varepsilon = \frac{P}{4\rho_0} \approx 2.4 \text{ erg/g.} \tag{6.15}$$

This value of ε is in the range between ε_2 and ε_1. It is clear that heating the material ahead of the front up to 350 °C will cause ε to approach ε_1. By increasing the initial density, one can reduce ε to ε_2. The desirable variations of ε can be provided by variations of the initial porosity at the same shock pressure. Therefore, it is fairly easy for porous silicon to fulfill the conditions for complete melting behind the shock front. Similar calculations can be performed for other materials.

To obtain a homogeneous melt, it is necessary that all the internal energy ε acquired by shock compression is uniformly distributed over the bulk. A sufficient condition for this is a small thermal relaxation time in comparison with the time of pressure action. This can be provided by a small value of the particle size. In general, under the shock compression of powder, the temperature is unevenly distributed over the volume. However, at pressures of about 100 kbar, even the powders with coarse particles (about 0.5 mm) are close to equilibrium in the shock front (time scale of temperature equilibration 10^{-7} s, see Chapter 5). This is the result of the high-strain-rate plastic deformation which particles undergo not only in the contact areas but also in bulk. This results in a more rapid establishment of thermal equilibrium in the shocked material for a comparatively coarse particle in comparison with the characteristic time of thermal diffusivity based on the initial particle size. This is, so to speak, a mechanically forced thermal equilibrium.

Moreover, the shock compression of a porous specimen is characterized by very high heating rates (about 10^{10} K/s, see Section 5.1), allowing for completion of the heating and quenching processes inside the time scale $\sim 10^{-6}$ s.

Nesterenko [1983a,b,c], [1984] performed experiments with silicon having an initial density approximately half of that for the monolithic material. The silicon particles were of the single-crystal type (initial specific resistance about 14 Ω cm) and had sizes of about 0.1 mm, a sample thickness of about 1 mm. To ensure the preservation of the shocked sample from post-shock crystallization, it was placed in a massive copper plate cooled to liquid nitrogen temperature (Nesterenko [1983c]). The powder was shock loaded with a flyer plate (Fig. 6.8). After shock compression, with a calculated pressure in the first shock 115 kbar,

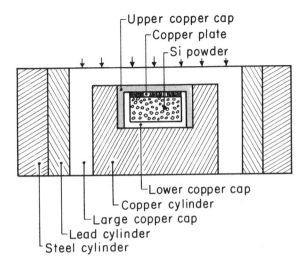

Figure 6.8. Experimental set-up for the shock loading and quenching of silicon powder.

the samples were recovered. The thickness of the shocked samples was about 0.5 mm that provided a relatively fast heat removal from the shocked material.

An X-ray diffraction pattern was recorded ((111) and (422) lines) with FeK_α radiation. The fine-structure characteristics of shocked samples were determined—the size of the coherent-scattering regions D and the microstrains $\Delta a/a$ (where Δa is the deviation and a is the lattice parameter) according to the procedure based on the line width (Gorelik, Rastorguev, and Skakov [1960]). The data obtained are $D \approx 1.33 \cdot 10^{-5}$ cm and $\Delta a/a \approx 0.46 \cdot 10^{-3}$. This means that a fine microcrystalline structure was received, due to the formation of a supercooled melt and subsequent rapid solidification. A similar technique was developed for high-pressure shock loading by a two-stage gas gun of the thin samples of different materials by Nellis et al. [1988], and was successfully applied to various materials.

This method—mechanically forced rapid melting near the triple point with a subsequent pressure release—can be used for the creation of microcrystalline and amorphous materials. The cylindrical geometry of shock loading with a central mandrel can also be used, and rapidly solidified films could be formed on the cylindrical surfaces from the melted zone near the central rod for the Mach reflection. The size of this zone can be varied by the parameters of loading as well as by changing a particle size of the powder (Section 2.5).

It is worthwhile mentioning that this method (Nesterenko [1983a,b,c], [1984]) for obtaining supercooled states is similar to the RPA (Rapid Pressure Application) method of creating a microcrystalline (or amorphous) state proposed by Sekhar, Mohan, Divaker, and Singh [1984] (see also Sekhar [1986]). The RPA method also replaces the quenching from heat diffusion by the pressure drop. This apparently can be performed more quickly and without dramatic restriction of the sample size. The influence of pressure (static) on the spectrum of the rapidly solidified phases was observed by Brajkin, Larichev, Popova, and Skrotskaya [1986].

Recently nanocrystalline silicon was synthesized using thermal spraying, in which silicon powder is injected into a high-energy flame where the particles melt and accelerate before colliding with the substrate (Goswami et al. [1999]). The available level of shock pressures under the particle's impact reported by the authors is up to 20 GPa (actually it is 10 GPa because the particle velocity behind the shock wave is about one-half of the velocity of the impacting object). This technique can be modified in a way that silicon particles are only preheated in the flame, and subsequently melted during the viscoplastic flow after collision under high-shock pressure. This process of melting can be facilitated if the initial silicon particles have some porosity. Then the pressure-induced supercooling described in this section may result in the amorphous silicon coating.

6.5   Sintering, Hot Isostatic Pressing of Shocked Powders

6.5.1   Sintering of Shocked Powders

It is necessary to apply high levels of shock pressure to the powder to get a relatively strong bonding between particles (see Section 6.1). These pressures are a few times higher than material strength. This provides a potential danger of the material's fracture in a rarefaction wave, especially taking into account its low ductility inherited due to heterogeneous deformation of the contacts under shock loading. The latter makes it impossible to get uniform bonding between particles throughout the sample bulk at plane shock loading. Additionally, in the most technologically convenient cylindrical geometry of loading, Mach reflection is created in the center of the sample, often accompanied by the quasistatic deformation of particles in bulk that cannot ensure strong bonding (Sections 2.5 and 6.1; Nesterenko [1992], [1995]).

One of the characteristic features of shock consolidation is that the deviations of the local values of pressures, strains, temperatures, and particle velocities, from averaged continuum values of these parameters, are comparable to or even larger than the values of the latter ones (Nesterenko [1992]). This is evident from the experimental data of contact temperatures (Section 5.1), the structural peculiarities of the compacted materials (Fig. 6.1), and the results of two-dimensional computer calculations (Figs. 5.13 and 5.14; Benson [1994], Benson, Nesterenko, and Jonsdottir [1995], [1996]). We can characterize this situation as highly heterogeneous behavior of the shocked powder.

To describe such behavior, it is very important to develop at least qualitative models for each structural level of the granular material. At the same time, from a mechanical and material point of view, these generic features of the process do not allow us to obtain the homogeneous final properties of the compacted material.

This makes absolutely inevitable the development of the technological scheme which will combine explosive loading with subsequent heat treatment, forging, rolling, or hot isostatic pressing (HIPing) (Nesterenko [1995]; Meyers, Benson, and Olevsky [1999]).

Raybould [1980] demonstrated that dynamically densified compacts from aluminum powder can be sintered at 300 °C, while temperatures of over 600 °C are required for the statically densified compacts. It was shown by Atroshenko [1984], Prummer [1987], and Graham [1993] that the explosive treatment of powders promotes the sintering process by activating it on the initial stage when defects created during the shock loading are not annealed. For example, after shock densification the activation energy of the sintering of nickel powder ПНК was 158 kJ/mol in comparison with the value of 265 kJ/mol after static densification (Atroshenko [1984]).

Bergmann and Barrington [1966] were the first to demonstrate that very fine particles of SiC and B_4C could be produced by explosive loading without introducing impurities. After shock treatment, powders were unusually responsive to sintering. They proposed using explosive shocking as a unique mechanism of the "cold-working" of ceramics that may permit subsequent control of their microstructure through primary recrystallization.

The structural refining and strength increase after explosive consolidation with subsequent sintering in comparison with traditional sintering were obtained by Bondar', Kostyukov et al. [1986] for the powder mixture TiC–TiNi. The recrystallization of explosively densified Ti-based powder with a creation of the fine-grain structure was observed by Oreshin, Pashkov, and Tkachev [1985]. Raybould [1987] found that bonding occurred during a low-temperature sintering below the α–β transition of shock-treated prealloyed Ti–6Al–4V powder. Activation of sintering is connected with extensive local shearing of the particle surfaces resulting in a high-dislocation density and removal of the surface oxides.

Properties of the explosively compacted high-temperature superconductors Y-123 also were essentially improved after proper annealing in the oxygen atmosphere (Nesterenko [1992]).

6.5.2 Shock Consolidation/Annealing (Sintering) of Microcrystalline Metallic Powders

The influence of shock compaction on the sintering of rapidly solidified powders (the most interesting case to be considered for application of shock consolidation or compaction) can differ from the case of powders with a normal microstructure due to the following reasons:

(1) Rapidly solidified materials are saturated by large numbers of lattice defects inherited from the liquid state. It is not clear whether there is any room for the additional defects generated by high-strain-rate plastic flow.

(2) The sintering process getting strong solid materials from rapidly solidified powders with a high strength usually demands high-temperature exposure of the powder for a relatively long time (usually about a few hours). Then the initial defect structure can be removed at the early stage of the densification process, for example, by the poligonization process.

Rapidly solidified stainless steel powder (ferrite–martensite type, EP-450) was successfully consolidated under plane shock loading at a pressure of 28 GPa (Nesterenko, Lazaridi, Pershin et al. [1989]; Lazaridi [1990]; Nesterenko [1992]).

The final tensile strength of samples was higher than half of the strength of solid material (0.64 GPa), but lower than the tensile strength of annealed and consolidated powder, see Figure 6.2. The samples for testing had a total length of 50 mm (gauge length 30 mm, crosssection area 3 · 5 mm²). The initial fine-grain microstructure of the particles was mainly preserved, but samples demonstrated no plasticity under tensile loading due to the lack of bonding on some interfaces (Fig. 6.3). The complete lack of plasticity of the explosively consolidated powders unfortunately should be considered as a characteristic feature for this method of consolidation, due to the practical impossibility of ensuring uniform bonding over the total contact area of particles (Nesterenko [1986], [1992]). Compacts with four particle sizes: +40–70 μm, +90–145 μm, +145–310 μm, and +310–440 μm were investigated. After shock consolidation, the samples had an average microhardness of about 7.70 GPa and a corresponding tensile strength depending on the particle size: 0.35 GPa (+90–145 μm), 0.5 GPa (+145–310 μm), and 0.4 GPa (+310–440 μm). X-ray analysis demonstrated an increase of the γ-phase from an initial 5–12% to 30–50%, after explosive consolidation.

Bondar', Nesterenko et al. [1990] and Nesterenko [1992] used heat treatment at a temperature of 600–1000 °C for 1 or 2 h in vacuum to increase plasticity of the compacts of this material. No essential changes in the material properties were observed at temperatures below 700 °C. In the interval 750–900 °C for 1 h the mechanical properties were essentially improved with decreasing the data spread with a temperature increase from the initial 20–30% to 5–20% at 900 °C, being the smallest for the smallest particle size. Tensile strength became correspondingly equal to 0.7 GPa (+40–70 μm), 0.6 GPa (+90–145 μm), 0.6 GPa (+145–310 μm), and 0.55 GPa (+310–440 μm), the largest for the smallest particle size; average microhardness decreased from 7.70 GPa to 2.20 GPa. This latter value of microhardness HV correlates well with the tensile strength σ_t according to the empirical relation $\sigma_t = HV/3$ for monolithic materials.

Even more important was the fact that the heat treatment resulted in reasonable plasticity. Relative elongation was correspondingly 12% (+40–70 μm), 8% (+90–145 μm), 6% (+145–310 μm), and 4% (+310–440 μm) after annealing during 1 h at 800 °C and 900 °C.

The increase in annealing temperature up to 1000 °C resulted in the increased tensile strength 0.92 GPa (+40–70 μm), 0.8 GPa (+90–145 μm), 1.1 GPa (+145–310 μm), and 0.6 GPa (+310–440 μm), but all the samples demonstrated brittle behavior without any plasticity under tension. So the optimal annealing conditions for atomized stainless steel powder EP-450 are 750–900 °C, for 1 h and they lead to reasonable strength and ductility.

Annealing of explosively consolidated rapidly solidified materials can be necessary, not only for their mechanical properties, but also for improvement of magnetic characteristics. Nesterenko, Lazaridi et al. [1989] demonstrated that annealed compacts from rapidly solidified (atomized in nitrogen) hard magnetic Mn–Al–C alloy have properties which allow us to use them as rotors in electrical devices. The initial microhardness of powder was 5.90 GPa, the coercitive force H_c = 224 kA/m, and induction B_r = 0.15 T. After shock densification, H_c decreased to 100 kA/m without a change of B_r. Optimized annealing (500 °C, 0.5 h, furnace cooling) resulted in the essential increase of B_r up to 0.28 T and a slight increase in H_c to 110 kA/m. Data for the explosively densified Mn–Al–C alloy powder prepared by different methods can be found in the paper by Vertman, Epanchintsev, Zvezdin, Nesterenko et al. [1989].

Dynamically densified samples from metastable alloys can be useful in research application. For example, explosively consolidated rapidly solidified samples of microcrystalline Zr–Nb alloy were used for the investigation of implanted profiles of He ions with energy in the interval of 2–3.5 Mev, with step 250 Kev, and doze 10^{17} cm^{-2} (Aksenov et al. [1989]).

6.5.3 Annealing of Consolidated Amorphous Metallic Materials

Detailed descriptions of the mechanical and magnetic properties of explosively consolidated amorphous powders and foils can be found in the papers by Nesterenko et al. [1987], [1988a,b], [1989], [1992], and Tabachnikova et al. [1990]. The compression strength of these materials can be higher or lower than the tensile strength of corresponding foils (about two to three times lower), due to the lack of bonding on the large part of the contacts. The magnetic properties of amorphous compacts can be of practical interest, especially when they are manufactured from amorphous foils.

Explosive compaction usually results in the degradation of properties of magnetically soft Co- and Fe-based alloy. The subsequent optimized annealing allows practically complete recovery of the magnetic properties typical for initial foils. Better results were obtained for compacts with a very small constant of magnetostriction (Co-based alloy, Zolotarev, Denisova, Nesterenko, and Pershin [1989]). For example, the annealing of a laminar cylindrical shield made from copper and soft magnetic Co-based foils with thicknesses of 15 μm and 26 μm (Fig. 6.4(c)), allowed increasing the shielding coefficient from 15 to 100.

It is interesting that the traditional annealing recommended for amorphous foils deteriorated the magnetic properties of explosive compacts. At the same time, the special regime of annealing at 500 °C for 1 h resulted in an essential improvement of the shielding coefficients (Nesterenko, Pershin et al. [1991]). The mechanical properties of compact from foil allow its use as a load-bearing element of the devices in combination with the shielding properties.

6.5.4 Sintering of Shocked Microcrystalline Ceramics

It is known that explosive working can be a valuable instrument as a method of ceramic powder modification, i.e., fragmentation of particles and their activation preceding traditional sintering (Graham [1993]).

Less established is the possibility of shock loading to activate sintering, or to change the recrystallization or phase transformation behavior of metastable materials. Beauchamp and Carr [1990] investigated the influence of shock treatment on the kinetics of the phase transformation of alumina powder produced by melting and quenching. They addressed the question as to whether phase transformation of the metastable alumina might be affected by the introduction of the defects during shocking additional to those inherited from the melt. The shock treatment has produced large changes in the kinetics of transformation from θ-phase aluminas to α-phase: the incubation period of approximately 1 h at 1050 °C, typical for unshocked powder, was completely eliminated. The main mechanism of this phenomenon was attributed to the increased concentration of nuclei in the θ-phase material by shock deformation.

Explosive loading can be effective for the activation of sintering of submicron nonocrystalline ceramics obtained by plasmochemical synthesis; this is difficult to accomplish through the ordinary methods of mechanical fragmentation. This was demonstrated by Nesterenko, Panin et al. [1992]. Kul'kov, Nesterenko et al. [1993] found that the dynamic loading of a nanocrystalline ZrO_2 powder results in a significant change in the powder morphology, without any increase in the crystallite size. Such loading leads to the generation of secondary microdeformations of the crystalline lattice of the initially tetragonal phase. As a result, the monoclinic phase is formed, whose appearance is caused by the reduction in the critical grain dimension.

It is worth mentioning that, compared to over ten different types of this powder (obtained using various synthesis methods), the explosively modified ZrO_2–Y_2O_3 has at least three specific features: the highest relative root-mean-squared microdeformation in combination with submicron particle size and a multiphase structure. The annealing of the powder produces a release of the stored deformation energy and, as a consequence, the disappearance of the monoclinic phase without any noticeable grain growth up to $T = 1100$ °C. The aforementioned peculiarities of powder morphology are able to activate the process of sintering. The same densities are indeed reached in a shock-modified powder, at several hundred degrees lower than in the unshocked powder.

The dynamic compaction of samples of rapidly solidified Al_2O_3–ZrO_2-based ceramics, with diameter 1 cm and initial thicknesses ranged from 400 microns to

900 microns, was investigated by Tunaboylu, McKittrick, Nellis, and Nutt [1994]. A 6.5 m long two-stage gas gun was used to generate plane shock loading with pressures ranging from 5.5 GPa to 41 GPa. Projectiles from different materials, weighing only a few grams, were used to reduce the amount of kinetic energy supplied to the powder and thus reduce the post-shock damage of the samples. It was reported that the compacts had good structural integrity without visible cracks and retained a fine-grain size for shock pressures below 12.6 GPa. A high-pressure phase of ZrO_2 was detected in compacts after shock loading at and above 6.3 GPa.

6.5.5   Shock Loading/Hot Isostatic Pressing of Rapidly Solidified Materials

6.5.5.1   HIPing of Shocked Ti–Al Alloy

Hot isostatic pressing (HIPing) is one of the most efficient methods of obtaining a high level of properties of consolidated powders. The HIPing of granular materials being a mature technological process also has some problems (German [1996]). Oxide and carbide inclusions on the particle surfaces under small intergranular shear deformation, typical of this technique, may result in relatively low-compact strength and high sensitivity to the stress concentrators. The subsequent plastic deformation of the HIPed compacts can result in breaking the surface defects, accelerating diffusion processes, and promoting a high strength and ductility of the material (German 1996).

From this point of view, dynamic compaction, being far from the technological level of development, has some advantages over hot isostatic pressing. It is accompanied by extensive shearing on the contact surfaces (see Fig. 6.1) in the dynamic regime of the particle's deformation, which can disrupt the surface inclusions like oxide or carbide films. Explosive pressing can considerably reduce the temperature that is necessary for complete densification in comparison with the HIP process. The phase and structural states are thus largely retained in rapidly solidified powders, and it is possible to obtain compacted materials with unique properties.

There is a possibility of combining both processes using their unique features and manufacturing materials with new qualities. Nesterenko, Lazaridi et al. [1985], Molotkov, Notkin, Elagin, Nesterenko, and Lazaridi [1987], [1991] and Nesterenko, Lazaridi, Molotkov, Notkin, and Elagin [1993] investigated the explosive compaction of Ti + 33%Al intermetallic rapidly solidified powder. Practical interest in this material is connected with the fact that Ti-based alloys with a high content of aluminum (15–36%) are comparable with Ni-based superalloys according to their heat resistance, and have a working temperature higher than 800 °C. The structure refining in these alloys by traditional methods of thermomechanical treatment (plastic deformation of the ingot with a subsequent recrystallization) is impossible because of very low plasticity. Elagin, Korobov, Molotkov et al. [1986] used a rapid solidification process and demonstrated that an increase of the quenching rate results in a fine-grain structure with high-defect density. The dispersed α_2-phase is uniformly distributed throughout the material bulk (γ-phase).

This rapidly solidified intermetallic powder, with its high strength and crystal structure saturated by defects, is a good example of the demonstration of the possible beneficial applications of explosive densification, combined with subsequent HIPing. The results given here were obtained in 1985–1987. It is interesting that independent research with a similar material Ti + 33%Al was performed by Meyers and Ferreira and Shang in New Mexico Tech approximately at the same time (Ferreira [1989]; Shang and Meyers [1991]).

The powder was made by centrifugal atomization in helium, grain size 63–415 μm, and the tapped density is 55% of the theoretical density. In our experiments, cylindrical capsules from titanium, with 15–20 mm diameter, wall thickness 1 mm, and 100–300 mm long, were packed with granules of Ti–34Al alloy. The capsules with powder inside were evacuated to a residual pressure less than $6.55 \cdot 10^{-3}$ Pa and were outgassed for 4 h at 400 °C, after which they were sealed in vacuum by electron beam welding. The "cold" explosive pressing at normal temperature, CEP, was performed in a cylindrical geometry and at explosive parameters (density, 1 g/cm^3, detonation speed, 4 km/s, and thickness about the radius of the capsule) which ensured the "regular" regime of loading. This means that the creation of a Mach stem in the center (and subsequent cavity) was mainly avoided, and density was uniform throughout the bulk of the sample. Buckling during the explosive pressing was practically absent and the capsule shrank only radially. This regime of loading did not destroy the vacuum inside the capsule. The detonation pressure was below 7 GPa (corresponding to the first shock in the powder with an amplitude of 3.5 GPa), while the loading time was about 5 μs.

The density obtained by cold explosive pressing is 95–98% of the theoretical. The granules retained their dendritic structure, being heavily deformed in the quasistatic regime (Fig. 6.9(a)). Cold explosive pressing (CEP) produced considerable work hardening, as is evident from the elevated etchability despite the small change of microhardness from the initial value of 5.00 ± 0.80 GPa to 5.80 ± 0.40 GPa. This can produce qualitative changes in the metastable defect structure in the original rapidly quenched granules, and may cause dramatic structural changes during the subsequent HIPing. The particles are deformed in the quasistatic regime. The transition to the dynamic regime demands pressure in the first shock 5 GPa (Section 2.5) although the regime is close to the beginning of the Mach reflection. The macrostructure is fairly homogeneous in the longitudinal and transverse sections. This is in qualitative agreement with our conclusion (Section 6.1) that transition to the Mach regime develops under a quasistatic type of particle deformation.

The hot isostatic pressing (HIP) of explosively densified material without capsule opening was performed at $T = 1180 \pm 20$ °C and $P = 130 \pm 10$ MPa during 4 h. The material structure after the HIPing of CEP-pressed compacts is presented in Figure 6.9(b). The subsequent HIPing brings the density of the

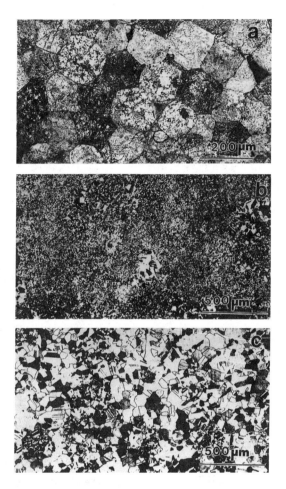

Figure 6.9. The structure of rapidly solidified Ti–33Al alloy after combined treatment: (a) CEP; (b) refined microstructure after CEP + HIP; (c) microstructure after HIP is presented for comparison.

cold explosively densified material close to the theoretical. The sample after traditional HIPing is shown in Figure 6.9(c) for comparison.

It is evident that the microstructure becomes substantially finer than in the direct traditional HIP, with the grain size being reduced from 50–100 to 3–5 μm (compare Figs. 6.9(b) and (c)). The mechanism of this structural change is probably connected with the work hardening during CEP which ensured the primary recrystallization during HIPing.

The tensile strength was determined using standard cylindrical specimens at room temperature. The combination of explosive densification with subsequent HIPing increased the yield strength from 0.47 ± 0.02 GPa (which is typical for HIPing of this alloy) to 0.67 ± 0.02 GPa (Molotkov et al. [1987]; [1991]; Nesterenko et al. [1992], [1993]). The most likely reason for this is reduction in grain size. The characteristics of plasticity for the two types of compacted specimens were similar, and the refining of the grain size unfortunately did not result in the increase of tensile ductility.

In connection with these results, it is necessary to mention that HIPing after explosive consolidation was performed during the time (4 h) typical for the existing industrial scale production process using the HIP unit at the former VILS (Ull-Union Institute for Light Alloys). At the same time, it is clear that the high density of the compact after explosive densification can drastically decrease the heating time in comparison with the powder, having an initial density 55% of solid. That is why the preliminary cold explosive densification can essentially decrease the time of the high-temperature exposure of metastable materials and can be used for rapidly solidified materials or for materials after mechanical alloying (Nesterenko [1989]).

The strong influence of explosive densification on the recrystallization behavior of the rapidly solidified granules Ti–34 Al was explained by creation of the defect system in the crystal lattice, during their high-strain-rate plastic deformation on the structural level, different from that which was originated during the quenching from the melt. Due to the quasistatic regime of particle deformation (Fig. 6.9(a)) the role of shock waves in the particle interior is not important. These experiments demonstrate the principle importance of the plastic deformation of rapidly solidified particles to control the recrystallization during subsequent HIPing.

6.5.5.2   Explosive Densification Versus Milling as Preconditioning for HIPing

It is very interesting to compare two different methods of particles plastic deformation before HIPing: explosive densification (Nesterenko, Lazaridi et al. [1985]; Molotkov et al. [1987], [1991]; Nesterenko et al. [1992], [1993]) and mechanical milling (Morris and Leboeuf [1997]). Some atomized powders in the latter case also contained carbon additives to prevent grain growth during HIPing due to the presence of fine carbides at grain boundaries.

Morris and Leboeuf observed between two and twenty times smaller grain sizes using HIPing of the atomized powders of γ-alloy based on the composition Ti–48Al–2Mn–2Nb after their milling during 10 min and 10 h, correspondingly, with the same carbon content (0.06C). If simple HIPing of these powders produced a 7.5 μm grain size, the additional milling, according to the mentioned regimes, resulted in the grain sizes 3.7 μm and 0.38 μm, correspondingly. The increase in initial carbon content (up to 0.6C) decreased the grain size to 3.9 μm after HIPing; the milling of this material before HIPing, according to the mentioned regimes, decreased the grain size to 2.2 μm (milled 10 min) and to

0.32 μm (milled 10 h). The main conclusion is that milling even for short periods (10 min) induces enough strain in the powders with low carbon content to accelerate recrystallization through the formation of new nuclei. After milling for 10 h the refinement of the grain size shows little dependence on the carbon content of the powders. According to Morris and Leboeuf, this indicates that the driving forces for nucleation and growth of new grains are dominated by the defect structure created during the milling process.

Explosive densification plus HIPing of atomized Ti–34Al powder also resulted in essential grain refinement (Molotkov, Notkin, Elagin, Nesterenko, and Lazaridi [1991]; Nesterenko, Lazaridi, Molotkov, and Notkin [1993]). This structural change increased the yield strength from 0.47 GPa to 0.67 GPa. Unfortunately, no ductility was observed in the tensile test after this combined treatment.

The compression flow stresses $\sigma_{0.2}$ at room temperature also increased dramatically after HIPing of the milled powders of the alloy Ti–48Al–2Mn–2Nb following the Hall–Petch-type relation with the constant $k_y = 1.2$ MPa m$^{1/2}$, similar to one found by other authors for fine γ-duplex microstructures. According to Morris and Leboeuf [1997], this demonstrates that the variations of the strength of these materials is totally determined by the grain size effect. The strain hardening was observed during a compression test of the materials produced from the milled powders, as well as in local plasticity near the cracks during the toughness tests. Further research to enhance the strength at the grain boundary necessary for slip transfer across the grains is desirable to increase the ductility of these materials.

The milling of the atomized powder Ti–48Al–2Mn–2Nb before HIPing resulted in a remarkable increase of hardness: from 2.50 GPa after HIPing to 3.04 GPa and 6.06 GPa after combined HIPing and milling during 10 min and 10 h for powders with a carbon content of 0.06C. The data for a carbon content of 0.6C are correspondingly 2.97 GPa versus 3.29 GPa and 6.36 GPa. The microhardness of the cast alloy with a similar composition is 1.97 GPa.

The electron microscope analysis (TEM) showed that the main part of the material Ti–33Al after traditional HIPing consists of grains of the γ-phase, which have high defect densities, mainly dislocations and stacking faults (Molotkov, Notkin, Elagin, Nesterenko, and Lazaridi [1991]). There are also small amounts of α_2-phase segregation, which varied in morphology (irregular particles, globules, disks, plates), and in size from hundredths of a micron to several microns. In contrast with traditional HIP, the application of the combined CEP + HIP approach leads to recrystallized γ-grains which are almost free from dislocations. At the same time, the α_2-phase is present as enlarged irregular segregations. The larger γ-grains (100–150 μm) occasionally showed a structure analogous to that obtained in direct HIP.

This behavior is in striking agreement with the result of Morris and Leboeuf [1997] which also obtained the dislocation-free γ-grains after HIPing of the atomized and ball-milled powders of the γ-alloy Ti–48Al–2Mn–2Nb for 10 min and 10 h.

Cold plastic deformation of rapidly solidified Ti–6Al–4V powders in a SPEX 8000 impact ball mill, prior to hot isostatic pressing, was employed by Nesterenko and Indrakanti [1999] to ensure the equiaxed final microstructure. The resulting increased compressive yield strength and the microhardness ensured a good ballistic performance of this material under the long-rod penetration test (Nesterenko, Indrakanti, Brar, and Gu [2000a,b]).

Comparison between the two methods of particle deformation before HIPing—explosive loading and mechanical milling of atomized particles—definitely points to the explosive loading method as more industrially preferable than milling, according to the cost and volume of material available after treatment. There are no restrictions to the explosive processing of powders on the scale of 100 kg or more per shot. Additionally, there is no need for a strong compact without cracks after explosive densification accompanied by HIPing. The quasistatic deformation of particles is sufficient to create a desirable defect structure, qualitatively changing the recrystallization behavior of atomized powders during HIPing. An additional benefit of explosive densification can be a decreased time of the high-temperature exposure, due to the higher thermal conductivity of the densified powders if they are HIPed after densification in the same capsule being vacuumed before the explosion (Nesterenko et al. [1993]).

The combination of the cold explosive densification with hot isostatic pressing can be useful for grain refining in some materials where traditional methods are ineffective, like the dispersion-hardened or intermetallic alloys (see, e.g., Grigorovich et al. [1986]).

6.5.6   HIPing of Explosively Densified Ceramics

Explosive modification can also be useful for the post-shock HIPing of ceramic powders. Shang and Meyers [1996] combined shock densification at a low pressure (in a quasistatic regime, just above the threshold for pore collapse) followed by the HIPing of SiC powder. Two particle sizes (44 μm and 7 μm) with and without the addition of the elemental mixture Si + C were investigated. The aim of this approach was to induce a local reaction to increase the temperature of the particle surfaces, thus enhancing a bonding with the help of the reaction products. For explosive densification, the cylindrical double-tube fixture (Meyers and Wang [1988]) was used, which is able to generate shock pressures in powder (6 GPa and 10 GPa) that can be several times higher than those generated with a contact explosion. The well-densified samples were obtained only from 7 μm SiC. They were subjected to subsequent HIPing at 1873 K and 2223 K for 2 h, and an argon pressure of 200 MPa.

A considerable degree of reaction was induced by this process at both temperatures. The HIPing at 1873 K could not collapse the pores resulting from shrinkage due to reaction and the average microhardness of the sample was 14.8 ± 5 GPa. The increase in temperature up to 2223 K resulted in a better bonding of SiC particles by the reaction products.

Explosively densified 7 μm SiC without a reactive mixture of silicon and carbon was HIPed at 1873 K and 2223 K. The HIPing annealed out the lattice defects in the interior of the SiC particles caused by the deformation during shock loading. Unfortunately, the residual intergranular voids with diameters around 0.2 μm were not eliminated by the HIPing. The average microhardness of the compact after HIPing was 26 ± 2 GPa (1873 K), comparable with the hot-pressed SiC (27 GPa, Cercom, Vista). The average microhardness of the compact after HIPing was 26 ± 2 GPa (1873 K), comparable with the hot-pressed SiC (27 GPa, Cercom, Vista). The average compressive strength for the shock/HIP SiC was approximately 300–500 MPa, being essentially less than that for the hot-pressed SiC (1.5–2.1 GPa) or for the sintered silicon carbide (1.2–1.6 GPa, Cercom, Vista). The lower compressive stress of the shock/HIP SiC was connected with a residual macrocracking.

6.6   Hot Explosive Compaction

There are two different reasons to combine preliminary heating of the samples with explosive compaction. The first one is connected with the decrease of the powder yield strength, due to the temperature increase. This can help to decrease pressure for complete densification, and to ensure the dynamic regime of particle deformations with the subsequent bonding (see Section 6.1). Industrial explosives with low-detonation velocities (and shock pressures in the sample) can be used. A second advantage of hot explosive consolidation is connected with the increase of ductility of high-strength materials, which can prevent material cracking at the stage of pore compaction and at the stage of unloading.

Cross [1959] was the first to suggest the combination of preheating with explosive compaction. The apparent technical difficulty of the hot explosive pressing which must be overcome is the potential danger of the contact of the hot sample and explosive. In many approaches, the powder in the sealed capsule is preheated in a high-temperature furnace and after that rapidly placed into an explosive for consolidation (Gorobtsov and Roman [1975]; Bhalla [1980]; Birla and Krishnaswamy [1981]; Taniguchi and Kondo [1988]; Yoshida, Yoshioka, Kimura, Hirabayashi, Sawaoka, and Prummer [1991]).

Wang, Meyers, and Szecket [1988] used the system of preheating the powder (sealed at a 10^{-5} torr vacuum) in the furnace up to 850 °C. Then a steel container with powder slides down to the explosive cylindrical assembly and initiate a detonation. The system used a bottom initiation and an armor plate to trap the specimen. One of the modifications to this approach was developed by Yu and Meyers [1992].

They preheated powder in the furnace, then the flyer plate with explosive and an installed detonator slides down to the assembly and detonates. This system uses the planar impact of the flyer plate to consolidate the powder, and can be applied in the conditions of the proving ground. A more advanced design

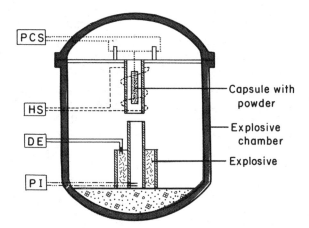

Figure 6.10. Set-up for hot explosive pressing: PCS, pulse current source; HS, heater supply; DE, detonator; PI, position indicator.

combined with double-tube explosive loading, which included initiation of the explosive at the top and the system acceleration downward, was developed by Ferreira, Meyers, Chang, Thadhani, and Kough [1991].

Kecskes and Hall [1995] used combustion synthesis for hot explosive consolidation, to fabricate a high-density sample of the W–Ti alloy suitable for scaling of the model penetrator applications. In this method, the heating is provided by the chemical reaction of a highly exothermic mixture of titanium and carbon powders surrounding the capsule with the W–Ti alloy powder. This allows for the elimination of the electrical furnace and system that moves the hot powder into contact with the explosive. Some disadvantages are connected with the use of a relatively thick thermal isolation between the explosive and powder with a "chemical furnace."

Hot explosive pressing (HEP) was provided in the explosive chamber (Fig. 6.10), allowing experiments to be repeatedly conducted in laboratory conditions (Nesterenko, Lazaridi et al. [1985]; Molotkov, Notkin, Elagin, Nesterenko, and Lazaridi [1991]). The set-up consisted of the explosion chamber, within which there were two coaxial cardboard tubes placed on a layer of sand, and between which the explosive was placed. The simple electrical heater (HS) was also placed in the explosive chamber.

Before compaction, the initially cold capsule, filled with powder containing the granules, was suspended on a wire within the heater, and was brought to the working temperature for the required time after closing the explosive chamber. Then the wire was burned with a current pulse (PCS) and the capsule fell inside the explosive charge, where it closed the contact in the position indicator at the bottom of the explosive charge. The explosive was detonated by applying a high voltage to the detonator (DE). In these experiments, the heater was undamaged by the explosion and can be used repeatedly; the conduction leads, PI and DE, merely require replacement at the end. The sand bed served to prevent the flight of the capsule after the explosion. This method also enables the use of various loading schemes (cylindrical, planar) which can be provided by appropriate shapes in the capsule and the cardboard holders.

Preheating up to 950 °C was used in these experiments. The maximum shock-wave pressure was below 10 GPa, while loading times were about 5 μs.

Analysis of the samples after HEP showed that the material structure demonstrated pronounced radial symmetry and indicated much higher temperatures in the central zones.

Granules close to the capsule's inner surface were represented by 5–10 μm equiaxial γ-grains, as after HIPing at the same temperature (950 °C), and within them there were a few α_2-phase segregations (Molotkov et al. [1991]; Nesterenko [1992]). There was a difference from the structure resulting after simple HIPing. The granules were enclosed in continuous rims of small grains, having a lower aluminum content, very low defect concentrations, and an ordered α_2-phase lattice as indicated by microdiffraction. The α_2-phase in that state is not characteristic of the γ-phase alloys and has not previously been observed in them. The aluminum segregation at the contacts is also observed, perhaps because of extensive shearing and high-velocity gradients in the interface area.

Comparatively larger (10–30 μm) grains appear toward the center (Fig. 6.11). They have the banded relief, characteristic of materials after β → γ-transformation. The transformation occurs in this material at 1350–1400 °C at atmospheric pressure. Such grains appear in the most highly deformed parts, namely at the junctions between three or more granules and in the surface layer (Fig. 6.11(a),(b)). This again demonstrates a highly nonuniform distribution of local temperatures even for a mainly quasistatic regime of particle deformation. Nearer the center, the grains with this relief represented the only structural component of the particles bulk (Fig. 6.11(c),(d)). These grains did not enlarge appreciably in the central part of the specimen, in comparison with the outside part.

Electron microscopy (Molotkov et al. [1991]) showed that the grain structure in conditions of HEP differed considerably from those obtained in the β → γ-transformation at atmospheric pressure. In the latter case, the banded relief was due to γ-twins, along which there were small amounts of thin α_2-layers. In contrast to HEP, the grains consisted almost completely of α_2-plates. The amount of γ-phase was reduced so much that it could not be observed in the grains, and its presence was confirmed only by weak reflections on the electron-diffraction patterns. All the α_2-platelets contained antiphase domains, visible in the dark-field and light-field pictures.

An explanation of the large amount of metastable α_2-phase can be connected to the pressure influence on the phase diagram, and with fast cooling which is too rapid to allow subsequent decomposition. It is possible to assume that the

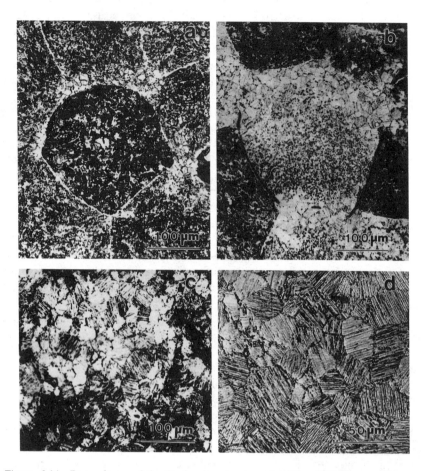

Figure 6.11. Dependence of the final microstructure for Ti + 34%Al granules on a radius of the capsule after hot explosive consolidation at 950 °C: (a), (b) middle point; (c), (d) central zone. Heterogeneous recrystallization of the extensively dynamically deformed granules in the triple points is evident in (a) and (b).

excess of the α_2-phase is formed by $\gamma \rightarrow \alpha_2$ transformation involving a martensite mechanism. Each homogeneous shearing region is bounded by the orientation of the γ-twin plate, so the twin boundaries are inherited by the α_2-phase (Molotkov et al. [1991]; Nesterenko [1992]).

The density obtained with HEP is close to theoretical. There is a considerable portion of intergranular fracture on the failure surface in the peripheral zones of the densified sample. The failure is intragranular at the center of the sample. The recrystallized α_2-rims around the granules show brittle failure, with cleavage in the grain bodies. Intercrystallite failure also predominates in parts having a fine-grained altered structure, and the fracture often shows signs of internal relief in the grains. Chains of α_2-segregations along the boundaries of the altered grains fail by brittle cleavage. A similar failure occurs in large transformed grains (Molotkov et al. [1991]).

These dynamic loading conditions produce considerable temperature gradients in the cross section, with estimated temperature differences of 600 °C for each 5–7 mm. If the alloy can show phase or structural transformations in the temperature range occurring over the radius, the structure will regularly vary along the radius. One can vary the diameter and loading conditions to regulate that temperature gradient, and preheating can be used to adjust the position on the temperature scale. Axisymmetric loading for cylindrical specimens by explosive can be recommended as a rapid method of determining the structures formed over a wide temperature range on pulse loading. A desirable structure type can be produced throughout the material bulk, by adjusting the loading scheme from cylindrical to planar and employing the required loading intensity.

6.7   Shock Densification of Composites

The interesting and unique capability of applying the shock compaction can be connected to composites where the traditional methods of densification are impossible or detrimental to the final properties, because of the chemical interaction between components.

Deribas, Staver, Nesterenko, Lihodid, and Mironov [1975] and Deribas, Nesterenko, and Staver [1976] used an explosive consolidation of a mixture of Ni–Cr alloy with BN to create dense composite layers on the inner surface of stainless steel tubes with a large ratio (up to 10) of length to diameter. Tubes with inner diameter of 50 mm to 500 mm were explosively manufactured. The use of the traditional methods of coating was impossible, according to the geometry, and was due to the large differences in the density of components. Heat treatment of the explosively densified tubes with a coating resulted in very good performance tested in engine. The heat treatment was provided with the steel rod inside, which was removed later.

The interesting features of the mixture's behavior, composed of light BN particles and heavy additives in the tangential shock in the geometry used for practical applications, were published by Deribas, Nesterenko, and Staver [1976].

The results of the material properties and technological steps developed by our group were never published, due to the classification of the patent in 1975, except for the brief discussion in a review paper by Deribas [1986, Fig. 19]. However, after the disintegration of the Soviet Union, the authors of the patent began to live in different countries (Russia and Ukraine) and attempts to declassify this patent and convert it into a normal one became technically impossible. Later the method was reproduced in Minsk in the Institute of Powder Metallurgy. This is one of the known successful applications of explosive consolidation, partially because the final product-part of the engine design (tube with coating) was obtained during the process. Definitely an attempt to manufacture the coating material without a substrate and then to adjust it to the inner surface of tube would fail.

Bhalla and Williams [1977] explore the possibility of manufacturing the stainless steel wire-reinforced aluminum composite by explosive compaction. No evidence was found of strong bonding between the wires and the matrix, probably due to the quasistatic regime of the aluminum particle's deformation. Evacuation of the powder before compaction improved the quality of the compact. The tensile and fatigue properties of the explosively manufactured composites were comparable with other aluminum matrix composites and with high-strength aluminum alloys.

Christman, Heady, and Vreeland [1991] and Tong, Ravichandran, Christman, and Vreeland [1995] processed SiC particulate reinforced titanium matrix composites using shock-wave consolidation. The high reactivity of titanium with most reinforcement materials makes the application of traditional methods based on hot pressing unreliable. Fully dense composite compacts free from interfacial reactions and macroscopic cracks have been obtained. Process variables (starting powder size, dispersion of SiC particles, chemical impurity level, and pre-compaction degassing and heat treatment) have been investigated. Proper post-consolidation annealing has also been explored to improve the impurity segregation and, thus, the ductility of the matrix. Controlled heat treatments of the densified composites can produce desirable interfacial reaction zones sizes. The processed materials are ideal for studying the effects of the interfacial properties on the mechanical behavior of particulate reinforced metal matrix composites. Again, it is necessary to emphasize that it is impossible to expect reliable mechanical properties from composites just after explosive densification, due to the intrinsic heterogeneous character of the deformation of the contact particles. This will definitely result in low ductility. But this example demonstrates that shock consolidation can be a very useful step in composite material's manufacturing.

6.8   Particle Comminution and Activation by Dynamic Loading

Explosive loading can also be very effective for the fragmentation of ceramics and may permit subsequent control of their microstructure through primary recrystallization (Bergmann and Barrington [1966]). These authors demonstrated that very fine particle-sized SiC and B_4C could be produced without introducing impurities, and an increase of surface area after treatment was up to 850% for the SiC particles (initial particle size is 165–406 μm). Another interesting feature mentioned in this paper is that dynamic comminution results in a favorable particle-size distribution permitting closer packing. The ball milling was used only as a method for the deagglomeration of the fine particles produced during the shocking.

The modification and activation of ceramic powders resulting from shock loading is analyzed by Graham [1993]. Significant changes in the specific surface at a relatively low amplitude of shock (e.g., three times for TiB_2), residual strains with values typical for cold-worked metals, reduction of the coherent scattering areas (crystallite size), and large concentration of point defects are typical features of shock-loaded materials. This behavior has a strong effect on the enhancement of chemical reactivity (zirconia + lead oxide), increasing catalytic activity (rutile, zinc oxide), enhanced dissolution rate (silicon nitride) and sinterability (aluminum nitride), and reduction in the transformation temperature (monoclinic to tertragonal conversion of zirconia).

Dynamic loading is especially attractive for the comminution of submicron ceramics. This is difficult to accomplish by the ordinary methods of mechanical fragmentation with a limit on the minimum particle size of 1 μm (Herbst and Sepulveda [1978]).

It was demonstrated by Panin, Nesterenko et al. [1991], Nesterenko, Panin et al. [1992], and Kul'kov, Nesterenko et al. [1993] that the dynamic loading of a nanocrystalline zirconium dioxide powder, obtained by plasmochemical synthesis with a particle size of about 0.3 μm, results in a significant change in the powder morphology and in a decrease of particle size without any increase in the crystallite size. Such loading leads to the appearance of secondary microdeformations of the crystalline lattice of the initially tetragonal phase. As a result, the monoclinic phase is formed, whose appearance is caused by the reduction in the critical grain dimension.

Dynamic comminution can result in submicron particle sizes and does not involve the contact of any particles with foreign materials as is typical for the traditional processes of grinding and milling.

The unique possibility of dynamic comminution can be realized using the optimized shock loading of granular materials with a subsequent high-strain, high-strain-rate plastic flow in a geometry similar to the TWC method test (see Chapter 4). Multiple fracture of ceramic particles (SiC, Alumina) is observed under these conditions especially inside the shear bands (see Figs. 4.20, 4.21, and 4.23). This means that the comminution of ceramic particles can be successfully achieved during confined high-strain flow without high shock pressures (stresses in the used geometry of the TWC method is about 10 kbar). Special precautions should be taken to avoid sintering of particles inside the shear bands instead of comminution (Figs. 4.24 and 4.25).

The introduction of large numbers of defects into the crystalline lattice of ceramic materials and residual strains, accompanied by the reduction of particle sizes, can result in the unique properties of powder unattainable by other methods for different applications.

6.9   Conclusion

The shock loading of powders may result in a simple increase in density, or can be accompanied by the appearance of physical bonding between the particles. The proposed criteria for the strong bonding of powders under shock loading were analyzed in this chapter. This approach, based on the microkinetic energy and scaling shock pressure by initial material strength, is able to explain the existing experimental data on the tensile strength of compacted materials, particularly its independence on initial particle size.

The specific features of the shock loading of powders require pressures a few times higher than the material strength to ensure strong bonding between particles. A strong compact can be obtained using explosive consolidation with tensile strength comparable with the solid. But a highly heterogeneous deformation of particles in the shock wave prevents uniform bonding over the interfaces of contacting particles. This results, as a rule, in a poor ductility of explosively consolidated powders. According to the presented analysis this is an inherent defect of this method, which cannot be eliminated.

The shock loading of powders is a useful instrument in material science to investigate the rapid solidification of different materials including high-pressure conditions. This method has the unique possibility of creating a pressure-quenched amorphous state for materials with an abnormal melting behavior (drop of melting temperature with pressure).

Dynamically densified samples from metastable alloys and composite powders can be useful in research applications.

The advantages of shock densification can be used to densify the initially amorphous materials with preservation of the amorphous state. The final mechanical and magnetic properties of bulk materials consolidated from amorphous foils (or with alternating layers of copper foils) have useful properties for practical application.

Finally, explosive treatment can be considered as a valuable "cold-working" process with an industrial scale possibility for the preconditioning of rapidly solidified powders or composites, before HIPing to refine the final microstructure in metallic alloys and ceramics or for the activation of ceramic powders for various applications including biomaterials.

Problems

1. Find the order of magnitude of the maximal thickness of amorphized alloy under a high-pressure shock wave if the duration of the shock pulse is equal to 1 μs. The thermal conductivity of the alloy is equal to 0.1 cm^2/s.

2. Describe the difference between diaplectic glass and glass obtained from the melt. Which type of glass could be obtained after ball milling?

3. Describe the main differences between pressure-induced glass states obtained in static and shock conditions.

4. What are the main differences between the shock compression of amorphous powder and a stack of amorphous foils, at the same initial densities and the same shock pressure?

5. Find the relation between the thickness of the critical melt in the consolidation of foils and spherical amorphous powders of the same material, where the retention of the amorphous state is possible.

6. Describe possible applications of dynamic densification for the processing of advanced materials.

7. Explain why dynamic densification alone cannot produce materials with a high ductility.

8. Explain why a transition to a dynamic regime of particle deformation is necessary for bonding between particles.

9. What are the main differences between bonding processes in explosive welding and in explosive consolidation?

10. Why is shear localization across a particle bulk rarely observed during the dynamic consolidation of rapidly solidified powders, despite the fact that material of the particle bulk is very susceptible to the localization process?

11. Explain the positive effect on mechanical properties of the initial preheating of powders before dynamic consolidation.

12. What are the advantages of the explosive treatment in comparison with milling, both considered as preconditioning methods before HIPing?

13. Find the critical shear strain resulting in the coalescence of carbide particles on the particle interface. Consider the state of strain as of pure shear with the main elongation along the diagonal and the grains inside the particles initially of cubic shape with carbide particles in the corners. The criterion can be established based on the requirement that the cube diagonal after shear deformation should be equal to the carbide particle diameter.

14. Rewrite the criterion represented by Eq. (6.6) in terms of the shock pressure P (α_0 is the initial porosity defined as the ratio of solid density to the initial density of the powder):

$$P \geq 6\rho_s \left\{ \frac{\kappa\left(c\Delta T_m\right)^2}{a\alpha_0\left(\alpha_0 - 1\right)^{3/2}} \right\}^{2/5}.$$

Neglect the potential energy and the temperature dependence of the particle's heat capacity c.

15. Calculate the shock pressure behind the Mach stem in copper powder with an initial porosity $\alpha_0 = 2$ in the cylindrical geometry of shock compaction without a central mandrel. Consider the wave speed of the Mach stem equal to the detonation speed (6 km/s). Is this enough shock pressure for the copper melting?

References

Ahrens, T.J., and Rosenberg, T. (1968) Shock Metamorphism: Experiments on Quartz and Plagioclase. In: *Shock Metamorphism of Natural Materials.* (Edited by B.M., French, and N.M., Short) Mono Book, Baltimore, pp. 59–81.

Ahrens, T.J., Kostka, D., Vreeland, T. Jr., Schwarz, R.B., and Kasiraj, P., (1984) Shock Compaction of Molibdenum Powder. In: *Shock Waves in Condensed Matter—1983: Proceedings of American Physical Society Topical Conference* (Edited by J.R. Asay, R.A. Graham, and G.K. Straub). Elsevier Science, Amsterdam, pp. 443–446.

Aksenov, P.V., Lizunov, Y.D., Ryazanov, A.I., Nesterenko, V.F., and Pershin, S.A. (1989) Investigation of Energy Dependence of Penetration Depth for He Ions in Microcrystalline Alloy Zr–Nb. *Voprosy Atomnoi Nauki i Tekhniki. Ser. Yaderno-Physichiskie. Issledovanya(Teoriya i Experiment),* 1989, no. 2, p. 107 (in Russian).

Altshuler, L.V. (1965) Application of Shock Waves in Shock Pressure Physics. *Uspehi Phisicheskih Nauk,* **85**, no. 2, pp. 197–258 (in Russian).

Anan'in, A.V., Breusov, O.N., Dremin, A.N., Pershin, S.V., and Tatsii, V.F. (1974) The Effect of Shock Waves on Silicon Dioxide. I. Quartz. *Fizika Goreniya I Vzryva,* **10**, no. 3, pp. 426–436.

Arndt, J. (1984) Shock Isotropization of Minerals. In: *Shock Waves in Condensed Matter—1983: Proceedings of American Physical Society Topical Conference* (Edited by J.R. Asay, R.A. Graham, and G.K. Straub). Elsevier Science, Amsterdam, pp. 473–480.

Asay, J.R., and Hayes, D.B. (1975) Shock-Compression and Release Behavior Near Melt States in Aluminium. *J. Appl. Physics,* **46**, no. 11, pp. 4789–4800.

Atroshenko, E.S. (1984) The Peculiarities of Sintering Kinetics after Explosive Compaction. *Physics and Chemistry of Materials Treatment,* no. 1, pp. 51–56 (in Russian).

Baer, M.R. (2000) Computational Modeling of Heterogeneous Reactive Materials at the Mesoscale. In: *Shock Compression of Condensed Matter—1999* (Edited by M.D. Furnish, L.C. Chhabildas, and R.S. Hixson). AIP, New York, pp. 27–33.

Batsanov, S.S. (1994) *Effects of Explosives on Materials.* Springer-Verlag, New York.

Beauchamp, E.K., and Carr, M.J. (1990) Kinetics of Phase Change in Explosively Shock-Treated Alumina. *J. of the American Ceramic Soc.,* **73**, no. 1, pp. 49–53.

Belyakov, L.V., Valitskii, V.P., Zlatin, N.A., and Mochalov, S.M. (1966) About Melting of Lead in Shock Wave. *Doklady Akademii Nauk SSSR,* **170**, no. 3, pp. 540–543 (in Russian).

Benson, D.J. (1994) An Analysis by Direct Numerical Simulation of the Effects of Particle Morphology on the Shock Compaction of Copper Powder. *Modelling Simul. Mater. Sci. Eng.,* **2**, pp. 535–550.

Benson, D.J., Nesterenko, V.F., and Jonsdottir, F. (1995) Numerical Simulations of Dynamic Compaction. In *Net Shape Processing of Powder Materials,* AMD-**216** (Edited by S. Krishnaswami, R.M. McMeeking, and J.R.L. Trasorras). ASME, New York, pp. 1–8.

Benson, D.J., Nesterenko, V.F., and Jonsdottir, F. (1996) Micromechanics of Shock Deformation of Granular Materials. In: *Shock Compression of Condensed Matter—1995, Proceedings of the Conference of the American Physical Society Topical Group on Shock Compression of Condensed Matter* (Edited by S.C. Schmidt and W.C. Tao). AIP Press, New York, pp. 603–606.

Benson, D.J., Nesterenko, V.F., Jonsdottir, F. and Meyers, M.A. (1998) Quasistatic and Dynamic Regimes of Granular Material Deformation Under Impulse Loading. *J. Mech. Phys. Solids*, **45**, nos. 11/12, pp. 1955–1999.

Bergmann, O.R., and Barrington, J. (1966) Effect of Explosive Shock Waves on Ceramic Powders. *J. of the American Ceramic Soc.*, **49**, no. 9, pp. 502–507.

Bhalla, A.K., and Williams, J.D. (1977) Production of Stainless Steel Wire-Reinforced Aluminium Composite Sheet by Explosive Compaction. *Journal of Materials Science*, **12**, no. 3, pp. 522–530.

Bhalla, A.K. (1980) Hot Explosive Compaction of Metal Powders. *Transactions of Powder Metal Association, India*, **7**, no. 9, pp. 1–8.

Binggeli, N., and Chelikowsky, J.R. (1991) Structural Transformation of Quartz at High Pressures. *Nature*, **353**, pp. 344–346.

Birla, N.C., and Krishnaswamy, W. (1981) Consolidation of Prealloyed Ti–6Al–2Sn–4Zr–2Mo Spherical Powders. *Powder Metallurgy*, **24**, no. 4, pp. 203–209.

Bondar', M.P., and Nesterenko, V.F. (1991) Contact Deformation and Bonding Criteria under Impulsive Loading. *Fizika Goreniya i Vzryva*, **27**, no. 3, pp. 103–117 (in Russian).
English translation: *Combustion, Explosion, and Shock Waves*, November 1991, pp. 364–376.

Bondar', M.P., Kostyukov, N.A., Ovechkin, B.B. et al. (1986) Effect of Explosive Compaction upon the Properties of TiC–TiNi Composition. In: *Proceedings of IX International Conference on High Energy Rate Fabrication* (Edited by V.F. Nesterenko and I.V. Yakovlev). Lavrentyev Institute of Hydrodynamics and Special Design Office of High-Rate Hydrodynamics, Novosibirsk, pp. 141–144 (in Russian).

Bondar', M.P., Nesterenko, V.F., Teslenko, T.S., and Lazaridi, A.N., (1990) Optimization of Heat Treatment Regimes of Explosive Compacts from Rapidly Solidified Granules of Chromium Steel. In: *Impulse Treatment of Materials* (Edited by A.F. Demchuk, V.F. Nesterenko, V.M. Ogolikhin, A.A. Shtertzer, Y.V. Kolotov, S.A. Pershin, and V.I. Danilevskaya). Special Design Ofice of High-Rate Hydrodynamics and Institute of Theoretical and Applied Mechanics, Novosibirsk, pp. 92–102 (in Russian).

Brajkin, V.V., Larichev, V.I., Popova, S.V., and Skrotskaya, G.G. (1986) Metallic Glasses and Amorphous Semiconductors Obtained by Melt Quenching under High Pressures. *Uspehi Phisicheskih Nauk*, **150**, no. 3, pp. 466–467 (in Russian).

Breusov, O.N. (1978) The Possible Role of Metastability in the Processes under Shock Compression. In: *Critical Phenomenon. Physico-Chemical Transformations in Shock Waves*. Institute of Chemical Physics, Chernogolovka, pp. 122–125 (in Russian).

Carslaw, H.S., and Jaeger, J.C. (1988) *Conduction of Heat in Solids*. Clarendon Press, Oxford.

Chaplot, S.L., and Sikka, S.K. (1993) Molecular-Dynamics Simulation of Pressure-Induced Crystalline-to-Amorphous Transition in Some Corner-Linked Polyhedral

Compounds. *Physical Review B (Condensed Matter)*, **47**, no. 10, pp. 5710–5714.

Chen, H., He, Y., Shiflet, G.J., and Poon, S.J. (1994) Deformation-Induced Nanocrystal Formation in Shear Bands of Amorphous Alloys. *Nature*, **367**, no. 6463, pp. 541–543.

Christman, T., Heady, K., and Vreeland, T., Jr. (1991) Consolidation of Ti-SiC Carticle-Reinforced Metal-Matrix Composites. *Scripta Metallurgica et Materialia*, **25**, no. 3, pp. 631–636.

Cross, A. (1959) Try Hot Explosive-Compaction for Sintered Powder Products. *Iron Age*, **184**, no. 26, December, pp. 48–50.

Decarli, P.S., and Jamieson, J.C. (1959) Formation of an Amorphous Form of Quartz under Shock Conditions. *J. Chem. Phys.* **31**, no. 6, pp. 1675–1677.

Deribas, A.A., Staver, A.M., Nesterenko, V.F., Lihodid, E.P., and Mironov, V.M. (1975) Unpublished results.

Deribas, A.A. (1986) Industrial Applications of Explosive Working. In: *Proceedings of IX International Conference on High Energy Rate Fabrication* (Edited by V.F. Nesterenko and I.V. Yakovlev). Lavrentyev Institute of Hydrodynamics and Special Design Office of High-Rate Hydrodynamics, Novosibirsk, pp. 13–39 (in Russian).

Deribas, A.A., Nesterenko, V.F., and Staver, A.M. (1976) The Separation of Components in Explosive Compaction of Multicomponent Materials. In: *Proceedings of III International Symposium on Metal Explosive Working*, Marianske Lazni, Chehoslovakia, Semtin, Pardubice, Chehoslovakia, Vol. **2**, pp. 367–372 (in Russian).

Deribas, A.A., and Staver, A.M. (1974) Shock Compression in Porous Cylindrical Bodies. *Fizika Goreniya i Vzryva*, **10**, no. 4, pp. 568–578 (in Russian).

Deribas, A.A. (1980) *Physics of Explosive Hardening and Welding*. Nauka, Novosibirsk.

Elagin, D.V., Korobov, O.S., Molotkov, A.B. et al. (1986) The Influence of Crystallization Conditions and Technical Treatment on Structure and Phase Content of Ti–Al Alloy. *Metals, Izvestiya of USSR Academy of Sciences*, no. 5, pp. 123–128 (in Russian).

Fan, G.J., Quan, M.X., Hu, Z.Q., Loser, W., and Eckert, J. (1999) Deformation-Induced Microstructural Changes in $Fe_{40}Ni_{40}P_{14}B_6$ Glass. *J. Mater. Res.* **14**, no. 9, pp. 3765–3774.

Ferreira, A. (1989) PhD thesis, New Mexico Tech., Soccorro.

Ferreira, A., Meyers, M.A., and Thadhani, N.N., Chang, S.N., and Kough, J.R. (1991) Dynamic Compaction of Titanium Aluminides by Explosively Generated Shock Waves; Experimental and Materials Systems. *Metallurgical Transactions*, **22A**, pp. 685–695.

Ferreira, A., Meyers, M.A., and Thadhani, N.N. (1992) Dynamic Compaction of Titanium Aluminides by Explosively Generated Shock Waves; Microstructure and Mechanical Properties. *Metallurgical Transactions*, **23A**, pp. 3251–3261.

Frenkel, J. (1955) *Kinetic Theory of Liquids*. Dover, New York, pp. 102, 130.

Friend, C.M., and MacKenzie, P.J. (1987) Fabrication of Multi-Laminar Metallic Glass/Aluminium Composites by Explosive Compaction. *J. Mater. Sci.*, **6**, no. 1, pp. 103–105.

Fujii, Y., and Kowaka, M. (1985) The Pressure Induced Metallic Amorphous State of SnI$_4$: I. A Novel Crystal—to Amorphous Transition Studied by X-Ray Scattering. *J. Phys. C (Solid State Physics)*, **18**, pp. 789–797.

German, R.M. (1996) *Sintering Theory and Practice*. Wiley, New York.

Gorelik, S.S., Rastorguev, L.P., and Skakov, Yu.A. (1960) *X-Ray and Electron Optic Analysis*. Metallurgiya, Moscow, p. 145 (in Russian).

Gorobtsov, V.G., and Roman, O.V. (1975) Hot Explosive Pressing of Powders. *Int. J. Powder Metal., & Powder Technology*, **11**, no. 1, pp. 55–60.

Goswami, R., Sampath, S., Herman, H., and Parise, J.B. (1999) Shock Synthesis of Nanocrystalline Si by Thermal Spraying. *J. Mater. Res.*, **14**, no. 9, pp. 3489–3492.

Gourdin, W.H., Echer, C.L., Cline, C.F., and Tanner, L.E. (1981) Microstructure of Expolosively Compacted Aluminum Nitride Ceramic. Preprint, Lawrence Livermore Laboratory, UCRL-8527.

Gourdin, W.H., (1984a) The Analysis of the Localized Microstructural Changes in Dynamically Consolidated Metal Powders. In: *Proceedings of International Conference on High Energy Rate Fabrication*, San Antonio, NM, pp. 85–92.

Gourdin W.H. (1984b) Energy Deposition and Microstructural Modification in Dynamically Consolidated Metal Powders. *J. Applied Phys.*, **55**, no. 1, pp. 172–181.

Gourdin, W.H. (1984c) Prediction of Microstructural Modification in Dynamically Consolidated Metal Powders. In: *Shock Waves in Condensed Matter—1983: Proceedings of American Physical Society Topical Conference* (Edited by J.R. Asay, R.A. Graham, and G.K. Straub). Elsevier Science, Amsterdam, pp. 379–382.

Graham, R.A. (1993) *Solids Under High-Pressure Shock Compression*. Springer-Verlag, New York.

Gratz, A.J., Nellis, W.J., Christie, J.M., Brocious, W., Swegle, W., and Cordier, P. (1992) Shock Metamorphism of Quartz with Initial Temperatures −170 to +1000 °C. *Phys. Chem. Minerals*, no. 19, pp. 267–288.

Grigorovich, V.K., Sheftel, E.P., Polyakova, I.R., and Mkrtumov, A.S. (1986) The Dispersion Hardening of Sendast. *Izvestiya of USSR Academy of Sciences (Metals)*, no. 4, pp. 144–138 (in Russian).

Hare, A.W., Murr, L.E., and Carlson, F.P. (1984) Implosive Consolidation of a Particle Mass Including Amorphouse Materials. Pat. 4.490.329 USA, IC B22F1/00; B22F1/02, Publ. 12.25. 84.

He, Y., and Schwarz, R.B. (1996) Bulk Amorphous Metallic Alloys: Synthesis by Fluxing Techniques and Properties. *Journal of Metals* (Abstracts of 1997 TMS Annual Meeting), November, p. 31.

Heczko, O., and Ruuskanen, P. (1993) Magnetic Properties of Compacted Alloy Fe73.5Cu7Nb3Si13.5B9 in Amorphous and Nonocrystalline State. *IEEE Transactions on Magnetics*, **29**, no. 6, pt. 1, pp. 2670–2672.

Hemley, R.J., Jephcoat, A.P., Mao, H.K., Ming, L.C., and Manghnani, M.H. (1988) Pressure-Induced Amorphization of Crystalline Silica. *Nature*, **334**, no. 6177, pp. 52–54.

Herbst, J.A., and Sepulveda, J.L. (1978) Fundementals of Fine and Ultrafine Grinding in a Stirred Ball Mill. In: *Proceedings of the International Powder and Bulk Solids Handling and Processing Conference, Rosemont, Illinois, May 16–18, Industrial and Scientific Conference Management*, Chicago, Illinois, pp. 452–470.

Inoe, A. (1996) Ferromagnetic Bulk Amorphous Alloys. *Journal of Metals* (Abstracts of 1997 TMS Annual Meeting), November, p. 31.

Ishakov, R.S., Kirko, V.I., Kuzovnikov, A.A. et al. (1984) Investigation of the Structure of the Bulk Amorphous Ferromagnetic Alloy $Co_{58}Ni_{10}Fe_5B_{16}Si_{11}$ Obtained by Explosive Compaction Based on the Local Magnetic Anisotropy Characteristics. Preprint 265f, Kirenskii Institute of Physics. Siberian Branch USSR Academy of Sciences, Krasnoyarsk, 31 pp. (in Russian).

Jain, M., and Christman, T. (1994) Synthesis, Processing, and Deformation of Bulk Nanophase Fe–28Al–2Cr Intermetallic. *Acta Metallurgica et Materialia*, 42, no. 6, pp. 1901–1911.

Jain, M., and Christman, T. (1996) Formation of Fine Nanocrystalline Microstructure in Bulk Fe–28Al–2Cr. *Nanostructured Materials*, no. 7, pp. 719–723.

Kasiraj, P., Vreeland, T. Jr., Schwarz, R.B., and Ahrens T.J. (1984a) Shock Consolidation of a Rapidly Solidified Steel Powder. *Acta Metall.*, 32, no. 8, pp. 1235–1241.

Kasiraj, P., Vreeland, T. Jr., Schwarz, R.B., and Ahrens, T.J. (1984b) Mechanical Properties of a Shock Consolidated Steel Powder. In: *Shock Waves in Condensed Matter—1983: Proceedings of American Physical Society Topical Conference* (Edited by J.R. Asay, R.A. Graham, and G.K. Straub). Elsevier Science, Amsterdam, pp. 439–442.

Kecskes, L.J., and Hall, I.W. (1995) Hot Explosive Consolidation of W–Ti Alloys. *Metallurgical and Materials Transactions A*, 26A, no. 9, pp. 2407–2414.

Kingma, K.J., Meade, C., Hemley, R.J., Ho-Kwang Mao, Veblen, D.R. (1993) Microstructural Observations of Alpha-Quartz Amorphization. *Science*, 259, no. 5095, pp. 666–669.

Kondo, K. (1997) Magnetic Response of Powders to Shock Loading and Fabrication of Nanocrystalline Magnets. In: *High-Pressure Shock Compression of Solids IV. Response of Highly Porous Solids to Shock Loading* (Edited by L. Davison, Y. Horie, and M. Shahinpoor). Springer-Verlag, New York, pp. 309–330.

Kormer, S.B., Sinitsyn, M.V., Kirillov, G.A., and Urlin, V.D. (1965) Exprimental Determination of Shock Temperatures of NaCl and KCl and Their Melting Curves up to Pressures 700 kbar. *J. Exp. Theor. Physics*, 48, no. 4, pp. 1038–1049 (in Russian).

Korth, G.E., and Williamson, R.L. (1995) Dynamic Consolidation of Metastable Nanocrystalline Powders. *Metall. Mater. Transactions A*, 26A, October, pp. 2571–2578.

Kostyukov, N.A., and Kuz'min, G.E. 1986) Criteria of Occurrence of "Central-Zone"-Type Macroinhomogeneities in the Shock-Wave Loading of Porous Media. *Fizika Goreniya i Vzryva*, 22, no. 5, pp. 87–96 (in Russian).
English translation: *Combustion, Explosion, and Shock Waves*, 1987, March, pp. 573–581.

Kruger, M.B. (1990) Memory Glass; an Amorphous Material Formed from $AlPO_4$. *Science*, 249, pp. 647–649.

Kul'kov, S.N., Nesterenko, V.F., Bondar', M.P., Simonov, V.A., Mel'nikov, A.G., and Korolev, P.V. (1993) Explosion Activation of Quench-Hardened ZrO_2–Y_2O_3 Ceramic Submicron Powders. *Fizika Goreniya i Vzryva*, 29, no. 6, pp. 66–72 (in Russian).
English translation: *Combustion, Explosion, and Shock Waves*, 1993, 29, no. 6, pp. 728–733.

Lazaridi, A.N. (1990) The Influence of Initial Characteristics of Steel Granules and Loading Regimes on the Compact Strength. In: *Impulse Treatment of Materials* (Edited by A.F. Demchuk, V.F. Nesterenko, V.M. Ogolikhin, A.A. Shtertzer, Y.V. Kolotov, S.A. Pershin, and V.I. Danilevskaya). Special Design Ofice of High-Rate Hydrodynamics and Institute of Theoretical and Applied Mechanics, Novosibirsk, pp. 70–86 (in Russian).

Lenon, C.R.A., Williams, J.D., and Bhalla, A.K. (1978) Explosive Compaction of Metal Powders. *Powder Metallurgy*, **21**, no. 1, pp. 29–34.

McQueen, R., and Marsh, S.P. (1960) Equation of State for Nineteen Metallic Elements from Shock-Wave Measurements to Two Megabars. *J. Appl. Phys.* **31**, no. 7, pp. 1253–1269.

Meyers, M.A., Gupta, B.B., and Murr, L.E. (1981) Shock-Wave Consolidation of Rapidly Solidified Superalloy Powders. *Journal of Metals,* October, pp. 21–26.

Meyers, M.A., and Wang, S.L. (1988) An Improved Method for Shock Consolidation of Powders. *Acta Metall.*, **36**, no. 4, pp. 925–936.

Meyers, M.A. (1994) *Dynamic Behavior of Materials.* Wiley, New York.

Meyers, M.A., Benson, D.J., and Olevsky, E.A. (1999) Shock Consolidation: Microstructurally-Based Analysis and Computational Modeling. *Acta Materialia*, **47**, no. 7, pp. 2089–2108.

Mineev, V.N., and Savinov, V.V. (1967) Viscosity and Melting Temperature of Al, Pb and NaCl under Shock Compression. *J. Exp. Theor. Physics*, **52**, no. 3, pp. 629–636.

Molotkov, A.V., Notkin, A.B., Elagin, D.V., Nesterenko, V.F., and Lazaridi, A.N. (1987) Investigation of the Microstructure and Mechanical Properties of Compacts after Explosive Compactioin of Ti–34Al Alloy. In: *Metallurgiya Granul, Extended Abstracts of the Second All Union Conference on Metallurgy of Granules,* Ull-Union Institute for Light Alloys, Moscow, pp. 201–203 (in Russian).

Molotkov, A.V., Notkin, A.B., Elagin D.V., Nesterenko V.F., and Lazaridi, A.N. (1991) Microstructure after Heat Treatment for Explosive Compacts Made from Granules of Rapidly Quenched Titanium Alloys. *Fizika Goreniya i Vzryva*, **27**, no. 3, pp. 117–126 (in Russian).
English translation: *Combustion, Explosion, and Shock Waves,* November 1991, pp. 377–384.

Morris, D.G. (1980) Compaction and Mechanical Properties of Metallic Glass. *Metal Sci.*, **14**, June, pp. 215–220.

Morris, D.G. (1981) Melting and Solidification During Dynamic Compaction of Tool Steel. *Metal Sci.,* March, pp. 116–124.

Morris, D.G. (1982a) Rapid-Solidification Phenomena. *Metal Sci.,* October, pp. 457–466.

Morris, D.G. (1982b) The Properties of Dynamically Compacted Metglass 2826. *J. Mater. Sci.*, **17**, pp. 1789–1894.

Morris, M.A., and Leboeuf, M. (1997) Grain-Size Refinement of γ–Ti–Al Alloys; Effect on Mechanical Properties. *Materials Science and Engineering*, **A224**, pp. 1–11.

Negishi, T., Ogura, T., Ishii, H. et al. (1985) Dynamic Compaction of Amorphouse $Ni_{75}Si_8B_{17}$ and $Pd_{78}Cu_6Si_{16}$ Alloys. *Mater. Sci.*, no. 20, pp. 299–306.

Nellis, W.J., Maple, M.B., and Geballe, T.H. (1988) Synthesis of Metastable Superconductors by High Dynamic Pressure. In: *Proceedings of the SPIE—The International Society for Optical Engineering, Multifunctional Materials*, Los Angeles, CA, 11–12 Jan. 1988, Vol. **878**, pp. 2–9.

Nesterenko, V.F. (1975) Electrical Effects under Shock Loading of Metals Contact. *Fizika Goreniya i Vzryva* **11**, 444–456 (in Russian).
English translation: *Physics of Explosion, Combustion and Shock Waves*, 1976, July, **11**, pp. 376–385.

Nesterenko, V.F. (1983a) Thermodynamics of Shock Compression of Porous Materials. In: *High Pressure in Science and Technology: Proceedings of IX AIRAPT International High Pressure Conference*, New York, pp. 195–198.

Nesterenko, V.F. (1983b) Scope for Supercooled Melts by a Dynamic Method. *Fizika Goreniya i Vzryva*, **19**, no. 5, pp. 145–149 (in Russian).
English translation: *Combustion, Explosion, and Shock Waves*, March 1984, pp. 665–667.

Nesterenko, V.F. (1983c) Method to Obtain Supercooled Melts by a Shock Pressure. USSR Patent 176567 (unpublished).

Nesterenko, V.F. (1984) On Possibility of Obtaining Supercooleed Melts by Dynamic Methods. In: *Proceedings of International Conference on High Energy Rate Fabrication*, San Antonio, NM, pp. 133–135.

Nesterenko, V.F. (1985) Potential of Shock-Wave Methods for Preparing and Compacting Rapidly Quenched Materials. *Fizika Goreniya i Vzryva*, **21**, no. 6, 85–98 (in Russian).
English translation: *Physics of Explosion, Combustion and Shock Waves*, 1986, May, pp. 730–740.

Nesterenko, V.F., and Muzykantov A.V. (1985) Evaluation of Conditions for Retaining an Amorphous Material Structure During Consolidation by Explosion. *Fizika Goreniya i Vzryva*, **21**, no. 2, pp. 120–126 (in Russian).
English translation: *Combustion, Explosion, and Shock Waves*, September 1985, pp. 240–245.

Nesterenko V.F., Lazaridi, A.N. et al. (1985) Explosive Consolidation of Powders of Ti–33Al Alloy. Final Report #GR 01850003818, Special Design Office of High-Rate Hydrodynamics, Siberian Branch of the USSR Academy of Sciences, Novosibirsk, 44 p.

Nesterenko, V.F. (1986) Heterogeneous Heating of Porous Materials at Shock Wave Loading and Criteria of Strong Compacts. In: *Proceedings of IX International Conference on High Energy Rate Fabrication* (Edited by V.F. Nesterenko and I.V. Yakovlev). Lavrentyev Institute of Hydrodynamics and Special Design Office of High-Rate Hydrodynamics, Novosibirsk, pp. 157–163 (in Russian).

Nesterenko V.F., Pershin, S.A., Hinskii, A.P. et al. (1987) Explosive Consolidation of Rapidly Solidified Materials for Electromagnetic Applications. Final Report #GR 01860022255, Special Design Office of High-Rate Hydrodynamics. Siberian Branch of the USSR Academy of Sciences, Novosibirsk, 62 pp.

Nesterenko, V.F. (1988a) Micromechanics of Powders under Strong Impulse Loading. In: *Computer Methods in Theory of Elasticity and Plasticity: Proceedings of X All-Union Conference* (Edited by F.M. Fomin). Institute of Theoretical and Applied Mechanics, Novosibirsk, pp. 212–220.

Nesterenko, V.F. (1988b) Influence of the Parameters of Powder Internal Structure on the Process of Explosive Compaction. In: *Proceedings of International*

Symposium on Metal Explosive Working. Semtin, Purdubice, Chehoslovakia, Vol. **3**, pp. 410–417 (in Russian).

Nesterenko, V.F. (1988c) Nonlinear Phenomena under Impulse Loading of Heterogeneous Condensed Media. Doctor in Physics and Mathematics Thesis, Academy of Sciences, Russia. Lavrentyev Institute of Hydrodynamics, Novosibirsk, Siberian Branch.

Nesterenko, V.F., Lazaridi, A.N., Belyev, N.V., and Pestov, A.M. (1989) Explosive Compaction of Electric Motor Rotors from Hard-Magnetic Materials. In: *Procdings of the X International Conference on High-Energy-Rate Fabrication*, (Edited by S. Petrovich). Yugoslavia, Ljubliana, pp. 86–91(in Russian).

Nesterenko, V.F., Pershin, S.A., Tabachnikova, E.D., and Gorshkov, N.N. (1989) Strength Characteristics of Bulk Amorphous Materials. In: *Procedeengs of International Seminar on High-Energy Working of Rapidly Solidified Materials and High-T_C Ceramics* (Edited by V.F. Nesterenko and A.A. Shtertzer). Special Design Office of High-Rate Hydrodynamics, Institute of Theoretical and Applied Mechanics, Novosibirsk, pp. 113–117 (in Russian).

Nesterenko, V.F., Lazaridi, A.N., Pershin, S.A., Miller, V.Ya., Feschiev, N.H., Krystev, M.P., Minev, R.M., and Panteleeva, D.B. (1989) Propeties of Compacts from Rapidly Solidified Granules of Different Sizes after Shock-Wave Consolidation. In: *Proceedeings of the International Seminar on High-Energy Working of Rapidly Solidified Materials and High-T_C Ceramics* (Edited by V.F. Nesterenko and A.A. Shtertzer). Special Design Office of High-Rate Hydrodynamics, Institute of Theoretical and Applied Mechaniucs, Novosibirsk, pp. 118–126 (in Russian).

Nesterenko, V.F., Avvakumov, E.G., Pershin, S.A., Kormilitsyna, Z.A, Lazaridi, A.N., and Yazvitskii, M.Yu. (1989) Shock-Wave Compaction of Mechanically Activated Powder of the System Fe–Nd–B. *Fizika Goreniya i Vzryva*, **25**, no. 5, pp. 148–150 (in Russian).
English translation: *Combustion, Explosion, and Shock Waves*, March 1990, pp. 656–658.

Nesterenko, V.F., and Lazaridi, A.N. (1990) Regimes of Shock-Wave Compaction of Granular Materials. In: *High Pressure Science and Technology*. Gordon and Breach, New York, Vol. **5**, pp. 835–837.

Nesterenko, V.F., Pershin, S.A., Farmakovskii, B.V., Khinskii, A.P., Zolotarev, S.N., Usishchev, N.A., and Novikov, S.N. (1991) Explosive Compaction of Electronic Device Components Made of Amorphous Alloys. *Fizika Goreniya i Vzryva*, **27**, no. 4, p. 104–109 (in Russian).
English translation: *Combustion, Exlosion, and Shock Waves*, January 1992, pp. 485–489.

Nesterenko, V.F. (1992) *High-Rate Deformation of Heterogeneous Materials*. Nauka, Novosibirsk (in Russian).

Nesterenko, V.F., Panin, V.E., Kulkov, S.N., and Melnikov, A.G. (1992) Modification of Submicron Ceramics under Pulse Loading. *High Pressure Research*, **10**, pp. 791–795.

Nesterenko, V.F., Lazaridi, A.N., Molotkov A. V., Notkin A.B., and Elagin D.V. (1993) Method of Fabrication Compacts from Titanium-Aluminum Alloy. Russian patent #1464378, 13 October.

Nesterenko, V.F. (1995) Dynamic Loading of Porous Materials: Potential and Restrictions for Novel Materials Applications. In: *Metallurgical and Materials Applications of Shock-Wave and High-Strain-Rate Phenomena, Proceedings of the 1995 International Conference EXPLOMET-95* (Edited by L.E. Murr, K.P. Staudhammer, and M.A. Meyers). Elsevier Science, Amsterdam, pp. 3–13.

Nesterenko, V.F., and Indrakanti, S.S. (1999) Tailoring of Microstructure of Ti–6Al–4V Alloy by Combined Cold Plastic Deformation and Hot Isostatic Pressing". *Constitutive and Damage Modeling of Inelastic Deformation and Phase Transformation, Proceedings of Plasticity'99* (Edited by A.S. Khan). Cancun, Mexico. NEAT Press, Fulton, MD, pp. 251–254.

Nesterenko, V.F., Indrakanti, S.S., Brar, S. and Gu, Y. (2000a) Long Rod Penetration Test of Hot Isostatically Pressed Ti-Based Targets. In: *Shock Compression of Condensed Matter—1999* (Edited by M.D. Furnish, L.C. Chabildas, and R.S. Hixson). AIP, New York, pp. 419–422.

Nesterenko, V.F., Indrakanti, S.S., Brar, S. and Gu, Y. (2000b) Ballistic Performance of Hot Isostatically Pressed (HIPed) Ti-Based Targets. In: *Key Engineering Materials*, Vols. **177–180**, Trans Tech Publications, Switzerland, pp. 243–248.

Olevsky, E.A., (1998) Theory of Sintering: from Discrete to Continuum. *Materials Science & Engineering: Reports*, **23**, no. 2, pp. 41–99.

Oreshin, N.V., Pashkov, O.P., and Tkachev, R.K. (1985) Investigation of the Structure and Properties of Compacts from Ti-based Powder after Shock Wave Consolidation and Subsequent Syntering. In: *Metallovedenie i Prochnost Materialov*. Volgograd Polytechnical Institute, Volgograd, pp. 59–65 (in Russian).

Otooni, M.A. (1998) *Elements of Rapid Solidification: Fundamentals and Applications*. Springer-Verlag, Berlin.

Panin, V.E., Nesterenko, V.F., Kulkov, S.N., and Melnikov, A.G. (1991) Crushing of Submicron Ceramic Powder by Impulse Loading, *Fizika Goreniya i Vzryva*, July–August, **27**, no. 4, p. 140 (in Russian).

Peker, A., and Johnson, W.L. (1993) A Highly Processable Metallic Glass: $Zr_{41.2}Ti_{13.8}Cu_{12.5}Ni_{10.0}Be_{22.5}$. *Appl. Phys. Lett.*, **63**, no. 17, pp. 2342–2344.

Pershin, S.A., and Nesterenko, V.F. (1988) Shear Strain Localization with Pulsed Compaction of Rapidly Quenched Alloy Foils. *Fizika Goreniya i Vzryva*, **24**, no. 6, pp. 120–123 (in Russian).
English translation: *Combustion, Explosion, and Shock Waves*, May 1989, pp. 752–755.

Prummer, R. (1983) Powder Compaction. In: *Explosive Welding, Forming and Compaction* (Edited by T.Z. Blazynski). London, Applied Science, pp. 381–400.

Prummer, R., and Klemm, W. (1986) Massive Parts of Metallic Glasses Made by Explosive Liquid Phase Sinter Treatment. In: *Horizons of Powder Metallurgy, Proceedings of International Powder Metallurgy Conference PM 86*, Dusseldorf, July 7–11, pp. 845–850.

Prummer, R. (1987) *Explosivverdichtung Pulvriger Substanzen (Grundlagen, Verfahren, Ergebnisse)*. Springer-Verlag, Berlin (in German).

Prummer, R. (1988) Explosive Compaction of Metallic Glass Powders. *Mater. Sci. and Engineering*, **98**, pp. 461–463.

Psahie, S.G., Korostelev, S.Yu., and Panin, V.E. (1988) About Forming of Domains with Disordered Structure under Shock Propagation in Crystal. *Letters to Journal of Technical Physics*, **14**, no. 12, pp. 1645–1648 (in Russian).

Rakhimov, A.E. (1993) Optical Microstructure of Explosively Compacted Ribbon Toroids from Fe-based Amorphous Alloy. *J. Mater. Sci. Lett.*, **12**, pp. 1891–1893.

Raybould, D. (1980) The Cold Welding of Powders by Dynamic Compaction. *Int. J. Powder Metallurgy and Powder Technology*, **16**, no. 1, pp. 1–10.

Raybould, D. (1981) The Properties of Stainless Steel Compacted Dynamically to Produce Cold Interparticle Welding. *J. Mat. Sci.*, **16**, pp. 589–598.

Raybould, D. (1987) Cold Dynamic Compaction of Pre-Alloyed Titanium and Activated Sintering. In: *New Perspective in Powder Metallurgy (Fundamentals, Methods, and Applications)*, Powder Metallurgy for Full Density Products. Metal Powder Industries Federation, Princeton, NJ, Vol. **8**, pp. 575–589.

Roman, O.V., Gorobtsov, V.G., and Mitin, V.S. (1982) Structure and Properties of Fe-based Amorphous Materials. *Poroshkovaya Metallurgiya* (Minsk), no. 6, pp. 8–13 (in Russian).

Roman, O.V., Bogdanov, A.P., Voloshin, Yu.N. et al. (1983) Structure and Properties of Amorphouse Powders after Explosive Loading. *Metallovedenie I Termicheskaya Obrabotka Metallov*, no. 10, pp. 57–59 (in Russian).

Roman, O.V., Gorobtsov, V.G., Pikus, I.M., Boltuts, D. Yu., and Mirilenko, A.P. (1986) The Analysis of the Processes of Shock Loading of Rapidly Solidified Materials. In: *Procedeengs of IX International Conference on High Energy Rate Fabrication* (Edited by V.F. Nesterenko and I.V. Yakovlev). Lavrentyev Institute of Hydrodynamics, Special Design Office of High-Rate Hydrodynamics, Novosibirsk, pp. 179–182 (in Russian).

Salli, I.V. (1972) *Crystallization at Ultrahigh Cooling Rates*. Naukova Dumka, Kiev (in Russian).

Schwarz, R.B., Kasiraj, P., Vreeland, T.Jr., and Ahrens T.J. (1984a) A Theory for the Shock-Wave Consolidation of Powders. *Acta Metall.*, **32**, no. 8, pp. 1243–1252.

Schwarz, R.B., Kasiraj, P., Vreeland, T. Jr., and Ahrens, T.J. (1984b) The Effect of Shock Duration on the Dynamic Consolidation of Powder. In: *Shock Waves in Condensed Matter—1983: Proceedings of American Physical Society Topical Conference* (Edited by J.R. Asay, R.A. Graham, and G.K. Straub). Elsevier Science, Amsterdam, pp. 435–438.

Sekhar, J.A., Mohan, M., Divaker, C., and Singh, A.K. (1984) Rapid Solidification by Application of High Pressure. *Scripta Metall*, **18**, no. 1, pp. 1327–1330.

Sekhar, J.A. (1986) Rapid Pressure Application During Solidification. In: *Metallurgical Applications of Shock-Wave and High-Strain-Rate Phenomena* (Edited by L.E. Murr, K.P. Staudhammer, and M.A. Meyers). Marcel Dekker, New York, pp. 1083–1084.

Sekine, T. (1997) Shock Synthesis of Materials. In: *High-Pressure Shock Compression of Solids IV. Response of Highly Porous Solids to Shock Loading* (Edited by L. Davison, Y. Horie, and M. Shahinpoor). Springer-Verlag New York, pp. 289–308.

Shang, S.-S., and Meyers, M.A. (1991) Shock Densification/Hot Isostatic Pressing of Titanium Aluminide. *Metallurgical Transactions A*, **22A**, no. 11, pp. 2667–2676.

Shang, S.-S., and Meyers, M.A. (1996) Dynamic Consolidation/Hot Isostatic Pressing of SiC. *Journal of Materials Science*, **31**, no. 1, pp. 252–261.

Shtertser, A.A. (1988) Transmission of Pressure in Porous Media under Explosive Loading. *Fizika Goreniya i Vzryva*, **24**, no. 5, pp. 113–119 (in Russian).

English translation: *Combustion, Explosion, and Shock Waves*, 1988, **24**, no. 5, pp. 610–615.

Shtertser, A.A. (1993) Effect of the Particle Surface State on the Particle Consolidation in Explosive Compacting of Powdered and Granular Materials. *Fizika Goreniya i Vzryva*, **29**, no. 6, pp. 72–78 (in Russian).

English translation: *Combustion, Explosion, and Shock Waves*, 1993, **29**, no. 6, pp. 734–739.

Sikka, S.K., and Sharma, S.M. (1992) Close Packing and Pressure-Induced Amorphization. *Current Science*, **63**, no. 6, pp. 317–320.

Sikka, S.K., Sharma, S.M., and Chidambaram, R. (1994) Steric Constraints: A Powerful Criterion to Predict the Onset of Phase Transitions in Molecular Solids under Pressure. In: *High-Pressure Science and Technology-1993* (Edited by S.C., Schmidt, J.W., Shaner, G.A., Samara, and M., Ross). AIP Press, New York, pp. 213–216.

Sivakumar, K., Bhat, T.B., and Ramakrishnan, P. (1996) Dynamic Consolidation of Aluminium and Al-20 V/o SiCp Composite Powders. *Journal of Materials Processing Technology*, **62**, pp. 191–198.

Somayazulu, M.S., Sharma, S.M., Sikka, S.K., Garg, N., and Chaplot, S.L. (1994) Molecular Dynamical Calculations of α-Quartz Implications for Shock Results. In: *High-Pressure Science and Technology—1993* (Edited by S.C., Schmidt, J.W., Shaner, G.A., Samara, and M., Ross). AIP Press, New York, pp. 815–818.

Staudhammer, K.P. (1999) Fundamental Temperature Aspects in Shock Consolidation of Metal Powders. In: *Powder Materials: Current Research and Industrial Practices*, (Edited by F.D.S. Marquis). The Minerals, Metals and Materials Society, Warrendale, PA, pp. 317–325.

Staver, A.M. (1981) Metallurgical Effects under Shock Compression of Powder Materials. In: *Shock Waves and High-Strain-Rate Phenomena in Metals. Concepts and Applications* (Edited by M.A. Meyers, L.E. Murr). Plenum Press, New York, pp. 865–880.

Stishov, S.M. (1968) Melting at High Pressures. *Uspehi Phisicheskih Nauk*, **96**, no. 3, pp. 467–496 (in Russian).

Stolyarov, V.V., Zhu, Y.T., Lowe, T.C., Islamgaliev, and R.K., Valiev, R.Z. (2000) Processing Nanocrystalline Ti and its Nanocomposites from Micrometer-Sized Ti Powder Using High Pressure Torsion. *Materials Science and Engineering A*, **A282**, nos. 1–2, pp. 78–85.

Susic, M.V., Minelic, B., and Uskokovic, D.P. (1990) Crystallization Kinetics and Thermal Stability of Shock-Compacted Amorphous Ni78P22. *Materials Letters*, **9**, nos. 5/6, pp. 215–218.

Tabachnikova, E.D., Diko, P., Miskuf, J., Csach, K., Nesterenko, V.F., and Pershin, S.A. (1990) Strength Characteristics and Special Features of Failure of Volume Amorphous Alloys in the Temperature Range 300–4.2 K. *Kovove Materialy*, 1990, **28**, no. 4, pp. 386–395 (in Czech).

English translation: *Metallic Materials*, 1990, **28**, no. 4, pp. 222–226.

Takagi, M., Kawamura, Y., Araki, M., Kuroyama, Y., and Imura, T. (1988) Preparation of Bulk Amorphous Alloys by Explosive Consolidation and Properties of the Prepared Bulk. *Materials Science and Engineering*, **98**, pp. 457–460.

Taniguchi, T., and Kondo, K. (1988) Hot Shock Compaction of Alpha–Alumina Powder. *Advanced Ceramic Materials*, **3**, no. 4, pp. 399–402.

Tong, W., Ravichandran, G., Christman, T., and Vreeland, T., Jr. (1995) Processing SiC-Particulate Reinforced Titanium-Based Metal Matrix Composites by Shock Wave Consolidation. *Acta Metallurgica et Materialia*, **43**, no. 1, pp. 235–250.

Tonkov, E.Yu. (1979) *Phase Diagrams of Elements at High Pressures*. Nauka, Moscow (in Russian).

Tse, J.S., and Klug, D.D. (1991) Mechanical Instability of α-Quartz; A Molecular Dynamics Study. *Phys. Rev. Lett.*, **67**, no. 25, pp. 3559–3562.

Tunaboylu, B., McKittrick, J., Nellis, W.J., and Nutt, S. (1994) Dynamic Compaction of Al_2O_3-ZrO_2 Compositions. *J. Am. Ceram. Soc.*, **77**, no. 6, pp. 1605–1612.

Valiev, R.Z., Lowe, T.C., and Mukherjee, A.K. (2000) Understanding the Unique Properties of SPD-Induced Microstructures. *Journal of Metals*, **52**, no. 4, pp. 37–40.

Vertman, A.A., Epanchintsev, O.G., Zvezdin, Y.I., Nesterenko, V.F., Pershin, S.A., Revdel, M.P., and Rodina, T.S. (1989) Preparation of High Coercivity Materials of the System Mn–Al–C by Explosive Compaction. *Fizika Goreniya i Vzryva*, Nov.-Decem. 1989, **25**, no. 6, pp. 120–124 (in Russian). English translation: *Combustion, Explosion, and Shock Waves*, May 1990, pp. 772–775.

Vreeland, T., Kasiraj, P., Ahrens, T., and Schwarz, R.B. (1983) Shock Consolidation of Powder—Theory and Experiment. In: *Proceedings of Mat. Res. Soc. Annual Meeting*, Boston, pp. 18–25.

Vukcevic, M., Glisic, S., and Uskokovic, D. (1993) Dynamic Compaction of Al–Li–X Powder Obtained by a Rotating Electrode Process. *Materials Science and Engineering*, **A168**, pp. 5–10.

Wang, S.L., Meyers, M.A., and Szecket, A. (1988) Warm Consolidation of IN 718 Powder. *J. Mat. Science*, **29**, pp. 1786–1804.

Wilkins, M.L. (1984) Dynamic Powder Compaction. In: *Proceedings of International Conference on High Energy Rate Fabrication*, San Antonio, NM, pp. 63–69.

Yoshida, M., Yoshioka, Y., Kimura, Y., Hirabayashi, H., Sawaoka, A.B., and Prummer, R.A. (1991) In: *Aerospace, Refractory and Advanced Materials. Advances in Powder Metallurgy*. Metal Powder Industries Federation, Princeton, NJ, Vol. **6**, pp. 199–209.

Yu, L.H., and Meyers, M.A. (1992) Shock Synthesis of Silicides. In: *Shock-Wave and High-Strain-Rate Phenomena in Materials, Proceedings of the 1995 International Conference EXPLOMET—90*, San Diego (Edited by M.A. Meyers, L.E. Murr, and K.P. Staudhammer). Marcel Dekker, New York, pp. 303–309.

Zhang, T., Inoue, and A., Masumoto, T. (1991) Amorphous Zr–Al–TM (TM–Co,Ni,Cu) Alloys with Significant Supercooled Liquid Region of Over 100 K. *Materials Transactions, JIM*, **32**, no. 11, pp. 1005–1010.

Zolotarev, S.N., Denisova, O.N., Nesterenko, V.F., Pershin, S.A., and Lazaridi, A.N. (1989) Explosive Compacts from Amorphous Soft-Magnetic Alloys. In: *Proceedeings of the International Seminar on High-Energy Working of Rapidly Solidified Materials and High-T_C Ceramics* (Edited by V.F. Nesterenko and A.A. Shtertzer). Special Design Office of High-Rate Hydrodynamics, Institute of Theoretical and Applied Mechaniucs, Novosibirsk, pp. 97–112 (in Russian).

Afterword

To conclude this book, I would like to outline what are the probable next steps in the corresponding areas of mesomechanics of heterogeneous materials in the attempt to predict further developments. Very often, such predictions do not materialize but can be of some help especially for young researchers.

Usually, the phenomena that are sensitive to the mesoscale cannot be straightforwardly predicted based solely on the conservation laws. The classical example illustrating this statement is a strongly nonlinear soliton in the 1-D granular material composed by collective motion of five particles resulting from impact by one particle. This qualitatively new wave (spatial size and shape are independent of amplitude, and initial sound speed does not determine the soliton parameters) is a stationary solution of a wave equation based on only one small parameter—ratio of particle size to characteristic wave length. Such an equation is more general than the weakly nonlinear Korteweg–de Vries equation, which is based on two small parameters representing a relatively small amplitude of long wave disturbance. This more general wave equation has stationary solutions with unique properties. A similar situation can be expected in other strongly nonlinear cases of wave motion, for example, waves in water channels, electrical lines, and plasma.

Strongly nonlinear strain solitons in granular materials were found in experiments, numerical calculations, and using the analytical approach. The uniqueness of the granular system, even in the simple 1-D chain, is that it provides the unusual possibility of demonstrating all type of waves: linear waves when amplitude is infinitesimally small in comparison with initial prestress, weakly nonlinear, KdV-type waves when they are comparable, and strongly nonlinear waves when there is no initial precompression. Solitary waves or shock waves can be generated depending on the duration of the initial pulse. A practically unexplored area is nonstationary wave transformation, both from the analytical continuum and the numerical approach. Especially interesting is the establishment of criteria for an impulse split into strongly nonlinear soliton trains and dependence of the number of solitons on boundary conditions.

The next promising step in the case of a highly nonlinear system is, first, the development of 3-D experiments and numerical calculations to understand the response of granular media to the dynamic loading. A very important task is to identify the similar phenomena in different materials in the areas of acoustic and electromagnetic waves.

Another challenge is to create such materials and systems, which do not exist in natural conditions, and to find practical applications for these new wave dynamics. Actually a relatively long 1-D chain of uncompressed elastic grains placed into a tube is the simplest example of a metamaterial. It just so happens that new wave dynamics was initially developed for this specific material. In the future, due to the development of experimental methods for the generating of ultrashort impulses (like laser-driven shocks), the emphasis may be shifted from the shock wave approach to the investigation of solitons in crystal lattices and in other materials with periodic structure.

Mesomechanics of porous materials needs a strongly coupled approach between 2-D and 3-D numerical simulations and experiments capturing material response on the mesolevel. Temperature measurements during shock for materials with calibrated dynamic properties may help to verify 3-D numerical modeling. A very interesting area is modeling in numerical calculations 2-D and 3-D interfacial phenomena on the contacts of solid and powder or between powders with different physical properties.

A new area is the impact phenomena in high-gradient materials containing macrocavities and heterogeneities with sizes comparable with the diameter of a projectile. Nonsymmetrical pore collapse and "forced" shear localization create unique damage pattern that can be interesting for ballistic applications. For example, such materials are able to deflect and to stop a long rod projectile within the target, reducing the absorbed linear momentum.

Strongly nonlinear shock wave dynamics in laminar materials should be considered not only as a part of shock wave research, but also as a part of highly nonlinear wave dynamics. The development of periodic structures with properties of "sonic vacuum" may be very interesting from a practical point of view.

Shear localization in heterogeneous materials can provide very interesting results in the area of structural patterning inside shear bands. Observation of vorticity inside shear bands, cascade of shear bands and fracture, chemical reactions and sintering of silicon carbide with small initial particle size, and other examples demonstrate that this area where material is in the extreme state can provide new results.

Shear band patterning is a relatively new area of intensive research. Patterning in the assemble of a large number of shear bands in simple geometry and in the uniform boundary conditions on the remote part of the sample still presents a challenge for experimental research and modeling. Only first steps have been made in this direction. One of the interesting problems here is the scaling of shear band spacing with a controlled strain rate and self-organization in the assemble of shear bands advancing from nuclei to well-developed shear bands.

Continuum models of shear band self-organization are justified for solid and granular materials in certain areas of grain size (particle size) and size of the material sample. At the same time, experimentally observed dramatic influence of material microstructure on the kinetics of softening may result in different shear band patterns.

Dynamic methods of materials treatment may help to synthesize new structures unattainable with the usual methods. Due to the inherent highly heterogeneous nature of contact deformation, simple shock (explosive) consolidation cannot provide materials with high strength and ductility. This dream of many years should be considered mirage. At the same time, shock loading may provide very unusual and useful properties as the precursor technique. One unexplored area is pressure-induced quenching of melts with the anomalous melting curve, which can be realized not only in the bulk but also in high-velocity plasma spray in coatings. The combination of dynamic loading with optimized HIPing can be a very promising direction for future research.

Index

"Abnormal" material, 101, 120
Activation, 488
Al–Cu laminate, 257
Alumina, 341
Amorphization, 465
Amorphous, 454, 455, 459, 475
Amplification, 275
Anharmonic approximation, 5
Annealing, 475
Armor, 340
Asymmetric, 393
Attenuation, 253, 261, 275
Averaged functions, 24

Backscattering, 78
Benson's model, 413
Bimodal, 361
Blast mitigation, 285, 295
BN, 222, 422, 425
Bonding, 451, 453
Boundary, 50, 76, 173, 212, 252
BRG model, 400, 403, 457

Carroll–Holt model, 138, 145, 165, 234, 461
Carrying ton, 118
Cavity, 224, 229, 341
Ceramic, 312, 482
Chain in liquid, 46

Characteristic times, 3
"Cold" boundary layers, 212, 217
Collapse, 309, 405
Collective behavior, 328
Collision of solitary waves, 33, 74
Comminution, 361, 488
Compact support, 20
Composites, 487
Compression, 211
Compression soliton, 17, 96, 115
Consolidation, 443, 451, 452, 473
Contact
    deformation, 198
    temperature, 389, 395
Continuum, 35, 98
Control procedures, 27
Convective derivative, 11, 20
Copper, 222
Critical
    amplitude, 63
    angle, 214
    thickness, 283
Cu + BN, 423
Cu–Ni, 390
Cylindrical, 414, 451

Damping, 274, 287, 290, 292
Dashpots, 5
Decoupling, 432
Deformable wall, 280

Densification, 443
Diatomic chain, 54, 56
Dielectric–metal transition, 208
Dimensionless analysis, 27
Discrete chain, 35, 98, 104
Disposable structures, 293
Dissipation, 2, 44, 162
Divergent form, 11
Dynamic deformation, 161, 166,
    183, 187, 189, 193, 205, 449

Elastic-plastic, 251
Elliptical particles, 73
Energy distribution, 418
Equation of motion, 4, 26
Exact solution, 154
Experiments, 65, 142, 154, 167,
    232, 262, 309, 314, 320, 333,
    386, 402, 423
Explosive chambers, 286
Explosive welding, 442, 452

Fermi–Pasta–Ulam Problem, 3
Fiber, 100
Final particle velocity, 272
Foam, 276, 278, 280, 299
Foil, 454
Force chains, 88
"Forced" localization, 316
Fracture, 224, 228
Fragmentation, 28, 30
Free surface, 224
"Free" volume, 466
Friction, 341

Gas–liquid system, 101
Glass powder, 265, 270
"Gradient catastrophe," 25
Grady–Kipp model, 328
Granular, 314, 340, 354, 366, 367,
    385
Gravity, 50
Groove, 227

Heat, 366, 385
Hertz law, 3
Heterogeneity, 198
Heterogeneous heating, 398, 455
High-gradient
    barriers, 284
    materials, 505
HIPing, 230, 472, 477
HMX, 158, 202, 203, 371
Hot explosive compaction, 483
Hot zones, 417
Hugoniot, 22, 210, 419

Implosion, 310
Impulse decomposition, 40, 61
Impurities, 77
Inert, 314
Inhomogeneity, 219
Initial compression, 46
Initiation, 323
Instability, 406
Interface, 221, 253, 396

Jetting, 229

KdV equation, 6, 109, 123, 504
KdV soliton, 7
Kinetic energy, 45

Lagrangian, 12, 23, 92
Laminar materials, 247, 262
Laminar model, 408
Laminate, 246, 253, 278
Leading shock, 253
Localized shear, 406
Long rod, 230
Long-wave approximation, 5, 91

Mach reflection, 451
Macroscales, 137
Macroshear, 453
Magnetic field, 211

Material modeling, 174
Maximal particle velocity, 272
Melting, 426, 444
Mesoscales, 137
Mesoshear, 369
Metallization, 205, 207
Metallographic, 205
Metamaterial, 108, 505
Microcrystalline, 473
Microhardness, 177
Microkinetic energy, 161, 183, 447
Microscales, 137
Microstructural, 386
Mixtures, 222, 267, 418
Molinari model, 332
Morse potential, 5

Nanocrystalline, 463
Nb–Si, 369
Neston, 118
Ni + BN, 423
Nonequilibrium, 385, 418, 456
Nonstationarity, 215
"Normal material," 96, 115
"Newton's cradle," 26, 31
Noncontact method, 207, 210
Nonlinearity
    strong, 1, 13, 20, 30, 39, 103,
        107, 504
    weak, 6, 39
Nonstationary waves, 47, 215

One-dimensional, 3, 87, 90, 104
Optical, 386, 393
Ordered packing, 169, 408
Oxide layers, 206

Penetration, 340
Perforated structures, 295
Periodic solution, 19
Perlite, 276
Phenomenological, 62
Piston impact, 41

Plastic deformation, 80
Pore, 405, 432
Pore collapse, 138
Porolon, 276, 279, 299
Porosity, 193, 421
Porous materials, 246
Porous target, 230
Potential energy, 14, 45, 93, 111,
    113
Powders, 385, 394, 416, 418, 465
Power-law, 90
Precompaction, 217
Precursor, 89, 217
Prefractured, 312, 348
Preservation, 455
Pressure-induced supercooling, 467

Quasistatic deformation, 161, 183,
    187, 189, 193, 205, 426

Random chain, 58
Random packing, 171, 411
Rapidly solidified, 477
Rarefaction shock, 122
Rarefaction soliton, 101, 120
RDX, 276, 284
Reaction, 367, 370, 426
Reactive, 314
Rectangular particles, 414
Reflected soliton, 52
Regularized equation, 12, 13, 92,
    108
Restrictions, 168
Rigid confinement, 286
Rigid wall, 278

Sawdust, 290
Self-organization, 307, 320, 506
Separation, 43, 222
Shear band
    patterning, 307, 320, 333, 340,
        505
    scaling, 340

spacing, 307, 329, 332
thickness, 366
Shear localization, 307
Shock
  absorbers, 90
  amplification, 275
  attenuation, 253, 261, 274
  front, 215, 269, 389, 401
  "long," 257
  mitigation, 283
  secondary, 394
  "short," 259
  stationary, 271
  "strong," 257
  transformation, 248
  wave, 21, 118, 180
  wave structure, 141
  wave width, 156, 159, 202, 215,
    269, 271
  "weak," 259
SiC, 312, 342, 358, 362, 364
Silicon, 469
Simple shear, 346
Sintering, 362, 472, 476
"Skin" model, 415, 457
Solitary wave
  weakly nonlinear, 7
  strongly nonlinear, 18, 28, 69,
    96, 117, 121
Soliton width, 19, 97, 117, 121,
  124
"Sonic vacuum," 7, 108
  equation, 10, 11, 12, 91, 107,
    108
  soliton, 39, 96, 117
Sound speed, 6, 93, 103, 210,
Speed of solitary wave, 16
Spin, 201
"Spontaneous" localization, 321
Stability, 22
Stainless steel, 324, 334, 387, 413
Stationary solutions, 13
Strain partitioning, 342
Strain rate, 187

Strength, 189
Striker impact, 67
"Strongly compressed" chain, 3
Supercooled, 464, 467
Superheated, 393
Supersonic, 17
Surface temperature, 450
Symmetrical, 405

Teflon–paraffin laminate, 255
Tensile, 204
Thermal relaxation, 415
Thermoelectric, 388
Thick-Walled Cylinder method, 308
Three-dimensional, 82, 87, 223
Ti–33%Al, 477
Ti–48Al–2Mn–2Nb, 481
Ti–6Al–4V, 325, 334, 473
Ti–Al laminate, 261
Ti-based, 231, 387
Ti + Si, 368
Titanium, 311
TNT, 285, 299
TNT + RDX, 249
Toda chain, 5
Transient soliton, 52
Transition, 187, 189, 193
Transmitted momentum, 279
Transmitting lines, 124
Transverse, 100
Two-dimensional, 82, 412
Two-particle chain, 53

Vermiculite, 287
Viscoplastic, 229, 401

W + BN, 422
"Weakly compressed" chain, 7
Weakly nonlinear equation, 6
Whitham method, 22
WKB method, 330
Wright–Ockendon model, 329